Progress in Physics

Volume 13

Functional Integration, Geometry and Strings

Proceedings of the
XXV Karpacz Winter School
of Theoretical Physics
Karpacz (Poland),
20 February – 5 March 1989

Edited by
Zbigniew Haba
Jan Sobczyk

1989

Birkhäuser Verlag
Basel · Boston · Berlin

Dr. Zbigniew Haba
Dr. Jan Sobczyk
Institute of Theoretical Physics
University of Wroclaw
Cybulskiego 36
50–205 Wroclaw
Poland

CIP-Titelaufnahme der Deutschen Bibliothek

Functional integration, geometry and strings: proceedings of the XXV Karpacz Winter School of Theoretical Physics, Karpacz (Poland), 20 February – 5 March 1989 / ed. by Zbigniew Haba; Jan Sobczyk. – Basel ; Boston ; Berlin : Birkhäuser, 1989
(Progress in physics ; Vol. 13)
ISBN 3-7643-2387-6 (Basel . . .)
ISBN 0-8176-2387-6 (Boston)
NE: Haba, Zbigniew [Hrsg.]; Zimowa Szkola Fizyki Teoretycznej <25, 1989, Karpacz>; GT

Printed in Germany on acid-free paper
ISBN 3-7643-2387-6
ISBN 0-8176-2387-6

CONTENTS

III. FUNCTIONAL INTEGRATION AND CONFORMAL FIELD THEORY

IV. QUANTIZATION OF CONSTRAINED SYSTEMS

* The lecturer

Foreword

It is already a quarter of a century since the first Winter School took place in Karpacz in 1964. Every year since then (except 1982) the School is organized by members of the Institute of Theoretical Physics of Wroclaw University. The topics of the School vary reflecting the development of modern theoretical physics as well as interests of people at our Institute.

The program of the present XXV Winter School was concentrated around the string model as the fundamental unified theory of Nature. As it came out from private conversations as well as from lively discussions after lectures and even more at a panel discussion this point of view has its ardent adherents as well as adversaries (who constituted a minority among the participants). In any case the string theory brings new ideas into physics. It stimulates our imagination what the fundamental theory can be. The development of string theory already enriched the conventional quantum field theory and mathematical physics. Let me only mention the development of conformal field theory and topological field theories. I think that these new perspectives are reflected by the contributions to these Proceedings. As string theory is a quickly developing field we made some effort to supply these Proceedings timely even at the cost of risking that some lecturers may be unable to comply with our early deadline.

The articles in this book reproduce quite accurately the programm of the School. There were some major lectures delivered by experts in the field continuing for a couple of days starting from an elementary level and reaching recent advanced developments. In addition, there were talks concerned with some special topics. The material is divided into sections. The major lectures are given preference in the sequence of articles in the section.

It is now my pleasure to thank these people who made it possible the School to take place. I would like to thank the authorities of the Wroclaw University, first of all the Rector Professor J. Ziolkowski for a generous financial support and for understanding of our problems. The School was organized in university houses in Karpacz. I wish to thank the staff for their contribution to a pleasurable atmosphere at the School. Last but not least I am grateful to my colleagues from the Institute for their generous help in preparation of the School. Among many colleagues whom I am indebted let me mention Ziemowit Popowicz and Zbigniew Jaskolski. Jan Sobczyk took the responsibility of the secretary. In fact, he was a coorganizer of the School. We have collaborated closely shaping the program and these Proceedings.

July 1989

Zbigniew Haba
Director of the School

FUN WITH STRINGS

MICHIO KAKU

Physics Dept., City College of the City Univ. of New York, N.Y. 10031

ABSTRACT

String theory has emerged as the leading candidate for a theory of all known forces, including a finite theory of quantum gravity. Unfortunately, the model was discovered quite by accident in 1968 and has been evolving backwards ever since. As a consequence, its underlying geometry is completely unknown. Unfortunately, the mathematics of the superstring has soared to dizzying heights. On one hand, this has unified various branches of mathematics and open up entirely new ones. On the other hand, this has tended to obscure the fact that the basic ideas underlying superstring theory are basically simple and intuitive.

Thus, the purpose of these lectures is not to blind the reader with dense equations, but to give an over-view of the simple, topological ideas that lie at the heart of string theory. The audience for these lectures is not necessarily those already familiar with the model, but for those still grappling with these concepts. These lectures have an underlying theme, that the enormous power of superstring theory arises from a new feature: topology. We will thus use intuitive concepts, rather than a blizzard of obscure (and often unenlightening) equations, to isolate the essential features of the model.

At the end of these lectures, we will end with string field theory, the only formalism in which all features of the model are manifest. In particular, we will stress that *closed string field theory is non-polynomial, as in Einstein's equations*.

I. THE SEARCH FOR UNIFICATION

In the twentieth century, two different formalisms have emerged which can, in principle, explain all known physical phenomena: general relativity, which gives us a compelling description of the beginning of the universe, and the quantum theory, which has unified the theory of sub- atomic particles. It seems bizarre that Nature, at its most fundamental level, would have two sets of laws, based on two sets of principles, two sets of equations, and operating on two different scales. It seems hard to believe that Nature could be so malicious as to create a world split in such an artificial fashion.

Some of the deepest thinkers of the century have struggled with the problem of quantizing gravity, and have failed. We are reminded of the famous comment by Niels Bohr after Wolfgang Pauli gave a long talk outlining his version of the unified field theory. Speaking for the audience, Bohr said, "We are all agreed that your theory is crazy. The question which divides us, 'Is it crazy enough?'"

The question remains today, "Is string theory crazy enough to unify all known forces?"

One important clue is the fact that the underlying theme behind general relativity and the quantum theory is symmetry. The gauge principle has emerged as probably the single most important formalism in all of high energy physics in the last several decades. If we chart the history of unification, we see that each theory is based on increasingly sophisticated Lie groups:

Theory	**Symmetry**
electro − mag.	$U(1)$
electro − weak	$SU(2) \times U(1)$
strong	$SU(3)$
GUT	$SU(5)$; $O(10)$
gravity	$Sp(4)$; $Gl(4)$
supergravity	$Osp(4/8)$
string	?

Symmetry, however, is not enough. By naively computing quantum corrections to a gravity theory, we find that the equations blow up in our face, a sure sign that the theory is incomplete.

Normally, we use Feynman's bag of tricks to juggle an infinite number of divergent diagrams, which all miraculously sum to zero. However, because κ has dimensions, this is impossible. We cannot add apples to oranges. Faced with a dimensionful Newton's constant G, the usual lore of renormalization theory collapses completely. Thus, we are faced with the daunting task of creating a *finite* theory of gravity, finite to every order in κ.

String theory[1−6], however, apparently solves this seemingly impossible problem because it introduces a new feature into quantum mechanics, topology. Thus, it provides a different mechanism in which to unite the trio of fundamental constants: c , Planck's constant, and the Planck mass. String theory therefore may ultimately answer Dirac's objection to renormalization theory. Dirac once said, "This (renormalization theory) is not sensible mathematics. Sensible mathematics involves neglecting a quantity when it turns out to be small - not neglecting it because it is infinitely great and you do not want it."

For example, the scattering of two open strings can create a single loop diagram. Unlike the point particle example, where shrinking a propagator yielded a divergence, you cannot shrink the hole to a point. In fact, the diagram is no longer ultraviolet divergent, but suffers from a much milder infrared divergence.

The irony of string theory is that, although it is a theory which unifies all known forces, the theory itself is so ununified. To the beginner, string theory, rather than proceeding from first principles, seems to be a loose collection of folklore, rules of thumb, and arbitrary conventions.

By contrast, think of general relativity, which was historically derived from geometry. First came Einstein's equivalence principle and the geometrical picture of general covariance. From the geometry came a unique action, the Einstein-Hilbert action. Then came the classical theory of curved manifolds (e.g. Schwarzschild's solutions, Robertson- Walker solutions, etc.). Last comes the yet unfinished quantum theory of gravitation. Thus, the historical sequence can be viewed as follows:

$$\textit{Geometry} \rightarrow \textit{Action} \rightarrow \textit{Classical Theory} \rightarrow \textit{Quantum Theory}$$

String theory, however, has a bizarre history, one of the strangest in the history of modern science. No where else has there been a theory proposed to solve the wrong problem, which then died an abrupt death, only to resurrected as a theory of everything!

Viewed in this fashion, we see that string theory has been evolving backwards, ever since its accidental discovery in 1968 by G. Veneziano and M. Suzuki at CERN when they were thumbing through a math book. Independently, they noticed that the Euler beta function, discovered in the last century, had almost all the correct properties of a hadronic S-matrix:

$$A(s,t) = B(s,t) = \int_0^1 dx\, x^{-\frac{1}{2}s-2}(1-x)^{-\frac{1}{2}t-2}$$

(Unfortunately, Suzuki, after listening to the advice of a senior physicist, was discouraged from publishing his result. There is a lesson here: never take too seriously the advice of your superiors.)

Thus, string theory was, from the very start, born as a quantum theory, a candidate which might satisfy the postulates of G. Chew's S- matrix program. The one postulate missing in the dual model was unitarity. To solve this problem, K. Kikkawa, B. Sakita, and M. Virasoro[7] proposed to treat the Beta function was a Born term for a perturbation expansion. This program was eventually solved for open bosonic strings by several groups (Kaku, Yu, Lovelace, Alessandrini [8–10]).

However, the simple classical picture underlying the dual model was finally isolated by Y. Nambu and T. Goto[11–12], who correctly saw that a vibrating string could reproduce the spectrum of the dual model. Then GGRT[13] and Mandelstam[14] made the link between the classical picture of Nambu-Goto and the quantum picture of KSV.

The action (in the light cone gauge) was finally discovered in 1974 (Kaku and Kikkawa[15]). Unfortunately, string theory died at around that time when it became apparent that the model only made sense in 26 dimensions (or 10 dimensions for the Neveu-Schwarz-Ramond model[16]). Who wanted to work in 26 dimensions when Yang-Mills theory and asymptotic freedom were blazing the trail for a theory of strong interactions? Scherk and Schwarz's suggestion[17] that string theory be reinterpreted as a theory of everything (where 6 of the 10 dimensions have curled up too small to be seen) fell like a lead balloon.

In 1984, Green and Schwarz's discovery[18] that string theory was free of all quantum anomalies set off a revival of interest in superstring theory. Quickly, new developments in compactification schemes created models which began to resemble the real world. Heterotic strings[19] with exceptional groups seemed to give us plenty of room for interesting phenomenology, and compactifications on Calabi-Yau spaces[20] and orbifolds[21] have given us entire new areas of research. Also, Witten's new open string theory[22] has taken the development of string field theory to a new height.

In fact, the "superstring stampede" has sorely distressed some senior figures in physics. One has suggested that superstring theory is like "American television commercials," (i.e. all advertisement and no content). Another has suggested that it be compared to former President Reagan's Star Wars program (untestable, enormously expensive, draining away all the top scientific talent).

Personally, I don't feel that the main stumbling block to superstring theory is the lack of a SSC which encircles our galaxy. I don't think bankrupting the GNP of our quadrant of the galaxy is the solution to the problem. The real problem is theoretical, rather than experimental. If we were a bit more clever, perhaps we could solve the theory non-perturbatively and then break the theory down from the Planck length down to present energies. Until this is done, it makes no sense to pontificate on the ultimate fate of the model, one way or the other.

Unfortunately, the problem that dogged Kaluza-Klein theory for the past 70 years is still with us: why did the higher dimensions suddenly curl up down to the Planck length? Today, we are no closer to an answer than we were in 1919.

Let us summarize briefly the positive and negative aspects of the model. On the positive side,

1) the model appears to be a finite theory of gravity to all orders in perturbation theory. Thus, string theory has no rivals.

2) all consistent string theories necessarily include gravity. Dropping gravity ruins all the beautiful properties of the theory.

3) with rather mild assumptions, one can obtain an astonishingly rich description of particle interactions, including GUT's, the minimal model, axions, chiral fermions, fermion generations, etc.

On the negative side, we have the following:

1) the model, defined at the Planck length of 10^{19} Gev/c^2, cannot be tested in the foreseeable future.

2) the model has *millions* of classical solutions, including solutions which describe the known world. Which is the correct vacuum?

3) the model is probably strongly coupled[23], meaning that perturbative methods are useless.

4) studies of the high energy behavior of the theory in perturbation theory shows that it is not Borel summable[24]; in fact, all classical vacua are probably unstable. (This can also be viewed as a positive development, because it rules out all the millions of wrong vacua that appear consistent to all orders in perturbation theory but have unbroken supersymmetry and massless dilatons.)

The usual approach to a non-perturbative calculation is, of course, to use field theory. However, string field theory has not lived up to its promise of a non-perturbative formulation of the model. In fact, progress in string field theory has come to a grinding halt because a successful closed string field theory is missing. Moduli space is either under-counted, or over-counted an infinite number of times. Because all the interesting string theories are based on closed strings, the lack of a closed string theory has prevented any progress in this direction.

Recently, however, we have proposed the solution to this nagging problem. Again, the answer lies in topology. The answer is that closed string field theory is *non-polynomial,* as in Einstein's theory (which is contained within it), with each interaction representing the topology of a *polyhedron.* We will comment more on this later.

Let us begin by discussing the simplest possible relativistic theory, the theory of point particles. A point particle is described by a vector which points from the origin to the location of the particle $x_\mu(\tau)$, where τ parametrizes the location of the point along the trajectory.

Similarly, a string is described by a vector which points from the origin to its location in space-time $X_\mu(\sigma,\tau)$. Its action is given by the area swept out by the string as it moves in space-time (see fig.1).

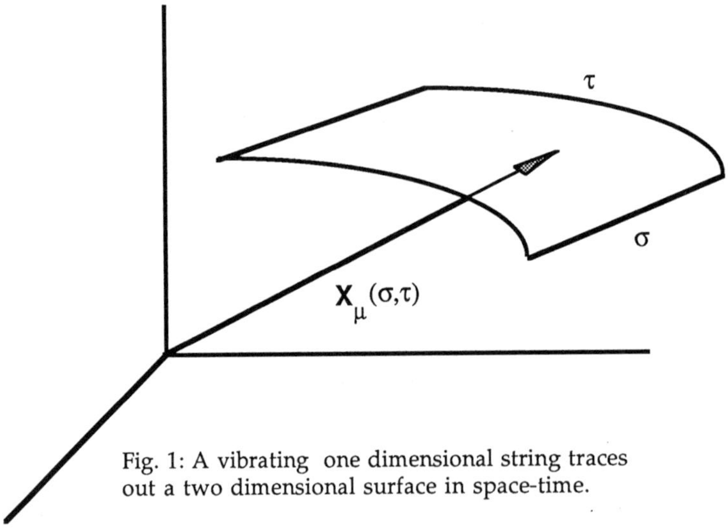

Fig. 1: A vibrating one dimensional string traces out a two dimensional surface in space-time.

The first quantized point particle action is given by the invariant length that it sweeps out moving in space-time (see ref. 1 for notations):

$$L = m \int d\tau \sqrt{\dot{x}_\mu(\tau)^2}$$

Notice that the theory is independent of how we parametrize the trajectory of the point particle, i.e. the action is invariant under:

$$\tau \to \tilde{\tau}(\tau) \; ; \qquad \delta x_\mu(\tau) = \epsilon \, \partial_\tau x_\mu(\tau)$$

This symmetry is reflected in the fact that not all canonical momenta p_μ are independent, i.e.

$$p_\mu \equiv \frac{\delta L}{\delta \dot{x}_\mu} \; ; \qquad p_\mu^2 + m^2 = 0$$

Lastly, the transition to the familiar second quantized Klein-Gordon equation is given by placing the constraint directly onto the Hilbert space of point particles:

$$(p_\mu^2 + m^2) \, |\phi\rangle = 0$$

The awkward feature of a first quantized point particle theory is that interactions must be imposed by hand on the path integral. If we take the Fourier transform of the path integral, then the transition amplitude is given by:

$$\sum_{Topologies} \langle \prod_j e^{ik_j x_j} \rangle = \sum_{Topologies} \int Dx \; e^{iS + \sum_j ik_j x_j}$$

Notice that the sum over topologies must be placed by hand in the path integral. Thus, unitarity must be tediously checked order by order.

The first quantized string theory progresses almost identically to the point particle case, except for a key difference.

We begin by postulating that the (Nambu-Goto) string action[11–12] is given by the two-dimensional area that a one dimensional string sweeps out when it moves in space-time:

$$L = \frac{1}{2\pi\alpha'} \left(\dot{X}_\mu^2 \, X'^2_\nu - (\dot{X}_\mu \, X'^\mu)^2 \right)^{1/2}$$

where the dot (prime) refers to derivatives with respect to τ (σ) and $\alpha' = 1/2$ is the string tension.

This action is invariant under two-dimensional reparametrizations of the world sheet:

$$\tau \to \tilde{\tau}(\tau, \sigma) \; ; \qquad \sigma \to \tilde{\sigma}(\tau, \sigma)$$

$$\delta X_\mu = \epsilon_\tau \, \partial_\tau X_\mu + \epsilon_\sigma \, \partial_\sigma X_\mu$$

Yet another way to write this action is to introduce a two dimensional metric tensor g_{ab}, so that the action is manifestly invariant under two-dimensional reparametrizations:

$$L = \frac{-1}{4\pi\alpha'} \sqrt{g} \; g^{ab} \, \partial_a X_\mu \, \partial_b \, X^\mu$$

(By eliminating the metric tensor, we obtain the Nambu-Goto action back again.)

As before, by taking calculating the canonical conjugate P_μ, we find that it is constrained:

$$P_\mu \equiv \frac{\delta L}{\delta \dot{X}_\mu}$$

$$(P_\mu^2 + \frac{X'^2}{(2\pi\alpha')^2}) = 0$$

$$P_\mu X'^\mu = 0$$

There are, however, two crucial differences which separate point particle and string dynamics, both originating from topology. First, notice that points interact over Feynman *graphs,* while strings interact over *manifolds*. Thus, there are an infinite number of Feynman rules compatible with point particles of arbitrary spin, but there are only a *finite* number of manifolds over which the string can interact. This severely restricts the number of consistent string theories. (In fact, only four are known.)

Second, notice that the constraints generate an algebra, called the Virasoro algebra, which generate reparametrizations. The generators are given by the Fourier moments of the constraints:

$$L_n = \frac{\pi}{4} \int_{-\pi}^{\pi} d\sigma\, e^{in\tau} [P_\mu + \frac{X'_\mu}{2\pi\alpha'}]^2$$

The algebra generated by these operators is:

$$[L_n , L_m] = (n - m)L_{n+m} + \frac{c}{12}(n^3 - n)\delta_{n,-m}$$

(c = 26). This algebra will play a crucial role at every stage in this theory.

Let us now quantize the theory. Notice that the path integral[25] is formally divergent:

$$\int DX\, e^{i\int L\, d\tau\, d\sigma} = \infty$$

because of the gauge symmetry. To eliminate this infinite gauge redundancy, we must fix the gauge. There are three ways in which to quantize the theory:

Gupta-Bleuler Quantization

In the Gupta-Bleuler formalism, we fix the gauge: $g_{ab} = \delta_{ab}$, and the resulting action breaks reparametrization invariance (but maintains the subgroup of *conformal transformations*):

$$L = \frac{1}{2\pi} \left(\dot{X}_\mu^2 - X'^2_\mu \right)$$

This conformal gauge action is easy to quantize. Let us define:

$$[P_\mu(\sigma), X_\nu(\sigma')] = -i\, \delta_{\mu\nu}\, \delta(\sigma - \sigma')$$

where we can decompose the fields in terms of their Fourier moments:

$$X_\mu = x_{0\mu} + \sum_{n=1}^{\infty} \frac{1}{\sqrt{n}} \left(a_{n\mu} - a_{n\mu}^\dagger \right) \cos n\sigma$$

$$P_\mu = \frac{1}{\pi} \left[p_{0\mu} + \sum_{n=1}^{\infty} \sqrt{n} \left(a_{n\mu} + a_{n\mu}^\dagger \right) \cos n\sigma \right]$$

The Lagrangian, of course, propagates negative metric ghosts associated with the time-like mode of X. To eliminate them, the Gupta-Bleuler quantization method states that the state vectors of the theory must satisfy:

$$L_n |\phi\rangle = 0 \; ; \quad n > 0$$

$$(L_o - 1) |\phi\rangle = 0$$

This means that the constraints are constructed to vanish on the Hilbert space. The solution of these constraints is quite difficult, but reveals that the Hilbert space is free of ghosts in 26 dimensions.

The state $|\phi\rangle$ is the second quantized field which we will discuss at the end of this article. We can decompose it in terms of its various harmonic oscillators:

$$|\phi\rangle = \phi(x)|0\rangle + A^\mu(x)\, a_{1\mu}^\dagger |0\rangle + h^{\mu\nu}\, a_{1\mu}^\dagger a_{1\nu}^\dagger |0\rangle + \cdots$$

Thus, while the first quantized X_μ describes the motion of a single string, the second quantized field formalism describes the composite, multi-string set of all possible modes within the theory. Notice that the lowest state is a tachyon, and the first massive move is a Maxwell or Yang-Mills field. Thus, string theory necessarily includes the standard gauge theories as a subset.

For closed strings, we must introduce two sets of harmonic oscillators a_μ and $\bar{a}_\mu$ (because there are cosine and sine modes simultaneously). Thus, the spectrum of the closed string contains a massless spin two field $h_{\mu\nu}$, (the graviton):

$$|\phi\rangle = \phi(x)|0\rangle + h^{\mu\nu}\, a_{1\mu}^\dagger \bar{a}_{1\nu}^\dagger |0\rangle + \cdots$$

Light Cone Quantization

In the light cone quantization program, we set the unwanted longitudinal modes to zero from the very beginning. We fix:

$$X^+ = 2\alpha' p^+ \tau$$

The advantage of this approach is that we can eliminate all the redundant ghost modes of the theory from the very start in terms of only the transverse components. The Virasoro constraints can be set identically to zero by solving for the longitudinal modes. The disadvantage is that the formalism is awkward and we must re-establish Lorentz symmetry at every step. The surprising feature is that the Lorentz generators of the broken theory only close in 26 dimensions. Specifically, the problem occurs with the commutator[13]:

$$[M^{-i}, M^{-j}] = \frac{-1}{p^{+2}} \sum_{m=1}^{\infty} [\alpha_{-n}^{i} \alpha_{n}^{j} - \alpha_{-n}^{j} \alpha_{n}^{i}] \Delta_n$$

where:

$$\Delta_n = \frac{n}{12}(26-D) + \frac{1}{n}\left[\frac{D-26}{12} + 2-2a\right]$$

In order for this commutator to vanish, we must have the dimension of space-time to be 26 and the intercept to be one.

BRST Quantization

The BRST method combines the best features of both approaches. We still maintain Lorentz covariance of the action, but the negative metric states do not worry us because they cancel against the ghosts circulating in the theory because of the Faddeev-Popov determinant. When we exponentiate the Faddeev-Popov determinant, we must introduce two anti-commuting ghost fields b and c. The Faddeev-Popov determinant is given by:

$$\Delta_{FP} = \det |\partial_z| = \int DbDc\; e^{i\int b\, \partial_z c\, d^2z \,+\, h.c.}$$

The resulting gauge-fixed action, including both the string field X and the ghost fields b and c, still has a residual symmetry generated by the BRST charge. The action is:

$$L = \frac{1}{2\pi}\left\{ \partial_z X_\mu\, \partial_{\bar z} X_\mu + b\,\partial_{\bar z} c + \bar b\,\partial_z \bar c \right\}$$

As before, we see that this is invariant under:

$$\begin{aligned} \delta X_\mu &= \epsilon[c\,\partial_{\bar z} X_\mu + \bar c\,\partial_z X_\mu] \\ \delta c &= \epsilon[c\,\partial_{\bar z} c + \bar c\,\partial_z c] \\ \delta b &= \epsilon[\bar c\,\partial_z b + c\,\partial_{\bar z} b \\ &\quad + 2\partial_{\bar z} c b + \partial_z X_\mu \partial_z X_\mu] \end{aligned}$$

The generator of this symmetry is given by[26]:

$$Q = c_o(L_o - a) + \sum_{n=1}^{\infty} [c_{-n} L_n + L_{-n} c_n] - \tfrac{1}{2} \sum_{n,m=-\infty}^{\infty} :c_{-m} c_{-n} b_{n+m}: (m-n)$$

Fixing:

$$Q^2 = 0$$

sets the intercept to be one and the dimension of space-time to be 26. The physical states of the theory are defined by:

$$Q\,|phy> = 0$$

Let us now discuss interactions, beginning with tree diagrams (where ghost modes decouple). As in point particle theory, interactions are introduced by summing over different topological configurations in the path integral. The scattering amplitude is given by the Fourier transform over two-dimensional complex Riemann surfaces[25]:

$$A_N(k_1, k_2, \cdots, k_N) = \sum_{Topologies} \int d\mu_N \int Dg_{ab} \int DX$$

$$\times\, \Delta_{FP} \exp i \int L d\sigma d\tau \left\{ \prod_{i=1}^{N} \sqrt{g} \exp ik_{i,\mu} X^{\mu i} \right\}$$

$$= \sum_{Topologies} \int Dg_{ab} \int d\mu < \prod_{i=1}^{N} \sqrt{g} \exp ik_{i,\mu} X^{\mu i} >$$

In the conformal gauge, the solution to the equation of motion is given by ordinary electrostatics, i.e. Poisson's equation (J_μ equals the sum over momenta):

$$\nabla^2 X_\mu = -i\pi J_\mu$$

Let us say that the topology over which we integrate is given by the upper half plane. Then the Neumann function defined over the upper half plane is given by the method of images:

$$N(z, z') = \ln|z - z'| + \ln|z - \bar{z}'|$$

Inserting this Neumann function into the path integral, we can solve for the N-tachyon scattering amplitude:

$$\int d\mu \int DX \exp\left\{ -\frac{1}{4\pi\alpha'} \int \partial_z X_\mu \partial_{\bar{z}} X^\mu d^2z + iJ_\mu X^\mu \right\}$$

$$= \int d\mu \exp\left\{ \alpha' \int J_\mu(z)\, G(z,z')\, J^\mu(z') dz dz' \right\}$$

$$= \int_0^1 \prod_{i=3}^{N-1} dz_i \;\; \theta(z_{i+1} - z_i) \prod_{2 \le i < j \le N} (z_i - z_j)^{2\alpha' k_i \cdot k_j}$$

where the z_i represent the points on the real axis which label the tachyons.

II. SUPERSYMMETRY

So far, we have only discussed the spectrum of the bosonic string. The introduction of spinors led to the discovery of supersymmetry, which first appeared in the string model.

Let us introduce a rather strange new field[16], ψ_μ, which is a *spinor in two dimensions* and a *vector in space-time*.

The action is given by[27]:

$$L = -\frac{1}{2\pi}\left(\partial_a X_\mu \, \partial^a X^\mu - i\psi^\mu \gamma^a \partial_a \psi_\mu\right)$$

(where the gamma matrices are the Dirac matrices in two dimensions). This action is invariant under global supersymmetry:

$$\delta X_\mu = \bar{\epsilon}\, \psi_\mu$$

$$\delta \psi_\mu = -i\gamma^a \epsilon \partial_a \, X_\mu$$

where we suppress the two dimensional spinor index of ψ_μ. (Notice that this two dimensional supersymmetry is not a physical one which exchanges space-time fermions with bosons. It is a supersymmetry of fields defined on the world sheet.)

There are several novel features of this simple but important action. First, because ψ_μ has two spinor components, there is an ambiguity as to how to decompose it in Fourier components. We can take either anti-symmetric boundary conditions (giving us the Neveu-Schwarz or bosonic sector) or symmetric ones (giving us the Ramond or fermionic sector):

$$\begin{cases} \textit{Ramond} & \psi_{0\mu}(\pi,\tau) = \psi_{1\mu}(\pi,\tau) \\ \textit{Neveu-Schwarz} & \psi_{0\mu}(\pi,\tau) = -\psi_{1\mu}(\pi,\tau) \end{cases}$$

Given two different types of boundary conditions for our spinor, we naturally have two different ways in which to power expand these two fields in Fourier components with integer (half-integer) modes. This is a feature entirely new in quantum field theory: the distinction between fermions and bosons is given by topology, i.e. their boundary conditions on a strip.

The second novel feature of this model comes from its symmetry algebra. From this action, we can extract the energy-momentum tensor:

$$\begin{aligned} T_{ab} &= \partial_a X^\mu \, \partial_b X_\mu \\ &\quad + \frac{i}{4}\bar{\psi}^\mu \gamma_a \, \partial_b \, \psi_\mu + \frac{i}{4}\bar{\psi}^\mu \gamma_b \, \partial_a \, \psi_\mu - (\textit{Trace}) \end{aligned}$$

The moments of T_{ab} give us the usual Virasoro generators L_n. The new feature, however, is that we can also extract the generator of two dimensional supersymmetry:

$$J_a = \tfrac{1}{2}\gamma^b \gamma_a \psi^\mu \partial_b X_\mu$$

The moments of J_a, in turn, give us two-dimensional supersymmetry generators (G_m for the NS model for half- integral m and F_m for the Ramond model for integral m):

$$NS: \begin{cases} [L_m, L_n] = (m-n)L_{m+n} + \dfrac{D}{8}(m^3-m)\,\delta_{m,-n} \\ [L_m, G_r] = (\tfrac{1}{2}m - r)\, G_{m+r} \\ \{G_r, G_s\} = 2L_{r+s} + \tfrac{1}{2}D\,(r^2-\tfrac{1}{4})\,\delta_{r,-s} \end{cases}$$

$$R: \begin{cases} [L_m, L_n] = (m-n)L_{m+n} + \dfrac{D}{8}m^3\,\delta_{m,-n} \\ [L_m, F_n] = (\tfrac{1}{2}m - n)F_{m+n} \\ \{F_m, F_n\} = 2L_{m+n} + \tfrac{1}{2}Dm^2\,\delta_{m,-n} \end{cases}$$

As usual, we demand that the Hilbert space of the theory respect these constraints:

$$(L_n, G_m)\,|\phi\rangle = 0\,; \quad m = integer$$

$$(L_n, F_m)\,|\phi\rangle = 0\,; \quad m = \tfrac{1}{2}\,integer$$

Solving these constraints, we find that the Hilbert space is free of ghosts if space-time is 10 dimensional.

Historically, the conformal gauge action with global two-dimensional supersymmetry was discovered first. Later, the complete action, with local two dimensional supersymmetry, was discovered. In addition to the usual supersymmetry pair of fields, it also contains a supersymmetry pair consisting of a two-dimensional vierbein e_a^b and its fermionic partner χ^a:

$$(X_\mu, \psi_\mu)\,; \qquad (e_a^b, \chi^a)$$

The full action is given by[28-9]:

$$L = \frac{-1}{4\pi\alpha'}\sqrt{g}\,[\,g^{\alpha\beta}\partial_\alpha X^\mu \partial_\beta X_\mu - i\bar{\psi}^\mu \gamma^\alpha \nabla_\alpha \psi_\mu$$

$$+ 2\bar{\chi}_\alpha \gamma^\beta \gamma^\alpha \psi^\mu \partial_\beta X_\mu + \tfrac{1}{2}\bar{\psi}_\mu \psi^\mu \bar{\chi}_\alpha \gamma^\beta \gamma^\alpha \chi_\beta\,]$$

(α, β are curved two dimensional indices.) This action is invariant under:

$$
\text{2D SUSY} \quad \begin{cases} \delta X^\mu = \bar{\epsilon}\, \psi^\mu \\ \delta \psi^\mu = -i\gamma^\alpha \epsilon(\partial_\alpha X^\mu - \bar{\psi}^\mu \chi_\alpha) \\ \delta e^a_\beta = -2i\bar{\epsilon}\gamma^a \chi_\beta \\ \delta \chi_\alpha = \nabla_\alpha \epsilon \end{cases}
$$

as well as Weyl, reparametrization, and local two dimensional Lorentz invariance.

Although the NS-R model is the most common formulation of the superstring, for many years it was not understood that the model actually contained a physical, 10 dimensional supersymmetry as well as the two dimensional supersymmetry on the world sheet. By projecting out by hand the odd "G-parity" modes (i.e. number of b oscillator modes) of the Fock space, one could eliminate the unwanted states, like the tachyon (which violates supersymmetry) and obtain a theory supersymmetric in space-time. This the GSO projection[30].

Obviously, what is required is a re-formulation in which 10 dimensional supersymmetry is manifest from the very beginning. A great advance was the formulation of the Green-Schwarz model, in which there is a physical 10 dimensional supersymmetry which links the X_μ with a genuine space-time fermion θ^A (with 32 complex components, A = 1,2).

The GS action[31] is:

$$
\begin{aligned} S = \frac{-1}{4\alpha'\pi} \int d\sigma\, d\tau \Big\{ & \sqrt{g}\; g^{\alpha\beta}\, \Pi_\alpha \cdot \Pi_\beta \\ & + 2\,i\, \epsilon^{\alpha\beta} \partial_\alpha X^\mu \,(\, \bar{\theta}^1\, \Gamma_\mu\, \partial_\beta \theta^1 - \bar{\theta}^2 \Gamma_\mu\, \partial_\beta\, \theta^2\,) \\ & - 2\, \epsilon^{\alpha\beta}\, \bar{\theta}^1\, \Gamma^\mu\, \partial_\alpha\, \theta^1\;\; \bar{\theta}^2\, \Gamma_\mu\, \partial_\beta\, \theta^2 \Big\} \end{aligned}
$$

where (Γ^μ are 10 dimensional Dirac matrices):

$$
\Pi^\mu_\alpha = \partial_\alpha X^\mu - i\bar{\theta}^A\; \Gamma^\mu\, \partial_\alpha\, \theta^A
$$

This action is invariant under:

$$
\begin{aligned} \delta\theta^A &= \epsilon^A \\ \delta X^\mu &= i\bar{\epsilon}^A\; \Gamma^\mu\, \theta^A \end{aligned}
$$

as well as:

$$
\begin{aligned} \delta\theta^A &= 2i\Gamma \cdot \Pi_\alpha\, \kappa^{A\,\alpha} \\ \delta X^\mu &= i\bar{\theta}^A\, \Gamma^\mu\, \delta\theta^A \end{aligned}
$$

$$\delta(\sqrt{g}\ g^{\alpha\beta}) = -16\sqrt{g}\,(P_{-}^{\alpha\gamma}\ \overline{\kappa}^{1\beta}\partial_\gamma\,\theta^1 + P_{+}^{\alpha\gamma}\ \kappa^{2\beta}\ \partial_\gamma\theta^2)$$

$$P_{\pm}^{\alpha\beta} = \tfrac{1}{2}(g^{\alpha\beta} \pm \epsilon^{\alpha\beta}/\sqrt{g})$$

The advantage of this GS formulation is that it is space-time supersymmetric to all levels. However, the price we pay for this is that covariant quantization of the theory is notoriously difficult, because it is a coupled system even at the free level. It's still not clear if we can actually quantize the model covariantly[31].

Therefore, we will go to the non-covariant light cone gauge, where the theory becomes linear:

$$\frac{-1}{4\pi\alpha'}\int d\sigma\, d\tau\, (\,\partial_a X^i\, \partial_a X^i\ -\ 2i\alpha'\,\bar{S}\,\gamma^b\partial_b\, S)$$

where S is the fermion field rescaled by $\sqrt{p^+}$ which satisfies $\Gamma^+ S = 0$. (The bosonic field has 8 real components, which is also the number of reduced components in the spinor field after we have applied all the constraints.) This action is manifestly space-time supersymmetric, with generators given by:

$$Q^a = (2p^+)^{1/2}\, S_0^a$$

$$Q^{\dot a} = (p^+)^{-1/2}\gamma^i_{a\dot a}\sum_{n=-\infty}^{\infty} S_{-n}^a\ \alpha_n^i$$

Lastly, we have the Type I theory if we have both open and closed strings. If we have θ^A with opposite (identical) chirality, we have the Type IIa (IIb) closed string theory.

III. HETEROTIC STRING

In 1919, when Kaluza wrote his celebrated article, he proposed that an unseen fifth dimension was curled up into a tiny ball. For example, the simplest possible "compactification" scheme involves one-dimension, where we make the identification that x-space is periodic: $x = x + 2\pi R$ A one dimensional scalar function must then be expanded in terms of periodic eigenfunctions: $\phi(x) = \sum_n \phi_n\, e^{ipx}$.

Demanding that $\phi(x) = \phi(x + 2\pi R)$ then quantizes the momentum: $p = n/R$. We see that momentum is quantized by taking periodic boundary conditions.

Likewise, the most successful superstring theory, the heterotic string[19], uses the compactification process in a crucial way. However, at first glance, it is a strange-looking string. Before, we noted that the closed string has two uncoupled oscillators, one for the cosine and one for the sine modes of a closed string. In the heterotic string, we demand that these two modes live in different dimensions: the left-moving string sector is purely bosonic and lives in 26 dimensions, where 16 of the dimensions have been "compactified" (i.e. rolled up into a ball), while the right-moving sector is in 10 dimensions and is supersymmetric. By consistency arguments, we can show that the 16 dimensional lattice upon which we will compactify the 16 left-moving co- ordinates must be the root lattice of $E_8 \times E_8$. The right-moving modes, by contrast, contains a supersymmetric multiplet. The fields are:

Left−Movers	**Right−Movers**
X_i	X_i
X_I	S_α

for i = 1,10 and I = 1,16. The final action, in the light cone gauge, is[19]:

$$S = \frac{-1}{4\pi\alpha'} \int d\tau \int_0^\pi d\sigma \Big(\partial_a X^i \, \partial_a X^i + \sum_{I=1}^{16} \partial_a X^I \, \partial_a X^I + i\bar{S}\, \gamma^- \, (\partial_\tau + \partial_\sigma) S \Big)$$

where we enforce the constraints:

$$(\partial_\tau - \partial_\sigma)\, X^I = 0$$
$$\gamma^+ S = \tfrac{1}{2}(1 + \gamma_{11})\, S = 0$$

This action is explicitly supersymmetric under the variation:

$$\begin{cases} \delta X^i = (p^+)^{-1/2}\, \bar{\epsilon}\, \gamma^i \, S \\ \delta S^a = i\,(p^+)^{-1/2}\, \gamma_- \, \gamma_\mu \, (\partial_\tau - \partial_\sigma)\, X^\mu \, \epsilon \end{cases}$$

This rather strange split between the left and right movers (which move in different dimensions) leads to a consistent theory. The zero mode sector of the theory, in fact, yields supergravity coupled to $E_8 \times E_8$ matter and gauge fields. Scattering amplitudes, both for trees and loops, have also been constructed. The scattering of four massless gauge bosons, for example, is described by:

$$A_4 = g^2 K(\rho_i, k_i)\cdot\epsilon \; \frac{\Gamma(-1-\frac{u}{8}+\frac{U}{2})\, \Gamma(-1-\frac{s}{8}+\frac{S}{2})\, \Gamma(-1-\frac{t}{8}+\frac{T}{2})}{\Gamma(1+\frac{u}{8})\, \Gamma(1+\frac{s}{8})\, \Gamma(1+\frac{t}{8})}$$

where $K(\rho_i, k_i)$ is a complicated function of the polarizations ρ_i and

$$S = (K_1 + K_2)^2$$
$$T = (K_2 + K_3)^2$$
$$U = (K_1 + K_3)^2$$
$$S + T + U = 8$$

and where K_i represents the lattice vector describing the isospin nature of the vector meson in the root space of $E_8 \times E_8$.

Let us count how many states there are in the theory, to check that the theory is really $E_8 \times E_8$ invariant. The left moving sector, for example, has a total of 73,764 states at the first massive level. The right- moving sector has 256 states. Thus, we have a total, of 256 x 73,764 = 18,883,584 states at this level! It is thus prohibitively difficult to construct the spectrum in terms of $E_8 \times E_8$ multiplets. The proof that the spectrum can be grouped according to $E_8 \times E_8$ multiplets involves using the *Kac-Moody algebras*[32]. The Kac-Moody generators have the following commutation relation:

$$[\, T^i_m , T^j_n \,] = i f^{ijl}\, T^l_{m+n} + km\, \delta^{ij} \delta_{m,-n}$$

Notice that the n=0 sector is just an ordinary Lie algebra. We can re- write this algebra in the Cartan-Weyl representation by smearing all operators around a circle parametrized by the angle θ:

$$[\, H_i(\theta), H_j(\theta) \,] = 0$$
$$[\, H_i(\theta), E_{\vec{\alpha}}(\theta') \,] = -2\pi\delta(\theta-\theta')\, \vec{\alpha}_i\, E_{\vec{\alpha}}(\theta)$$

and:

$$[\, E_{\vec{\alpha}}(\theta), E_{-\vec{\alpha}}(\theta) \,] = 2\pi\delta(\theta-\theta') \sum_i \vec{\alpha}_i H_i(\theta) + 2\pi i \delta'(\theta-\theta')$$

and

$$[\, E_{\vec{\alpha}}(\theta), E_{\vec{\beta}}(\theta') \,] = \begin{cases} 2\pi\delta(\theta-\theta')\, E_{\vec{\alpha}+\vec{\beta}}(\theta) & \text{if } \vec{\alpha}+\vec{\beta} \in \Gamma \\ 0 & \textit{otherwise} \end{cases}$$

Specifically, the representation of the Kac-Moody generators used by the heterotic string are:

$$p_I = H_I \; ; \quad E(K) = \int \frac{dz}{2\pi i z} : e^{2iK^I X^I_L(z)} : C(K)$$

where K is a lattice vector and C(K) is a cocycle.

Using vertex operators, we can explicitly construct a representation of the Kac-Moody algebra. Thus, since the vertex operators generate the spectrum of the theory, the spectrum itself must be invariant under Kac-Moody transformations. Therefore, even though the spectrum is not manifestly $E_8 \times E_8$ symmetric, we have established that it actually is symmetric under that group.

IV. MULTI-LOOPS

After the original discovery of the Euler beta function by Veneziano and

Suzuki, KSV proposed to treat it as a Born term; higher loops could then be generated by the unitarity relation:

$$\text{Im}\, T_{ij} = -\tfrac{1}{2} \sum_{n} < i\,|T\,|n> < n\,|T^{\dagger}\,|j>$$

Fortunately, the mathematicians have already calculated the Neumann functions on surfaces with holes. They are particular examples of automorphic functions.

For the first loop, the Neumann function for a Riemann surface with a hole is given in terms of:

$$\ln \psi(x, w) = \ln (1-x) - \tfrac{1}{2}\ln x + \frac{\ln^2 x}{2\ln w}$$

$$+ \left[\sum_{n=1}^{\infty} \ln (1 - w^n x) + \ln (1 - w^n/x) - 2\ln (1 - w^n) \right]$$

(This function preserves the property that the point z is identified with the point wz, where w is complex.) Thus, the first loop N-tachyon scattering amplitude for closed strings is equal to:

$$A = \int \prod_{i=1}^{N} d^2 z_i \; |w|^{-4} \; |f(w)|^{-48} \left(\frac{-4\pi}{\ln |w|} \right) \prod_{i<j} \chi_{ij}^{\frac{1}{2}k_i \cdot k_j}$$

where: $\nu_j = (2\pi i)^{-1} \ln z_1 z_2 \cdots z_j$; $\nu_{ji} = \nu_j - \nu_i$;
$\tau = \nu_M = (2\pi i)^{-1} \ln w$; $w = z_1 z_2 \cdots z_N$; $c_{ji} = z_{i+1} z_{i+2} \cdots z_j$; and:

$$\chi (z, w) = \exp \frac{(\ln^2 |z|)}{2 \ln |w|} \left| z^{-\frac{1}{2}}(1-z) \prod_{m=1}^{\infty} \left\{ \frac{(1-w^m z)(1-w^m/z)}{(1-w^m)^2} \right\} \right|$$

$$\chi_{ij} = 2\pi \exp \left(\frac{-\pi(\text{Im}\, \nu_{ji})^2}{\text{Im}\, \tau} \right) \left| \frac{\Theta_1(\nu_{ji} \mid \tau)}{\Theta'_1(0 \mid \tau)} \right|$$

Let us re-write this as:

$$A = \int d^2\tau \, (\text{Im}\, \tau)^{-2} \, C(\tau) \, F(\tau)$$

where:

$$C(\tau) = 4(\tfrac{1}{2}\, \text{Im}\, \tau)^{-12} \, e^{4\pi\, \text{Im}\, \tau} \, |f(e^{2\pi i \tau})|^{-48}$$

$$F(\tau) = \pi^N \, \text{Im}\, \tau \int \prod_{i=1}^{N-1} d^2 \nu_i \prod_{i<j} (\chi_{ij})^{\frac{1}{2}k_i \cdot k_j}$$

A careful analysis of the divergences shows that the amplitudes of string theory are not ultraviolet divergent at all. If we analyze singularities in $q = e^{2\pi^2/\ln w}$, then we find infrared divergences correspond to squeezing the interior holes of the Riemann surface, i.e. infrared divergences lie at the boundaries of the integration region corresponding to a change in topology of the Riemann surface. The single loop open string graph diverges when the interior hole vanishes. This corresponds, by conformal invariance, to extracting a closed loop with zero momentum. Thus, the divergences correspond to a tachyon or a dilaton scattering into the vacuum.

We find that the Type I superstring only has q^{-1} poles, and hence slope renormalization can absorb this infinity. A q^{-3} divergence never occurs because the boson and fermion internal legs cancel against each other.

We also find that the Type II superstring single loop amplitude is actually finite. Also, there are no 2 and 3 point single loop graphs, and hence self-energy insertions on external legs are simply forbidden.

At first, it may appear trivial to write down the N-loop amplitude: the recipe is to calculate the Neumann function of the disk with N-holes (first written down by Burnside in the 1890s) and insert it into the path integral. However, the problem, for closed loops, is that the resulting answer is infinite. There is a new, hidden symmetry group for the closed loop surfaces that needs to be factored out, which is called *the modular group*.

For example, take the torus. Let the a-cycle be any closed loop drawn around the torus, and a b-cycle any closed loop which encircles the interior hole of the torus. If we take a razor blade, slice the torus along these cycles, and then open up the resulting surface, we find a parallelogram with opposite sides identified (which has the topology of a space-war video game; moving our rocket off the right edge, we emerge from the left). The horizontal width of the parallelogram is 1 and τ represents the complex co- ordinates of one corner. Obviously, any point in this plane z is identical with a point z + 1, as well as z + τ. But we also want our plane to reflect the fact that it must be invariant under interchange of the a- and b-cycles Thus, the plane has the symmetry:

$$\tau \rightarrow \tau + 1$$

$$\tau \rightarrow \frac{-1}{\tau}$$

This transformation, in turn, can be re-written as:

$$\tau = \frac{a\tau + b}{c\tau + d}$$

where a,b,c, and d are *integers* and ad - bc = 1 . This group is called the modular group SL(2, **Z**).

Naively integrating over all τ thus integrates over an infinite number of copies of the same thing. The single-loop amplitude must integrate over one and only one modular invariant region of the z plane, called the fundamental region.

For example, the integrand of the single loop amplitude is invariant under the modular transformation[33]:

$$\chi\left(\frac{\nu}{c\tau + d}, \frac{a\tau + b}{c\tau + d}\right) = |c\tau + d|^{-1} \chi(\nu, \tau)$$

so that $F(\tau) = F(\tilde{\tau})$ is invariant.

For the multi-loop amplitude, the situation becomes worse and, in fact, the region of integration is poorly understood for genus 2 and higher. Mathematicians over the past century have made little headway in finding co-ordinates which integrate over one copy of the fundamental region. This represents one of the most difficult problems facing multi-loop calculations.

Let us define the problem more precisely. Let us define the "intersection number" of any two cycles. If two cycles never meet (see fig. 2), then their intersection number is zero. If they cross each other, then their intersection number is ± 1 (depending on the orientation of their cross). For the torus, for example, the intersection numbers for the a- and b-cycle form a matrix:

$$\begin{pmatrix} 0 & 1 \\ -1 & 0 \end{pmatrix} \leftrightarrow \begin{cases} (a, a) = (b, b) = 0 \\ (a, b) = -(b, a) = 1 \end{cases}$$

What is the group which leaves the intersection matrix invariant? Notice that (by definition) the group Sp(2, **Z**) leaves this metric invariant. Notice also that Sp(2, **Z**) is precisely isomorphic to the modular group Sl(2, **Z**), as expected.

Now let us calculate the intersection matrix for a genus g Riemann surface. Because cycles around the ith hole do not cross cycles around the jth hole, the intersection matrix is simple: just place the above 2 X 2 matrix along the diagonal of a 2g X 2g matrix. The group which leaves this metric invariant is thus Sp(2g, **Z**), which

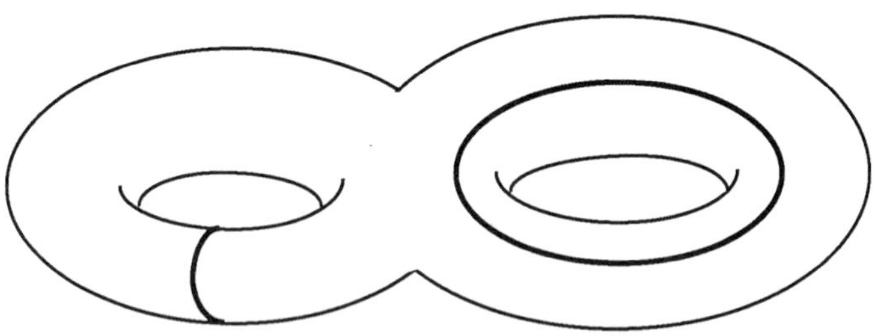

Fig. 2: Non-intersecting a-cycles (on the left) and b-cycles (on the right) for a genus two Riemann surface.

is the modular group for the genus g torus.

The simplest way to generate the modular group Sp(2g, **Z**) is to use the method of *Dehn twists*. Take a genus g Riemann surface and cut it along one of its cycles. Then twist one end of the cut by 360 degrees, and resew the two cut ends back again. Notice that we have now made a *global* transformation in the Riemann surface which cannot be smoothly mapped back to the identity, i.e. when Dehn twists D_a and D_b cut a torus around the a- or b-cycles, we find that cycles transform as $D_a(a) = a \;\; ; \;\; D_a(b) = a + b$.

The whole point of this topological discussion is that we must divide out by the group generated by Dehn twists, called the Mapping Class Group, which contains Sp(2g, **Z**) as a sub-group, when we construct the multi-loop amplitude. Let us call, loosely speaking, *Teichmuller space* as the space generated by integrating over all the parameters of the multi-loop amplitude. z_i and τ, for example, are Teichmuller parameters. Each point in Teichmuller space corresponds to an entire Riemann surface. This space, of course, is too big, because we over-count those surfaces which are related to each other by Dehn twists. In practice, we wish to integrate over just one copy of the fundamental domain, which spans *moduli space*. Thus, the fundamental problem facing all of multi-loop calculations is to construct moduli space, defined as:

$$\frac{\textit{Teichmuller Space}}{\textit{Mapping Class Group}} = \textit{Moduli Space}$$

Each loop requires three moduli to parametrize it. Thus, the dimension of moduli space is equal to 3g - 3. Let us now review the four main parametrizations of the multi-loop amplitude in terms of moduli:

1) Schottky groups[8-10].

2) Theta functions[34-7].

3) Weil-Petersson co-ordinates[38].

4) Light cone co-ordinates[13-15].

Let us begin with the first parametrization, in terms of *Schottky groups*, which was the first parametrization to be discovered in 1970[8-10]. As we have seen, the string amplitudes are all projectively invariant. A projective transformation can be re-written as the following:

$$P(z) = \frac{az + b}{cz + d} \quad \to \quad \frac{P(z) - z_1}{P(z) - z_2} = X \, \frac{z - z_1}{z - z_2}$$

Instead of the numbers a,b,c,d, we now have the number X, which is called the "multiplier" of the projective transformation, and the numbers z_1 and z_2, which are fixed points or invariant points (i.e. $P(z_1) = z_1$). These three parameters constitute the three moduli for the each loop within the multi-loop amplitude. For the single loop amplitude, the multiplier is equal to w and the invariant points are equal to $w^{\pm\infty} = 0, \infty$. Notice that the successive action of a projective transformation on a circle divides up the complex plane into a sequence of concentric circles, which converge on the fixed points.

The N-loop amplitude is constructed from N projective transformations P_i. The final answer is given by:

$$A_N = \int \prod_{\alpha=1}^{2N+1} d^{26}k_\alpha \, d\mu \prod_{\{P\}} \prod_{\{\tilde{P}\}} (1 - X_{\{\tilde{P}\}})^{-24}$$
$$\times \prod_{1=i<j=M} (z_i - \{P\}z_j)^{k_i \cdot k_j} \prod_{\beta=1}^{N} dX_\beta \, X_\beta^{-\alpha(k_\alpha)-1}$$
$$\times \prod_{i=1}^{M} \left(\frac{z_i - \{P\}x_\lambda^{(2)}}{z_i - \{P\}x_\lambda^{(1)}} \right)^{k_i \cdot k_\lambda}$$
$$\times \left\{ \frac{x_\beta^{(1)} - \{P\}x_\lambda^{(1)}}{x_\beta^{(2)} - \{P\}x_\lambda^{(1)}} \, \frac{x_\beta^{(2)} - \{P\}x_\lambda^{(2)}}{x_\beta^{(2)} - \{P\}x_\lambda^{(2)}} \right\}^{\frac{1}{2}k_\beta \cdot k_\lambda}$$

where:

$$d\mu = \prod_{i=1}^{M} dz_i dV_{abc}^{-1} \prod_{\alpha=1}^{N} dx_\alpha^{(1)} \, dx_\alpha^{(2)} \, (x_\alpha^{(1)} - x_\alpha^{(2)})^{-2}$$

where $\{P\}$ represents the product of all distinct projective transformations P_i, where the Roman letters refer to the external lines, and the Greek letters represent the loops. Notice that the region of integration lies between the limit points of the set $\{P\}$. We must have these limit points forming a *discrete* region on the real axis (which is the region of integration for the multi-loop amplitude). Projective groups which satisfy this non-trivial property are called Schottky groups.

For open loops, this gives us the exact answer for the multi-loop amplitude. However, for closed loops we must truncate the region of integration in a way that is unknown.

Second, let us turn our attention to another promising approach, the *theta function*[34–5] method, which is very elegant (but ultimately faces the Schottky problem). Because mathematicians have proven that moduli space is a complex space, it is natural to use complex co-ordinates. Specifically, we will use the complex *period matrix* Ω. Any Riemann surface of genus g has g Abelian integrals ω_i which can be defined on it. The normalization of these integrals is arbitrary, but usually taken so that its integrals over the various cycles are as follows:

$$\int_{a_i} \omega_j = \delta_{ij} \; ; \quad \int_{b_i} \omega_j = \Omega_{ij}$$

This period matrix generalizes the complex variable τ found for the torus. It can be shown rather easily that the period matrix is symmetric and has a positive definite imaginary part. In general, the $\frac{1}{2} g(g+1)$ elements of a symmetric $g \times g$ matrix (with positive definite imaginary part) generates a space called the Siegel upper half plane.

Under a Sp(2g,Z) transformation, the period matrix transforms as:

$$\Omega' = (A\Omega + B)(C\Omega + D)^{-1}$$

where the A,B,C,D are each symplectic $g \times g$ matrices. Thus, two different period matrices may describe the same Riemann surface if they are related by a Sp(2g,Z) transformation.

Now that we have a mathematical description of the mapping class group in terms of Sp(2g,Z) defined on Dehn twists, the next task is to write down functions defined over a Riemann surface with g handles. If we take the single loop Θ function, and then force it to have these periodicity properties over the Riemann surface with g handles, we naturally add an additional set of infinite terms to the summation. Eventually, this summation approaches:

$$\Theta(\mathbf{z}|\Omega) = \sum_{n \in Z^g} e^{i\pi \mathbf{n}\cdot\Omega\cdot\mathbf{n} + 2\pi i \mathbf{n}\cdot\mathbf{z}}$$

where the sum over the vector $\mathbf{n}$ is taken over a g dimensional lattice. The Θ function is defined not over the complex variable z, but over the vector $\mathbf{z}$ defined by:

$$\mathbf{z} = \int_{p_o}^{z} \boldsymbol{\omega}$$

where p_o is an arbitrary point on the surface. Since ω_i has g components, then z_i is also a vector with g components.

It also possible to construct Θ functions which have *spin structures* defined over them, i.e. if we take the NS-R spinor ψ_μ and move it around the a- and b-cycles, we pick up factors of ± 1. The spin structure $[\alpha]$ of a manifold is defined by specifying complete set of all ± 1 factors, denoted by the vectors **a** and **b**, which we pick up by moving the spinor around all cycles.

Then we can now define the generalized Θ function with spin structure:

$$\Theta[\alpha](\mathbf{z}|\boldsymbol{\Omega}) = \sum_{n \in Z^g} e^{i\pi(\mathbf{n}+\mathbf{a})\cdot\boldsymbol{\Omega}\cdot(\mathbf{n}+\mathbf{a}) + 2\pi i(\mathbf{n}+\mathbf{a})\cdot(\mathbf{z}+\mathbf{b})}$$

$$= e^{i\pi\mathbf{a}\cdot\boldsymbol{\Omega}\cdot\mathbf{a} + 2\pi i\mathbf{a}(\mathbf{z}+\mathbf{b})}\, \Theta(\mathbf{z}+\boldsymbol{\Omega}\,\mathbf{a}+\mathbf{b}|\boldsymbol{\Omega})$$

It is now a straightforward task to calculate how these generalized Θ functions transform under a mapping class group transformation. Under Sp(2g,Z), we find:

$$\Theta[\alpha](\mathbf{z}'|\Omega') = \epsilon\, e^{-i\pi\phi} \det(C\Omega + D)^{1/2}\, \Theta(\mathbf{z}|\Omega)$$

where

$$\mathbf{z}' = (C\Omega + D)^{-1}\cdot\mathbf{z}$$

and

$$\begin{bmatrix}\mathbf{a}'\\ \mathbf{b}'\end{bmatrix} = \begin{pmatrix} D & -C\\ -B & A\end{pmatrix}\begin{bmatrix}\mathbf{a}\\ \mathbf{b}\end{bmatrix} = \frac{1}{2}\begin{bmatrix}(CD^T)_d\\ (AB^T)_d\end{bmatrix}$$

and the phase factor is given by:

$$\phi = \mathbf{a}D^T B\mathbf{a} + \mathbf{b}C^T A\mathbf{b} - 2\mathbf{a}B^T C\mathbf{b} - (\mathbf{a}D^T - \mathbf{b}C^T)(AB^T)_d$$

where T represents the transpose and d represents taking the diagonal element. The symbol ϵ represents the eighth root of unity (with a certain restriction which does not worry us here).

Now that we have defined Θ functions with the proper periodicity on the Riemann surface, our next task is to actually calculate the measure of the multi-loop amplitude. Here, we are greatly aided by a remarkable result of Belavin and Knizhnik[36], which simply states that the measure of the multi-loop amplitude, because it is the product of left-moving and right-moving modes, is the square of a certain holomorphic function (up to zero modes):

$$\textit{Holomorphic Factorization}: \quad Z = \int_F \frac{\eta \wedge \bar{\eta}}{(\det \operatorname{Im}\Omega)^{13}}$$

where F is the fundamental region of the surface, and η is a 3g-3 form:

$$\eta = \prod_{i=1}^{3g-3} dy_i\, F(y)$$

where y_i are a parametrization of the Teichmüller parameters, and F(y) is holomorphic, has no zeroes, and has second order poles at the boundary where the surface degenerates.

As a result, almost by inspection, we can write down the measure of the multi-loop amplitude by examining the complex structure of theta functions!

Unfortunately, the elegance and beauty of the theta function approach must be balanced against the fact that the period matrix does not have the correct number of dimensions. As we have seen, there are $\frac{1}{2}g(g+1)$ co-ordinates in the period matrix, but the dimension of moduli space is 3 g - 3. Thus, they only agree for two and three loops. For loops higher than g=4, the period matrix no longer adequately parametrizes moduli space. This is the *Schottky problem,* i.e., determining when a matrix in the Siegel upper half plane is the period matrix of a Riemann surface. Thus, the two important stumbling blocks for the theta function approach are the Schottky problem and the fact that the fundamental domain for moduli space in this formulation is unknown. (Only within the past few years have mathematicians solved the Schottky problem[37], but the answer is so highly non-linear that its application to string theory is not clear.)

Finally, let us say a few words about the last two parametrizations. The *Weil-Petersson parametrization*[38] is attractive because it has a nice, visual appeal. Imagine a pair of pants. Obviously, by chopping up a genus g Riemann surface along its various a-cycles, we wind up with a bunch of disconnected pairs of pants. Thus, the pants represents, in some sense, the elemental unit of string interaction. For the 6 real moduli for each loop, take the geodesic circumference l_i of each of the three legs of the pants, and twist each leg by the angle θ_i. Then the invariant measure for the multi-loop amplitude is simply:

$$Measure = \prod_{i=1}^{3g-3} l_i \, dl_i \, d\theta_i$$

The simplicity of this result must be balanced against the fact that this only provides a parametrization of Teichmuller space, not moduli space. Thus, this parametrization over-counts moduli space an infinite number of times.

Finally, there are *light cone co-ordinates,* which have solved the century-old problem of explicitly finding the fundamental domain of moduli space. Along with Penner co-ordinates, light-cone co-ordinates are the only way in which to successfully triangulate moduli space.

Light cone perturbation theory is guaranteed to give us precisely one cover of moduli space, without having to artificially truncate higher string amplitudes by hand. Thus, string field theory has solved a mathematical problem that dates back to the time of Riemann, a successful triangulation of moduli space.

Although a rigorous proof of the finiteness of the superstring is still lacking, the mathematicians have given us a way in which to rigorously reformulate this question. In particular, the potential singularities of the multi-loop amplitude can be re-expressed in terms of the Selberg zeta function[39]:

$$Z(s) = \prod_{l \text{ primitive}} \prod_{p=0}^{\infty} \left[1 - \nu(\gamma) \, e^{-(s+p)l} \right]$$

Analysis of the Selberg functions confirms the fact that the measure has a singularity where the topology of the genus g surface degenerates. If we now squeeze any of the loops and deform the topology, the Selberg zeta function tells us that the singularities factorize, just as we expected.

Much work has been on multi-loop superstring amplitudes, but the theory is much less well developed than the bosonic theory. There is an ambiguity[40] in defining the superstring amplitudes which has caused some confusion (but it appears to be a total derivative in moduli space). Furthermore, superstring amplitudes (excluding the light cone parametrization) suffers from a super-Schottky problem.

V. CLASSICAL SOLUTIONS AND CONFORMAL FIELD THEORY

So far, we have the esthetically pleasing result that there are only four consistent string theories in 10 dimensions, the Type I, IIa and b, and the heterotic string. Now, we wish to compactify the remaining 6 dimensions to produce a physically realistic four dimensional string theory. Although the model is reasonably constrained in 10 dimensions, the problem, however, is that literally *millions* of classical solutions are possible in four dimensions. With a little bit of work, we can even produce realistic models with 3 to 4 generations of quarks and leptons and the minimal symmetry group SU(3) X SU(2) X U(1). Let us study some of these vacua (keeping in mind the true non- perturbative vacuum does not have to resemble any of these perturbative classical vacua).

Our first choice is to compactify the ten dimensional space on the six torus, leaving us with four dimensional space-time. However, compactifying on this surface transforms N=1 supersymmetry in 10 dimensions into N=4 supersymmetry in 4 dimensions, which is phenomenologically undesirable.

Thus, we must search for new compactification schemes which will be more phenomenologically acceptable. There are four major types of acceptable classical solutions that have been intensely investigated:

1) Calabi-Yau manifolds[20,41]
2) Orbifolds[21,42]
3) Free fermions and free bosons[43]
4) Asymmetric orbifolds[44].

We begin our discussion with *Calabi-Yau manifolds* (keeping in mind that that there are large overlaps among the four sets of solutions). We start with the following set of reasonable assumptions which we want in a phenomenologically acceptable model:

1) the 10 dimensional universe has compactified to four dimensional space times

some unknown manifold K^6:

$$M^{10} \to M^4 \times K^6$$

where M^4 is a maximally symmetric space,

$$R_{\mu\nu\alpha\beta} = \frac{R}{12} (g_{\mu\alpha} g_{\nu\beta} - g_{\mu\beta} g_{\nu\alpha})$$

and K is compact.

2) we assume that an N = 1 local supersymmetry remains unbroken after the compactification.

3) we assume that some of the bosonic fields can be set to zero:

$$H = d\phi = 0$$

These are rather mild assumptions, but we find that they provide an enormously powerful set of restrictions on the six dimensional manifold K. In particular, the assumption that N=1 supersymmetry survives the compactification is a surprisingly stringent assumption. It implies that the variation of the spin 3/2 gravitino ψ_A found in supergravity (the massless sector of the string) must be zero:

$$\delta\psi_A = \frac{1}{\kappa} D_A \epsilon = 0$$

This is non-trivial: the existence of a covariantly constant spinor ϵ places an enormous number of constraints on the manifolds. In particular, if we differentiate the equation once again:

$$D_i \epsilon = 0 \to [D_j, D_i] \epsilon \to R_{ijkl} \Gamma^{kl} \epsilon = 0$$

Thus, we can show that the space is Ricci flat, $R_{ij} = 0$.

Let us define the *holonomy group* as the group generated when we take a spinor around successive closed paths around a point. Normally, when going around a closed loop, spinors accumulate a phase factor which, in six dimensions, generates the holonomy group SO(6), or equivalently SU(4). But if we constrain ϵ to be covariantly constant, the holonomy group becomes a subgroup of SU(4). To see this, note that it transforms as:

$$\epsilon \to U \epsilon$$

By a SU(4) transformation, any spinor can be brought into the column vector (0 , 0 , 0 , ϵ_o). This, in turn, means that the U matrix only rotates the three zero indices in the column matrix. Thus, U must be a member of SU(3), and so the space K^6 has *SU(3) holonomy*.

Fortunately, Kähler manifolds that are Ricci flat (and hence have vanishing first Chern class) can be shown to be equivalent to spaces of SU(3) holonomy and are easy to construct (and are called Calabi-Yau manifolds).

The problem, however, is that now we are flooded with potentially tens of thousands of manifolds preserving N=1 supersymmetry!

Next, we wish to break $E_8 \times E_8$ down to E_6. This can be accomplished consistently by embedding the spin connection into the gauge field, thereby mixing two different manifolds. Fortunately, this gives rise to desirable results. We know that E_8 by itself does not have chiral representations, and hence is not an acceptable candidate for model building, but E_6 does. In fact, the decomposition under this breaking of the 248 elements of the adjoint of E_8 become:

$$\mathbf{248} = (\mathbf{3}, \mathbf{27}) + (\overline{\mathbf{3}}, \overline{\mathbf{27}}) + (\mathbf{8}, \mathbf{1}) + (\mathbf{1}, \mathbf{78})$$

This is gratifying, because the **27** is the most suitable representation for the quarks and leptons in GUT theories with E_6.

Next, we want to break the model even further to the standard model $SU(3) \times SU(2) \times U(1)$ without destroying N=1 supersymmetry. Normally, this is quite difficult. Breaking schemes which break the group down to the standard model also break N=1 supersymmetry. One solution to this problem is to use the method of Wilson lines. For *non-simply connected manifolds,* the Wilson loop:

$$U = P \exp \int_C A_\mu \, dx^\mu$$

does not necessarily equal 1 even if the curvature tensor vanishes. This is because we cannot always shrink closed loops into points. Thus, we can break E_6 symmetry down to a subgroup G by choosing an element of U and forcing G to commute with this element. For example, if we take an element of U which is $Z_5 \times Z_5$, we find:

$$\frac{E_6}{Z_5 \times Z_5} = SU(3)_C \times SU(2)_L \times SU(2)_R \times U(1) \times U(1)$$

Unfortunately, we see that, in addition to arriving at the standard model, we also have unwanted U(1) groups.

In addition, topological arguments allow us to calculate the number of quark-leptons in the theory. Since the number of massless modes on a manifold is related to the number of harmonic forms that can exist on that manifold, we have a simple relation between generation number and topological invariants:

$$\textit{Generation Number} = \tfrac{1}{2} \, |\chi(M)|$$

Unfortunately, the Euler number χ of Ricci flat Kähler manifolds is usually quite high, but we can always reduce this number drastically by taking a submanifold. For example, we can divide out by a discrete group which preserves a certain polynomial in the co-ordinates. A good example is CP_4 divided by $Z_5 \times Z_5$, which is not simply connected. This manifold has four generations.

Calabi-Yau manifolds are thus rich in detail and can give reasonably acceptable phenomenology. However, they are difficult to work with because the fields are highly coupled and non-linear. The remaining set of compactification schemes are simpler because they use *free fields*.

Orbifolds[21,42], for example, are probably the limiting case of a Calabi-Yau space. Orbifolds are created by taking a torus T_6 and dividing out by some discrete group Z_n, which allows fixed points.

$$\textit{Orbifold} : \ \frac{T_6}{Z_n}$$

Orbifolds are not true manifolds, because they have fixed points. (For example, think of a plane divided by Z_2. This is formed by wrapping up the plane into a cone. The tip of the cone is a fixed point. A cone is thus not a manifold.) Fortunately, these fixed points apparently do not spoil the properties of the string model. Non-trivial constraints are placed on orbifolds by modular invariance, which mixes the boundary conditions non-trivially. The boundary conditions:

$$X(\sigma_1 + 2\pi, \sigma_2) = h\, X(\sigma_1, \sigma_2)$$
$$X(\sigma_1, \sigma_2 + 2\pi) = gX(\sigma_1, \sigma_2)$$

(where h and g are elements of Z_n) can break the symmetries of the string model. The subgroup of the space-time group and the internal group which survives the process is the subgroup which commutes with g and h.

Phenomenologically, there are many solutions to this compactification, some with three generations and the group $SU(3) \times SU(2) \times U(1)^n$, which yields too many U(1) factors. Unfortunately, the method of Wilson lines does not change the rank of the group, so in general we will have too many U(1) factors.

The next method is the use of *free fermions and bosons*[43]. One step in this direction is to use modular invariance and the absence of tachyons and anomalies to systematically derive all possible solutions. We begin with the amplitude A written as a sum over spin structures $[\mathbf{a}\ \mathbf{b}]$. The total amplitude is then:

$$\sum_{\mathbf{a},\mathbf{b}} C \begin{bmatrix} \mathbf{a} \\ \mathbf{b} \end{bmatrix} A \begin{bmatrix} \mathbf{a} \\ \mathbf{b} \end{bmatrix}$$

Now we demand that the coefficients C's factorize properly, have modular invariance, and yield models which have no tachyons or anomalies. Remarkably, it is possible to solve the constraints on C and obtain solutions. The simple ones reproduce some of the known compactifications, but thousands of other compactifications are still being investigated. This yields the hope that we may be able to exhaust many possibilities and obtain a realistic model.

Lastly, there are asymmetric orbifolds where (as in the heterotic string) we have left and right-movers which are compactified on different manifolds. Again, after compactification, we demand that our amplitudes be modular invariant, which is sufficient to define a large class of these asymmetric orbifolds.

Clearly, at this point we are faced with the unpleasant task of haphazardly categorizing millions upon millions of classical solutions. Obviously, what is needed is a systematic way in which to make sense of this jungle of classical solutions.

The most elegant method may be to recognize that each classical solution corresponds to a *conformal field theory*[45-6], and hence the problem is to systematically write down all conformal field theories.

The philosophy of conformal field theory is that one can reconstruct all Green's functions of a two dimensional conformal theory by knowing its short distance behavior. For example, if a field ϕ_l, when it is close to the energy-momentum tensor T(z), satisfies the following short-distance behavior:

$$T(z)\phi_l(z_1,\bar{z}_1) = \frac{\Delta_l}{(z-z_1)^2}\,\phi_l(z_1,\bar{z}_1) + \frac{1}{z-z_1}\,\frac{\partial}{\partial z}\,\phi_l(z_1,\bar{z}_1)$$

then we say that it a primary field with conformal dimension Δ_l. Given this simple relation, we can then derive Ward identities for all Green's functions:

$$\sum_{j=1}^{N}\left[z_j^{n+1}\frac{\partial}{\partial z_j} + (n+1)\,\Delta_j z_j^n \right] < \phi_1(z_1,\bar{z}_1)\cdots\phi_N(z_N,\bar{z}_N) > \; = 0$$

Thus, without knowing the Lagrangian, we can extract almost all the information of the theory once the conformal properties of the fields are known.

In addition to primary fields, we have an infinite tower of secondary fields associated with each primary field. For example, consider the infinite number of states composed of all possible products of raising operators L_{-n} operating on the vacuum:

$$|\{\lambda\}> \;\equiv L_{-l_1}^{\lambda_1}\;\cdots\; L_{-l_n}^{\lambda_N}\,|0>$$

(This forms the universal enveloping algebra of a Lie algebra.) We call this collection of states a *Verma module*. To form an irreducible representation of the conformal group, we must have:

$$\det < \{\lambda\} \mid \{\lambda'\}> \;\neq 0$$

Remarkably, this determinant can actually be calculated[47], once we know the central term c of the algebra. From this, we can compute a discrete series of unitary (ghost free) representations for[48]:

$$c = 1 - \frac{6}{m(m+1)}$$

where m is a positive integer and $c < 1$.

The calculation of the unitary representations for the N=1 and N=2 superconformal algebras are also being intensely investigated. For example, for N=2 world sheet supersymmetry, one can show that another discrete unitary series exists for $c < 3$ if[49]:

$$c = \frac{3n}{n+2}, \quad n = 1,2, \cdots$$

Normally, a construction of representations of the N=2 superconformal algebra would be not of physical interest. However, it has been shown that N=2 world sheet supersymmetry on the background manifold creates N=1 space-time supersymmetry in four dimensions after compactification[50]. Thus, N=2 world sheet supersymmetry gives us a handle on constructing large classes of classical solutions with N=1 space-time supersymmetry.

We recall that N=1 space-time supersymmetry in four dimensions was the essential ingredient which selected out Calabi-Yau manifolds. We expect, therefore, that N=2 superconformal spaces may be related to six dimensional Calabi-Yau spaces.

Recently, Gepner[51] has actually been able to construct exact solutions for these N=2 superconformal spaces, giving us Calabi-Yau spaces which generalize the approximate results we presented earlier. What is remarkable about this construction is that mathematicians do not know the metric tensor on these spaces. Also, the construction is highly non-linear, in contrast to the simpler free field constructions mentioned earlier.

There are several hundred such Calabi-Yau spaces we can construct in this fashion. If we include all the various twists and reductions, we can obtain over a million Calabi-Yau manifolds. Because these results are exact, it is now possible to use this as a laboratory to test many of the conjectures and approximate results obtained on Calabi-Yau manifolds.

Gepner's construction is surprisingly simple, but powerful. First, we use the equivalence of the N=2 superconformal fields with the parafermion fields of Zamolodchikov and Fateev[52] (which generalize the fields of the Ising model) and the SU(2) WZW model[53] to write down the primary fields and the characters of their conformal families. Then one enforces modular invariance on these characters. Finally, one takes tensor products of various representations, preserving the relation[51]:

$$c = \sum_{i=1}^{l} \frac{3k_i}{k_i + 2} = \frac{3}{2}(10 - D)$$

where D = 4. Then we check to see which conformal field theories correspond to which Calabi-Yau manifold. In this fashion, one can reproduce the Calabi-Yau manifolds that have appeared in the literature and generate many more. For example, tensoring five copies of the k = 3 theory yields the quintic CP^4 Calabi- Yau manifold.

VI. STRING FIELD THEORY

Up to now, we have discussed the perturbative approach to string theory. However, as we have stressed, the perturbation theory is not Borel summable and the coupling constant is probably strong, so that the millions of classical vacua that have been found cannot be taken that seriously yet. Also, we have stressed that the first quantized approach taken so far seems *ad hoc*. Many of the rules for string theory seem arbitrary, contrived, and, to a beginner, certainly bewildering.

We now discuss the field theory of strings which (in principle) may give us a clue as to how to formulate the theory non-perturbatively, and where all the properties can be derived from an action, without appealing to folklore and rules of thumb. We discuss the light cone theory, the BRST theory, and the newer geometric theory, which predicts that closed string field theory is non- polynomial.

We begin the second quantized theory by postulating the existence of a string field $\Phi(X)$ which can describe a multi-string theory. The field functional Φ is not a function of σ. It is a functional of a string variable X which is defined over a region of σ :

$$\Phi(X) = \langle X|\Phi\rangle = \Phi[X(\sigma_1), X(\sigma_2), X(\sigma_3), \cdots X(\sigma_N)]$$

We postulate a Schrodinger equation[15]:

$$L(\Phi) = \Phi^*(i\partial_\tau - H)\Phi$$

$$H = \frac{\pi}{2p^+}\int d\sigma \left(P_i^2 + \frac{X'^2_i}{\pi^2} \right)$$

Notice that this action reproduces precisely the Green's functions for free strings found in the first quantized approach:

$$\Delta_{1,2} = \langle X_1, \tau_1|X_2, \tau_2\rangle$$

$$= < X_1 | e^{-iH\tau} | X_2 >$$

$$= \int_{X_1}^{X_2} DX \; e^{i\int_{\tau_1}^{\tau_2} L(\sigma,\tau)\, d\sigma\, d\tau}$$

$$= \int D^2\Phi \; \Phi^*(X_1)\, \Phi(X_2)\; e^{i\int_{X_1}^{X_2} \int d\tau\, L(\Phi)\, DX}$$

where:

$$L(X) = \frac{1}{2\pi} \int d\sigma \;\; (\dot{X}_i^2 - X'^2_i)$$

Thus, at the free level, we can describe the theory either in the first or second quantized language. There is no difference.

We can reproduce all the identities found in field theory. In particular, we can show that the Green's function is the matrix element of two fields:

$$\Delta_{12} = << 0 | \Phi^\dagger_{p^+}(X_1, \tau_1)\; \Phi_{q^+}(X_2, \tau_2) | 0 >>$$

$$= \delta(p^+ - q^+) \int DX \; e^{i\int L\, d\sigma d\tau} \prod_{\sigma} \delta(\vec{X}(\sigma,\tau_1) - \vec{X}_1(\sigma))$$

$$\times \prod_{\sigma'} \delta(\vec{X}(\sigma,\tau_2) - \vec{X}_2(\sigma))$$

where $|0>>$ is the vacuum of the second quantized string.

Strings interact in the simplest way possible: they join at their endpoints. The interactions is given by a simple Dirac delta function:

$$S_3 = \int dp^{+r}\; \delta(\sum_{r=1}^{3} p^{+r}) \int DX_{123}\; \Phi^\dagger(X_3)\; \Phi^\dagger(X_1)\; \Phi(X_2) \delta_{123} + h.c.$$

where δ_{123} is the overlap Dirac delta function:

$$\delta_{123} = \prod_{0<\sigma_1<\pi\alpha_1} \delta\,(X^3_\mu(\sigma_3) - X^1_\mu(\sigma_1)\,) \prod_{0<\sigma_2<\pi\alpha_2} \delta\,(X^3_\mu(\sigma_3) - X^2_\mu(\sigma_2)\,)$$

where: $\sigma_3 = \pi|\alpha_3| - \sigma_1$ and $\sigma_3 = \pi\alpha_2 - \sigma_2$ and the α_i sum to zero. Thus, the three string interaction is actually the simplest possible string interaction that one can postulate, a delta function. To actually construct an operatorial expression for δ_{123}, a convenient approach was pioneered by Mandelstam. Notice that the following conformal map converts the complex upper half z-plane into the ρ plane, which looks like a series of strips which collide with each other, as in a light cone diagram[14]:

$$\rho(z) = \tau + i\sigma = \sum_{n=1} \alpha_i \ln (z - x_i)$$

The points x_i on the x-axis are sent to infinity in the ρ plane, and hence correspond to strings coming in from infinity. As z hops over z_i, the ρ plane jumps by $\pi\alpha_i$, which is the width of a string. Thus, Mandelstam's map correctly reproduces the scattering of strings in the ρ plane.

For the closed string theory, the interaction is purely cubic. Surprisingly, Einstein's theory of general relativity (which is necessarily non-polynomial in the metric tensor) becomes cubic when expressed as a light cone string field theory! (This is because the higher non-polynomial interactions are generated as effective terms by the higher point string interactions. As the mass of the intermediate states go to infinity, these higher string interactions become contact terms.)

For the open string theory, however, there are actually five interactions altogether (see fig. 3). Surprisingly, the closed string interaction necessarily appears as part of the open string theory. There is a simple way to see that five interactions are

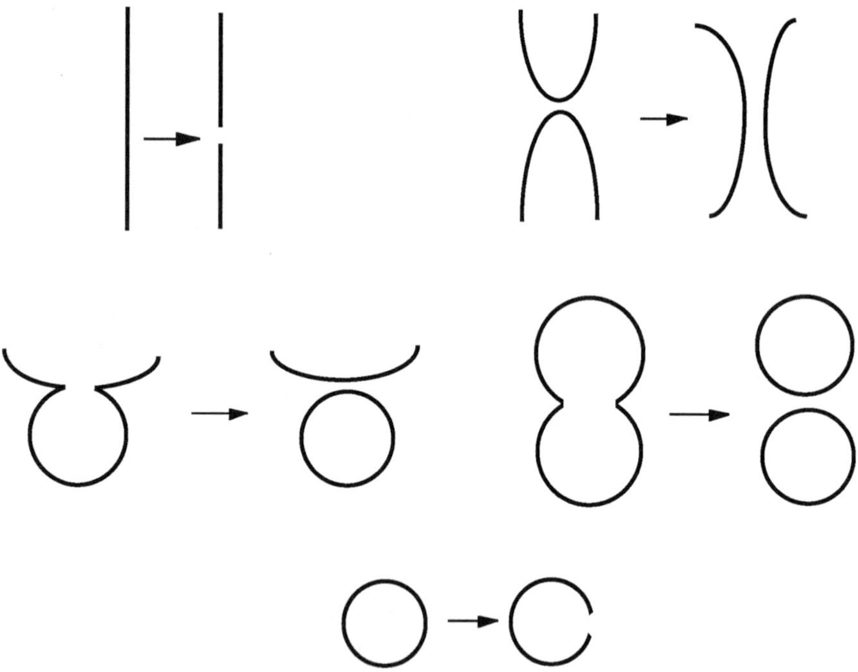

Fig. 3: The five interactions of light cone string field theory.

necessary for the open string case, and only three interactions are necessary for the closed string case.

To see this, take the real part of both sides of the above equation:

$$\tau = \sum_{n=1} \alpha_i \ln |z - x_i|$$

Notice that τ is precisely the electrostatic potential function created by fixed charges sitting at x_i with charges α_i. In particular, the equipotential lines of the diagram are obtained by keeping τ fixed. However, keeping fixed τ also means taking a snapshot of the string topology at any given time. Thus, equipotential lines of the z-plane have the same topology as interacting strings!

For example, examine the equipotential lines corresponding to two positive and two negative charges placed on opposite sides of the Koba- Nielsen disk. The equipotential lines collide at an interior point near the center of the disk. But this is precisely the topology mapped out by the four string interaction (fig. 4).

Fig. 4: The equipotential lines trace out the time evolution of the four string interactions.

Likewise, the presence of closed-open string interactions can also be reproduced by equipotential lines. Thus, using only simple electrostatics, we can completely map out the entire set of elementary string interactions in the field theory.

If we let Φ symbolically represent an open string field, and Ψ a closed string field, then the interactions of the open string light cone theory have the structure[15]:

$$\Phi^3 + \Psi^3 + \Phi^4 + \Phi\,\Psi + \Phi^2\Psi$$

while the closed string theory has the structure Ψ^3.

VII. BRST STRING FIELD THEORY

The disadvantage of the previous approach is that it is clearly not covariant. A major advance was the formulation of a truly covariant model, where the action has a geometrical interpretation. The key to the covariantized approach is to construct a theory invariant under a gauge transformation[54]:

$$\delta |\Phi> = \sum_{n=1}^{\infty} L_{-n} \; |\Lambda_n>$$

We postulated this invariance based on an analogy with the Yang-Mills theory, which is also invariant under a field which rotates into ghost fields. In fact, if we decompose the previous equation, the L_{-1} generator gives us precisely the linear part of the variation of the Yang-Mills theory. Similarly, for the closed string, the L_{-2} generator gives us the linear part of the variation of the graviton.

To generalize this discussion, notice that BRST quantization gives us an operator Q which is nilpotent, as in the d operator found in the theory of forms. Thus, a first guess which would be to postulate the following free action:

$$L = < \Phi | Q | \Phi>$$
$$\delta |\Phi> = Q \, |\Lambda>$$

Because Q is nilpotent, the action is invariant under this variation. Notice that the field $|\Phi>$ is now necessarily a composite of ghost modes as well as X-modes, i.e. it contains all possible combinations of b- and c-ghost creation operators, with the restriction that we only take ghost number $-\frac{1}{2}$ states (c has ghost number +1, and b has ghost number -1).

To introduce interactions, we follow the analogy with the theory of forms, treating Q as the counterpart of the nilpotent operator d. In the theory of forms, Witten[22] constructed an action by postulating the following rules, based on Q and a new multiplication rule for strings, which we call *:

1) Existence of a nilpotent derivation

$$Q^2 = 0$$

2) associativity of the * product

$$[A * B] * C = A * [B * C]$$

3) Leibnitz rule

$$Q[A * B] = QA * B + (-1)^A \, A * QB$$

4) the product rule:

$$\int A * B = (-1)^{AB} \int B * A$$

5) the integration rule:

$$\int QA = 0$$

With these rules, we can now show that the invariant action is:

$$L = \int \Phi * Q\Phi + \frac{2}{3}\Phi * \Phi * \Phi$$

which is a Chern-Simons term.

In order to construct this operation *, we want a totally symmetric configuration of three interacting strings. However, the string interaction given for the light cone theory is not symmetric and hence does not satisfy our requirement. We thus postulate a new theory of strings in which strings interact at their midpoints (see fig. 5). We need to postulate a new type of symmetric interaction, with the overlap delta function given by:

$$\delta_{123} = \prod_{i=1}^{3} \prod_{0<\sigma_i<\frac{1}{2}\pi} \left(X_\mu^i(\sigma_i) - X_\mu^{i-1}(\pi - \sigma_i) \right)$$

where all strings now have *equal* parametrization length π. Thus, strings interact at precisely their midpoints.

Does this unorthodox choice reproduce the Veneziano model? The answer lies in creating a conformal map which can take us from the upper half plane to the strip where strings interact symmetrically. The solution is to use Riemann cuts, with the map:

$$\rho(z) = \frac{(z+1)(z-2)(z-\frac{1}{2}) - (z^2 - z + 1)^{3/2}}{z(1-z)}$$

or an equivalent map which is multi-valued[55]. With proper normalization, we can take the Riemann cut between $\frac{1}{2} + i\frac{1}{2}\sqrt{3}$ and its complex conjugate. By drawing the equipotential lines found for both types of maps, we find precisely the topology of the midpoint interaction. With a bit more work (using the theory of third elliptic integrals), we can actually show that this interaction successfully reproduces the Veneziano model.

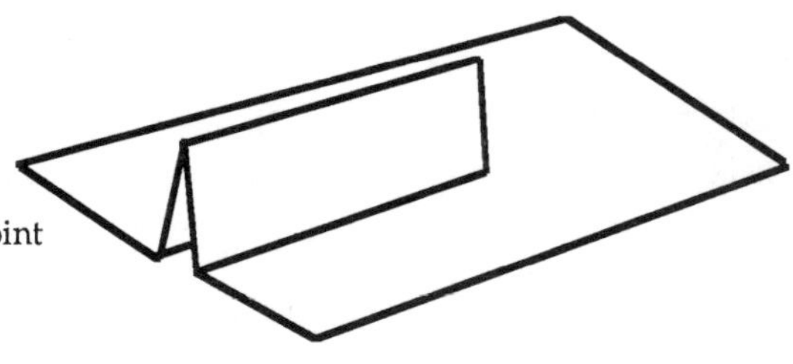

Fig. 5: Three-string scattering in the midpoint gauge.

VIII. GEOMETRIC STRING FIELD THEORY

Buoyed by the elegance of this construction, numerous attempts were made to extend this theory to closed strings and superstrings. However, the results have been disappointing. Using conformal field theory methods, Witten constructed an open superstring generalization of the bosonic theory, but it apparently has an anomaly. The vertex contains a "picture changing" operator $X(\frac{1}{2}\pi)$ defined at the midpoint. Because the algebra requires that we multiply several of these vertices together, we find X^2, which is formally infinite[56]. Various attempts have been made to eliminate this disastrous infinity[57], but none is yet universally accepted. Second, attempts to generalize the theory to closed strings have met with failure. Moduli space is severely under-counted by this midpoint geometry. The Shapiro-Virasoro closed string amplitude cannot be obtained by these midpoint interactions, so a new approach must be tried.

Furthermore, the previous theory also raises as many questions as it answers. For example, why is the older light cone theory so dramatically different from the covariant theory? By eliminating longitudinal states from Witten's theory, we certainly *do not* obtain the light cone theory, because the topology of the graphs are totally different, and the light cone theory has closed strings as elementary fields, while the covariant theory has closed strings as bound states.

Attempts were made at a "covariantized light cone" theory[58] based on putting ghosts in the older theory, but this theory has unwanted, unphysical parameters ("string length") which violate unitarity and give an infinite over-counting of states. Worse, this theory does not even reproduce the old dual model. Because proper time is lacking in this theory, it predicts 4g-3 moduli parameters for a genus g surface, which is unacceptable. The Riemann surfaces this theory generates for higher loops does not resemble anything found in the dual model.

As a result, we have recently proposed an entirely new approach to the problem. One difficulty has been that a blizzard of unnecessary gauge artifacts (ghosts, ghost counting, ghost insertions, midpoints, etc.) have clouded the essential string dynamics, which must ultimately be simple.

Geometric string theory[59−63] was created in direct analogy with general relativity and Yang-Mills theory, which can be uniquely derived from two simple group theoretical statements:

Global Symmetry: the theory must propagate pure (i.e. ghost-free) fields A_μ^a and $g_{\mu\nu}$ transforming as irreducible representations of SU(N) and the Lorentz group.

Local Symmetry: the theory must be locally invariant under SU(N) and general coordinate transformations.

The unique solution to these two simple principles is general relativity coupled to a Yang-Mills theory. Notice that the five axioms of cohomology are consequences of these two simple principles. The five axioms of cohomology by themselves have no physical content.

By direct analogy, we will postulate an entirely new gauge group, which we call *the universal string group (USG)* from which the multiplication and integration rules will be exactly determined from Clebsch-Gordonry and the tensor calculus. We therefore postulate the following two principles, from which we shall uniquely derive all of string field theory:

Global Symmetry: the theory must propagate pure fields A_σ^α transforming as an irreducible representation of Diff(S).

Local Symmetry: the theory must be locally invariant under the universal string group.

99% of the work is in defining the tensor calculus of the USG. Only 1% of the work is involved in writing down the action, which only takes a few lines.

To define the USG, let us first define a group based on *unparametrized* strings defined in physical space-time, rather than the unphysical one-dimensional parameter space of σ. Let {C} be the set of all physical unparametrized oriented open strings. Let us define a *triplet* of physical strings if they can be arranged as in fig. 6. (Notice that the strings do not touch at their midpoints, because there is not such thing as a midpoint for unparametrized strings). Let us define an *anti-triplet* of

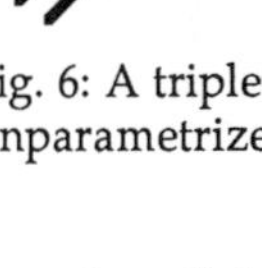

Fig. 6: A triplet of three unparametrized strings.

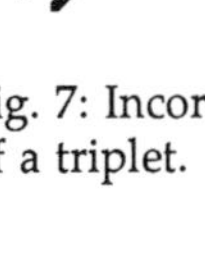

Fig. 7: Incorrect definition of a triplet.

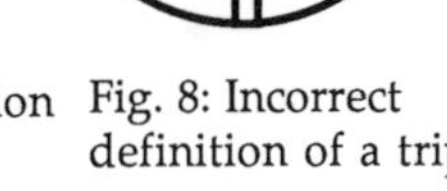

Fig. 8: Incorrect definition of a triplet.

three strings if their orientations are reversed. Let us define the string $\overline{C}$ conjugate to C if $\overline{C}$ appears in the same triplet at C. Then we define a subgroup of the USG as the group which maps the points in C into itself and all its conjugates:

$$USG : \begin{cases} C \to C \\ C \to \overline{C} \end{cases}$$

Let us now define:

$$f_{C_1C_2C_3} = \begin{cases} +1 \text{ for } \textit{triplets} \\ -1 \text{ for } \textit{anti-triplets} \\ 0 \textit{ otherwise} \end{cases}$$

Then the structure constants for the string group are:

$$f_{C_1C_2}^{C_3} = f_{C_1C_2\bar{C}_3}$$

$$f_{C_1C_2}^{C_3} = -f_{C_2C_1}^{C_3}$$

Let us define L_C to be an operator associated with a physical string C. The index C is a member of the nondenumerable set given by Cantor's transfinite numbers. Then the algebra for the string group is defined as:

$$[L_{C_1}, L_{C_2}] = f_{C_1C_2}^{C_3} L_{C_3}$$

(We sum over repeated indices. Notice that the joining points of several strings do not necessarily have to meet). We call this a *transfinite algebra*.

The essential point is that the Jacobi identities are exactly satisfied for this algebra. For example,

$$[L_{C_{[1}}, [L_{C_2}, L_{C_{3]}}]] = 0$$

When we analyze the Jacobi identity, we find: we find $1 - 1 + 0 = 0$. We can now define how covariant ϕ_C and contravariant fields ϕ^C transform under the string group:

$$\delta\phi_{C_1} = -f_{C_1C_2}^{C_3} \phi_{C_3} \Lambda^{C_2}$$

$$\delta\phi^{C_1} = f_{C_2C_3}^{C_1} \phi^{C_2} \Lambda^{C_3}$$

Notice that we now have two invariants under the string group:

$$\delta(\phi^C \phi_C) = 0$$

$$\delta(\phi^{C_1}\phi^{C_2}\phi^{C_3} f_{C_1C_2C_3}) = 0$$

Now observe that a theory based on fig. 7 does not obey the Jacobi identities. By checking the Jacobi identity for these three strings, we see that $1+0+0 \neq 0$ Furthermore, we find that the ϕ^3 interaction is not gauge invariant:

$$\delta(\phi^3) \neq 0$$

There is a way out of this problem, which is to add a four-string interaction, such that:

$$\delta(\phi^3+g\phi^4)\sim 0$$

This, in fact, is the origin of the four-string interaction.

The important observation is that when we generalize these results to the closed string case, we find precisely the *opposite* situation, that the endpoint gauge closes without any four string interaction.

Let us naively choose fig. 8 as a triplet. Then:

a) triplets are their own anti-triplets. It is easy to show that, by rotating a triplet of closed strings in physical space, we can convert it into the anti-cyclic ordering. Thus, the algebra collapses.

b) it is impossible to construct constant anti-symmetric tensors like $f_{C_1C_2}^{C_3}$ in loop space for symmetric configurations. By rotating a triplet, we can always invert the first and second strings and thus show that the tensor is symmetric. Thus, anti-symmetric structure constants are impossible to define.

c) the Jacobi identities do not close for symmetric triplet configurations. Because this configuration is fully symmetric, it produces four-string diagrams in the Jacobi identity with the topology of a tetrahedron, as in fig. 9. Because these tetrahedron graphs are symmetrical, we find that $1+1+1\neq 0$

Thus, we are now forced to adopt a new approach to closed strings.

In order to cure the problem with Jacobi identities, we are now forced to take diagrams such that $|L_1|+|L_2|=|L_3|$ as in fig. 10. This new definition of a triplet will guarantee that diagrams like fig. 9 never appear in the Jacobi identity. Notice that we can define an outer and inner string in fig. 10 because the configuration is not totally symmetric in three strings. Then:

Fig. 9: Tetrahedron graph created by Jacobi identities.

Fig. 10: Correct definition of a triplet of closed strings.

$$f_{C_1C_2C_3} = \begin{cases} +1(-1) \text{ for } \textit{triplets} \text{ if } C_3 = \textit{outer (inner) string} \\ 0 \textit{ otherwise} \end{cases}$$

Notice, as we have stressed, that only symmetric matrices in C_1 and C_2 are possible for closed strings. Now define:

$$f_{C_1C_2}^{C_3} = f_{C_1C_2\tilde{C}_3}$$

where $\tilde{C}$ is a closed string with reversed orientation from C.

Our string algebra is now defined as:

$$\{L_{C_1}, L_{C_2}\} = \hat{f}_{C_1C_2}^{C_3} L_{C_3}$$

There are several surprising features to this equation. Notice that we are forced to introduce *anti-commutators* rather than commutators. Thus, the generators are Grassman-like. In order for the right hand side to have the same statistics as the left hand side, we are forced to consider $\hat{f}$ as a Grassman operator. Thus, the algebra is not a Lie algebra in the usual sense. (It becomes a standard Lie algebra only when we introduce a parametrization.)

We will take the following choice:

$$\hat{f}_{C_1C_2}^{C_3} = f_{C_1C_2}^{C_3} \hat{\iota}_{C_1 \cap C_2 \cap C_3}$$

Notice that $\hat{\iota}$ is a Grassman operator which is defined only at the interaction point of three strings (the endpoint) in a triplet as in fig. 10. If we now calculate the Jacobi identities, we find that 1 - 1 + 0 = 0. The crucial minus sign does not come from the anti-symmetry of the structure constants, because the structure constants are symmetric. It comes from moving the Grassman variable defined at the interaction points past each other. (This is the origin of the "ghost insertion operator" found in BRST theories.)

We can show that

$$\delta(\phi^C \phi_C) = 0$$

$$\delta(\phi_{C_1} \phi_{C_2} \phi_{C_3} \hat{f}_{\tilde{C}_1\tilde{C}_2}^{C_3}) = 0$$

where ϕ^C and ϕ_C are independent fields. Thus, ϕ^2 and ϕ^3 terms are again invariant.

If, however, we were to take the fully symmetric choice of three closed strings as our basic triplet, then we would need an infinite number of polyhedra to close the algebra. This, in fact, is the origin of the non-polynomial nature of the USG in the midpoint gauge. *Thus, the four open string interaction and the non-polynomial closed string interactions are just non-linear realizations of the USG.*

Now that we have given the generators L_C of the string group (SG), let us give the structure of the gauge group of string field theory, the universal string group (USG):

$$SG = \frac{USG}{\text{Diff(S)} \times \text{Weyl} \times \text{general covariance}}$$

i.e. the USG, divided out by the symmetries of the first quantized string theory, give us the unparametrized string group. (For simpicity, we will assume flat space and delete a discussion of the important Weyl sector of the theory[62]).

To incorporate reparametrization invariance, we must necessarily introduce a larger set of fields, including vierbeins and connections. We want a theory which is manifestly invariant under local changes of the parametrizations from σ to $\bar{\sigma}$:

$$\bar{X}_\mu(\bar{\sigma}) = X_\mu(\sigma) \; ; \quad \bar{\sigma} = \sigma + \zeta^\sigma(X)$$

By functionally integrating over ζ^σ (which will eventually be embedded in the vierbein), we functionally integrate over all possible parametrizations of a string.

A field $\Phi(X)$ transforms as:

$$\delta\, \Phi(X) = \epsilon^\sigma(X)\, X'^{\mu\sigma}\, \partial_{\mu\sigma}\, \Phi(X)$$

and all structure constants implicitly contain the $\zeta^\sigma(X)$ field:

$$f_{X_1 X_2 X_3} = \prod_{i=1}^{3} \prod_{0<\sigma_i<a_{i,i-1}} \delta\left(X^i_\mu(\sigma_i) - X^{i-1}_\mu(\sigma_{i-1}) \right)$$

where a_{ij} is the parametrization distance between the ith and jth strings. Notice that by functionally integrating over $\zeta^\sigma(X)$ which appears within the structure constants, the theory functionally integrates over strings which can interact at their endpoints, midpoints, or any point in between. Thus, previous tring field theories are treated as gauge choices (called the *endpoint gauge* and the *midpoint gauge*). Also, we have a class of gauges, called the *interpolating gauge,* which allows us to smoothly interpolate between these two. In fig. 11, we see that the USG (the unified string group when we include supersymmetry) uniquely determines geometric string theory, which in turn yields the interpolating, endpoint, and midpoint gauges.

Now let us be concrete. Let L_σ represent the Virasoro generators smeared over the string. Let L_X represent the generator L_C of the string group where a physical string C is represented by a parametrized string X. The algebra of the USG is then given by:

$$[\, L_\sigma, L_\rho \,] = f^{\,\omega}_{\sigma\rho}\, L_\omega$$

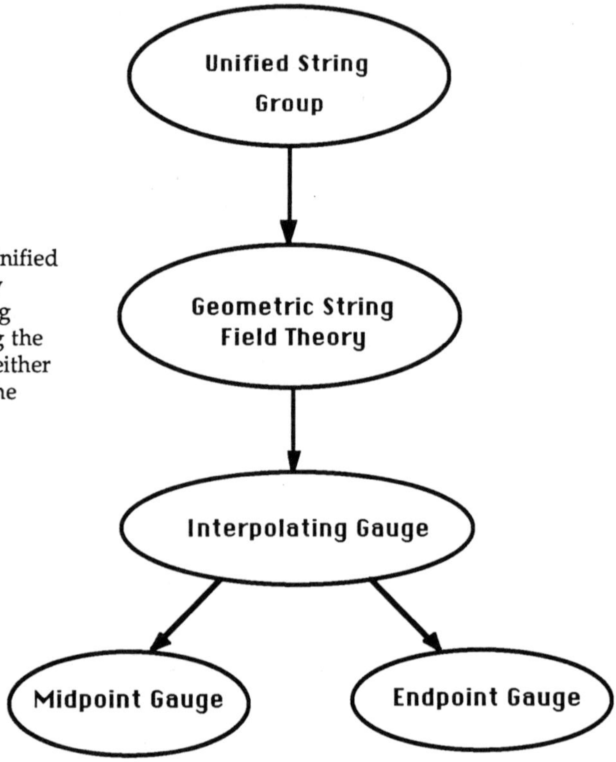

Fig. 11: Gauging the Unified String Group uniquely yields Geometric String Field Theory. By fixing the gauge, we can derive either the midpoint gauge, the enpoint gauge, or the interpolating gauge.

$$[\, L_\sigma , L_X \,] = X'^{\mu\sigma} \, \partial_{\mu\sigma} \, L_X$$

$$[\, L_X , L_Y \,] = f^Z_{XY} \, L_Z$$

The first line is the usual Virasoro algebra, where σ, ρ, ω are continuous indices. The second line states that the USG is the semi-direct product of the string group with the reparametrization group.

Anticipating the introduction of general covariance, we introduce a *tangent space* into the theory, spanned by irreducible Verma modules, labelled by α, β. If $|\alpha>$ represents an element within a Verma module, then the action of a Virasoro generator is to transform it into another member:

$$L_\sigma \, |\alpha> \; = \sum_\beta f^{\,\beta}_{\sigma\alpha} \; |\beta>$$

where the f's are Clebsch-Gordon coefficients.

Let us now introduce connection fields, covariant derivatives, and curvatures. We work in analogy with Yang-Mills theory coupled to gravity, where we have a connection field A^i_μ for SU(N), a connection field ω^{ab}_μ for reparametrizations, and a vierbein. Similarly, we will introduce three fields, the smallest necessary to gauge the USG: (a) the connection field A^{α}_{σ}, where α represents a Verma module (b) the connection field $\omega^{\alpha}_{\mu\sigma,\beta}$ for reparametrizations (c) the vierbein $e^{\nu\rho}_{\mu\sigma}$.

Thus, we have the analogy:

$$A^{i}_{\mu} \rightarrow A^{\alpha}_{\sigma}$$

$$\omega^{ab}_{\mu} \rightarrow \omega^{\alpha}_{\mu\sigma,\beta}$$

$$e^{\nu}_{\mu} \rightarrow e^{\nu\beta}_{\mu\sigma}$$

The connection field A^{α}_{σ} transforms as:

$$\delta A^{\alpha}_{\sigma} = \epsilon^{\sigma} f^{\alpha}_{\sigma\beta} A^{\beta}_{\sigma} + \epsilon^{\omega} f^{\rho}_{\omega\sigma} A^{\alpha}_{\rho}$$

From this field, we can construct the covariant derivative from the Virasoro generators:

$$D_{\sigma} = \partial_{\sigma} + A_{\sigma}$$

(where we suppress Verma module indices; see ref. 1 for notations).

Similarly, the connection field for reparametrizations will be represented by $\omega^{\beta}_{\mu\sigma,\alpha}$, so that we can define a covariant derivative of a Verma module ϕ^{α}:

$$\nabla_{\mu\sigma} \phi^{\alpha} = \partial_{\mu\sigma} \phi^{\alpha} + \omega^{\alpha}_{\mu\sigma,\beta} \phi^{\beta}$$

Also, the introduction of a vierbein $e^{\nu\rho}_{\mu\sigma}$ allows us to write down a measure which is invariant under reparametrizations:

$$| e | DX = \det |e^{\nu\rho}_{\mu\sigma}| \prod_{\mu} \prod_{n} dX^{\mu}_{n}$$

From these covariant derivatives, we can now construct curvature tensors defined on the string group and the reparametrization group:

$$[D_{\sigma}, D_{\rho}] = F_{\sigma,\rho}$$

$$[\nabla_{\mu\sigma}, \nabla_{\nu\rho}]^{\alpha}_{\beta} = R^{\alpha}_{\mu\sigma,\nu\rho,\beta}$$

Notice that the vierbein still has more than enough components to gauge reparametrizations, which only require a vector field $\zeta^{\sigma}(X)$. Thus, we must fix the constraints:

$$R^{\alpha}_{\mu\sigma,\nu\rho,\beta} = 0$$

$$\nabla_{\mu\sigma} e^{\lambda\omega}_{\nu\rho} = 0$$

which fixes the vierbein in terms of the vector field and the connection in terms of the vierbein.

Now let us construct the action. Naively, we might postulate $F_{\sigma\rho} F_{\sigma\rho}$, but the Kronecker delta $\delta_{\sigma\rho}$ is *not* a constant tnesor under the USG. We also know that D_σ has weight 2, while an invariant action must have weight 1, so we are forced to introduce a new constant tensor which resembles a Dirac matrix. The Dirac γ^μ, of course, can be viewed as a Clebsch-Gordon coefficient between spinors and vectors. Similarly, we need a new Clebsch-Gordon coefficient γ^σ which takes us from vectors ϕ_σ to Verma modules ϕ^α. Products of these matrices are also constant tensors:

$$\epsilon^{\sigma_1 \sigma_2 \cdots \sigma_N} = \prod_{i=1}^{N} \gamma^{\sigma_i}$$

The final (and unique) action is therefore:

$$L = \det |e^{\nu\rho}_{\mu\sigma}| \left[\epsilon^{\sigma\rho\theta} [A_\sigma \times \nabla_\rho \times A_\theta + \frac{2}{3} A_\sigma \times A_\rho \times A_\theta] \right]$$

To fix the gauge, we notice that the theory is invariant under:

$$\delta A^\alpha_\sigma = \Sigma^\alpha_\sigma \, ; \quad (\gamma^\sigma \Sigma_\sigma) = 0$$

which allows us to replace A^α_σ with a Verma module ϕ^α. The action then becomes:

$$\Phi \times \gamma^\sigma \nabla_\sigma \Phi + \frac{2}{3} \Phi \times \Phi \times \Phi$$

where the vierbein is still contained within the action. Lastly, we have the gauge degree of freedom to fix the vierbein to:

$$e^{\nu\rho}_{\mu\sigma} = \delta^\nu_\mu \, \delta^\rho_\sigma$$

for open strings in the midpoint gauge and closed strings in the endpoint gauge.

The crucial point is that, for open strings in the endpoint gauge and closed strings in the midpoint gauge, the vierbein does not disappear and has to be carefully removed as a gauge degree of freedom. The removal of the vierbein is nontrivial, and in fact will generate all the higher order string interactions as gauge artifacts. In particular, we will see that closed string field theory in the midpoint gauge is nonpolynomial.

IX. NONPOLYNOMIAL GEOMETRIC STRING FIELD THEORY

In this section, we will briefly show how the four-string interaction for open strings[62] and the non-polynomial structure[63] of closed string interactions can be derived in the same way that the instantaneous four-fermion Coulomb term can be derived in ordinary QED. The origin of the Coulomb term is that the propagator is not invariant under a gauge transformation. For example, in a covariant gauge the

A_o field may propagate as $\frac{1}{\Box}$ while in the Coulomb gauge the propagator for A_o becomes ∇^{-2}. Thus, because the A_o no longer has any time derivatives on it, we may functionally integrate out over A_o which immediately leaves us with the four-fermion interaction:

$$\bar{\psi}\gamma_o\psi \, \nabla^{-2} \, \bar{\psi}\gamma_o\psi$$

An identical situation happens with the geometric string field theory. The propagator for closed strings:

$$D = \int_{|z|<1} z^{L_o - 1}\bar{z}^{\bar{L}_o - 1} d^2 z$$

is not gauge invariant under a transformation generated by the universal string group. We find:

$$UDU^{-1} \neq D$$

In fact, as in QED, we can actually gauge rotate the theory so that the propagator term $z^{L_o}\bar{z}^{\bar{L}_o}$ becomes the number 1. Instead of eliminating the A_o field, as in QED, we eliminate the string vierbein $e_{\mu\sigma}^{\nu\rho}$ that exists in the theory. *The elimination of the vierbein creates the closed four-string interaction. Since the vierbein couples to the three-vertex, the four-string interaction is just the square of the three-vertex in the interpolating gauge.*

Similarly, for closed strings, the removal of the vierbein in the midpoint gauge produces an infinite sequence of polyhedra, each of which is precisely the product of interpolating vertices.

Specifically, let the operator $U(\zeta^\sigma)_{\alpha_1}^{\alpha^2}$ change the parametrization length of a string from α_2 to α_1. Then the U operator, which is a function of the vierbein, allows us to go from interpolating vertices to endpoint and midpoint vertices:

$$\prod_{i=1}^{3} U_{\alpha}^{\alpha_i} \, |V_{\alpha_1\alpha_2\alpha_3}>_{interpolating} = |V_{\alpha\alpha\alpha}>_{midpoint}$$

$$\prod_{i=1}^{3} U_{\bar{\alpha}_i}^{\alpha_i} \, |V_{\alpha_1\alpha_2\alpha_3}>_{interpolating} = |V_{\bar{\alpha}_1\bar{\alpha}_2\bar{\alpha}_3}>_{endpoint}$$

where $\sum\bar{\alpha}_i = 0$ and α_i are arbitrary.

For open strings, at for the on-shell scattering amplitude for $1 + 2 \to 5 \to 3 + 4$, we have shown that[62]

$$< V_{125}| \, D_5 \, |V_{534}>_{midpoint} = < V_{125}| \, D_5 \, |V_{534}>_{endpoint} + \int d\alpha_5 \, \mu < V_{125}|V_{534}>$$

Thus, when removing the vierbein by making the transition from the midpoint gauge to the endpoint gauge, the propagator D_5 is not invariant and, in fact, can be sent to 1, thereby producing a contact term. Notice that the four-string interaction. is precisely the square of the interpolating vertex.

For closed strings, we have[61–3]:

$$< V_{125}| D_5 |V_{534}>_{endpoint} = < V_{125}| D_5 |V_{534}>_{midpoint}$$
$$+ \int d\alpha_5 \, d\theta_5 \, \mu < V_{125}|V_{534}>$$

Making the transition from the endpoint gauge to the midpoint gauge thus produces the tetrahedron graph, which again is the square of the interpolating vertex.

Now, repeat this same process in the action. In point particle field theory, if we functionally integrate over one of one set of fields in an action, we often produce higher order interactions in the action, including non-polynomial terms. However, these interactions are not contact terms, and hence the non-polynomial nature of the

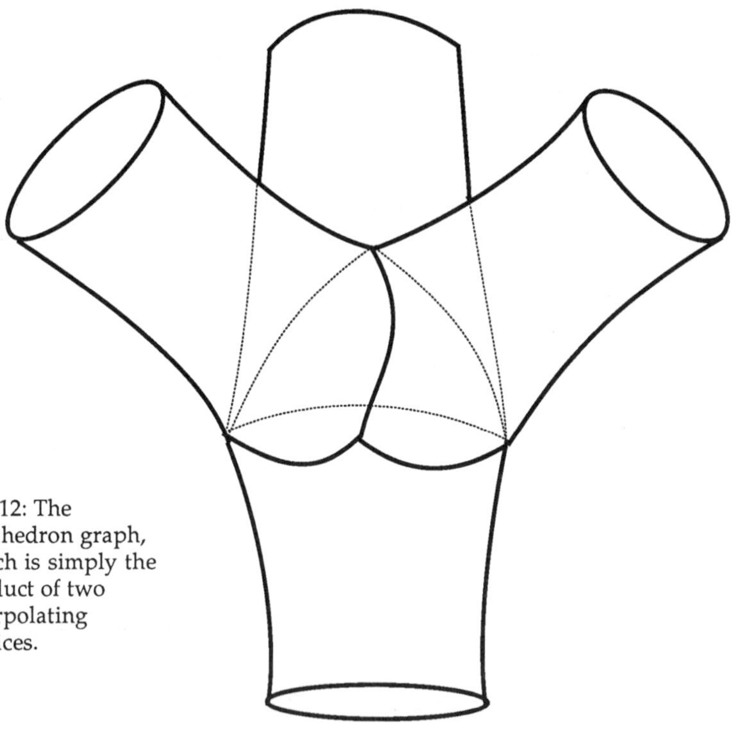

Fig. 12: The tetrahedron graph, which is simply the product of two interpolating vertices.

action is an illusion. In string theory, however, this is not the case. The propagator is not invariant under the USG, so some of the higher order interactions collapse into genuine contact terms, thus producing four-string interactions in the action and even non-polynomial terms.

If we carefully remove the vierbein and extract all higher order contact terms produced by this process, we find that geometric string field theory for open strings, which is cubic in the the midpoint gauge, generates precisely the five interactions found in the light cone theory. Repeating this process for closed strings, we find that geometric string field theory, which cubic for the endpoint gauge, becomes non-polynomial for the midpoint gauge.

We can summarize the way in which geometric string field theory, which is cubic, generates new interactions whenever the gauge is fixed (Φ represents open string fields and Ψ represents closed string fields):

$$(\Phi^3)_{geometric} = \begin{cases} \left(\Phi^3 + \Phi^4 + \Phi^2\Psi + \Psi^3 + \Psi\Phi\right)_{endpoint} \\ (\Phi^3)_{midpoint} \end{cases}$$

$$(\Psi^3)_{geometric} = \begin{cases} (\Psi^3)_{endpoint} \\ \left(\Psi^3 + \Psi^4 + \Psi^5 + \cdots\right)_{midpoint} \end{cases}$$

It may seem surprising that closed string field theory in the midpoint gauge is non-polynomial, but this can actually be seen at the level of complex maps. For example, the map:

$$\frac{d\rho}{dz} = N \, \frac{\prod_{i=1}^{2(N-2)} (z - z_i)^{1/2}}{\prod_{r=1}^{N} (z - \gamma_r)}$$

takes us from the complex plane to series of horizontal cylinders representing the collision of N closed strings. By carefully examining the range of integration, we can analytically show that the Koba-Nielsen plane cannot be covered if string interactions are purely cubic. In fact, using the method of equipotential lines, we can show that the topology of the missing region corresponds exactly to N-faced polyhedra.

Consider the Shapiro-Virasoro amplitude. In fig. 13, we find that a large triangular region (which includes the north and south pole and most of the Riemann sphere) is missing when we only use strings interacting at their midpoints. How-

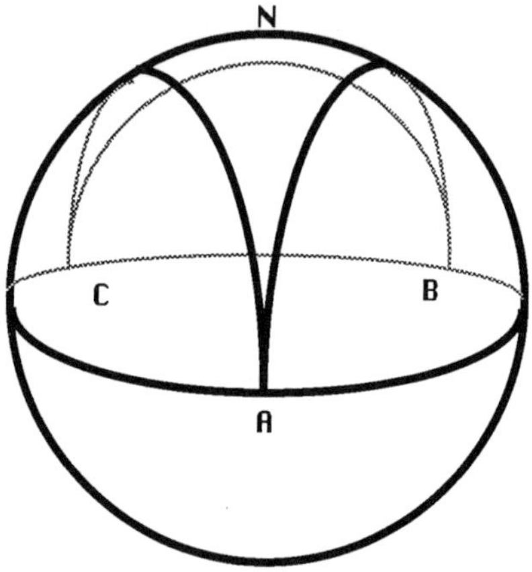

Fig. 13: The triangular region surrounding the north and south pole is the missing region of the Shapiro-Virasoro model. It is precisely filled by the tetrahedron graph.

ever, the equipotential lines of the charges in the missing region form exactly the topology of a tetrahedron.

Think of a tetrahedron as in fig. 14, with each of the four faces having circumference equal to 2π. Let a_{ij} equal the parametrization distance linking the ith and jth face. There are 6 edges labelled by a_{ij} and four constraints coming from the circumference, so there are two arbitrary parameters which fixes the tetrahedron. Notice that this is precisely the number of moduli necessary to describe the Shapiro-Virasoro model.

For a general polyhedron with N faces, there are 3(N-2) edges a_{ij}. Once we fix the circumference of each face, there are therefore 3(N-2) - N = 2N - 6 undetermined parameters, which is precisely the number of Koba-Nielsen variables or moduli in a Riemann surface describing N string scattering. Thus, we have the

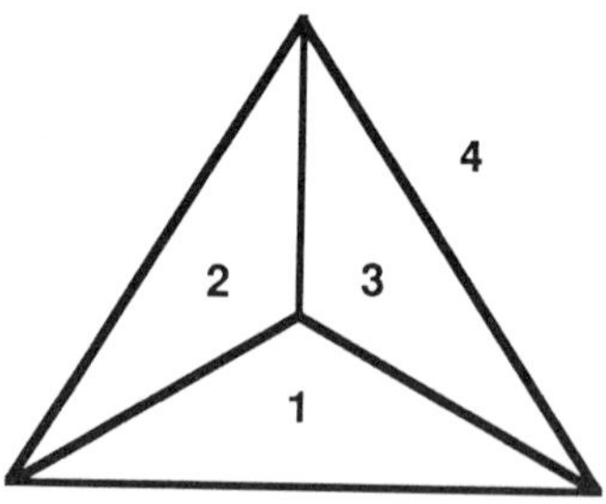

Fig. 14: The tetradhedron graph., representing the collision of four closed strings.

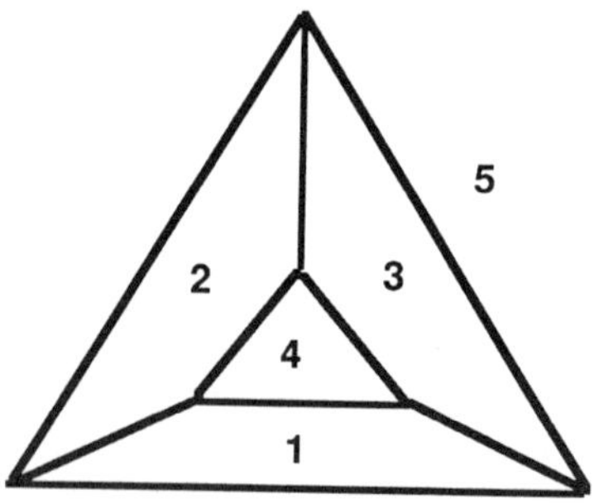

Fig. 15: The prism graph, representing the collision of five closed strings.

correct dimensions to fill any missing region of the complex plane.

For the general case, we have found that the boundary conditions on the missing region are that (a) the circumference of any face must be 2π and (b) the perimeter of any closed path which encircles several faces must be greater than 2π. For the tetrahedron, this becomes $a_{ij} \leq 2\pi$. Since there are are 2N - 6 = 2 independent variables, we will choose a_{12} and a_{13}. It is now a simple matter to solve for the region of integration:

$$\textit{Missing Region}: \quad \begin{cases} a_{12},\ a_{13} < \pi \\ a_{12} + a_{13} > \pi \end{cases}$$

In the $a_{12}-a_{13}$ plane, this missing region corresponds to a right triangle. This, in turn, can be mapped onto the northern hemisphere of the Riemann sphere, and we find that the tetrahedron precisely fills the missing region and, in fact, occupies most of the complex sphere.

Notice that, inside the missing region, closed strings *do not interact at their midpoints*. In fact, the overlap distance of two closed strings is less than π, i.e. *the missing region is simply the square of the interpolating vertex.*

For N=5 as in fig. 15, we now have 2N - 6 = 4 independent variables. The a_{ij} now satisfy:

$$\sum_{j=1;\, j\neq i}^{5} a_{ij} = 2\pi \qquad (i = 1,2,3)$$

$$a_{14} + a_{24} + a_{34} = 2\pi$$

$$a_{15} + a_{52} + a_{53} = 2\pi$$

Let us carefully choose our independent variables. We choose $a_{12},a_{24},a_{13},a_{35}$ (A naive choice will actually include unwanted dependent variables because of hidden identities.)

Setting the fact that all $a_{ij} < \pi$, we now find the complete description of the missing region:

$$\textit{Missing Region}: \quad \begin{cases} a_{12},\ a_{24},\ a_{13},\ a_{35} < \pi \\ a_{12} < a_{35} < a_{12} + a_{24} \\ a_{13} < a_{24} < a_{13} + a_{35} \end{cases}$$

The previous derivation used only a non-linear realization of the USG. However, a direct proof, by brute force, can also generate the same non-polynomial action. We can power expand our action as follows:

$$L = <\Psi|Q\Psi> + \sum_{n=3}^{\infty} \alpha_n <\Psi^n>$$

which we force to be invariant under:

$$\delta|\Psi> = Q|\Lambda> + \sum_{n=1}^{\infty} \beta_n |\Psi^n \Lambda>$$

This expansion is complicated by the fact that there are more than one n-faced polyhedra, which we label by the index i, i.e.

$$<\Psi^n> \equiv \sum_i c(n)_i <\Psi^n>_i$$

where:

$$(n!)^{-1} <\Psi_{[1}|<\Psi_2| \cdots <\Psi_{n]}|V>_i = <\Psi_1\Psi_2 \cdots \Psi_n>_i$$

(where the brackets represent all possible symmetrizations of the indices). By setting the variation of the action equal to zero, we wish to calculate the three sets of unknowns: α_n, β_n, $c(n)_i$.

After a long calculation, we can deduce the precise value of all coefficients. Let $s(n)_i$ represent the number of ways that that the ith polyhedron with n faces can by symmetrically rotated into itself. Then we define:

$$<\Psi^n>_{SYM} = \sum_{i=1} \frac{1}{s(n)_i} <\Psi^n>_i$$

The final result is then[63]:

$$L = <\Psi|Q\Psi> + 2^2 g <\Psi^3>_{SYM} + 2^2 g^2 <\Psi^4>_{SYM}$$

$$+ 2^3 g^3 <\Psi^5>_{SYM} + \cdots + 2^{n-2} g^{n-2} <\Psi^n>_{SYM} + \cdots$$

X. NEW DIRECTIONS

In summary, there is a wealth of problems that have opened up concerning string theory that will certainly keep physicists (and mathematicians) busy for years to come! Hopefully, we have introduced the reader to some of them. Let us, however, end with a discussion of where research might possibly take us in entirely different directions. Although speculation often outraces details, these new avenues may eventually open up new insights into string theory:

1) Universal Moduli Space[64]. Instead of string field theory, perhaps the modular properties of Riemann surfaces of infinite genus are strong enough to determine the structure of the entire perturbation series taken as a whole. This would be a radical departure from the present direction of string field theory. It tries to exploit the beautiful modular properties of the perturbation theory, which are broken by string field theory.

2) Renormalization Group[65]. Perhaps the conformal condition $\beta = 0$ should be taken as an equation of motion using renormalization group arguments. Starting with the space of "all possible two dimensional theories," perhaps the classical equation of motion of such a theory would be the known conformal field theories.

4) Knots[66]. Perhaps the arduous classification of possible conformal field theories (and hence all possible classical vacua of the string) can be simplified by leaping into the third dimension via knot theory. From the three dimensional vantage point of knot theory, many of the "miracles" of two dimensional conformal field theory can be simply explained.

5) Beyond the Planck Length[67]. Perhaps there is a deconfining phase transition at the Hagedorn temperature, and a new phase opens up. Perhaps the string theory is a "low energy" phase of a higher theory. Arguments from high temperature physics and high energy behavior of string amplitudes seem to indicate that newer symmetries and newer degrees of freedom may be opening up at those fantastic energies and temperatures. Perhaps topological theories give us a glimpse of this new theory, which is defined in a realm where the conventional notions of space and time fade away.

ACKNOWLEDGMENTS

We would like to thank B. Sakita, L. Hua, J. Lykken, and S. Giddings for extensive discussions. This work was partially funded by the US/Japan Cooperative Science Program, NSF-INT-8715626, and also NSF-PHY-86-15338 and CUNY-FRAP-13873.

REFERENCES

1) M. Kaku, *Introduction to Superstrings,* Springer-Verlag, New York, 1988.

2) M. Green, J. Schwarz, and E. Witten, *Superstring Theory,* Cambridge Univ. Press, Cambridge, 1987.

3) M. Green and D. Gross, eds., *Unified String Theories,* World Scientific, Singapore, 1986.

4) M. Jacob, ed., *Dual Theory,* North Holland, Amsterdam, 1974.

5) J. H. Schwarz, ed., *Superstrings: the first 15 years of superstring theory,* World Scientific, Singapore, 1985; Phys. Rep. 89, 223 (1982).

J. Scherk, Rev. Mod. Phys. 47, 1213 (1975).

6) P. Frampton, *Dual Resonance Models,* Benjamin, New York, 1974.

7) K. Kikkawa, B. Sakita, M.A.Virasoro, Phys. Rev. 184, 1701 (1969).

8) M. Kaku and L.P. Yu, Phys. Lett. 33B, 166 (1970); Phys. Rev. D3, 2992, 3007, 3020 (1971).

M. Kaku and J. Scherk, Phys. Rev. D3, 430, 2000 (1971).

9) C. Lovelace, Phys. Lett. 32B, 703 (1970); 34B, 500 (1971).

10) V. Alessandrini, Nuovo Cim., 2A, 321 (1971).

11) Y. Nambu, Lectures at the Copenhagen Summer Symposium, 1970.

12) T. Goto, Prog. Theor. Phys. 46, 1560 (1971).

13) P. Goddard, J. Goldstone, C. Rebbi and C.B. Thorn, Nucl. Phys. B56, 109 (1973).

14) S. Mandelstam, Nucl. Phys. B64, 205 (1973); B69, 77 (1974).

15) M. Kaku and K. Kikkawa, Phys. Rev. D10, 1110, 1823 (1974).

16) A. Neveu and J. Schwarz, Nucl. Phys. B31, 816 (1971); Phys. Rev. D4, 1109 (1971).

P. Ramond, Phys. Rev. D3, 2415 (1971).

17) J. Scherk and J. Schwarz, Nucl. Phys. B81, 118 (1974); T. Yonea, Prog. Theor. Phys. 51, 1907 (1974).

18) M. Green and J. Schwarz, Phys. Lett. 149B, 117 (1984); 151B, 21 (1985).

19) D. J. Gross, J.A. Harvey, E. Martinec, and R. Rohm, Nucl. Phys. B256, 253 (1986); B267, 75 (1986).

20) P. Candelas, G. Horowitz, A. Strominger and E. Witten, Nucl. Phys. B258, 75 (1985).

21) L. Dixon, J. Harvey, C. Vafa, and E. Witten, Nucl. Phys. B261, 678 (1985); B274, 285 (1986).

22) E. Witten, Nucl. Phys. B268, 253 (1986).

23) M. Dine and N. Seiberg, Nucl. Phys. B301, 357 (1988); Phys. Rev. Lett. 57, 2625 (1986); Nucl. Phys. B306, 137 (1988).

24) D. Gross and V. Periwal, Phys. Rev. Lett. 60, 2105 (1988); D. Gross and P. Mende, Phys. Lett. B197, 129 (1987).

25) C.S. Hsue, B. Sakita, and M.A. Virasoro, Phys. Rev. D2, 2857 (1970).

J.L. Gervais and B. Sakita, Nucl. Phys. B34, 632 (1971).

26) M. Kato and K. Ogawa, Nucl. Phys. B212, 443 (1983).

27) J.L. Gervais, and B. Sakita, Nucl. Phys. B34, 632 (1971).

28) S.Deser and B. Zumino, Phys. Lett. 65B, 369 (1976).

29) L. Brink, P. Di Vecchia, and P. Howe, Phys. Lett. 65B, 471 (1976).

30) G. Gliozzi, J. Scherk, and D. Olive, Nucl. Phys. B122, 253 (1977).

31) M. Green and J. H. Schwarz, Phys. Lett. 136B, 367 (1984); Nucl. Phys. B198, 252, 441 (1982).

See also: R. Kallosh and M. Rakmanov, Phys. Lett. B209, 233 (1988); R. Kallosh and A. Morozov, Phys. Lett. B207, 164 (1988); Int. J. Mod. Phys. A3, 1943 (1988).

32) V. Kac, *Infinite Dimensional Lie Algebras,* Birkhauser, Boston, 1983.

33) J. Shapiro, Phys. Rev. D5, 1945 (1972).

34) L. Alvarez-Gaume, G. Moore, and C. Vafa, Comm. Math. Phys. 106, 1 (1986). J. Fay, *Theta Functions on Riemann Surfaces,* Lecture Notes in Mathematics, Vol. 352, Springer-Verlag, Berlin, 1973.

D. Mumford, *Tata Lectures on Theta,* Birkhauser, Basel, 1983.

35) G. Moore and P. Nelson, Nucl. Phys. B266, 58 (1986).

A. Morozov and A. Perelomov, Phys. Lett. 183B, 296 (1987).

36) A.A. Belavin and V.G. Knizhnik, Phys. Lett. 168B, 201 (1986).

37) N. Ishibashi, Y. Matsuo and H. Ooguri, Mod. Phys. Lett. A2, 119 (1987); L. Alvarez-Gaume, C. Gomez, and C. Reina, Phys. Lett. 55B, 55 (1987).

38) E. D'Hoker and D.H. Phong, Rev. Mod. Phys. 60, 917 (1988) and references therein.

39) A. Selberg, J. Indian Math. Soc. 20, 47 (1987).

E. D'Hoker and D.H. Phong, Nucl. Phys. B269, 205 (1986).

40) E. Verlinde and H. Verlinde, Phys. Lett. B192, 95 (1987).

J.J. Atick, G. Moore, and A. Sen, Nucl. Phys. B307, 221 (1988).

41) H. Nishino and S. Gates, Phys. Lett. B189, 45 (1987); M. Cvetic, Phys. Rev. Lett. 59, 2829 (1987); P. Candelas, A. Dale, C.Lutken, R. Schimmrigk, Nucl. Phys. B298, 493 (1988); J. Ellis, K. Enqvist, S. Kalara, D. Nanopoulos, K. Olive, Nucl. Phys. B306, 445 (1988).

42) L. E. Ibanez, Phys. Lett., B181, 269 (1986); K. Kobayashi, Phys. Rev. Lett. 58, 2507 (1987); D. Bailin, A. Love, S. Thomas, Phys. Lett. B188, 193 (1987); L. Ibanez, J. Kim, H.P. Nilles, F. Quevedo, Phys. Lett. B191, 292 (1987); P.K. Mohapatra, Phys. Lett. B192, 108 (1987); H. Sato, Phys. Lett. B201, 78 (1988);

I. Senda, Phys. Lett. 209, 221 (1988).

43) H. Kawai, D. Lewellen, H. Tye, Phys. Rev. Lett. 57, 1832 (1986); W. Lerche, D. Lust, and A. Schellekens, Nucl. Phys. B287, 477 (1987); L. Antoniadis, C. Bachas, and C. Kounnas, Nucl. Phys. B289, 87 (1987).

44) K. Narain, Phys. Lett. B169, 41 (1986); K.S. Narain, M.H.Sarmadi, and C. Vafa, Nucl. Phys. B288, 551 (1987).

45) A.A. Belavin, A.M. Polyakov, and A.B. Zamolodchikov, Nucl. Phys. B241, 333 (1884).

V.G. Knizhnik and A.B. Zamolodchikov, Nucl. Phys. B247, 83 (1984).

46) D. Friedan, E. Martinec, S. Shenker, Nucl. Phys. B271, 93 (1986).

47) V. Kac, Lect. Notes in Phys, 94, 441 (1979).

48) D. Friedan, Z. Qui, and S. Shenker, Phys. Rev. Lett. 52, 1575 (1984).

49) W. Boucher, D. Friedan, and A. Kent, Phys. Lett. 172B, 316 (1986).

50) C. Hull and E. Witten, Phys. Lett. 160B, 398 (1985).

A. Sen, Nucl. Phys. B278, 289 (1986).

T. Banks, L. Dixon, D. Friedan and E. Martinec, Nucl. Phys. B299, 613 (1988).

51) D. Gepner, Nucl. Phys. B296, 757 (1988); Phys. Lett. 199B, 380 (1987).

52) A.B. Zamolodchikov and V.A. Fateev, Sov. Phys. JETP 62, 215 (1985).

53) E. Witten, Comm. Math. Phys. 92, 451 (1984).

V. Knizhnik and A.B. Zamolodchikov, Nucl. Phys. B247, 83 (1984).

54) M. Kaku, Nucl. Phys. B267, 125 (1985).

T. Banks and M. Peskin, in *Symposium Anomalies, Topology, and Geometry,* World Scientific, Singapore, 1985.

55) D. Gross and A. Jevicki, Nucl Phys. B283, 1 (1987); B287, 225 (1987).

S. Samuel, Nucl. Phys. B296, 187 (1988).

E. Cremmer, C.B. Thorn, and A. Schwimmer, Phys. Lett. 179B, 57 (1986).

S. Giddings, Nucl. Phys. B278, 242 (1986).

S. Giddings and E. Martinec, Nucl. Phys. B278, 91 (1986).

56) C. Wendt, SLAC-PUB 4442 (1987).

57) O. Lechtenfeld and S. Samuel, Phys. Lett. B213, 431 (1988).

A.R. Bogojevic, A. Jevicki and G.W. Meng, BROWN-HET-672.

58) H. Hata, K. Itoh, T. Kugo, H. Kunitomo, and K. Ogawa, Phys. Rev. D34, 2360 (1986).

59) M. Kaku, CCNY-HEP-14 (Aug. 1986); CCNY-HEP-3 (Mar. 1987); CCNY-HEP-7 (1988). (See ref. 1, Chapter 8).

60) M. Kaku, Phys. Lett. 200B, 22 (1988); Int. J. Mod. Phys. A2,1 (1987).

61) M. Kaku, Phys. Rev. D38, 3052 (1988).

M. Kaku and J. Lykken, Phys. Rev. D38, 3067 (1988).

L. Hua and M. Kaku, "Deriving the Shapiro-Virasoro Amplitude From Geometric String Field Theory," CCNY-HEP-NYU/CCNY preprint.

62) M. Kaku, "Deriving the Four-String and Open/Closed String Interactions From Geometric String Field Theory," CCNY-HEP-88-7 (to appear in Int. J. of Mod. Physics).

M. Kaku, "Unifying Light-Cone, Witten's, and Osp(26,2/2) String Field Theories," CCNY-HEP-89-3.

63) M. Kaku, "Non-polynomial Geometric String Field Theory: I," CCNY-HEP-89-6.

64) D. Friedan and S. Shenker, Phys. Lett. B175, 287 (1986).

65) T. Banks and E. Martinec, Nucl. Phys. B293, 733 (1987); V. Periwal, PUPT-1988; S. Das, G. Mandal, and S. Wadia, TIRF-TH- 88/33 (1988); J. Hughs, J. Liu, and J. Polchinski, Univ. of Texas preprint (1988).

66) E. Witten, Comm. Math Phys. (to appear).

67) J.J. Atick and E. Witten, Nucl. Phys. B310, 291 (1988).

E. Witten, Nucl. Phys. B311, 46 (1988).

D.J. Gross, Phys. Rev. Lett. 60, 1229 (1988).

Prof. Michio Kaku, Physics Dept., City College of New York of the City University of New York, 138th St. at Convent Ave., New York, N.Y. 10031, USA.

I. STRINGS: SUMMING OVER SURFACES

BORDERED SURFACES, OFF-SHELL AMPLITUDES, SEWING, AND STRING FIELD THEORY

Steven Carlip
The Insitute for Advanced Study
Princeton, NJ 08540 U.S.A.

The Polyakov path integral in string theory is a beautiful perturbation expansion, with many deep and elegant mathematical properties. But it is nevertheless a perturbation expansion, and over the past few years it has become increasingly evident that we do not know the underlying theory it is an expansion for. Not too long ago, this did not seem to be a very urgent problem: there was no shortage of interesting questions which could be addressed perturbatively. As those questions have been answered, however, nonperturbative issues have become increasingly important. In particular, the properties which characterize realistic models — the particle content, coupling strengths, etc. — depend on the choice of vacuum, which can only be determined nonperturbatively.

On a more philosophical level, we do not yet have a simple underlying principle or geometrical picture to provide a foundation for string theory. Unlike general relativity, in which spacetime and gravity are unified, string theory describes objects propagating in a prescribed background. The background is not arbitrary — consistent string propagation provides strong constraints — but neither is it fully unified with the string.

This is not a happy situation, and much work has gone into developing a nonperturbative string theory. The resulting efforts fall into two general categories. The first, "top down," approach looks for basic principles which might underlie such a theory. Work on the geometry of loop space, the study of topological quantum field theories, and the search for general symmetries of string field theory are inspired by this approach.

The second, "bottom up," approach looks more directly for elements of perturbation theory which can be extended to a nonperturbative theory. The attempt to systematize and classify conformal field theories, which are candidate string vacua, falls in this category. So does the study of "sewing," that is, the effort to construct Polyakov amplitudes for complicated world sheets from simple building blocks.

These lectures will deal with the current status of the sewing problem. The rationale for this approach is that any nonperturbative string theory must reproduce the Polyakov path integral as a perturbation series. If our experience in ordinary field theory is a guide — and admittedly it may not be — the terms in such a perturbation series, like Feynman diagrams, are likely to be built up from simple "vertices" and "propagators," which can themselves be represented as (off-shell) Polyakov amplitudes. Hence an understanding of how to put together simple components into more complicated world sheet amplitudes is likely to give us much-needed information about the structure of nonperturbative string theory.

To understand sewing, we must first understand the building blocks, off-shell Polyakov amplitudes. This is the subject of my first lecture. Next we will explore the sewing of conformal field theories at a fixed conformal structure, that is, the reconstruction of correlation functions for a fixed surface Σ from those on a pair of surfaces Σ_1 and Σ_2 obtained by cutting Σ along a closed curve. We will then look at the problem of sewing *amplitudes*, integrals of correlation functions over moduli space. This will necessitate an understanding of how to build the moduli space of a complicated surface from simpler moduli spaces. Finally, we will briefly examine vertices and string field theories.

Much of the work reported here was done in collaboration with various combinations of Steven Blau, Minot Clements, and Stephen and Vincent Della Pietra [1–6]. The idea of sewing is an old one, dating back to work by Bardakci, Halpern, and Shapiro [7] in the early days of string theory. In the newer incarnation of string theory, a number of authors have done related work, among them Morozov and Rosly [8–10], Sonoda [11], Ohrndorf [12–14], and many others; there are also connections to the "operator sewing" of Vafa *et al.* [15–16]. I will restrict my attention in these lectures to the closed oriented bosonic string; for the beginnings of an extension to the superstring, see [17] and [18].

1. Bordered Surfaces and Off-Shell String Amplitudes

To understand how to build up complicated Polyakov amplitudes from simple pieces, we must first understand the pieces, path integral for surfaces with boundaries. The idea that such path integrals can be interpreted as off-shell string amplitudes is an old one, dating back to the pioneering paper of Cohen, Moore, Nelson, and Polchinski [19], who calculated the amplitude for the surface with the topology of a cylinder and showed that it could be understood as the tree level string propagator. Progress in understanding more complicated amplitudes has been slow, however, in part because of technical difficulties (now largely solved) and in part because of problems in interpreting the boundary parameterizations necessary to calculate the path integral.

1.1. The Polyakov Path Integral

The action for the closed oriented bosonic string is

$$S = \frac{1}{2} \int_{\Sigma} \sqrt{g} g^{ab} \partial_a X^\mu \partial_b X_\mu \tag{1.1}$$

where Σ is an oriented surface with boundary $\partial\Sigma$ and Riemannian metric g. The Polyakov path integral is*

$$Z[\bar{X}] = \frac{\int [dg][dX] e^{-S}}{\mathrm{Vol}(\mathit{Diff} \times \mathit{Weyl})} \tag{1.2}$$

where $\mathrm{Vol}(\mathit{Diff} \times \mathit{Weyl})$ is the volume of the symmetry group of S, the group of diffeomorphisms and Weyl transformations, and the path integral is over all configurations with fixed boundary values $X|_{\partial\Sigma} = \bar{X}$. More precisely, to fix the boundary values of X, we must choose a parameterization $\mathcal{P} : S^1 \to \partial\Sigma$ for each component of $\partial\Sigma$, and set

$$X|_{\partial\Sigma} = \bar{X} \circ \mathcal{P}^{-1} \tag{1.3}$$

The Polyakov amplitude thus depends on both $\bar{X}$ and $\mathcal{P}$. The dependence on $\mathcal{P}$ is crucial; we will return to this point later.

* See [20] for an introduction to the path integral.

As in the case of the path integral for a surface without boundaries, the integral over metrics in (1.2) can be split into an integral over diffeomorphisms and Weyl transformations and an integral over the moduli space

$$\mathcal{M}(\Sigma) = \text{Metrics}/\mathit{Diff} \times \mathit{Weyl} = T(\Sigma)/G(\Sigma) \tag{1.4}$$

where $T(\Sigma)$ is the Teichmüller space of Σ and $G(\Sigma)$ is the mapping class group (see section 1.3). As usual, there is a Jacobian (or Faddeev-Popov determinant) proportional to $(\det P^\dagger P)^{\frac{1}{2}}$, where P is the differential operator which takes vectors to symmetric traceless tensors,

$$(Pv)^{ab} = \nabla^a v^b + \nabla^b v^a - \frac{1}{2} g^{ab} \nabla_c v^c \tag{1.5}$$

The integration over diffeomorphisms and Weyl transformations then cancels the volume Vol(*Diff* × *Weyl*) in (1.2), leaving an integral over $\mathcal{M}(\Sigma)$ with a measure $d\mu(\mathcal{M})$ which can be written in terms of quadratic and Beltrami differentials on Σ. (In the constant negative curvature gauge, $d\mu(\mathcal{M})$ is the Weil-Petersson measure [1].)

As in the case of a closed surface, we can write the Jacobian $(\det P^\dagger P)^{\frac{1}{2}}$ as a ghost path integral,

$$\int [db][dc] \exp\{ \int_\Sigma \sqrt{g}\, b_{ab}\, (Pc)^{ab} \} \tag{1.6}$$

The only new feature when Σ has boundaries is that we must now specify boundary conditions for b and c. The appropriate set of boundary conditions was first proposed by Alvarez [21]:

$$b_{nt} = c^n = 0 \tag{1.7}$$

where n and t denote normal and tangent components at $\partial\Sigma$. These conditions can be most easily understood as arising from the fact the diffeomorphisms must map $\partial\Sigma$ to itself. The ghost c comes from the generator of infinitesimal diffeomorphisms, and the condition $c^n = 0$ means that infinitesimal deformations of the boundary are tangent to the boundary; $b_{nt} = 0$ is the adjoint boundary condition, in a sense that will be discussed in the next lecture.

Returning to the amplitude (1.2), we can now perform the X integral, since the action is quadratic in X. The procedure is standard: we shift X by the classical solution $X_{c\ell}$, where

$$\begin{aligned} \Delta X_{c\ell} &= 0 \\ X_{c\ell}|_{\partial\Sigma} &= \bar{X} \end{aligned} \tag{1.8}$$

Then $S[X] = S[X_{c\ell}] + S[Y]$, where $Y|_{\partial\Sigma} = 0$, and the Y integral simply gives the determinant of the Laplacian Δ with Dirichlet boundary conditions. Combining this result with the Jacobian $(\det P^\dagger P)^{\frac{1}{2}}$, we find

$$Z_\Sigma[\bar{X}] = \int\limits_{\mathcal{M}(\Sigma)} \sigma(\mathcal{M}(\Sigma);\bar{X}) \tag{1.9}$$

where

$$\sigma(\mathcal{M}(\Sigma);\bar{X}) = (\frac{\det{}'\Delta_\Sigma}{\int_\Sigma \sqrt{g}})^{-13}(\det(P^\dagger P)_\Sigma)^{\frac{1}{2}} e^{-S_{c\ell}[\bar{X}\circ\mathcal{P}]} d\mu(\mathcal{M}) \tag{1.10}$$

is the Polyakov measure and $S_{c\ell}$ is the classical action,

$$S_{c\ell}[\bar{X}\circ\mathcal{P}] = S[X_{c\ell}] = \frac{1}{2}\int\limits_{\partial\Sigma} dx \int\limits_{\partial\Sigma} dx' \bar{X}\circ\mathcal{P}(x)\partial_n\partial_{n'}G(x,x')\bar{X}\circ\mathcal{P}(x') \tag{1.11}$$

with $G(x,x')$ the Dirichlet Greens function for Σ.

1.2. Determinants

To make sense of (1.9), we must understand the integrand — the determinants and Greens function — and the region of integration — the moduli space for a bordered surface. Let us begin with the integrand.

Much progress has been made over the past few years in understanding the corresponding determinants on closed surfaces (see for instance [22]); as Morozov has explained at this school, these can be written in terms of simple geometric quantities such as the prime form and Abelian differentials. To extend these results to bordered surfaces, we can make use of techniques, first suggested by Lovelace [23], based on the existence of a closed double $\overline{\Sigma}$ for the surface Σ. The double $\overline{\Sigma}$ of an arbitrary surface with boundary Σ is formed by gluing together two copies of Σ along their common boundaries. While a metric on Σ will not ordinarily extend smoothly to $\overline{\Sigma}$, any metric on Σ is conformal to one which extends smoothly; that is, any complex structure on Σ extends to the double.

The double $\overline{\Sigma}$ thus consists of two copies of Σ, together with an involution $I\colon \overline{\Sigma} \to \overline{\Sigma}$ which sends a point on one copy to the corresponding point on the other. The fixed points of I are boundary curves of Σ, and one can write $\Sigma = \overline{\Sigma}/I$. The fundamental characteristic of I is that it is anticonformal, that is, that it anticommutes with the complex structure:

$$I_* J = -J I_* \tag{1.12}$$

As a consequence, I behaves nicely with respect to differential operators on $\overline{\Sigma}$; for example,

$$I_* \partial = \overline{\partial} I_* \tag{1.13}$$

Geometrical objects on $\overline{\Sigma}$, such as eigenfunctions of differential operators, thus break up naturally into pieces even and odd under I. As we shall see, this makes it possible to write determinants on Σ in terms of corresponding determinants on $\overline{\Sigma}$. But $\overline{\Sigma}$ is a closed surface, for which such determinants have simple expressions, so such a construction goes a long way towards allowing us to understand the integrand in (1.9).

Consider first the ghost determinant $(\det P^\dagger P)_{\overline{\Sigma}}$. Under the involution I, the eigenspace of $P^\dagger P$ splits into even and odd subspaces. It is not hard to see that the even subspace consists of eigenvectors whose normal components vanish and whose tangent components have vanishing normal derivatives at $\partial\Sigma$. But these conditions are precisely the boundary conditions (1.7). Hence, writing $(P^\dagger P)^\pm = \frac{1}{4}(1 \pm I_*)P^\dagger P(1 \pm I_*)$, we have

$$\det(P^\dagger P)_\Sigma = \det(P^\dagger P)^+_{\overline{\Sigma}} \tag{1.14}$$

At the same time, the complex structure J maps even vectors into odd vectors: by (1.12), if $I_* v = v$, then $I_*(Jv) = -Jv$. This means that the even and odd eigenspaces of $(P^\dagger P)_{\overline{\Sigma}}$ are isomorphic, and that the determinants of $(P^\dagger P)^\pm$ are equal. But it is clear that $\det(P^\dagger P)_{\overline{\Sigma}} = \det(P^\dagger P)^+_{\overline{\Sigma}} \det(P^\dagger P)^-_{\overline{\Sigma}}$. Hence

$$\det(P^\dagger P)_\Sigma = (\det(P^\dagger P)_{\overline{\Sigma}})^{\frac{1}{2}} \tag{1.15}$$

For the scalar determinant $(\det \Delta)$, the situation is somewhat more complicated. An eigenfunction of Δ on $\overline{\Sigma}$ which is odd under I now corresponds to an eigenfunction on Σ which vanishes on the boundary, i.e., a Dirichlet eigenfunction, while an eigenfunction on $\overline{\Sigma}$ which is even under I corresponds to a Neumann eigenfunction on Σ. But in contrast with the ghosts, the complex structure J has no natural action on scalars, and the even and odd eigenspaces of Δ are not isomorphic. Nevertheless, it is possible to relate $\det \Delta^+$ and $\det \Delta^-$, for instance by examining the dependence of their ratio on the moduli of $\overline{\Sigma}$. It can then be shown that

$$\det \Delta_\Sigma = \left(\frac{\det' \Delta_{\overline{\Sigma}}}{\int_\Sigma \sqrt{g}} \right)^{\frac{1}{2}} R^{-\frac{1}{2}} \tag{1.16}$$

where R can be calculated in terms of the even and odd parts of the period matrix of $\overline{\Sigma}$ [1,6,8].

The remaining geometrical elements of the integrand of (1.9)— the Greens function in $S_{c\ell}$ and the measure $d\mu(\mathcal{M})$ — can also be expressed in terms of quantities on the double $\overline{\Sigma}$. Details are given in [1]; the same results are discovered by different methods in [8-9] and [24].*

To make sense of (1.9), however, it is not sufficient to understand the integrand: we must also know the region of integration. We now turn to this question.

1.3. Moduli Spaces

Let us first recall a few definitions. Let Σ be a topological two-manifold, with or without boundary. Denote by $\mathit{Diff}(\Sigma)$ the group of diffeomorphisms of Σ, and by $\mathit{Diff}_0(\Sigma)$ the subgroup of diffeomorphisms which are isotopic to the identity. The mapping class group $G(\Sigma)$ is the quotient space $\mathit{Diff}(\Sigma)/\mathit{Diff}_0(\Sigma)$ of equivalence classes of diffeomorphisms which cannot be smoothly deformed to the identity. $G(\Sigma)$ is generated by Dehn twists, diffeomorphisms induced by cutting Σ along a closed curve, twisting one boundary by 2π, and rejoining the two boundaries.

The (reduced) Teichmüller space $T(\Sigma)$ is the quotient of the space of Riemannian metrics on Σ by $\mathit{Diff}_0(\Sigma) \times \mathit{Weyl}(\Sigma)$, where $\mathit{Weyl}(\Sigma)$ is the group of Weyl transformations $g \to e^\phi g$. The moduli space $\mathcal{M}(\Sigma)$ is the quotient of the space of metrics by the full group $\mathit{Diff}(\Sigma) \times \mathit{Weyl}(\Sigma)$, including the mapping class group, so $\mathcal{M}(\Sigma) = T(\Sigma)/G(\Sigma)$. In general, Teichmüller space has a simple structure — if Σ is of genus g with b boundary components and n punctures, $T(\Sigma)$ is a

---

⋆ An error in [1] is discussed in [25]; the final results are not affected, however.

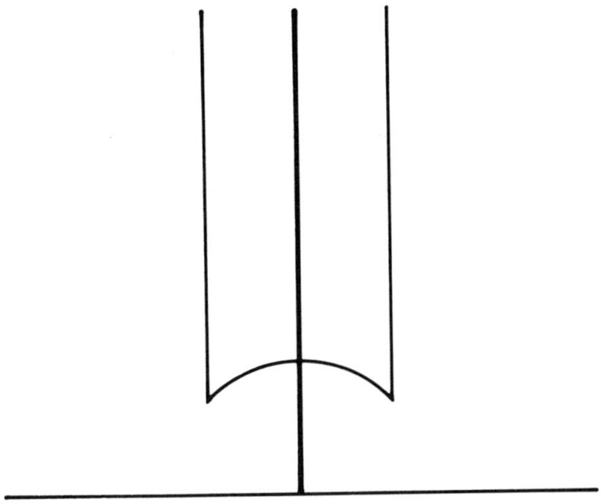

Figure 1. Moduli and Teichmüller spaces for the cylinder and the torus.

$(6g-6+2n+3b)$-dimensional ball — while moduli space is much more complicated. But in string theory, it is moduli space rather than Teichmüller space that occurs naturally: the action is invariant under all diffeomorphisms, not only those which can be deformed to the identity.

To understand the moduli spaces of bordered surfaces, let us begin with a simple example, the cylinder, whose double is the torus. The moduli space of the torus is well known: it is the "keyhole" region shown in Figure 1. The moduli space of the cylinder, on the other hand, is the entire positive imaginary axis (it is determined by one parameter, the length of the cylinder), and it is not a subspace of the moduli space of the torus. What has happened is that the torus' mapping class group is too large — it includes elements which are not in the mapping class group of the cylinder — so its moduli space is too small.

The Teichmüller space of the torus, on the other hand, is the entire upper half plane, and the Teichmüller space of the cylinder — again, the positive imaginary axis — is now a subspace. This subspace has a simple description. The Teichmüller space of the torus inherits an involution, $\tau \mapsto -\bar{\tau}$, from the involution I, and the Teichmüller space of the cylinder is precisely the fixed point set of this involution. It follows from the work of Earle [26] and Sepälä [27] that this description generalizes to an arbitrary surface. The Teichmüller space of any bordered surface Σ is precisely the fixed point set of the Teichmüller space of its double $\overline{\Sigma}$ under the action of the involution

$$J \mapsto -I_* \circ J \circ I_* \tag{1.17}$$

where J is the complex structure of $\overline{\Sigma}$, and thus parameterizes $T(\overline{\Sigma})$.

As in the case of the cylinder, to obtain the correct region of integration in (1.9), one must still take the quotient of this fixed point set by the action of the mapping class group $G(\Sigma)$, which can be viewed as a subgroup of $G(\overline{\Sigma})$. This can be difficult to do in practice. Up to this complication, however, we have now succeeded in expressing the entire path integral for Σ in terms of quantities which characterize $\overline{\Sigma}$.

1.4. Parameterizations

The amplitude (1.9) differs from an amplitude for a compact world sheet in one important way: through the classical action S_{cl}, $Z[\bar{X}]$ depends implicitly on the parameterization $\mathcal{P}$. Any two parameterizations differ from each other by diffeomorphisms of a circle, so at first sight it appears that we must fix an element of *Diff* S^1 for each boundary component of Σ to make sense of Z.

In fact, the situation is more complicated. The parameterization $\mathcal{P}$ must be chosen for each point in the moduli space of Σ, and for a generic surface, there is no canonical way of comparing the parameterization for different points of $\mathcal{M}$. We must thus pick $\mathcal{P}$ for each point of $\mathcal{M}$; more precisely, we must specify a section of a $(\mathit{Diff}\ S^1)^b$ bundle (b is the number of components of $\partial\Sigma$) over $\mathcal{M}$.

We will see later that the crucial problem in sewing Polyakov amplitudes is the determination of an appropriate parameterization for the boundaries being sewn. Unfortunately, the meaning of such a parameterization is not well understood. In the original paper of Cohen *et al.* [19] on the off-shell propagator, it was proposed that one should integrate over all reparameterizations of the boundaries. Such an integral is extremely hard to perform explicitly, however, and the resulting expressions are intractable.

Redlich [28–29] has suggested instead that the choice of parameterization should be viewed as a choice of gauge in a string field theory. This is an attractive idea, and it has been verified in some special cases. In particular, we understand a good deal about *free* string field theory, and a number of parameterizations of the boundaries of the cylinder have been shown to give the same free string field theory propagator in different gauges. For the interacting theory, the situation is more complicated, however, since the choice of section described above depends on the moduli space $\mathcal{M}(\Sigma)$, which does not ordinarily appear in a string field theory.

It would be interesting to try to relate the change in the Polyakov amplitude under a boundary reparameterization to the BRST charge Q. Ideally, it might be possible to show that amplitudes with different parameterizations differ by Q-exact quantities; this would relate reparameterizations to gauge transformations in the BRST theory. So far, however, little progress has been made in this area.

1.5. The Ghost Boundary Term

One remaining feature of the amplitude (1.9) deserves discussion. So far in the derivation, we have assumed that the ghost fields b and c vanish at the boundary $\partial\Sigma$. This is certainly a reasonable assumption for physical amplitudes. But for sewing to work, we will also need to admit amplitudes with nonvanishing ghost boundary values. This is not surprising: if we consider the inverse of sewing, "cutting," there is no reason to believe that the ghost fields should always vanish at the curve along which we cut Σ.

For the scalar field X, we saw that Dirichlet boundary *conditions* for the determinant $(\det\Delta)$ came from the choice of Dirichlet boundary *values* — the specification of $\bar{X}$ — on $\partial\Sigma$. Correspondingly, Alvarez boundary conditions (1.7) for the ghost determinant $\det P^\dagger P$ should come from the specification of $\bar{b}_{nt}$ and $\bar{c}^n$ at the boundary. (For simplicity, I am being somewhat careless here about the parameterization dependence of $\bar{b}$ and $\bar{c}$, which is complicated by the fact the b and c are not scalars, but tensors, with tensor transformations under reparameterizations. With a little work, the boundary conditions for b and c can be expressed more precisely, but we will not need such details here.)

The ghost action in (1.6) is linear in b and c, so at first sight we might expect no boundary contribution to the path integral: even with nonvanishing $\bar{b}$ and $\bar{c}$, it seems that the classical action should vanish. When we try to evaluate the classical action explicitly, however, we find a problem. The ghost action

$$S_{gh} = \int_\Sigma \sqrt{g}\, b_{ab}\,(Pc)^{ab} \tag{1.18}$$

has *no* extrema. In fact, if we split b and c into classical fields and fluctuations,

$$\begin{aligned} & b = b_{c\ell} + b' \\ & P^{\dagger} b_{c\ell} = 0, \quad (b_{c\ell})_{nt} = \bar{b}_{nt}, \quad b'_{nt} = 0 \\ & c = c_{c\ell} + c' \\ & P c_{c\ell} = 0, \quad (c_{c\ell})^n = \bar{c}^n, \quad c'^n = 0 \end{aligned} \tag{1.19}$$

we find that

$$S_{gh} = \int_{\Sigma} \sqrt{g}\, b'_{ab} (Pc')^{ab} + 2 \int_{\partial\Sigma} \bar{b}_{nt} c'^t \tag{1.20}$$

The last cross-term means that $b_{c\ell}$ and $c_{c\ell}$ are not true extrema of the action: the variation $\delta S_{gh}(\bar{b}, \bar{c})$ does not vanish to first order.

The cure to this dilemma was first proposed by Birmingham and Torre [30], who added a boundary term

$$S'_{gh} = -2 \int_{\partial\Sigma} b_{nt} c^t \tag{1.21}$$

to the ghost action. With this addition, it is easy to see that

$$S_{gh} + S'_{gh} = \int_{\Sigma} \sqrt{g}\, b'_{ab} (Pc')^{ab} - 2 \int_{\partial\Sigma} \bar{b}_{nt} \bar{c}^t \tag{1.22}$$

which no longer has a cross-term between the classical fields and the fluctuations, but instead has a classical boundary contribution, as desired.

Actually, life is more complicated. For all but the simplest world sheet topologies, b has zero-modes h^i ($i = 1, \ldots, \dim\mathcal{M}(\Sigma)$), which satisfy $P^{\dagger} h^i = 0$, ${h^i}_{nt} = 0$. The presence of these zero-modes has two implications:

1. $c_{c\ell}$ doesn't exist: $\int_{\Sigma} h^i (P c_{c\ell}) = 2 \int_{\partial\Sigma} {h^i}_{nn} \bar{c}^n \neq 0$, so $P c_{c\ell}$ cannot vanish.

2. $b_{c\ell}$ is not unique: it is determined only up to terms proportional to the h^i.

The first problem can be solved by defining $c_{c\ell}$ by

$$(P c_{c\ell})^{ab} = \sum_i \left(2 \int_{\partial\Sigma} {h^i}_{nn} \bar{c}^n \right) h^{iab} \tag{1.23}$$

which always has a (unique) solution. To deal with the second problem, we can write b as the sum of three components,

$$b = b_{c\ell} + b' + \sum \lambda_i h^i$$
$$\int_\Sigma b_{c\ell}^{ab} {h^i}_{ab} = \int_\Sigma b'^{\,ab} {h^i}_{ab} = 0 \tag{1.24}$$

We then find that

$$S_{gh} + S'_{gh} = \int_\Sigma \sqrt{g}\, b'_{ab} (Pc')^{ab} + 2 \int_{\partial\Sigma} \bar{b}_{nn} \bar{c}^n + 2\lambda_i \int_{\partial\Sigma} {h^i}_{nn} \bar{c}^n \tag{1.25}$$

As in the case of the scalar field, we can write the boundary contribution in (1.25) in terms of a Greens function ${F^a}_{a'b'}(x, x')$ for the differential operator P:

$$\int_{\partial\Sigma} \bar{b}_{nn} \bar{c}^n = - \int_{\partial\Sigma} \int_{\partial\Sigma} \bar{b}_{nt} {F^t}_{n'n'} \bar{c}^{n'} \tag{1.26}$$

which has a structure much like the scalar boundary term (1.11). As for the remaining term in (1.25) involving the b zero-modes, the path integral over b must now include an integral over the λ_i in (1.24), which will give a delta function

$$\delta\Big(\int_{\partial\Sigma} {h^i}_{nn} \bar{c}^n\Big) \tag{1.27}$$

in the amplitude. We shall see later that this delta function, discovered independently by Ohrndorff [13], is precisely what is needed to make the ghost contribution to the amplitude sew correctly.

To summarize our results so far,

1. The path integral for a bordered surface can be evaluated fairly explicitly in terms of well-understood quantities on its double.
2. Getting the correct region of integration, the moduli space of Σ, is hard.
3. All fields in the theory, including the ghosts, give boundary contributions to the path integral; when zero-modes are present, these boundary contributions become somewhat more complicated.
4. There is a huge "gauge symmetry" coming from the possibility of reparameterizing the boundary curves, which is not fully understood.

2. Sewing Conformal Field Theories

Now that we know something about off-shell Polyakov amplitudes, we can ask how to sew them together. The amplitude (1.9) is determined by the integrand, the Polyakov measure, and the region of integration, moduli space. Correspondingly, the problem of sewing naturally splits into two issues: how to combine measures, and how to build up complicated moduli spaces from simple ones. In this lecture, we shall address the first of these questions.

More precisely, we shall investigate what can be called "sewing at a fixed conformal structure." We begin with a world sheet Σ with a given metric (or a given conformal structure — as usual, the Weyl dependence cancels in the end), and cut it along a dividing curve C to form a pair of new surfaces Σ_1 and Σ_2 (see Figure 2). The metric on the original world sheet induces metrics on the cut surfaces; for now, we hold these fixed, working at only a single point in the moduli space of each surface. Observe that the parameterization of the curve C determines parameterizations of the boundaries $\partial\Sigma_1$ and $\partial\Sigma_2$; there is no ambiguity in the "gauge choice" at this stage.

Note that we could instead cut Σ along a nondividing curve — cutting open a handle — to form a single new surface with two additional boundary components. The generalization to this case is straightforward, but the notation is somewhat more complicated; see [3] for details.

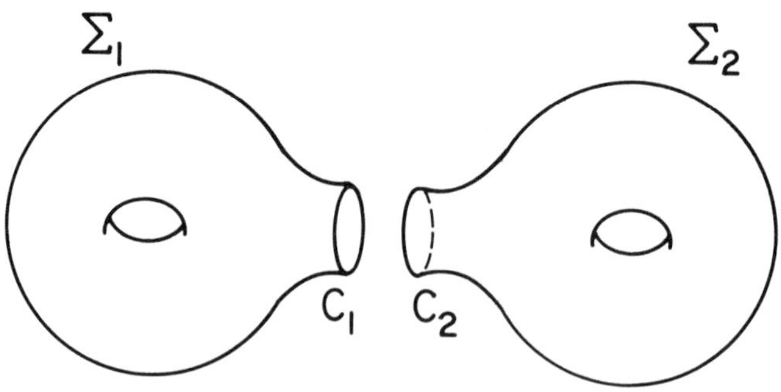

Figure 2. Sewing Σ_1 to Σ_2 along C gives the surface Σ.

We can now attempt to reconstruct the Polyakov measure on Σ from the measures on Σ_1 and Σ_2 by functionally integrating over the boundary values of the fields along C. Since the measure is made up of determinants and Greens functions, this will require an understanding of the relationships between such quantities on the two surfaces. The goal of this lecture is to outline the proof of such relationships.

To understand what must be shown, consider a single scalar field with action (1.1), with g now fixed. As before, the path integral over X can be performed, giving

$$Z_{\Sigma_1}[\bar{X}] = (\det \Delta)^{-\frac{1}{2}} e^{-S_{cl}[\bar{X}]} \tag{2.1}$$

with a corresponding equation for Σ_2. For sewing to hold, we must have

$$Z_\Sigma = \int [d\bar{X}] Z_{\Sigma_1}[\bar{X}] Z_{\Sigma_2}[\bar{X}] \tag{2.2}$$

Substituting the expression (1.11) for the classical action into (2.2) and performing the integral over $\bar{X}$, we can see that (2.2) requires that

$$\det \Delta_\Sigma = \det \Delta_{\Sigma_1} \det \Delta_{\Sigma_2} \det(\partial_n \partial_{n'} G_{\Sigma_1}(x,x') + \partial_n \partial_{n'} G_{\Sigma_2}(x,x')) \tag{2.3}$$

Such a relation between determinants is far from obvious, but we shall see that it is true.

The kernel $\partial_n \partial_{n'} G_{\Sigma_1}(x,x')$ in (2.3) has an interesting interpretation which we will find helpful later. Consider specifying $\bar{X}$ on $\partial\Sigma_1$, extending this Dirichlet boundary data harmonically to the interior of Σ_1, and then calculating the *Neumann* boundary data $\partial_n X$ for this extension. $\partial_n \partial_{n'} G(x,x')$ is then the map from the original Dirichlet data to the corresponding Neumann data. Forman [31] calls such a map between different boundary data a Poisson map; we will see that this map occurs quite generally.

The full proof of relations like (2.3) is rather complicated, and will involve a number of new concepts. The principles, however, are easy:

1. Look at the variation

$$\delta \log \det \Delta = \operatorname{Tr} \delta\Delta\, \Delta^{-1} = \operatorname{Tr} \delta\Delta G \tag{2.4}$$

2. Express G_Σ in terms of G_{Σ_1} and G_{Σ_2}: for x and x' on Σ_1, for instance,

Green's theorem implies that

$$G_\Sigma(x,x') = G_{\Sigma_1}(x,x') + \int\limits_C dy \int\limits_C dy' \partial_n G_{\Sigma_1}(x,y) G_\Sigma(y,y') \partial_{n'} G_{\Sigma_1}(y',x') \tag{2.5}$$

2.1. General Ingredients

Rather than restricting ourselves to the fields which occur in ordinary bosonic string theory, let us consider a more general Gaussian conformal field theory, with an action on Σ_1 of the form

$$S = \int\limits_{\Sigma_1} b\, Lc \tag{2.6}$$

where $b \in V^-$ and $c \in V^+$ are sections of vector bundles over Σ_1 and L: $V^+ \to V^-$ is a linear differential operator of order n. For now, let us assume for simplicity that there are no zero modes, i.e., that L has no kernel or cokernel. The notation of (2.6) suggests a standard b-c system, but we can also take $b = c = X$, for instance, and $L = -\frac{1}{2}\Delta$, to recover the scalar action (1.1). By considering such a general theory, we can make our proofs somewhat simpler, although at the expense of some added notation. The derivations here are sketchy; for details, see [3].

As in the scalar case discussed above, we must begin by giving boundary data for b and c. To do so, we first introduce a projection π of V^+ onto the Cauchy data (the boundary value and first $n-1$ derivatives) for c. A suitable set of boundary data for c consists of half of its Cauchy data (e.g., Dirichlet data for a scalar field, or the positive frequency part of a spinor field when L is a first order operator). We label such a choice of boundary data by a capital letter, for instance A, and define a new projection π_A of c onto its A boundary data. An associated projection π_{1-A} projects c onto the other half of its Cauchy data, and defines another set of boundary data; if A denotes the Dirichlet data for a scalar field, for instance, then the $1-A$ data is the Neumann data.

We saw in the first lecture that the specification of Dirichlet *data* for X lead to determinants of the Laplacian with Dirichlet boundary *conditions* $X|_{\partial\Sigma_1} = 0$. A similar phenomenon occurs quite generally, arising from the separation of the fields into classical extrema and fluctuations. We therefore define the A boundary condition for L as the restriction of L to $\ker\pi_A$; we denote the resulting operator as L_A.

Given a choice A of boundary data for c, we must now determine suitable adjoint boundary data for b. We again define a projection π of b onto its Cauchy data, and consider the functional

$$\int_{\Sigma_1} (b\,Lc - L^\dagger b\,c) = W(\pi b, \pi c) \tag{2.7}$$

where W depends on the Cauchy data for b and c. We can now fix the adjoint boundary data $A^\dagger$ by the requirement that

$$W(\pi_{A^\dagger} b, \pi c) = W(\pi b, \pi_{1-A} c) \tag{2.8}$$

The data $A^\dagger$ are adjoint to A in the sense that $(L^\dagger)_{A^\dagger} = (L_A)^\dagger$. For the ghost fields b_{ab} and c^a considered in the last lecture, for instance, b_{nt} and c^n are adjoint data. Note that if the fields b and c are identical, as in the scalar case, we must choose self-adjoint boundary data, $A = A^\dagger$. For a scalar field, for example, $W(\pi X, \pi Y) = \int_C (-X\partial_n Y + Y\partial_n X)$, and both Dirichlet and Neumann data are self-adjoint.

We next need a Greens function for L. Assuming that L_A is invertible, we define the Greens function G_A with A boundary conditions by

$$L\,G_A = 1, \qquad G_A L|_{\ker\pi_A} = 1|_{\ker\pi_A}, \qquad \pi_A G_A = 0 \tag{2.9}$$

With such a Greens function, we can harmonically extend boundary data $\bar{c}$, that is, we can solve the boundary value problem

$$Lc = 0 \quad \text{with} \quad \pi_A c = \bar{c} \tag{2.10}$$

We thus have an extension map P_A from $\bar{c}$ to its extension by L, obeying the relationships

$$LP_A = 0, \qquad \pi_A P_A = 1 \tag{2.11}$$

It is not hard to show that such an extension map satisfies

$$P_A \pi_A = 1 - G_A L \tag{2.12}$$

2.2. Boundary Terms in the Action

We saw previously that for the bosonic string ghosts, the action (1.18) was not adequate for a bordered surface; a boundary term had to be added. The same is true for the more general action (2.6). Indeed, if we again write $b = b_{c\ell} + b'$, $c = c_{c\ell} + c'$, it is easy to check that $S[b,c] = S[b',c'] + W(\pi b_{c\ell}, \pi c')$, so the classical fields again fail to decouple from the fluctuations. To eliminate the cross-term, we must add to the action (2.6) a boundary contribution of the form

$$S_1 = -W(\pi_{A^\dagger} b, \pi_{1-A} c) \tag{2.13}$$

We then find that

$$(S + S_1)[b,c] = S[b',c'] - W(\pi_{A^\dagger} b_{c\ell}, \pi_{1-A} c_{c\ell}) \tag{2.14}$$

so the classical fields and fluctuations now decouple.

For the bosonic ghost system, the boundary contribution (2.13) is precisely the term S'_{gh} found in the last lecture. For the scalar field with Dirichlet boundary data, S_1 may also be present: if we write the action as $S = -\frac{1}{2}\int_\Sigma X\Delta X$, we find that $S_1 = \frac{1}{2}\int_{\partial\Sigma} X\partial_n X$, so $S + S_1$ gives the standard action (1.1).

Finally, using the extension map P_A we can write $b_{c\ell}$ and $c_{c\ell}$ in terms of their (specified) A and $A^\dagger$ boundary values $\bar{b}$ and $\bar{c}$. This is the generalization of the procedure which led to equation (1.26). We find

$$(S + S_1)[b,c] = S[b',c'] - W(\bar{b}, \pi_{1-A} P_A \bar{c}) \tag{2.15}$$

2.3. Sewing Measures

The term in (2.15) involving W is the generalization of the classical action (1.11) for the scalar field. Like (1.11), it is bilinear in the boundary data, with a kernel involving a Greens function (implicit in the extension map P_A). The map $\pi_{1-A} P_A$ is, in fact, a Poisson map: it is the projection onto $1 - A$ boundary data of the extension by L of A boundary data. If we denote the Poisson map between two sets of boundary data A and B by

$$\phi_{B,A} = \pi_B P_A, \tag{2.16}$$

we see that

$$Z_{\Sigma_1}[\bar{b}, \bar{c}] = \int [db][dc] e^{-(S+S_1)[b,c]} = (\det L_{\Sigma_1})^{\pm 1} \exp\{W(\bar{b}, \phi_{1-A,A}\bar{c})\} \tag{2.17}$$

where the sign in the exponent is negative if b and c are commuting, positive if they are anticommuting.

Equation (2.17) is the generalization of (2.1). For sewing to hold, the equivalent of (2.2) must still be satisfied, with the integral now over $\bar{b}$ and $\bar{c}$ rather than $\bar{X}$. To find the generalization of (2.3), we again integrate over boundary values. Since W is bilinear, the integral is Gaussian, and can be done explicitly. We find as the condition for sewing

$$\det L_\Sigma = \det L_{\Sigma_1} \det L_{\Sigma_2} \det(\phi^1_{1-A,A} + \phi^2_{1-A,A}) \tag{2.18}$$

where ϕ^1 and ϕ^2 are the Poisson maps for Σ_1 and Σ_2.

We now outline the proof of (2.18). First note that our sewing relationship involves determinants on different surfaces, so a direct comparison is difficult. We can avoid this problem, however, by considering the determinant on Σ to be a determinant on the cut surface $\Sigma' = \Sigma_1 \cup \Sigma_2$, but with "sewing boundary conditions," i.e., with the *entire* Cauchy data of b and c continuous across C. (There is a minor technical detail here: while we require that $\pi^1_A = \pi^2_A$, we need $\pi^1_{1-A} = -\pi^2_{1-A}$. The relative minus sign comes essentially because the outward normals change direction between Σ_1 and Σ_2; see [3]).

To proceed, let A and B be two arbitrary choices of boundary data. We first note that by (2.12), $\phi_{B,A}\phi_{A,B} = \pi_B P_A \pi_A P_B = \pi_B(1 - G_A L)P_B = \pi_B P_B = 1$. Hence an arbitrary variation of $\det \phi_{B,A}$ is

$$\delta \log \det \phi_{B,A} = \operatorname{Tr} \phi^{-1}_{B,A} \delta\phi_{B,A} = \operatorname{Tr} \pi_A P_B \pi_B \delta P_A \tag{2.19}$$

Further, $\delta(LP_A) = 0 = (\delta L)P_A + L(\delta P_A)$; acting on the left with G_A, we see that

$$\delta P_A = -G_A \delta L \, P_A \tag{2.20}$$

We also note that G_A is the inverse of L_A, so

$$\delta \log \det L_A = \operatorname{Tr} G_A \delta L \tag{2.21}$$

Combining (2.19)–(2.21), we see that

$$\begin{aligned}\delta \log(\det \phi_{B,A} \det L_A) &= \operatorname{Tr}[G_A \delta L - \pi_A P_B \pi_B G_A \delta L\, P_A] \\ &= \operatorname{Tr}[G_A - P_A \pi_A P_B \pi_B G_A] \delta L \\ &= \operatorname{Tr}[G_A - (1 - G_A L)(1 - G_B L) G_A] \delta L \\ &= \operatorname{Tr} G_B \delta L = \delta \log \det L_B\end{aligned} \tag{2.22}$$

and thus, up to a possible overall integration constant,

$$\det L_B = \det \phi_{B,A} \det L_A \tag{2.23}$$

This relation was first shown in a more special context by Forman [31].

We now apply this general result to relate A boundary conditions to sewing boundary conditions, $\pi_A^1 = \pi_A^2$ and $\pi_{1-A}^1 = -\pi_{1-A}^2$. These boundary conditions come from boundary data given by the projection

$$\pi_B = \begin{pmatrix} \pi_A^1 - \pi_A^2 \\ \pi_{1-A}^1 + \pi_{1-A}^2 \end{pmatrix} \tag{2.24}$$

L_B then corresponds to L on Σ, while L_A is L on $\Sigma_1 \cup \Sigma_2$ with A boundary conditions. If we denote by P_A^1 and P_A^2 the extension maps for Σ_1 and Σ_2, we find the Poisson map

$$\begin{aligned}&\phi_{B,A} = \pi_B P_A \\ &= \begin{pmatrix} (\pi_A^1 - \pi_A^2) P_A^1 & (\pi_A^1 - \pi_A^2) P_A^2 \\ (\pi_{1-A}^1 + \pi_{1-A}^2) P_A^1 & (\pi_{1-A}^1 + \pi_{1-A}^2) P_A^2 \end{pmatrix} = \begin{pmatrix} 1 & -1 \\ \phi_{1-A,A}^1 & \phi_{1-A,A}^2 \end{pmatrix}\end{aligned} \tag{2.25}$$

Thus

$$\det \phi_{B,A} = \det(\phi_{1-A,A}^1 + \phi_{1-A,A}^1) \tag{2.26}$$

Comparing (2.23) and (2.26), we have succeeded in proving the desired sewing relation (2.18).

2.4. Generalizations

Several generalizations of this proof are possible. First, although we have considered only the partition function $Z[\bar{b}, \bar{c}]$, it is not hard to show that arbitrary correlation functions sew properly. Indeed, if we add source terms for b and c to construct a generating function for correlators, the arguments of the last section can be repeated with minimal changes. By the same reasoning, sewing also holds for theories with both quadratic and linear couplings, such as scalar field theories with ϕR terms in the action.

The proof becomes somewhat more complicated if L or $L^\dagger$ have zero-modes. We have already seen this for the bosonic string ghosts; recall that the classical action included a zero-mode term $\int h_{nn}\bar{c}^n$, which must be generalized. The zero-modes show up in a number of places in the analysis: the operator L is no longer invertible, the Greens functions must be defined with appropriate projections onto spaces of zero-modes, and, as in the bosonic ghost case, the classical field equations either have an infinite number of solutions or none. Nevertheless, it is shown in [3] that if we are careful to keep track of the zero-modes everywhere they occur, the correct generalization of (2.2) still holds.

The procedure described here for sewing bordered surfaces is closely related to the operator techniques of Vafa *et al.* [15–16] and Ohrndorf [14]. Operator methods are normally applied to surfaces with punctures rather than finite boundaries. But to define sewing, one must also choose a coordinate z in the neighborhood of each puncture, and such a choice is equivalent to the choice of a parameterized boundary curve $|z| = 1$. The boundary data used by Vafa *et al.* at the punctures can then be understood as coherent state (Atiyah-Patodi-Singer [32]) boundary data for the sewing technique developed here.

Finally, a note for the experts. The relation (2.23) can be understood in terms of determinant line bundles: it is in fact an isomorphism between determinant lines, which is an isometry of the Quillen metrics.

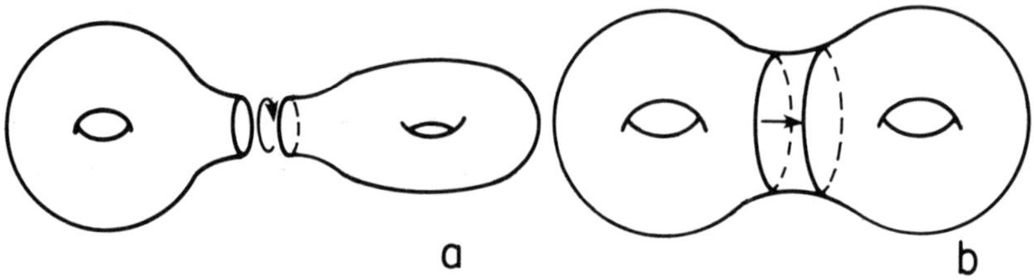

Figure 3. The moduli of Σ depend on a twist (3a), but a height parameter is redundant (3b).

3. Sewing Amplitudes

We have now seen that the *measures* in the Polyakov path integral sew together properly at a fixed metric. While this result is reassuring, it is not terribly surprising. The Polyakov measure is, after all, given by a path integral for a field theory on Σ, and we certainly should be able to perform such an integral by first integrating over fields with fixed values on the curve C and saving the integral over these boundary values until the end. The harder question, to which we now turn, is how to sew together the full *amplitudes*, complete with their integrations over moduli space.

The basic issue is now how to construct the moduli space of a sewn surface Σ from the moduli spaces of its component parts Σ_1 and Σ_2. The dimensions of the moduli spaces add correctly: $\dim(\mathcal{M}(\Sigma)) = \dim(\mathcal{M}(\Sigma_1)) + \dim(\mathcal{M}(\Sigma_2))$. Unfortunately, this does not mean that the moduli of Σ_1 and Σ_2 can serve as moduli for Σ. There are three basic problems:

1. The moduli of Σ depend on the parameterizations of the curves $\partial\Sigma_1$ and $\partial\Sigma_2$ being sewn together. Indeed, if we cut Σ along C, reparameterize one of the boundaries (say, $\partial\Sigma_1$), and sew Σ back together with the new parameterization, the generic effect will be to change the complex structure of Σ. This is the two-dimensional version of Kodaira-Spencer deformation theory [33].

2. We are missing a modulus of Σ: the complex structure of Σ depends not only on that of Σ_1 and Σ_2, but also on the "twist" angle at which they are sewn (see Figure 3a).

3. We have an extra "height" modulus: if we cut off a section of the neck around $\partial\Sigma_1$ (i.e., an annular neighborhood of the boundary) and glue it onto $\partial\Sigma_2$, we will have changed the moduli of Σ_1 and Σ_2 without changing the complex structure of Σ (see Figure 3b).

The last two points are perhaps most familiar to physicists in Fenchel-Nielsen or "length-twist" coordinates on Teichmüller space [34]. In these coordinates, a surface is parameterized by lengths of geodesics in a constant negative curvature metric and by twists along those geodesics. In sewing, the lengths of the two geodesics being identified must be equal, so one length parameter is redundant, while the extra twist must be put in by hand. As we shall see below, a similar mismatch of coordinates occurs in the coordinates for moduli space based on the decomposition of Σ into flat cylinders.

3.1. Eliminating Lengths

Let us begin with the third problem, the redundant length parameter. This redundancy has a simple interpretation. A Polyakov amplitude for a surface with boundary is analogous to a Feynman diagram in a quantum field theory with external propagators. It is not surprising that in order to combine two such diagrams, we must replace pairs of external propagators by single internal lines. This suggests that the removal of the redundant parameter should involve the inverse propagator.

The propagator — the Polyakov amplitude for the cylinder — has been calculated by Cohen *et al.* for the case of vanishing ghost boundary values. The inclusion of a nonzero $\bar{b}$ and $\bar{c}$ is straightforward, and it is not hard to show (for instance, by working with Fourier modes) that the inverse of the propagator is the operator

$$H_\gamma = \int\limits_\gamma dt\Big[-\frac{1}{2}(\frac{\delta}{\delta X})^2 + \frac{1}{2}(\partial_t X)^2 - \frac{\delta}{\delta b_{nt}}\partial_t\frac{\delta}{\delta c^n} + b_{nt}\partial_t c^n + 1\Big] \tag{3.1}$$

where γ is one of the boundary curves of the cylinder, with a parameterization induced from the flat metric. Not surprisingly, H may be recognized as the first quantized string Hamiltonian.

For the cylinder, it is fairly easy to show that

$$H_\gamma \sigma(\lambda; \bar{X}, \bar{b}, \bar{c}) = -\frac{d}{d\lambda}\sigma(\lambda; \bar{X}, \bar{b}, \bar{c}) \tag{3.2}$$

where σ is the Polyakov measure (1.10) (with the ghost classical action now included) and λ is the modulus. Essentially the same relation holds for an arbitrary surface [4]: for any surface Σ_1 with a boundary component γ_1,

$$H_{\gamma_1}\Big(\sigma_{\Sigma_1}(\mathcal{M}(\Sigma_1); \bar{X}, \bar{b}, \bar{c})\Big) = -\int_{\gamma_1} dt(n^a n^b - t^a t^b)(\frac{1}{\sqrt{g}}\frac{\delta}{\delta g^{ab}})\sigma_{\Sigma_1}(\mathcal{M}(\Sigma_1); \bar{X}, \bar{b}, \bar{c}) \tag{3.3}$$

where n^a and t^a are normal and tangent vectors at γ. In other words, the action of the inverse propagator on the boundary values at γ is equivalent to a change in the metric of Σ_1. (See also [13] for a similar argument.) While such a relation may seem surprising, it can be understood as the statement that the energy-momentum tensor calculated in terms of canonical variables (i.e., boundary values of the fields) is the same as the energy-momentum tensor calculated by varying the metric.

In a neighborhood of γ conformal to a flat cylinder, the functional derivative in (3.3) can be interpreted as the generator of the quasiconformal map which shortens that cylinder. The inverse propagator thus does what we hoped: it truncates an external cylinder, and thus removes a height modulus from the measure for Σ_1 or Σ_2. To proceed further, however, we must identify precisely what cylinder is being removed. This will require a detour into the theory of quadratic differentials on Riemann surfaces.

3.2. Quadratic Differentials

The importance of quadratic differentials in string theory was first observed by Giddings and Martinec [35]. A quadratic differential gives a natural foliation of the string world sheet, providing a definition of time and a means of time-slicing. At the same time, such a differential determines a Euclidean structure on the world sheet, breaking a surface Σ into flat cylinders joined along arcs at their boundaries.

Let us briefly review some of the theory of quadratic differentials [36–37]. A holomorphic quadratic differential on a surface Σ is a holomorphic (2,0)-form, i.e., a differential which can be written locally in the form $\phi(z) = f(z)dz^2$. If Σ is bordered, we require also that ϕ be real on $\partial\Sigma$; this ensures that ϕ extends to the double $\overline{\Sigma}$.

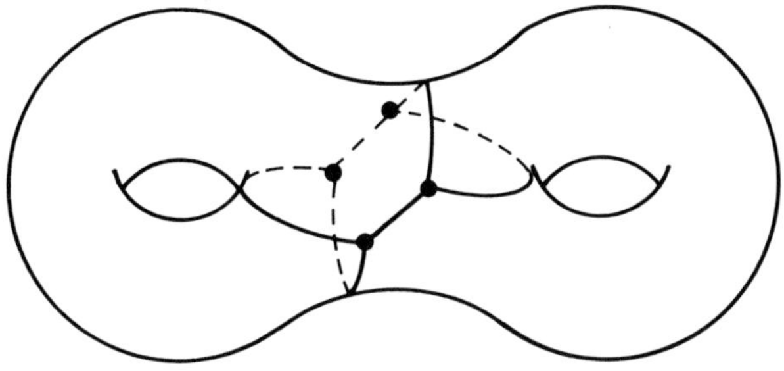

Figure 4. The critical graph Γ of a Jenkins-Strebel differential divides Σ into cylinders.

A horizontal trajectory of ϕ is a curve along which ϕ is real and positive. In a coordinate patch in which $\phi(z) = dz^2$, this is a curve Imz = const.; this equation is preserved in the transition from one patch to another. Similarly, a vertical trajectory is a curve along which ϕ is real and negative. On the plane, horizontal and vertical trajectories are simply horizontal and vertical lines; on a Riemann surface Σ, they are (open or closed) curves which foliate the surface.

A critical trajectory is a horizontal trajectory which joins critical points, i.e., zeros, of ϕ. A zero of order n will be the endpoint of $n + 2$ critical arcs, which may all be segments of distinct critical trajectories or may include both ends of a single trajectory (see Figure 4). A Jenkins-Strebel differential is a holomorphic quadratic differential whose noncritical horizontal trajectories are closed on Σ, or, if Σ is bordered, are closed on the double of Σ.

It can be shown that the critical trajectories of a Jenkins-Strebel differential ϕ form a compact graph Γ which divides Σ into cylinders, called ring domains (Figure 4).* If γ is a closed curve which wraps once around the ring domain R, R is said to be of homotopy type $[\gamma]$. Each domain R_i has a natural metric given by $|\phi(z)|$; in this metric, R_i is a flat cylinder with some circumference a_i and height b_i. A Jenkins-Strebel differential thus gives a decomposition of Σ into flat cylinders. Conversely, a collection of flat cylinders sewn together by boundary identifications to form a Riemann surface determines a Jenkins-Strebel differential: simply take $\phi(z) = dz^2$ in the flat coordinates on each cylinder.

* If Σ is bordered, its double will be divided into such ring domains; on Σ itself, some domains may be strips — i.e., half-cylinders — rather than cylinders.

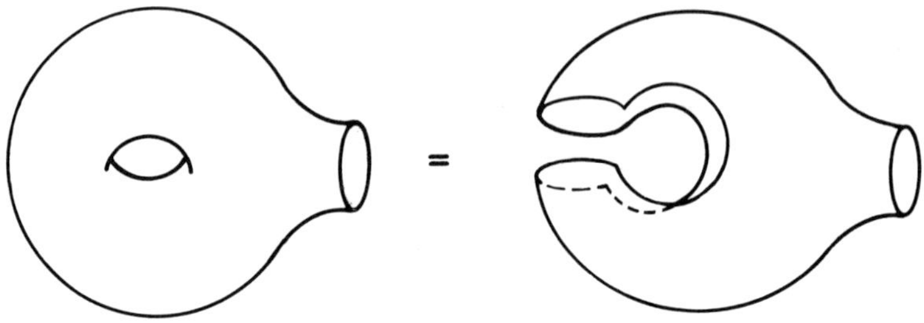

Figure 5: Σ can be represented as a single cylinder with arc identifications.

It is sometimes necessary to find Jenkins-Strebel differentials with specified ring domains. To do so, we choose a set S of free homotopy classes of curves $\{\gamma_i,\ i = 1, \ldots, n\}$ on Σ which represent circumferences of the ring domains we are interested in. Let us define the moduli space $\mathcal{M}(\Sigma, S)$ of the pair (Σ, S) as the space of equivalence classes of surfaces under diffeomorphisms and Weyl transformations which preserve the homotopy classes in S. $\mathcal{M}(\Sigma, S)$ is larger that the moduli space $\mathcal{M}(\Sigma)$, since we are not factoring out the diffeomorphisms which change the homotopy classes $[\gamma_i]$, but it is smaller than the Teichmüller space $T(\Sigma)$.

The known existence theorems for quadratic differentials then take the same general form: given a set S and certain additional data, there exists a unique Jenkins-Strebel differential with n ring domains of homotopy type $[\gamma_i]$ satisfying conditions determined by the additional data. For example, Strebel [36] has shown that as the additional data one may specify the heights b_i of the ring domains R_i, or the ratios of their moduli (as cylinders). These data determine a unique Jenkins-Strebel differential, whose ring domains give a decomposition of Σ into cylinders with the chosen heights or ratios of moduli.

Given such a cylinder decomposition, one can reconstruct Σ by sewing the cylinders back together. An existence theorem for quadratic differentials thus determines a set of coordinates for $\mathcal{M}(\Sigma, S)$. Note that there may be more than one topologically distinct critical graph Γ allowed by an existence theorem; that is, different points in the moduli space of Σ may be represented by topologically distinct ways of sewing together the same set of flat cylinders. This means that we actually have a cell decomposition of $\mathcal{M}(\Sigma, S)$, with different cells corresponding to different topologies of Γ [38]. Points in a cell are parameterized by the unprescribed data for the ring domains (e.g., the circumferences if the one fixes the heights), together with the sewing data (the lengths and twists of arcs which are to be identified) which determines how the cylinders are attached along Γ. Lower dimensional cells appear when segments of Γ shrink to points, i.e., when two zeros of the relevant quadratic differential coincide.

As an example of such a coordinate system, imagine specifying a single homotopy class $[\gamma]$ on Σ (see Figure 5). By a theorem of Strebel [36], there is a unique Jenkins-Strebel differential ϕ on Σ with a single ring domain of homotopy type $[\gamma]$ with unit circumference. Σ can thus be represented as a single flat cylinder with arcs identified along the boundaries. The moduli space $\mathcal{M}(\Sigma, [\gamma])$ is parameterized by the height b of this cylinder, a twist parameter θ which determines the relative orientations of its two boundary components, and a set of coordinates $\{m\}$ which describe the arc identifications. These latter arc identification coordinates can be identified with the moduli of the noded surface $\dot{\Sigma}$ formed by shrinking γ to a point [39].

For γ homotopic to a component of the boundary $\partial\Sigma$, the twist coordinate θ is not needed. In this situation, the moduli space $\mathcal{M}(\Sigma, [\gamma])$ is identical to the ordinary moduli space $\mathcal{M}(\Sigma)$. Indeed, the difference between $\mathcal{M}(\Sigma, [\gamma])$ and $\mathcal{M}(\Sigma)$ is that in $\mathcal{M}(\Sigma, [\gamma])$ we do not factor out the diffeomorphisms which change the homotopy type of γ. But if γ is a boundary curve, any diffeomorphism of Σ takes γ to itself, and thus does not change its homotopy type.

When $\partial\Sigma$ has two or more components $\{\gamma_1, \gamma_2, \ldots\}$, there exists another useful cylinder decomposition: Σ can be expressed as the union of two ring domains, of circumference 1 and homotopy type $\{[\gamma_1], [\gamma_2]\}$. The existence of such a representation follows from an existence theorem for quadratic differentials suggested by Kerckhoff and proven in [5]. The moduli for Σ now consist of two cylinder heights and a set of arc identification parameters. For three or more ring domains, no corresponding theorem is known, and there does not seem to be a natural coordinate system for $\mathcal{M}(\Sigma)$ in which more than two heights serve as coordinates.

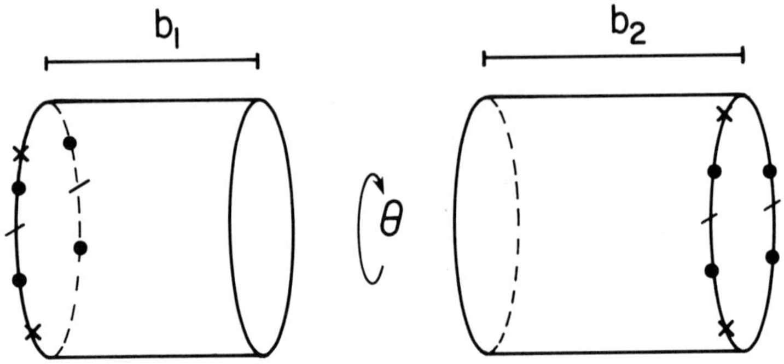

Figure 6. Sewing Σ_1 and Σ_2 as cylinders.

3.3. Sewing Moduli

Having completed our detour into the theory of quadratic differentials, let us return to the problem of sewing moduli spaces. Recall that we began by asking what cylinder on a string world sheet was cut off by the inverse propagator. We can now answer that question.

Let H act on the boundary component γ_1 of Σ_1, and let ϕ_1 be the unique quadratic differential with a single ring domain of unit circumference and homotopy type $[\gamma_1]$. ϕ_1 determines coordinates for $\mathcal{M}(\Sigma_1)$ consisting of a height b_1 and a set of moduli $\{m_1\}$ which describe the arc identifications. ϕ_1 also determines a parameterization of γ_1: in a neighborhood of γ_1, we can choose coordinates such that $\phi(z_1) = -\left(\frac{dz_1}{z_1}\right)^2$ and use $\arg(z_1)$ to parameterize γ_1. With this coordinate system for moduli space and this parameterization of γ_1, it is not hard to see that

$$H_{\gamma_1} = -\frac{d}{db_1} \tag{3.4}$$

H thus generates the quasiconformal transformation which collapses the single ring domain we have chosen to zero height, leaving a circle with a set of identified arcs;

its action on the Polyakov amplitude for Σ_1 is

$$H_{\gamma_1} \int_{\mathcal{M}(\Sigma_1)} \sigma_{\Sigma_1}(\mathcal{M}(\Sigma_1); \bar{X}, \bar{b}, \bar{c}) = \int_{\mathcal{M}(\Sigma_1)} \delta(b_1) \sigma_{\Sigma_1}(\mathcal{M}(\Sigma_1); \bar{X}, \bar{b}, \bar{c}) \tag{3.5}$$

We can now sew Σ_1 to Σ_2. To do so, we also represent Σ_2 as a single cylinder of unit circumference with boundary γ_2, with moduli b_2 and $\{m_2\}$. By attaching our truncated surface Σ_1 to γ_2, we obtain a sewn surface Σ, represented as a single cylinder of homotopy type $[\gamma] = [\gamma_1] = [\gamma_2]$, of height $b = b_2$, with arc identifications parameterized by the moduli $\{m_1, m_2\}$ (see Figure 6). These are almost the coordinates needed to parameterize the moduli space $\mathcal{M}(\Sigma, [\gamma])$; we are missing only the twist coordinate θ. But observe that our quadratic differential ϕ_1 did not really fix the parameterization of γ_1 uniquely: the coordinate z_1 in the neighborhood of γ_1 is determined only up to a phase. The choice of relative phase between z_1 and the corresponding coordinate z_2 in a neighborhood of γ_2 is precisely the twist coordinate we need.

To summarize, we have sewn together Σ_1 and Σ_2 by representing each surface as a flat cylinder of unit circumference with arcs identified along boundaries. We know how to sew cylinders; the procedure is unique up to a constant twist, which becomes one of the moduli of the sewn surface. This leaves us with one extra height — the sewn surface depends only on the sum $b_1 + b_2$ of the heights of Σ_1 and Σ_2 — but we eliminate the redundancy by using the inverse propagator to set b_1 to zero. More symmetrically, we could apply H to both surfaces and sew by inserting a propagator, but it is easy to see that the result would be identical.

Our Jenkins-Strebel differential has thus provided an answer to the three problems raised at the beginning of this lecture. ϕ determines the cylinder which must be truncated to eliminate the extra length parameter; it provides a natural parameterization of $\partial\Sigma$, up to a constant; and it determines the needed twist parameter by the remaining ambiguity in that parameterization. Combining this analysis of the moduli spaces with the previous results for sewing at fixed conformal structure, we can now sew together Polyakov amplitudes for the two surfaces Σ_1 and Σ_2 along boundary components γ_1 and γ_2. If we let T_θ denote a twist by angle θ, i.e., $f \circ T_\theta(s) = f(s+\theta)$, we see that

$$\int [d\bar{X}][d\bar{b}][d\bar{c}]\Big(\int\limits_{\mathcal{M}(\Sigma_2)} \sigma_{\Sigma_2}(b_2, \{m_2\}; \bar{X}\circ T_\theta, \bar{b}\circ T_\theta, \bar{c}\circ T_\theta)\Big)$$
$$\cdot H_{\gamma_1}\Big(\int\limits_{\mathcal{M}(\Sigma_1)} \sigma_{\Sigma_1}(b_1, \{m_1\}; \bar{X}, \bar{b}, \bar{c})\Big)$$
$$= \int\limits_{\mathcal{M}(\Sigma,[\gamma])} \int [d\bar{X}][d\bar{b}][d\bar{c}]\sigma_2(b_2, \{m_2\}; \bar{X}\circ T_\theta, \bar{b}\circ T_\theta, \bar{c}\circ T_\theta)\cdot \sigma_1(b_1 = 0, \{m_1\}; \bar{X}, \bar{b}, \bar{c})$$
$$= \int\limits_{\mathcal{M}(\Sigma,[\gamma])} \sigma_\Sigma(b_2, \theta, \{m_1, m_2\}) \tag{3.6}$$

which is essentially the sewing formula we want.

The relation (3.6) applies if we are sewing two disjoint world sheets along a single closed curve. It can also be generalized to the case of sewing two boundary components of a single surface Σ. In that case, we can use the second set of coordinates for moduli space described in the last section, based on the existence of a two-cylinder decomposition of Σ. The argument is the same: applying the inverse propagator to one boundary component eliminates the redundant length parameter, the appropriate Jenkins-Strebel differential fixes the parameterization of the boundary curves up to a constant, and the twist parameter is inserted to account for the remaining ambiguity in the parameterization.

Now consider sewing two surfaces along some arbitrary number n of boundary curves. The obvious generalization of the procedure discussed here would require a representation of each surface as a set of n cylinders of unit circumference, which would then be identified in pairs after the appropriate application of inverse propagators. Unfortunately, such a representation does not exist for $n > 2$: while it is possible to specify the appropriate homotopy classes, the existence theorem for quadratic differentials we would need is false. We can, however, proceed stepwise, first sewing our surfaces along a single boundary to form a single world sheet, then sewing boundary components of that world sheet two at a time. Such an iterative procedure allows us to sew any combination of Polyakov amplitudes.

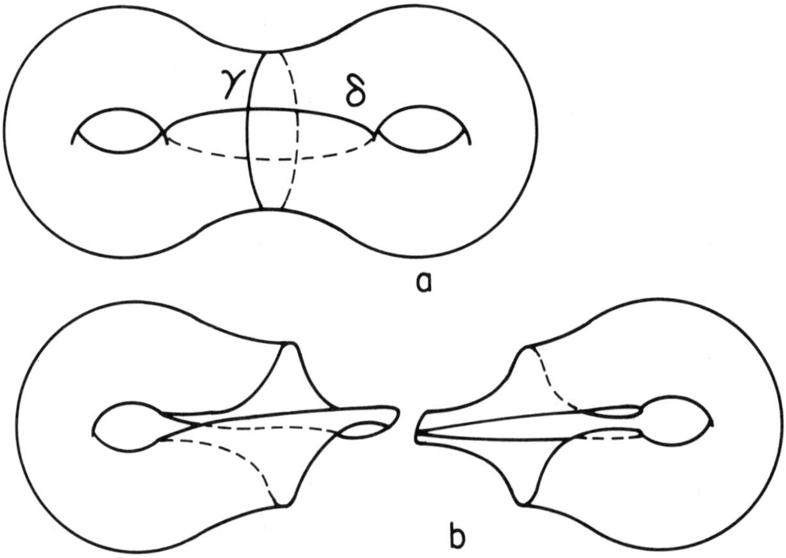

Figure 7. Cutting a genus two surface before (7a) and after (7b) a Dehn twist.

3.4. Troubles

We have now developed a procedure for sewing Polyakov amplitudes. Unfortunately, as it now stands, this technique has two serious drawbacks. First, observe that the integral on the right hand side of equation (3.6) is over the moduli space $\mathcal{M}(\Sigma, [\gamma])$ rather than $\mathcal{M}(\Sigma)$. This is too large a region of integration. The problem is exactly the same as that discussed in the first lecture when we investigated the moduli space of a surface and its double. The sewn surface has a mapping class group which is larger than the mapping class group of its components — diffeomorphisms which change the homotopy class of γ exist on Σ, but not on Σ_1 or Σ_2 — and the region of integration $\mathcal{M}(\Sigma)$ is consequently smaller than the region naturally obtained by sewing.

Consider, for example, the genus two surface of Figure 7a obtained by joining two one-holed tori along γ. A Dehn twist around the curve δ changes the homotopy class $[\gamma]$. If one cuts along the new curve obtained from γ by this Dehn twist, one obtains the two surfaces shown in Figure 7b, which are again topologically one-holed tori. (This is not obvious, but stare at the pictures; or note that the double of each component of Figure 7b is a genus two surface.) The world sheets in Figures 7a and 7b give identical contributions to the amplitude for the genus two surface, leading to an overcounting in (3.6). By considering higher order Dehn twists, one easily sees that the moduli space of the genus two surface will be counted an infinite number of times.

It may be that this problem is inevitable. In ordinary field theories, the contribution of any given Feynman diagrams must be divided by a "symmetry factor" to account for the overcounting of identical configurations. The overcounting of moduli space in sewing may be an indication of the need for a similar factor, now infinite, but still calculable (it is the volume of an appropriate quotient of the mapping class group). Alternatively, the equal contributions of Figures 7a and 7b may be a sign of a genuine new symmetry in string field theory which we do not yet understand. Yet another possibility is that the overcounting reflects an inadequate understanding of the measure in string field theory; in light cone string field theory, where a similar problem occurs, it may be curable if the measure is treated carefully [40].

A second problem with this technique is more troubling. Equation (3.6) seems to no longer refer to the quadratic differential ϕ — it is a relation between integrals over moduli spaces, and the coordinates on moduli space should be irrelevant. This is, unfortunately, an illusion: the Polyakov amplitudes for Σ_1 and Σ_2 depend on boundary parameterizations, which are determined by ϕ_1 and ϕ_2.

If these boundary parameterizations were determined once and for all, this dependence would not be a problem — we would have a gauged-fixed theory, but

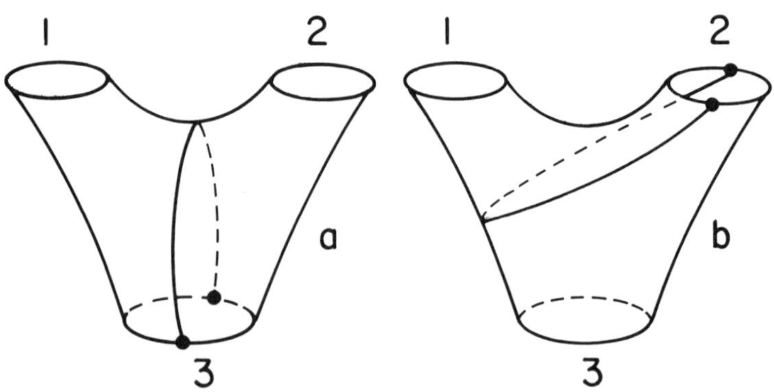

Figure 8. Different decompositions of a three-holed sphere into two ring domains.

that would still be a major step forward. Unfortunately, the choice of quadratic differential ϕ is not fixed, but depends on what is being sewn to what. In the three-holed sphere of Figure 8, for example, we choose one set of ring domains if we wish to sew boundary 1 to boundary 2, and another if we wish to sew boundary 1 to boundary 3. We thus have two Jenkins-Strebel differentials, and two parameterizations of boundary 1. As Ohrndorf [12] has shown explicitly for the case of two cylinders, the wrong parameterization will give the wrong integration region for the sewn surface. It is as if we were trying to calculate amplitudes in QCD with a propagator in light cone and axial gauge, a three-gluon vertex in axial and Lorentz gauge, and a four-gluon vertex in Lorentz and light cone gauge — we could still write down Feynman diagrams, but only by mixing and matching our components.

The basic problem is that we do not understand the gauge symmetry which underlies closed string theory. We are therefore stuck with peculiar gauge-fixed quantities which disguise any fundamental structure. A better understanding of the role of boundary parameterizations seems to be a prerequisite for further progress in the sewing problem.

4. Vertices and Other Miscellany

Finally, let me say a few words about the implications of our sewing procedure for string field theory vertices. It is widely believed that closed string field theory should involve only a single three-string vertex. This is certainly the case in open string field theory, where Witten's three-string midpoint interaction is sufficient to reproduce all amplitudes. Open string amplitudes can be constructed by sewing strips of unit width along the Witten vertex (Figure 9); an existence theorem for quadratic differentials is then used to show that one obtains a single covering of the relevant open string moduli space [41–42].

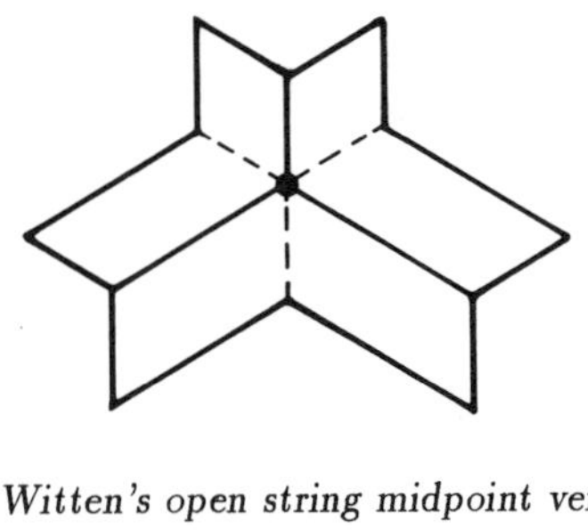

Figure 9. Witten's open string midpoint vertex.

By analogy, we can study possible closed string vertices by examining the sewing of three-holed spheres ("pairs of pants"). In particular, if we could find a represention of the three-holed sphere as three flat cylinders of equal circumference, we would have a good candidate for a closed string vertex. Such a representation would imply the existence of a quadratic differential with three ring domains of circumference 1 on an arbitrary three-holed sphere. Unfortunately, such an existence theorem is false. It is shown in [5] that the relevant cell decomposition of the moduli space of the three-holed sphere consists of *two* highest dimensional cells; while one cell consists of spheres which can be built from three flat cylinders of unit circumference (Figure 10a), the other consists of spheres which must be built from two cylinders and a strip (Figure 10b). Only by combining both cells do we obtain the entire moduli space of the three-holed sphere.

This means that the "generalized Witten vertex" obtained from Figure 10a by collapsing the heights of the three cylinders is not sufficient to give the moduli space of three-holed spheres. One can use the various known existence theorems for quadratic differentials to search for other vertices, but no likely candidates have been found. The one possible exception is the light cone decomposition of Giddings and Wolpert [43], which is related to the covariantized light cone field theory of the Kyoto group [44–45]. It is difficult to discuss sewing in such light cone coordinates for moduli space, however: string circumferences which are arbitrary in simple diagrams are promoted to moduli when these diagrams are sewn together, making it hard to keep track of moduli.

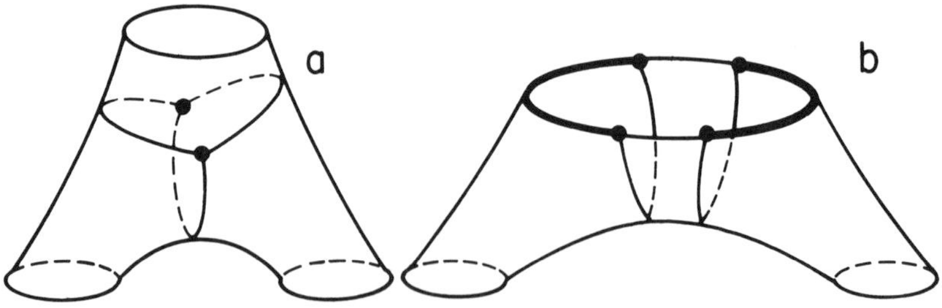

Figure 10. The closed string midpoint vertex (10a) and the remaining cell (10b) needed to cover the moduli space of the three-holed sphere.

One very interesting possibility, proposed by Saadi and Zwiebach [46] following an idea of Kaku's [47], is that one may need a nonpolynomial theory, with an infinite number of vertices, to correctly reproduce moduli space in string field theory. This would make string theory more closely analogous to general relativity, which is certainly nonpolynomial, and it might perhaps be an expected feature of a theory containing gravity. Saadi and Zwiebach have explicitly constructed the four- and five-string vertices which, together with the three-string vertex of Figure 10a, give the correct moduli spaces for the tree level four- and five-point functions. A recent preprint by Kugo, Kunitomo, and Suehiro [48] suggests the generalization to arbitrary tree diagrams. Such a nonpolynomial theory appears not to give the correct moduli spaces for loop diagrams, but this may reflect an inadequate understanding of the string field theory measure rather than an incorrect Lagrangian [40].

Where do we go from here? There are a number of important issues in sewing which are not yet resolved. Most important among them, I think, are the question of the role of boundary parameterizations and the search for new parameterizations of moduli space. These two issues may be related: we have seen that the quadratic differentials which specify coordinates for moduli space also give particular boundary parameterizations. Progress on either of these questions would represent an important advance in our understanding of the structure of string theory.

Acknowledgements

This work was supported in part by the U.S. Department of Energy under grant DE-AC02-76ERO2220.

REFERENCES

1. S. Blau, S. Carlip, M. Clements, S. Della Pietra, and V. Della Pietra, *Nucl.Phys.* **B301** (1988), 285
2. S. Carlip, M. Clements, and S. Della Pietra, in *Superstrings*, P.G.O. Freund and K.T. Mahanthappa, eds. (Plenum Press, New York, 1988)
3. S. Carlip, M. Clements, S. Della Pietra, and V. Della Pietra, "Sewing Polyakov amplitudes I," Insitute for Advanced Study preprint IASSNS-HEP-88/53 (1988)
4. S. Carlip, *Phys.Lett.* **209B** (1988), 464
5. S. Carlip, *Phys.Lett.* **214B** (1988), 187
6. S. Della Pietra and V. Della Pietra, "Analytic torsion and finite group actions," Institute for Advanced Study preprint IASSNS-HEP-89/28 (1989)
7. K. Bardakci, M.B. Halpern, and J.A. Shapiro, *Phys.Rev.* **185** (1969), 1910
8. A. Morozov and A. Rosly, *JETP Lett.* **45** (1987), 210
9. A. Morozov and A. Rosly, *Phys.Lett.* **195B** (1987), 554
10. A. Morozov and A. Rosly, "Strings and open Riemann surfaces," ITEP preprint (February 1988)
11. H. Sonoda, *Nuclear Physics* **B311** (1988), 401,417
12. T. Ohrndorf, *Phys.Lett.* **194B** (1987), 490
13. T. Ohrndorf, *Nucl.Phys.* **B301** (1988), 460
14. T. Ohrndorf, "Covariant operator formalism of the open bosonic string and Polyakov's path integral on bordered Riemann surfaces," Heidelberg preprint HD-THEP-88-37 (1988)
15. C. Vafa, *Phys.Lett.* **190B** (1987), 47
16. L. Alvarez-Gaumé, C. Gomez, G. Moore, and C. Vafa, *Nucl.Phys.* **B303** (1988), 455
17. L. Mezincescu, R.I. Nepomechie, and P.K. Townsend, "Path integral calculation of the closed spinning string propagator," University of Miami preprint UMTG-147 (1988)
18. A.V. Marshakov, "The path integral representation of the fermionic string propagator," P.N. Lebedev Physical Institute preprint, Moscow (1988)
19. A. Cohen, G. Moore, P. Nelson, and J. Polchinski, *Nucl.Phys.* **B267** (1986), 143
20. M.B. Green, J.H. Schwarz, and E. Witten, *Superstring theory* (Cambridge University Press, Cambridge, 1987)
21. O. Alvarez, *Nucl.Phys.* **B216** (1983), 125
22. Yu. Manin, *Phys.Lett.* **172B** (1986), 184
23. C. Lovelace, *Phys.Lett.* **32B** (1970), 703
24. A.S. Losev, *JETP Lett.* **48** (1908), 330

25. G. Collins, "Comment on the use of doubled surfaces," Stony Brook preprint ITP-SB-88-56 (1988)

26. C.J. Earle, in *Advances in the theory of Riemann surfaces*, 1969 Stony Brook Conference (Princeton University Press, Princeton, 1971)

27. M. Sepälä, *Teichmüller spaces of Klein surfaces*, Annales Academice Scientarium Fennica, Mathematica Dissertationes (Helsinki, 1978)

28. N. Redlich, *Phys.Lett.* **205B** (1988), 295

29. N. Redlich, *Nucl.Phys.* **B304** (1988), 129

30. D. Birmingham and C.G. Torre, *Class.Quant.Grav.* **4** (1987), 1149

31. R. Forman, *Inv.Math.* **88** (1987), 447

32. M.F. Atiyah, V.K. Patodi, and I.M. Singer, *Math.Proc.Camb.Phil.Soc.* **77** (1975), 43

33. K. Kodaira, *Complex manifolds and the deformation of complex structures* (Springer-Verlag, Berlin, 1985)

34. E. D'Hoker and D.H. Phong, *Nucl.Phys.* **B269** (1986), 205

35. S.B. Giddings and E. Martinec, *Nucl.Phys.* **B278** (1986), 91

36. K. Strebel, *Quadratic differentials* (Springer, Berlin, 1984)

37. F. Gardiner, *Teichmüller theory and quadratic differentials* (Wiley, New York, 1987)

38. J.L. Harer, *Inv.Math.* **84** (1986), 157

39. A. Marden and H. Masur, *Math.Scand.* **36** (1975) 211; **39** (1976) 232

40. H. Hata, *Phys.Lett.* **217B** (1989), 438, 445

41. S.B. Giddings, E. Martinec, and E. Witten, *Phys.Lett.* **176B** (1986), 362

42. W. Ogura, *Prog.Theor.Phys.* **79** (1988), 936

43. S.B. Giddings and S.A. Wolpert, *Commun.Math.Phys.* **109** (1987), 177

44. H. Hata, K. Itoh, T. Kugo, H. Kunitomo, and K. Ogawa, *Phys.Lett.* **172B** (1986), 195

45. H. Hata, K. Itoh, T. Kugo, H. Kunitomo, and K. Ogawa, *Phys.Rev.* **D35** (1987), 1356

46. M. Saadi and B. Zwiebach, "Closed string theory from polyhedra," MIT preprint MIT-CTP-1704 (1989)

47. M. Kaku and J. Lykken, *Phys.Rev.* **D38** (1988), 3067

48. T. Kugo, H. Kunitomo, and K. Suehiro, "Non-polynomial closed string field theory," Kyoto preprint KUNS 965 HE(TH)89/01 (1989)

Dr. Steven Carlip, School of Natural Sciences, Institute for Advanced Study, Princeton, NJ 08540, USA.

Operator Formalism for Strings on Riemann Surfaces

Theophil Ohrndorf*

Institut für Theoretische Physik
der Universität Heidelberg, FRG

1. Introduction

In the last few years various groups have proposed different operator formalisms for string theory [1-6]. Typically these approaches culminate in the construction of string vertices, multi-Reggon vertices, or string states. In fact all these names may be considered as being synonymous. Unfortunately only a few authors take care to relate their own approach to those already existing.

These lectures are intended to fill in this gap. We hope to convince the reader that string vertices may be understood as "quantum mechanical" transition amplitudes (TA) on bordered Riemann surfaces (RS), where the boundaries of the underlying RS have to be considered as parametrized curves. The TA will be defined explicitly with the help of the path integral. The various approaches mainly differ in the conventions used to parametrize the boundaries. These lectures are mainly based on the work of the author. A full account has been given in [7]and the reader is referred to this paper for more information.

Before presenting any details in the following sections, let us briefly outline the role played by the operator formalism in string theory. Already soon after the discovery of the Veneziano amplitude it was recognized [8] that the factorization properties of the generalized Veneziano amplitudes, which allow to interpret these amplitudes as scattering amplitudes, could be displayed in an elegant fashion by introducing propagators and vertex functions. These objects were represented with the help of a first quantized operator field, the so called Fubini-Veneziano field [9]. In this way an efficient algorithm employing "Feynman-like rules" was developed.

This algorithm allowed to represent the Veneziano like amplitudes in a way that their factorization properties were manifest just like in the Born approximation of ordinary field theory. Moreover this formalism provided an explicit representation of the Virasoro algebra and of the assiciated physical state conditions. Having such a representation is of crucial importance to demonstrate the decoupling of negative norm states. The operator formalism may therefore be understood as being an integral part of the definition of string theory.

Unfortunately the quantum field theoretic interpretation of the resulting theory was obscure, as it violated unitarity in just the same way as the Born approximation in field theory does, i.e. there are only one-particle intermediate states. The cure of this disease was readily found by imitating the behaviour of field theory. It was proposed to include loop corrections into the theory. Such loops may be formed by identifying the external states of a tree diagram. The implementation of this program made it necessary to construct generalizations of the scattering amplitudes, which do not only describe the scattering of the individual modes of the string, but of a full string state, usually called a Reggeon. Such amplitudes are therefore known as multi-Reggeon vertices. To represent these vertices explicitly the operator formalism had to be generalized. One has to introduce a separate Fubini-Veneziano field for each Reggeon.

All these developments occured rapidly after the discovery of the Veneziano amplitude in 1969. Since that time string theory has seen several important new developments. Most dramatically, its commmon interpretation has changed from being a phenomenological theory of hadrons to providing a potential candidate of a quantized theory of gravity. This has not only to be considered as being simply a change of point of view. To understand the gravitational nature of string theory it is necessary to find a formulation of this theory which allows to make contact with general relativity explicitly. Conceptually this is very important, because otherwise it seems to be almost impossible to even formulate questions about "how strings manage to quantize gravity", not to mention the problem of finding adequate answers to such questions. In fact in the last few years there have been numerous attempts to find a new formulation of string theory, usually occuring under the name of string field theory or non-perturbative string theory.

Unfortunately many of these attempts proceed simply by try-and-error. As so far no satisfactory solution has been found, it appears worthwile to approach this problem more systematically. Such a systematic way of searching for a good string field theory is provided by studying the various ways in which an arbitrary RS can be build out of only a few simple pieces by gluing these pieces together in a suitable manner. Let us outline the procedure more closely.

One of the basic obstacles which any field theoretic formulation of string theory has to overcome, is to find a solution to the so called "duality problem". Naively one might be tempted to interpret each Feynman like diagram associated with a dual amplitude as a Feynamn diagram of a string field theory. Unfortunately this does not work, as e.g. the four-point amplitude has poles in both the s-channel and in the t-channel simultaneously, a behaviour which cannot result from the contribution of a single Feynman diagram. The only way to circumvent this problem is by first splitting up the range of the integral representing the dual amplitude into several pieces, which are then associated with individual Feynman diagrams. In fact this problem is only a particular aspect of a much more general problem.

One of the advantages of modern string theory is the availabiliy of Polyakov's path integral formulation. Having performed the gauge fixing, any amplitude can be represented as a finite dimensional integral over the moduli space of an assiciated RS. The region of integration has to cover moduli space exactly once. To obtain a field theory from the path integral, according to the general philosophy one has to split up the RS into cylinders, which are connected by n-holed spheres ($n \geq 3$). Cylinders represent propagators, whereas the n-holed spheres represent vertices. A particular elegant way of decomposing a RS can be obtained by employing quadratic differentials with closed trajectories [10]. In general one identifies the moduli of the cylinders with the Schwinger parameters of the associated field theory. Similarly one may also, at least in principle, incorporate the moduli associated with the RS representing the vertices into a string field theory. The crucial criterion, which any acceptable field theory has to satisfy, is that the integrals over the moduli associated with the individual building blocks of string field theory have to combine in such a way that moduli space is covered precisely once. Interpreting the positions of the punctures on a RS as well as moduli, the relation to the "duality problem" becomes obvious.

We emphasize this problem here, although we have nothing to contribute to its solution, because it is in fact the central problem which any string field theory has to solve, if it wants to be taken seriously.

In these lectures we shall concentrate on the construction of the TA associated with only a single, fixed though arbitrary RS. We shall deal with the closed bosonic string and shall include both the X-coordinate as well as the $b-c$ ghost system. It will turn out that in the process of the construction of the TA the generalized Fubini-Veneziano field and its ghost analogues will occur quite naturally.

The TA which we shall construct may be interpreted as a (unintegrated) contribution to a correlation function of a string field theory in fixed order of perturbation theory. In particular once the problem of finding a suitable slicing of moduli space has been solved, the vertices of the associated field theory will be given by our TA. Moreover, as the TA only depend on the moduli and on the boundary parametrizations of the underlying RS, the calculation of any string field theory amplitude to fixed order of perturbation theory amounts to a determination of just these data. These facts may simplify the calculation of particular contributions enormously, compared with the naive generalization of standard point particle field theory techniques to strings.

2. First quantization and the path integral

In this section we shall work out the first quantization of the purely bosonic string coordinate X. As the TA is explicitly defined in terms of the path integral, the procedure will be a straightforward generalization of the path integral quantization of e.g. the harmonic oscillator, as it may be found in the famous textbook of Feynman and Hibbs [11].

The importance of this procedure hides in the fact, that we can derive an operator formalism explicitly from the path integral. In other words we do not have to resort to any reconstruction theorem, to introduce an operator formalism, as is usually done in conformal field theory. In this way we alleviate the problem of finding a reconstruction theorem valid on higher genus surfaces.

The important result will be formula (2.9). This formula is powerful enough to allow the construction of the TA in abstract Fock space almost completely. This construction will be carried out in the second half of this lecture.

To make these lectures more transparent, we shall deal with the TA of the X-coordinate and the TA of the ghost coordinates independently. In fact due to the presence of the Liouville mode only their product is well defined.

Our starting point is the following TA

$$K[X_1(\sigma),\cdots,X_N(\sigma)] = \int DXe^{-S} \tag{2.1}$$

with

$$S = \frac{1}{2\pi}\int_M d^2z\partial_z X^\mu \partial_{\bar{z}} X_\mu, \tag{2.2}$$

where the path integral includes only those configurations $X(z,\bar{z})$, which satisfy $X(z,\bar{z}) = X_i(\sigma)$ if $z \in C_i$. We shall assume that the boundary ∂M of the world sheet M consists out of N components C_i $i = 1,...,N$, which are just closed loops. The string coordinate $X(z,\bar{z})$ is required to satisfy Dirichlet boundary conditions along the boundary of M.

As the action (2.2) is quadratic in X (2.1) can be evaluated explicitly. It is a simple exercise to show that

$$K[X_1(\sigma),\cdots,X_N(\sigma)] = (det\Delta)^{-d/2}exp\{-S^{CL}[X(\sigma)]\}, \tag{2.3}$$

with the classical action

$$S^{CL}[X(\sigma)] = -\frac{1}{2\pi}\int_{\partial M} d\sigma X_\mu(\sigma)\partial_n X^{CL,\mu}(z(\sigma),\bar{z}(\sigma)), \tag{2.4}$$

where$X^{CL}(z,\bar{z})$ is the solution of the Dirichlet problem on M

$$X^{CL}_\mu(z,\bar{z}) = \int_{\partial M} d\sigma'\partial'_n G(z',z)X_\mu(\sigma'), \tag{2.5}$$

$\partial_n X^{CL}_\mu(z(\sigma),\bar{z}(\sigma))$ is its normal derivative and $z(\sigma)$ demarcates the boundary of M. $G(z',z)$ is the Dirichlet Green function on M. For simplicity we have suppressed its $\bar{z},\bar{z}'$ dependence.

Since the classical action can be written in the form (2.4), acting with the functional derivative $\overleftarrow{\delta}/\delta X^r(\sigma)$ on $K[X_1(\sigma),...,X_N(\sigma)]$ reproduces just $\partial_n X(\sigma)$. This property is important, because it allows to introduce an operator field $\partial_z\hat{X}^r$, which may be understood as a generalization of the old Fubini-Veneziano field. This operator has the property that it reproduces precisely the classical configuration $\partial_z X^{CL}_\mu$, the derivative of (2.5), when acting on $K[X_1(\sigma),...,X_N(\sigma)]$.

In order to introduce $\partial_z\hat{X}^r$ let us pick a particular component of ∂M, let's say r, and introduce a functional derivative operator $\overleftarrow{\delta}/\delta X^r(\sigma)$, which satisfies

$$[X^r_\mu(\sigma), \frac{\overleftarrow{\delta}}{\delta X^s_\nu}(\sigma')] = \delta^{rs}\eta_{\mu\nu}2\pi\delta(\sigma-\sigma'). \tag{2.6}$$

Letting $\overleftarrow{\delta}/\delta X^r(\sigma)$ act to the left is simply a convenient convention. The operator $\partial_z\hat{X}^r$, which reproduces the classical configuration $\partial_z X^{CL}$, is then defined by

$$\partial_z\hat{X}^r = \frac{1}{\dot{z}}\left(\frac{1}{2}\dot{X}^r(\sigma) - \frac{i}{4}\frac{\overleftarrow{\delta}}{\delta X^r}(\sigma)\right). \tag{2.7}$$

We have

$$\dot{z}(\sigma)\partial_z = \frac{1}{2}(\partial_\sigma - i\partial_n) \tag{2.8}$$

where ∂_n denotes the normal derivative across ∂M. As $\overleftarrow{\delta}/\delta X$ reprodues $\partial_n X$ by acting on (2.3), (2.7) and (2.8) imply that

$$K[X_1(\sigma),\cdots,X_N(\sigma)]\partial_z\hat{X}^r = K[X_1(\sigma),\cdots,X_N(\sigma)]\partial_z X^{CL} \tag{2.9}$$

Note that (2.9) is strictly valid only if z lies on C_r. Fixing the parametrization of the boundary of M with the help of a conformal mapping in the next paragraph, we shall represent $\partial_z \hat{X}^r$ in terms of a Laurent expansion, where the coefficients are just harmonic oscillators. This representation allows to analytically continue (2.9) away from C_r, to be valid in at least a small annulus type neighborhood around C_r. However, this representation breaks down as soon as the expansion of $\partial_z X^{CL}$ induced by the action of $\partial_z \hat{X}^r_\mu$ ceases to converge. In fact there is no way of making sense of $\partial_z \hat{X}^r$ beyond such a singular point. Therefore (2.9) is valid only if z is near C_r, but in general not all over M.

Let us next turn to the parametrization of the boundary ∂M. We shall induce a parametrization of C_r by conformally mapping the simply connected region of the z-plane, which is encircled by C_r, but which does not belong to M itself, onto a unit disk with coordinate ς_r. Choosing the angular coordinate to parametrize the boundary of the unit disk in the ς_r-plane, a parametrization of C_r is induced. All this has just the effect, that expanding $X_r(\sigma)$ in an ordinary Fourier series, where σ is the induced parameter on C_r

$$X_r(\sigma) = \sum_n X^r_n e^{-in\sigma}, \tag{2.10}$$

$\partial_{\varsigma_r} \hat{X}$ becomes

$$\partial_{\varsigma_r} \hat{X}^\mu_r = -\frac{1}{2} i \sum_n \varsigma_r^{-n-1} \alpha^{\mu,r}_n, \tag{2.11}$$

with

$$\alpha^r_n = -inX^r_{-n} + \frac{1}{2} i \frac{\overleftarrow{\partial}}{\partial X^r_n}. \tag{2.12}$$

(2.12) implies the standard commutation relations

$$[\alpha^{r,\mu}_n, \alpha^{s,\nu}_m] = \delta^{rs} \eta^{\mu\nu} n \delta_{n,-m}. \tag{2.13}$$

To illustrate the role of the conformal mappings ς_r we shall briefly list several examples, which are related to different operator approaches proposed in the literature. All the approaches known to the author proposed around 1970 [12, 13] work implicitly with ς_r being a projective mapping. A different mapping is employed by Mandelstam [14]. In terms of his famous $\rho(z)$ mapping it is simply $\varsigma_r \sim \; exp\{\rho(z)/\alpha_r\}$. Another mapping may be obtained by replacing $\rho(z)$ by Witten's mapping [15]. Although these mappings are of no concern as long as we use the TA to calculate only correlation functions of physical states directly, they become very important, if we begin to built up TA by composing them out of TA defined on submanifolds of M, because the information about the moduli space is hidden in these mappings.

Let us finally come to the Fock representation K_X of $K[X_1(\sigma), ..., X_N(\sigma)]$. We shall simply define K_X to be a Fock state, which reproduces $K[X_1(\sigma), ..., X_N(\sigma)]$ by forming inner products with $X_r(\sigma)$ eigenstates $|X_r(\sigma)>$

$$K[X_1(\sigma), \cdots, X_N(\sigma)] = K_X \prod_{r=1}^{N} |X_r(\sigma)> \tag{2.14}$$

The states $|X_r(\sigma)>$ may be understood as an infinite product of ordinary harmonic oscillator eigenstates associated with the individual oscillators α^r_n.

To construct K_X explicitly we have to employ a set of meromorphic functions $F^r_m(z)$. To define these functions, it is useful to introduce a new RS M_∞, which is just M augmented by the "boundary disks". By definition $F^r_m(z)$ is required to have an m-fold pole at $\varsigma_r(z) = 0$ and to be a regular function of z everywhere else on M_∞. As M_∞ is compact, this fixes $F^r_m(z)$ uniquely modulo the freedom to add an arbitrary constant. We shall assume in the rest of this section that g, the genus of M_∞, vanishes. Then functions F^r_m exist for all values of m.

Suppose that z lies near C_r, then $F^r_m(z)$ has the following expansion

$$F^r_m(z) = \delta_{rs}\varsigma_r^{-m}(z) - mN^{rs}_{mn}\varsigma^n_s(z) \qquad \begin{matrix} m \geq 1 \\ n \geq 0 \end{matrix} . \tag{2.15}$$

It is not too difficult, to express N^{rs}_{mn} as a double contour integral. A general formula will be given in the next section.

By multiplying the relation (2.9) with $F^r_m(z)$ and integrating along ∂M (2.9) becomes

$$K_X\{\alpha^r_{-m} - mN^{rs}_{mn}\alpha^s_n\} = 0. \tag{2.16}$$

(2.16) represents a set of relations which are solved by

$$K_X \sim \delta(\sum p_i) < 0|exp(\frac{1}{2}\alpha^r_m N^{rs}_{mn}\alpha^s_n). \tag{2.17}$$

The energy-momentum conserving δ-function arises by replacing $F^r_m(z)$ by just a constant. Obviously (2.16) does not fix the normalization of K_X. We shall determine it in the last lecture.

The result (2.17) may be used in two essentially different ways. Either one may saturate (2.17) with physical states to obtain scattering amplitudes. It can be shown that this procdure corresponds exactly to filling out the "boundary disks" with operator inserted disks. Alternatively one may use (2.17) as a building block to create new amplitudes by via gluing.

3. Loop corrections

If g, the genus of the world sheet M_∞, does not vanish, we have to modify the construction of K_X, which has been given at the end of the foreging section for $g = 0$ in several ways.

The fundamental reason that the $g = 0$ procedure cannot simply be taken over to $g \geq 1$ follows from the Weierstrass gap theorem. This theorem states that if the point at which the functions F^r_m are required to become singular is not a so called "Weierstrass point", then there exist no functions F^r_m for $g \geq m$. For a proof of this theorem we refer the reader to the standard mathematical literature [16, 17]. For simplicity we shall assume in the following that the point at which F^r_m becomes singular is not a Weierstrass point.

The missing of the functions F^r_m for $g \geq m$ implies that the relations (2.16) are no longer sufficient to fix the oscillator dependence of K_X completely. To circumvent this problem we shall simply drop the requirement so far imposed on F^r_m to be a single-valued function on M. Instead we shall allow F^r_m to change by a constant (period), when going around some of the nontrivial cycles of M. Introducing a canonical homology basis on M, we have the freedom to normalize the F^r_m in such a way that all the a-periods vanish. Expanding the F^r_m just like in (2.15) it is possible to represent the "Neumann coefficients" N^{rs}_{mn} as a double contour integral of the prime form $E(z_1, z_2)$ on M_∞

$$N^{rs}_{mn} = \oint_{C_r} d\varsigma_r(z_1)\varsigma_r^{-m-1} \oint_{C_s} d\varsigma_s(z_2)\varsigma_s^{-n-1} log E(z_1, z_2) \qquad n, m \geq 1. \tag{3.1}$$

Before we can generalize the set of relations (2.16) to the case $g > 0$ we have first to find a way of incorporating the non-vanishing periods of $\partial_z X$. Note that the single-valuedness of $X^{CL}(z, \bar{z})$ does not imply that $\partial_z X$ has non-vanishing periods.

To handle the non-vanishing periods of F^r_m and of $\partial_z X$ we shall introduce $2g$ new operators. In order to represent these operators again as differential operators, we shall allow $X(z, \bar{z})$ to become multi-valued, i.e.

$$\int_{a_i, b_i} d\sigma \partial_\sigma X = -2\pi Y_{a_i, b_i} \qquad i = 1, \cdots, g. \tag{3.2}$$

We can now use the fact that due to (3.2) $K[X_1(\sigma), ..., X_N(\sigma)]$ becomes a function of the periods Y_{a_i} and of Y_{b_i} to introduce the operators

$$\hat{p}_{a_i} = \frac{1}{2\pi i}\left(\frac{\overleftarrow{\partial}}{\partial Y_{b_i}} - 4\pi i Y_{a_i}\right) \tag{3.3}$$

$$\hat{p}_{b_i} = -\frac{1}{4}\left(\frac{\overleftarrow{\partial}}{\partial Y_{a_i}} + 4\pi i Y_{b_i}\right), \tag{3.4}$$

which satisfy

$$-\frac{1}{4} i\hat{p}_{a_i} \cong \frac{1}{2\pi i}\int_{a_i} dz \partial_z X \tag{3.5}$$

$$-i\hat{p}_{b_i} \cong \int_{b_i} dz \partial_z X. \tag{3.6}$$

Here the symbol $\cong$ indicates that these relations are only valid when acting with the operators on $K[X_1(\sigma), ..., X_N(\sigma)]$. $\hat{p}_{a_i}$ and $\hat{p}_{b_i}$ satisfy the following commutation relations

$$[\hat{p}_{a_i}, \hat{p}_{b_j}] = \delta_{ij} \tag{3.7}$$

We would like to point out that $\hat{p}_{a_i}$ and $\hat{p}_{b_i}$ provide a concrete realization of a set of operators previously proposed in [18]. To work with eigenstates of $\hat{p}_{a_i}$ and its complex conjugate we shall first perform the following transformation

$$K[X_1(\sigma),\cdots,X_N(\sigma),Y_{a_i},Y_{b_i}] = \prod_{i=1}^{g}\{\int dp_{a_i} d\bar{p}_{a_i} \delta(4Y_{a_i} + p_{a_i} - \bar{p}_{a_i}) \\ \times exp(2\pi i Y_{b_i}(p_{a_i} + 2Y_{a_i}))\} K[X_1(\sigma),\cdots,X_N(\sigma),p_{a_i},\bar{p}_{a_i}], \tag{3.8}$$

and then we shall represent $K[X_1(\sigma),...,X_N(\sigma),p_{a_i},\bar{p}_{a_i}]$ as a Fock state. In this way it is easy to demonstrate that (2.16) has to be replaced by

$$\begin{aligned} &\alpha^r_{-m} - mN^{rs}_{mn}\alpha^s_n - \frac{1}{2}\hat{p}_{a_k} a^r_{k,m} \cong 0 \\ &m > 0 \qquad n \geq 0 \\ &s = 1,\cdots,N \qquad k = 1,\cdots,g \end{aligned} \tag{3.9}$$

and by

$$\begin{aligned} -\tilde{a}^r_{k,n}\alpha^r_n + 2\hat{p}_{b_k} - \frac{1}{2}\hat{p}_{a_i}\tau_{ik} \cong 0 \\ n \geq 0 \qquad k,i = 1,\cdots,g \end{aligned} \tag{3.10}$$

where the coefficients $a^r_{k,m}$ are the Taylor coefficients of the analytic abelian differentials $\omega^k_{\varsigma_r}$ on M_∞

$$\omega^k_{\varsigma_r} = \sum_{n=1}^{\infty} a^r_{k,n}\varsigma_r^{n-1} \qquad k = 1,\cdots,g \tag{3.11}$$

and τ_{ik} is the period matrix on M_∞. If $n \geq 1$,

$$\tilde{a}^r_{k,n} = \frac{1}{n}a^r_{k,n}, \tag{3.12}$$

whereas $\tilde{a}^r_{k,0}$ is an undetermined constant. There is a set of relations similar to (3.9) and (3.10) for the complex conjugate of α^r_n. Note that with the help of the operators $\hat{p}_{a_i}$ and $\hat{p}_{b_i}$ holomorphic factorization is preserved, at least if we do not carry out the integral over p_{a_i} and $\bar{p}_{a_i}$, in contrast to the approach persured in [3, 6]. Solving (3.9) and (3.10) one obtains

$$\begin{aligned} K_X \sim \delta(\sum p_i) < 0|exp\{\frac{1}{8}p_{a_i}\tau_{ik}p_{a_k} + \frac{1}{2}p_{a_k}\tilde{a}^r_{k,m}\alpha^r_m \\ + \frac{1}{2}\alpha^r_m N^{rs}_{mn}\alpha^s_n + \text{c.c.}\} \\ m,n \geq 0. \end{aligned} \tag{3.13}$$

Reinserting (3.13) into (3.8) and setting $Y_{a_i} = Y_{b_i} = 0$, i.e. letting $X(z,\bar{z})$ again become single-valued, we finally get

$$\begin{aligned} K_X \sim \delta(\sum p_i) < 0|exp\{-\frac{1}{4}[\tilde{a}^r_{i,m}\alpha^r_m - \text{c.c.}](Re(\tau_{ij})^{-1} \\ \times [\tilde{a}^s_{j,m}\alpha^s_m - \text{c.c.}] + \frac{1}{2}\alpha^r_m N^{rs}_{mn}\alpha^s_n + \text{c.c.}\} \end{aligned} \tag{3.14}$$

4. The $b-c$-ghost system

The construction of the ghost TA $K[c_1(\sigma), b_1(\sigma), ..., c_N(\sigma), b_N(\sigma)]$ can be carried out in a manner almost parallel to the construction of $K[X_1(\sigma), ..., X_N(\sigma)]$. Let us therefore only emphasize those aspects, in which the ghost system differs substantially from the X-coordinate.

Instead of (2.3) one obtains the following TA for the ghosts

$$K[c_1(\sigma), b_1(\sigma), \cdots, c_N(\sigma), b_N(\sigma)] = (det\ P_1^+ P_1)^{+1/2} e^{-S_{gh}^{CL}} \delta, \tag{4.1}$$

where δ is a product of fermionic δ-functions. If $N \geq 3$ it is given by

$$\delta = \prod_{k=1}^{6g+3N-6} (\int_{\partial M} d\sigma h_{zz}^k \dot{z}^2 c(\sigma)), \tag{4.2}$$

with h_{zz}^k an analytic quadratic differential on M. This δ-function performs an important task. The classical action appearing in (4.1) is a functional of the classical configuration of the $b-c$ fields c^z and b_{zz}. Due to their equations of motion these configurations have to be analytic functions on M. In our formalism $c(\sigma)$ and c^z are related by

$$c(\sigma) = -\sqrt{2} Im(c^z/\dot{z}) \qquad z \in \partial M \tag{4.3}$$

and similarly for the b field. For generic values of $c(\sigma)$ it is impossible to find solutions to the equations of motion such that (4.3) is satisfied. This is only possible if we restrict our attention to such boundary data, which satisfy

$$\int_{\partial M} d\sigma h_{zz}^k \dot{z}^2 c(\sigma) = 0. \tag{4.4}$$

But this is just the condition imposed effectively by (4.2).

The role of the functions F_m^r is now taken over by either an inverse meromorphic differential ($g = 0$) or if $g \geq 1$ by a meromorphic quadratic differential $H_{m,zz}^r$ with the following expansion if z is near to C_r

$$H_{m,\varsigma_s\varsigma_s}^r = \delta_{rs}\varsigma_r^{-m-2} - \tilde{V}_{mn}^{rs}\varsigma_s^{n-2} \qquad m \geq -1 \qquad n \geq 2. \tag{4.5}$$

(4.5) leads to the following Fock representation of the TA if $g \geq 1$

$$K_{gh} \sim < 0|\tilde{V}_{gh}\tilde{\delta} \times (c.c.), \tag{4.6}$$

where

$$\tilde{V}_{gh} = exp(-b_m^r \tilde{V}_{mn}^{rs} c_n^s) \qquad m \geq -1, \quad n \geq 2, \tag{4.7}$$

and

$$\tilde{\delta} = \prod_{k=1}^{3g-3} (\tilde{H}^k_{n,s} c^s_n). \tag{4.8}$$

In contrast to (4.1) in (4.8) the holomorphic quadratic differentials on M_∞,

$$\tilde{h}^k_{\varsigma_r \varsigma_r} = \sum_{m=2}^{\infty} \tilde{H}^k_{m,r} \varsigma_r^{m-2} \qquad k = 1, \cdots, 3g-3 \tag{4.9}$$

appear. A formula similar to (4.6) holds for the $g = 0$ case.

There are several important remarks concerning the result (4.6). At first it can be shown [19] that (4.6) induces an amount of ghost number violation, which is given by -3 times the Euler characteristic of M. Secondly (4.6) is BRST invariant, i.e. it is annihilated by the operator

$$\sum_{r=1}^{N} Q_r \tag{4.10}$$

where Q_r refers to the BRST operator of an individual boundary component. Finally the precise prescription to obtain an amplitude from (4.6) by saturation with physical states must include the action of the operator

$$\int \prod_{i=1}^{3g-3+N} [d^2 \tau_i \int_M d^2 z_i \mu_z^{iz} b_{zz}] \times [c.c] \tag{4.11}$$

where μ_z^{iz} denotes the Beltrami differentials and we count the positions of the vertex operators as moduli.

5. Boundary variations and the normalization of the states

The method which we have employed so far to construct K_X and K_{gh} leaves the normalization of these states open. As it has to be expected that this normalization depends on the moduli of M_∞, we cannot simply ignore it. In this section we shall give a rough outline of how the normalization of these states can be found. We shall refrain from giving all the details, as the corresponding calculations are quite involved.

The general idea is first to establish a set of variational equations characterizing the response of (2.3) and of (4.1) to both variations of the boundary of M as well as to variations of the moduli of M_∞. We shall then go on and solve these variational equations in Fock space. A detailed calculation shows that the product $K[X_1(\sigma), \cdots, X_N(\sigma)] K[c_1(\sigma), b_1(\sigma), \cdots, c_N(\sigma), b_N(\sigma)]$ changes under a shift of ∂M by an amount measured by the vector field $\delta^z(\sigma)$ according to

$$\begin{aligned}&\hat{\delta}(K[X_1(\sigma), \cdots, X_N(\sigma)] K[c_1(\sigma), b_1(\sigma) \cdots, c_N(\sigma), b_N(\sigma)]) \\ &= K[X_1(\sigma), \cdots, X_N(\sigma)] K[c_1(\sigma), b_1(\sigma), \cdots, c_N(\sigma), b_N(\sigma)] Re\{-\frac{i}{\pi} \int_{\partial M} dz \delta^z : \hat{T} :\}\end{aligned} \tag{5.1}$$

where

$$\hat{T} = \hat{T}_X + \hat{T}_{gh} \tag{5.2}$$

and

$$T_X = -2(\partial_z X)^2 \tag{5.3}$$

$$T_{gh} = 2(\partial_z c)b_{zz} + c^z \partial_z b_{zz}. \tag{5.4}$$

The result (5.1) carries over to variations of the "isometric circles" of M_∞ provided we only replace ∂M appropriately.

We can calculate the variation of the product of Fock states $K_X K_{gh}$ under variations of ∂M in two different ways. Either by using their explicit form as being given by (3.14) and by (4.6) or by using (5.1). Matching both variations we get a set of relations among the different Neumann coefficients, which are identically satisfied for all coefficients apart from N_{00}^{rs}, which so far has not been determined. Solving the corresponding variational equation we find that (3.1) extends to $m = n = 0$ provided $r \neq s$. If $r = s$, the variational equation is solved by

$$N_{00}^{rr} = -log\acute{\varsigma}_r(z^r), \tag{5.5}$$

where z^r is the inverse image of $\varsigma_r = 0$, i.e.

$$\varsigma_r(z^r) = 0. \tag{5.6}$$

Similarly it follows that

$$\tilde{a}_{i,0}^r = \int^{z^r} dz \acute{\varsigma}_r \omega_{\varsigma r}^i. \tag{5.7}$$

Preceding in the same fashion, however, focussing on the variations of the moduli of M_∞ we find that the product of the RHS of (3.14) and (4.6) has to be multiplied by

$$(det'\Delta_\infty / \int_{M_\infty} d^2z\sqrt{g})^{-d/2} det^{+1/2}(P_1^+ P_1)_\infty \tag{5.7}$$

to give a normalized state. g denotes here the determinant of the world sheet metric and the subscript ∞ indicates that the corresponding determinants are defined on M_∞. Note that the result (5.7) had to be expected, as saturating $K_X K_{gh}$ with vacuum states should reproduce the vacuum energy on M_∞.

References

[1] A. Neveu and P. West, Nucl. Phys. **B 278** (1986) 601
[2] P. di Vecchia, K. Hornfeck, M. Frau, A. Lerda and S. Sciuto, Phys. Lett. **B 206** (1988) 643 and related work of this group
[3] C. Vafa, Phys. Lett. **B190** (1987) 47
[4] A. Le Clair, M.E. Peskin and C.R. Preitschopf, Slac-Pub 4306/4307 Jan. 1988
[5] J.L. Peterson and J. Sidenius, Nucl. Phys. **B301** (1988) 247
[6] L. Alvarez-Gaumé, C. Gomez, G. Moore and C. Vafa, Nucl. Phys. **B 303** (1988) 455
[7] Th. Ohrndorf, "Covariant Operator Formalism of the Bosonic String and Polyakov's Path Integral on Bordered Riemann Surfaces" Heidelberg-Preprint HD-THEP-1988-37
[8] S. Fubini, D. Gordon, and G. Veneziano, Phys. Lett. **B29** (1969) 679
[9] S. Fubini and G. Veneziano, Nuovo Cimento **67A** (1970) 29
[10] S. Carlip, these proceedings
[11] R.P. Feynman and A.P. Hibbs, "Quantum Mechanics and Path Integrals", Mac Graw-Hill, New York (1965)
[12] C. Lovelace, Phys. Lett. **B 32** (1970) 490
[13] D. Olive, Nuovo Cimento **3A** (1971) 399
[14] S. Mandelstam, Nucl. Phys. **B64** (1973) 205
[15] E. Witten, Nucl. Phys. **B268** (1986) 253
[16] H.M. Farkas and I. Kra, Riemann Surfaces, Springer 1980
[17] O. Forster, Riemannsche Flächen, Springer 1977
[18] R. Dijkgraaf, E. Verlinde and H. Verlinde, Comm. Math. Phys. **115** (1988) 649
[19] Th. Ohrndorf, Phys. Lett. **B 215** (1988) 280

TH. OHRNDORF
INSTITUT FÜR THEORETISCHE PHYSIK
DER UNIVERSITÄT HEIDELBERG
PHILOSOPHENWEG 16, D-6900 HEIDELBERG, FRG

HOW TO DERIVE THE OPEN STRING PROPAGATOR ?

Zbigniew Jaskólski
Institute of Theoretical Physics.
University of Wroclaw
Cybulskiego 36
PL-50-205 Wroclaw.POLAND

1. INTRODUCTION

The problem of the functional integral representation of the string propagator is a part of the off-shell formulation of the Polyakov string (see Carlip's lectures - this volume). The first approach to this problem based on O.Alvarez's paper [1] has been proposed by Cohen et al. [2]. The basic idea of this approach is to consider the path integral over metrized surfaces connecting prescribed contours in the target space. The resulting functional contains an averaging over boundary reparametrizations which in contrast to other functional measures in the Polyakov theory remains undetermined and is an untractable formal symbol. The expression for the propagator is therefore well defined only for the point-like strings where this averaging decouples [2,3]. In several papers [4-8] it was pointed out that the off-shell amplitudes in the string theory should be gauge dependent and therefore the averaging over boundary reparametrizations seems to be spurious. The arguments of Ref.[4,5] are mainly of the string field theory nature, while the derivations in Refs [6,7] are based on the BRST extention of the proper time representation of the propagator. Another derivation of the off-shell string propagator based on the Batalin-Fradkin-Vilkoviski phase space path integral has been proposed by Karchev [8]. In all these cases however the original geometrical interpretation of the Polyakov path integral is lost.

The aim of this lecture is to present a new derivation of the covariant path integral representation of the string propagator. We will use two kinds of arguments: the quantum mechanical ones based on the generalization of the Feynman functional quantization and the geometrical ones related to the requirement of the consistency of the Faddeev-Popov procedure. For the simplicity we will consider the open string case.

2. THE COVARIANT FUNCTIONAL QUANTIZATION OF THE GAUGE SYSTEMS

Within the Feynman path integral quantization of nondegenerate lagrangians the space of wave functions is the space of complex valued functions defined on the space $\mathbb{R}\times Q$ which can be regarded as a space of boundary conditions for trajectories of a system. (We use the position representation; in general one can take an arbitrary half of the Cauchy data). In this space one introduces the inner product defined by the volume form $d^n q$ on Q. The basic object is the integral kernel of the evolution operator expressed in terms of the functional integral over trajectories.

In the case of the gauge systems with a Lie algebra of the first class constraints the space $\mathbb{R}\times Q$ divides into classes of physically equivalent boundary conditions. We say that (t,q), $(t',q') \in \mathbb{R}\times Q$ are physically equivalent if there exist two trajectories connected by a gauge transformation starting at (q,t) and (q',t') respectively and ending at a common point (q'',t''). In many cases the equivalence classes can be described as orbits of some group $\tilde{\mathcal{G}}$ (acting on $\mathbb{R}\times Q$) induced by the group $\mathcal{G}$ of the gauge transformations (acting on the space of trajectories). For gauge systems without reparametrization invariance all constraints are linear in momenta and the group $\mathfrak{G}$ has a general structure $\{id_{\mathbb{R}}\}\otimes\bar{\mathfrak{G}}$. The $\mathfrak{G}$-invariant functions on $\mathbb{R}\times Q$ form in this case the space of physical states $\mathfrak{H}_{ph}$ (i.e. the space of states annihilated by constraints operators). The inner product in $\mathfrak{H}_{ph}$ can be constructed from a $\bar{\mathfrak{G}}$-invariant volume form on Q by the Faddeev–Popov procedure. The quantization is completed by evaluating the integral kernel of the evolution operator as a path integral over classes of the gauge equivalent trajectories.

The important feature of the gauge systems with the reparametrization invariance is the appearence of the first class quadratic in momenta constraints. Moreover the gauge transformations connect in this case trajectories starting at different times and the group $\mathfrak{G}$ has the structure $T_{\mathbb{R}}\otimes\bar{\mathfrak{G}}$ where $T_{\mathbb{R}}$ denotes the group of the time translations. The wave functions corresponding to the physical states are therefore time independent. In fact due to the vanishing of the canonical Hamiltonian the Schrödinger equation is trivial and the whole dynamic is described by the constraints equations. The space $\mathfrak{H}^{off}_{ph}$ of the $\bar{\mathfrak{G}}$-invariant functions on Q corresponds in this case to the space of the states annihilated by all the linear in momenta constraints. The evolution operator is simply the identity, but there exists the corresponding object defined by the path integral over trajectories – the propagator. It can be interpreted as the integral kernel of the operator inverse to the operator of a quadratic in momenta constraint. In the following we will describe how this general scheme of the covariant functional quantization can be applied in the case of the open bosonic string.

3. THE SPACE OF THE OPEN STRING WAVE FUNCTIONALS

In the euclidean formulation the space of the string trajectories consists of the triplets (M,g,x) where M is a

2-dim rectangle-like manifold, g is a riemannian metric on M and x is a map from M to the target space $\mathbb{R}^d$. The space of the general gauge transformations consists of the Weyl rescalings of the metric and the diffeomorphisms (which can change the manifold M). The first stage of the gauge fixing consists in the choice of some fixed manifold M. The space of the string trajectories and the space of the gauge transformations are reduced in this gauge to the space $\mathcal{M}_M \times \mathcal{E}^d_M$ and to the group $\mathcal{D}_M \circ W_M$ resp. (see Ref.[4] for another gauges).

Let us call two pairs of the opposite sides of M the "timelike"(T) and the "spacelike"(S) boundaries. For the x variables we have to impose the string edge conditions along the T-boundaries:

$$n^a_g \partial_a x^\mu = 0 \qquad (1)$$

where n_g is the normal vector to ∂M with respect to g. As was pointed out by O. Alvarez [1] in order to obtain the well defined F-P determinant one has to impose the mixed boundary conditions on ∂M for vector fields $\delta f \in \mathcal{T}_{id}\mathcal{D}_M$ and traceless tensors $\delta h \in \mathcal{T}_g \mathcal{M}_M$:

$$n_a \delta f^a = 0 \qquad t^a n^b (P\delta f)_{ab} = 0 \qquad (2)$$

$$n^a t^b \delta h_{ab} = 0 \qquad n_a (P^+ \delta h)^a = 0 \qquad (3)$$

(In addition we require that the boundary components of M be perpendicular one to another at all corners of M with respect to the metrics $g \in \mathcal{M}_M$ [9].). The boundary conditions (2,3) mean that in the path integral representation of the string propagator one cannot integrate over the whole space $\mathcal{M}_M$ but rather over a submanifold the tangent space of which is determined by (3). Similarly the conditions (2) lead to some restriction of $\mathcal{D}_M$. The integral submanifolds of the distributions (2),(3) can be constructed by means of the doublings M_D and M^D of M with respect to the T- and S-boundaries respectively, and some fixed involutions $\imath$ on M_D and $\jmath$ on M^D. The appropriate space of the string trajectories is then defined as the cartesian product

$\mathcal{M}^{ij}_{M}\times\mathcal{E}^{id}_{M}$ of the space $\mathcal{M}^{ij}_{M}$ of the metrics g on M admitting the smooth i- and j-symmetric extention on M_D and M^D resp., and of the space $\mathcal{E}^{id}_{M}$ of the maps $x: M \longrightarrow \mathbb{R}^d$ satisfying the condition (1). The gauge group is reduced to the semi-direct product $\mathcal{D}^{ij}_{M}\circledS\mathcal{W}^{ij}_{M}$ of the group $\mathcal{D}^{ij}_{M}$ of the diffeomorphisms admitting the smooth i- and j-symmetric extention on the doublings and the group $\mathcal{W}^{ij}_{M}$ of the functions $\phi: M \longrightarrow \mathbb{R}$ with the Neumann boundary conditions along ∂M. It can be shown that the whole formulation is independent of the choice of the involutions i, j.

Now we should determine the boundary conditions along the "initial" Σ_i and the "final" Σ_f components of the S-boundary of M. In the case of the point-like particle [2,10] it is possible to build up the covariant quantum theory on the space of the wave functions depending on the equivalence classes, (the orbits of the group $\tilde{\mathcal{G}}$ of the induced gauge transformations), i.e. with all the linear constraints solved. In the string case the classes of the physically equivalent boundary conditions form the space $\mathcal{K}^d$ of oriented contours in $\mathbb{R}^d$. Note, however, that in the canonical quantization of the string the solution of all the linear in momenta constraints requires the noncovariant light-cone gauge. On the other hand in the covariant orthogonal gauge a part of the linear constraints remains unsolved. It seems that it is the origin of the difficulties with the covariant formulation based on the space $\mathcal{K}^d$ [2]. It motivates us to consider some larger space of the boundary conditions.

Let us consider the space $\mathcal{M}_L\times\mathcal{E}_L$ of the pairs $(e,\tilde{x})$ where e is an einbein on the (arbitrary fixed) interval $L=[0,1] \subset \mathbb{R}$ and $\tilde{x}$ denotes a map $\tilde{x}: L \longrightarrow \mathbb{R}^d$ with the Neumann boundary conditions at the ends of L. The direct product $\mathcal{D}_L\times\mathbb{R}_+$ of the group $\mathcal{D}_L$ of the orientation-preserving diffeomorphisms of L and the constant rescalings of einbeins acts on $\mathcal{M}_L\times\mathcal{E}_L$ by:

$$\mathcal{M}_L\times\mathcal{E}_L \ni (e,\tilde{x}) \xrightarrow{(\gamma,\lambda)\,\in\,\mathcal{D}_L\times\mathbb{R}_+} (\lambda\gamma^*e,\gamma^*\tilde{x}) \in \mathcal{M}_L\times\mathcal{E}_L .$$

This action induces the principal fibre bundle structure:

$$\mathcal{M}_L\times\mathcal{E}_L \xrightarrow{\Pi_{\overline{\mathcal{E}}}} \overline{\mathcal{E}}^d_L = (\mathcal{M}_L\times\mathcal{E}_L)/(\mathcal{D}_L\times\mathbb{R}_+). \qquad (4)$$

We choose the base space $\overline{\mathcal{E}}^d_L$ of (4) as the space of boundary conditions. For every trajectory $(g,x)\in\mathcal{M}^{ij}_M\times\mathcal{E}^{id}_M$ we define the initial $\bar{c}_i$ and the final $\bar{c}_f$ boundary conditions by:

$$\bar{c}_i = \Pi_{\overline{\mathcal{E}}}(\sigma^*_i e_i,\sigma^*_i x_i) \quad , \quad (i\longrightarrow f) \qquad (5)$$

where $\sigma_i : L\longrightarrow\Sigma_i$, $(i\longrightarrow f)$ are arbitrary diffeomorphisms and

$$e^2_i = I^*_i g \ , \qquad x_i = I^*_i x \quad , \quad (i\longrightarrow f)$$

($I'_i:\Sigma_i\longrightarrow M$, $(i\longrightarrow f)$ are embeddings of the initial and final boundary into M). Note that the definition (5) of $\bar{c}_i, \bar{c}_f\in\overline{\mathcal{E}}^d_L$ is independent of the choice of σ_i, σ_f. The space of the "off-shell" string wave functionals defined on $\overline{\mathcal{E}}^d_L$ we will denote by $\overline{\mathfrak{H}}^{off}$.

In order to complete the construction of the space of string states one should determine the residual induced gauge transformations in $\overline{\mathcal{E}}^d_L$ (related to the unsolved linear constraints) and introduce an appropriate inner product in $\overline{\mathfrak{H}}^{off}$. The both problems are discussed in [11], here we will restrict ourselves to the path integral representation of the string propagator in the space $\overline{\mathfrak{H}}^{off}$.

4. THE OFF-SHELL OPEN STRING PROPAGATOR

According to the ideas of Sect.2 the string propagator $P[\bar{c}_i,\bar{c}_f]$ is defined as the path integral over the space $\overline{\mathcal{F}}_M(i,f)\subset\mathcal{M}^{ij}_M\times\mathcal{E}^{id}_M$ of the string trajectories starting at the "initial" class of boundary conditions $\bar{c}_i\in\overline{\mathcal{E}}^d_L$ and ending at the "final" one $\bar{c}_f\in\overline{\mathcal{E}}^d_L$:

$$P[\bar{c}_i|\bar{c}_f]= \int\limits_{\overline{\mathcal{F}}^{ij}_M(i,f)} \mathcal{D}(g,x)\left(\int\limits_{\mathcal{D}^{ij}_M}\mathcal{D}f\right)^{-1}\left(\int\limits_{\overline{\mathcal{W}}^{ij}_M}\mathcal{D}\phi\right)^{-1} e^{-S[g,x]} \qquad (6)$$

Since the conditions (5) are not invariant under the

action of the whole group $\mathcal{W}_M^{ij}$, the gauge invariance which should be taken into account in (6) is described by the subgroup $\mathcal{D}_M^{ij} \circ \overline{\mathcal{W}}_M^{ij} \subset \mathcal{D}_M^{ij} \circ \mathcal{W}_M^{ij}$ where $\overline{\mathcal{W}}_M^{ij}$ consists of the functions $\phi \in \mathcal{W}_M^{ij}$ which are constant on the S-boundaries of M.

The trajectories $(g,x) \in \overline{\mathcal{T}}_M(i,f)$ with fixed g form the affine space $\mathcal{E}_M^d(g|i,f)$ of mappings $x : M \longrightarrow \mathbb{R}^d$ fulfilling the boundary conditions :

$$x_i = \tilde{X}_i \circ \gamma_i[g,\hat{e}] \quad , \qquad (i \longrightarrow f),$$

where for every g the diffeomorphisms $\gamma_i[g,\hat{e}] : \Sigma_i \longrightarrow L$, $(i \longrightarrow f)$ are uniquely determined by the equations:

$$\mathrm{const}_i \times e_i = \gamma_i^*[g,\hat{e}] \quad , \qquad (i \longrightarrow f),$$

and $(\hat{e},\tilde{X}_i)$, $(\hat{e},\tilde{X})$ denote representants of the contours $\bar{c}_i$, $\bar{c}_f$ resp.:

$$\bar{c}_i = \Pi_{\mathcal{G}}(\hat{e},\tilde{X}) \quad , \qquad (i \longrightarrow f).$$

The integral over $\mathcal{E}_M^d(g|i,f)$ is gaussian and can be evaluated:

$$\int\limits_{\mathcal{E}_M^d(g|i,f)} \mathcal{D}x \; e^{-S[g,x]} = (\det_{DN} \mathcal{L}_g)^{-\frac{d}{2}} \, e^{-S[g,x_i^f[g]]} \qquad (7)$$

where $\mathcal{L}_g$ denotes the 2-dim. Laplace-Beltrami operator acting on the space of the scalar functions on M with the Dirichlet boundary conditions along the S-boundaries and the Neumann ones along the T-boundaries of M ; $x_i^f[g]$ is the solution of the boundary value problem

$$\mathcal{L}_g x_i^f[g] = 0 \quad ,$$
$$x_i^f[g] \circ I_i = \tilde{X}_i \circ \gamma_i[g,\hat{e}] \quad , \qquad (i \longrightarrow f) \; ,$$
$$(n^{\perp})^a \, \partial_a x_i^f[g] = 0 \qquad \text{on} \quad \partial M \setminus (\Sigma_i \cap \Sigma_f) \; .$$

The r.h.s. of the formula (7) is the $\mathcal{D}_M^{ij}$-invariant functional of g so one can apply the standard Faddeev-Popov method [10] for the conformal gauge slice $\mathcal{S}_\Theta^c = \{ g \in \mathcal{M}_M^{ij} : g = \exp(\phi) g^t , \; \phi \in \mathcal{W}_M^{ij}, \; t \in T_M^R \}$ determined by the global section $\Theta : T_M^R \ni t \longrightarrow g^t \in \mathcal{M}_M^{ij}$ in the bundle:

$$\mathcal{M}_M^{ij} \longrightarrow T_M^R = \mathcal{M}_M^{ij}/(\mathcal{D}_M^{ij} \circledcirc \mathcal{W}_M^{ij}) \qquad (8)$$

over the relative Teichmüller space T_M^R [12]. Taking into account the corner anomaly [9] one can obtain in the critical dimension d = 26 :

$$P[\bar{c}_i, \bar{c}_f] = \int_0^\infty dt \; M_{g^t}(\delta\psi|\delta\tilde{\chi}) \left(\frac{\det_{AA} P_{g^t}^+ P_{g^t}}{H(P_{g^t}^+)}\right)^{\frac{1}{2}} (\det_{DN} \mathcal{L}_{g^t})^{-13} \times \qquad (9)$$

$$\times \int_{\mathcal{W}_M^{ij}} \tilde{\mathcal{D}}\phi \left(\int_{\bar{\mathcal{W}}_M^{ij}} \mathcal{D}\phi \right)^{-1} \exp(\tfrac{5}{8} \sum_i \phi(z_i) - S[e^\phi g^t, x_i^i[e^\phi g^t]])$$

where we use the standard notations [1,2,10]; z_i denote the corners of M and the index AA in the symbol of the determinant of $P_{g^t}^+ P_{g^t}$ means that the mixed boundary conditions (2,3) on ∂M have been chosen.

As follows from the formula (9) even at the critical dimension the integration over "conformal factor" does not decouple. There are two reasons for this difficulty. First of all, the maximal symmetry group of the "classical action" $S[g,x_i^f[g]]$ is $\mathcal{D}_M^{ij} \circledcirc \bar{\mathcal{W}}_M^{ij}$ (rather than $\mathcal{D}_M^{ij} \circledcirc \mathcal{W}_M^{ij}$) . Secondly, due to the corner anomaly the symmetry group of the integrand in (9) is reduced further to the group $\mathcal{D}_M^{ij} \circledcirc \bar{\bar{\mathcal{W}}}_M^{ij}$ where $\bar{\bar{\mathcal{W}}}_M^{ij}$ consists of the functions $\phi \in \bar{\mathcal{W}}_M^{ij}$ satisfying the condition $\Sigma\phi(z_i)=0$. Let us observe that the problem disappeares if we consider in the formula (6) instead of the space of trajectories $\bar{\mathcal{T}}_M(i,f)$ the subspace:

$$\bar{\mathcal{T}}_M(i,f|\mathcal{B}_M) = \{ (g,x) \in \bar{\mathcal{T}}_M(i,f) : g \in \mathcal{B}_M \}$$

determined by the $\mathcal{D}_M^{ij} \circledcirc \bar{\bar{\mathcal{W}}}_M^{ij}$-subbundle $\mathcal{B}_M$ of the bundle (8). In fact for every subbundle $\mathcal{B}_M$ the functional:

$$P[\bar{c}_i, \bar{c}_f|\mathcal{B}_M] = \int_{\bar{\mathcal{T}}_M^{ij}(i,f|\mathcal{B}_M)} \mathcal{D}(g,x) \left(\int_{\mathcal{D}_M^{ij}} \mathcal{D}f \right)^{-1} \left(\int_{\bar{\bar{\mathcal{W}}}_M^{ij}} \mathcal{D}\phi \right)^{-1} e^{-S[g,x]} \qquad (10)$$

can be evaluated (=defined) in the critical dimention by

gaussian integrals and the F-P procedure.

In the case of the $\mathcal{D}^{ij}_{M}\circ\overline{\overline{\mathcal{W}}}^{ij}_{M}$-subbundle $\mathcal{B}^{0}_{M}(1)$ constructed by the action of the group $\overline{\overline{\mathcal{W}}}^{ij}_{M}$ from the subspace $\mathcal{M}^{ij}_{MO}(1)$ of the metrics with the zero scalar curvature and with the intrinsic lengths of the initial and the final boundary equal to 1, one can obtain [2,11]:

$$P[\overline{c}_i,\overline{c}_f\,|\mathcal{B}^0_M(1)]=\text{const}\int_0^\infty dt\; e^{\pi t}\,[\prod_{n>0}(1-e^{i2\pi nt})]^{-12}\; t^{-13} e^{-W[\overline{c}_i,\overline{c}_f\,|t]}$$

where $\quad W[\overline{c}_i,\overline{c}_f\,|t] =$ (11)

$$\frac{1}{4\pi\alpha'}\Big[\frac{(\overline{c}^{\,0}_i-\overline{c}^{\,0}_f)^2}{t}+\sum_{n=1}^{\infty}\frac{2\pi n}{sh(\pi nt)}[((\overline{c}^{\,n}_i)^2+(\overline{c}^{\,n}_f)^2)ch(\pi nt)-2\overline{c}^{\,n}_i\overline{c}^{\,n}_f]\Big]$$

For $\overline{c}_i = \Pi_{\overline{\mathcal{G}}}(e_i,\tilde{X}_i)$, $(i\longrightarrow f)$ the modes $\overline{c}^{\,n}_i$, $\overline{c}^{\,n}_f$, $n = 0,1,\ldots$ are defined by:

$$\overline{c}^{\,n\mu}_i = \ell_i^{-1/2}\int_L e_i\,\psi^{e_i}_n\,\tilde{X}^{\mu}_i\,ds\quad,\qquad \ell_i = \int_L e_i\,ds\quad,\quad (i\longrightarrow f).$$

where $\{\psi^{e_i}_n\}_{n>0}$, $(i\longrightarrow f)$ denote the complete sets of the orthonormal eigenfunctions of the 1-dim. Laplace-Beltrami operators related to the metrics e_i^2, e_f^2 on L resp..

For the representations $(\hat{e},\tilde{X}_i)$, $(\hat{e},\tilde{X}_f)$ of the contours $\overline{c}_i,\overline{c}_f \in \overline{\mathcal{C}}^{26}_L$ with $\hat{e} \equiv 1$, proceeding as in Ref.[2], one can rewrite the result (11) as

$$P[\tilde{X}_i,\tilde{X}_f\,|\mathcal{B}^0_M(1)]=\text{const}\int_0^\infty dt\; e^{\pi t}\prod_{n>0}(1-e^{-2\pi nt})\,\langle\tilde{X}_i\,|e^{-Ht}\,|\tilde{X}_f\rangle \qquad (12)$$

where H denotes the open string hamiltonian equal to the zero mode of the density $W(s) \approx P^2(s) + x'^2(s)$ of the quadratic constraints. By the introduction of the b-c ghosts and by an "exponentiation" of the F-P determinant one can construct [6] the BRST extension of the functional (12) :

$P[\tilde{X}_i,\hat{\theta}_i,\theta_i;\tilde{X}_f,\hat{\theta}_f,\theta_f\,|\mathcal{B}^0_M(1)] =$ (13)

$$\text{const}\int_0^\infty dt\,\langle\tilde{X}_i,\hat{\theta}_i,\theta_i\,|e^{-\pi t H_{BRST}}|\tilde{X}_f,\hat{\theta}_f,\theta_f\rangle.$$

$\hat{\theta}_i, \hat{\theta}_f$ and θ_i, θ_f denote the appropriately defined boundary values of the b-c ghosts variables and H_{BRST} is the BRST-extended open string hamiltonian. This form of the BRST-extended string propagator has been also proposed in Refs [7,8].

The formula (13) offers the quantum mechanical interpretation of the functional $P[\bar{c}_i, \bar{c}_f | \mathcal{B}^o_M(1)]$ (11) determined by the subbundle $\mathcal{B}^o_M(1)$. It can be regarded as the 'body' of the integral kernel of the operator inverse to the BRST extended operator of the zero mode $W_0 \sim \int(\wp^2(s)+x^2(s))ds$ of the quadratic in momenta constraints. Let us note that since the canonical string hamiltonian weakly vanishes there is a large freedom in the choice of the hamiltonian governing the evolution in the intrinsic (world sheet) time. Roughly speaking this freedom correspnds to the freedom in the choice of the $\mathcal{D}^{ij}_M \oplus \bar{\bar{W}}^{ij}_M$-subbundle $\mathcal{B}_M$ determining the functional $P[\bar{c}_i, \bar{c}_f | \mathcal{B}_M]$ (10).

5. CONCLUSIONS

It was shown that the generalized Feynman scheme of the covariant functional quantization leads to the consistent path integral representation of the string propagator in the space $\bar{\mathfrak{H}}^{off}$ of the string wave functionals defined on the space $\bar{\mathcal{E}}^{26}_L$. It turns out that this representation requires the choice of the additional geometrical data concerning the global geometry of the space of the allowed string trajectories. This leads to the whole family of the functionals $P[\bar{c}_i, \bar{c}_f | \mathcal{B}_M]$ (10). The important question is the BRST extention of the formulation presented above. The part of this extention concerning the propagator $P[\bar{c}_i, \bar{c}_f | \mathcal{B}_M(1)]$ in the special gauge ($\hat{e} = 1$) is rather well understood (11-13), [4-8]. The complete formulation should contain the construction of the inner product in the BRST-extended space of states and the cohomological description of the residual gauge invariance. Finally let us note that the generalization of the path integral representation of the propagator to the case of more complicated topologies of the world sheet M yields the expressions for the off-shell multiloop open string amplitudes [11]. The new interesting aspect of this construction appears. For every topological type of M one should choose some $\mathcal{D}^{ij}_{M}\odot\overline{\overline{\mathcal{W}}}^{ij}_{M}$-subbundle of $\mathcal{M}^{ij}_{M}$. It seems that the sewing property of the off-shell amplitudes leads to strong restrictions on the possible choices.

ACKNOWLEDGMENT

The author would like to thank S.Carlip for discussion and critical remarks.

REFERENCES

[1] O.Alvarez : Nucl.Phys. B216 (1983) 125

[2] A.Cohen,G.Moore,P.Nelson,J.Polchinski:Nucl.Phys. B267 (1986)143; Proc.Workshop on Unified String Theory, eds M.Green & D.Gross (World Scientific, 1986)

[3] A.Cohen,G.Moore,P.Nelson,J.Polchinski: Nucl.Phys. B281 (1986) 127;
S.K.Blau, M.Clements : Nucl.Phys. B284 (1987) 118
S.K.Blau, M.Clements, B.McClain : Univ. of Texas preprint UTTG-29-86 (1986)
S.K.Blau, S.Carlip, M.Clements, S.DellaPietra, V.DellaPietra : Nucl.Phys.B301 (1988) 285

[4] A.N.Redlich : Phys.Lett.205B (1988) 295; Inst. of Advanced Study preprint IASSNS/HEP-87/49 (1987)

[5] T.Ohrndorf : Heidelberg preprint HD-THEP-87-12 (1987)

[6] D.Birmingham, C.G.Torre :Phys.Lett.205B (1988) 289

[7] T.Lee : Ann.Phys. 183 (1988) 191

[8] N.I.Karchev : Phys.Lett. B 212 (1988) 158

[9] C.Varughese,W.I.Weisberger : Phys.Rev. D 37 (1988) 981

[10] Z.Jaskólski : Commun.Math.Phys.111 (1987) 439, J.Math.Phys.29 (1988) 1034

[11] Z.Jaskólski: "The Polyakov Path Integral over Bordered Surfaces (The Open String Amplitudes)" to appear

[12] C.J.Earle : Moduli of surfaces with symmetries, in Advances in the Theory of Riemann Surfaces ,1969 Stony Brook Conference, ed. L.Ahlfors (Princeton University Press , 1971)
M.Sepälä : Teichmüller Spaces of Klein Surfaces, Annales Academice Scientiarum Fennica, Mathematica Dissertaciones (Helsinki , 1978)

MULTILOOP (SUPER) STRING AMPLITUDES AND OTHER APPLICATIONS TO CFT OF THE OPERATOR FORMALISM ON ARBITRARY GENUS RIEMANN SURFACES

A.R.Lugo and J.Russo
International School for Advanced Studies
Strada Costiera 11, Trieste, Italy

Abstract:

Bosonic scattering amplitudes are computed for closed string and superstring theories at arbitrary genus by using a global operator formalism. The properties of the vertex operators are analyzed in this context. The VEV of the energy-momentum tensor is calculated in the case of a scalar field and b-c systems. All is in agreement with standard results.

Introduction

In the last few years there was an increasing interest in developing operator methods for the computation of correlation functions in CFT at arbitrary genus [1]. An important step in this direction was made by Krichever and Novikov in refs. [2,3]. They realized that, by simply using the Riemann-Roch theorem, one can perform an explicit construction of the bases for the spaces of holomorphic tensors of weight λ on Riemann surfaces with two punctures. These bases find a natural application in conformal theories.

Here we will work with the operator formalism developed in refs.[4,5,6]. A brief review of it is given in section 1. In section 2 we construct vertex operators and study their properties. These vertices are used in section 3 for the calculation of scattering amplitudes. In section 4 we investigate the connection between the path integral energy momentum tensor and the one corresponding to the present formalism. We show that it is a schwartzian connection.

1. Operator quantization of string theory.

A closed string can be parametrized by an angular coordinate σ and a time parameter τ. The world sheet swept out by the string can be conformally mapped to a Riemann surface Σ without two points , which will be denoted by P_+ and P_-, corresponding to the times $\tau\to\mp\infty$. The time parameter can be defined at arbitrary genus as follows [2,7]

$$\tau(P)=\mathrm{Re}\int_{P_0}^{P} dk \tag{1.1}$$

where dk is the differential of the third kind on Σ with simple poles at the points P_+ and P_- with residues +1 and -1 respectively. By using the Riemann-Roch theorem, one can see that this fixes dk uniquely up to the addition of (holomorphic) abelian differentials. A further requirement, namely that the periods of dk around the canonical homology basis (α_i,β_i) be pure imaginary (and thus $\tau(P)$ is a well-defined function) determines it unambiguously. The level curves of the function $\tau(P)$ are conventionally called C_τ.

The equation of motion for the matter field of the bosonic string theory is

$$\partial\bar\partial X^\mu=0 \quad ; \qquad P\neq P_\pm \tag{1.2}$$

This equation follows from a Hamiltonian (which is the generator of τ-translations) [5]. This implies that ∂X^μ $(\bar\partial X^\mu)$ is holomorphic (antiholomorphic) everywhere except $P_\pm$. A basis $\{\omega_n\}$ for the space of meromorphic 1-forms, holomorphic outside $P_\pm$, has been constructed in [2]. We can thus write the solutions to (1.2) as follows

$$\partial X^\mu(Q)=i/\sqrt{2}\sum_n \alpha_n^\mu\,\omega_n(Q) \quad ; \quad \bar\partial X^\mu(Q)=i/\sqrt{2}\sum_n \bar\alpha_n^\mu\,\bar\omega_n(Q) \tag{1.3}$$

Now the requirement of single-valuedness of X^μ implies the following relations

$$0=\oint_{C_\tau} dX^\mu=\oint_{\alpha_i} dX^\mu=\oint_{\beta_i} dX^\mu \tag{1.4}$$

By inserting the expansions (1.3) into (1.4), and solving these constraints, one arrives by integration to the following expression

$$X^\mu(P) = x^\mu - ip^\mu\tau(P) + i/\sqrt{2}\sum_{n<-g/2} (\alpha_n^\mu\,\phi_n(P,P_+) + \bar\alpha_n^\mu\,\bar\phi_n(P,P_+)) +$$

$$+ i/\sqrt{2}\sum_{n>g/2} (\alpha_n^\mu\,\phi_n(P,P_-) + \bar\alpha_n^\mu\,\bar\phi_n(P,P_-)) \tag{1.5}$$

where $\phi_n(P,Q_0)$ are harmonic (single-valued) functions given by

$$\phi_n(P,Q_0) = \int_{Q_o}^{P} (\omega_n - \sum_{j=1}^{g} (F_{nj}\eta_j + \bar{G}_{nj}\bar{\eta}_j)) \tag{1.6}$$

with

$$F_{nj} = i/2 \sum_{i=1}^{g} ((\tau_2^{-1}\bar{\tau})_{ji} a_n^i - (\tau_2^{-1})_{ji} b_n^i) \tag{1.7a}$$

$$G_{nj} = i/2 \sum_{i=1}^{g} ((\tau_2^{-1}\bar{\tau})_{ji} \bar{a}_n^i - (\tau_2^{-1})_{ji} \bar{b}_n^i) \tag{1.7b}$$

Here $\{\eta_i,\ i=1,..,g\}$ are the abelian differentials with the standard normalization [8], τ_{ij} is the period matrix, and a_n^i, b_n^i are the periods of ω_n around α_i, β_i respectively.

The quantization of the theory is performed by promoting the coefficients of the expansions to operators acting on a Fock space. The commutation rules follow as usual from the canonical commutation relations ($\alpha'=1$):

$$[\alpha_n^\mu,\alpha_m^\nu] = \gamma_{nm}\,\eta^{\mu\nu} , \qquad [\bar{\alpha}_n^\mu,\bar{\alpha}_m^\nu] = \bar{\gamma}_{nm}\,\eta^{\mu\nu}$$

$$[\alpha_n^\mu,\bar{\alpha}_m^\nu] = 0 , \qquad [p^\mu,x^\nu] = -i\eta^{\mu\nu} \tag{1.8}$$

where $\gamma_{nm}=1/2\pi i\oint dA_n A_m$, and A_n are the dual functions to ω_n [2]. The vacuum state $|0\rangle$ is defined to be the state annihilated by p^μ and α_n^μ, $\bar{\alpha}_n^\mu$ with $n>g/2$, while the dual vacuum $\langle 0|$ similarly with $n<-g/2$. The normal order product is defined as usual by arranging annihilation operators on the right. It is useful to introduce the coefficients $\tilde{\gamma}_{nm}$ and $\tilde{s}_{nm}$ by the formulas (see ref.[10] for explicit expressions)

$$:\alpha_n^\mu\alpha_m^\nu: = \alpha_n^\mu\alpha_m^\nu - \eta^{\mu\nu}\,\tilde{\gamma}_{nm} \tag{1.9a}$$

$$:\alpha_n^\mu\bar{\alpha}_m^\nu: = \alpha_n^\mu\bar{\alpha}_m^\nu - \eta^{\mu\nu}\,\tilde{s}_{nm} \tag{1.9b}$$

The generalized Virasoro operators L_n and $\bar{L}_n$ are defined as the coefficients in the expansion of T and $\bar{T}$, that is

$$T(Q) \equiv \sum_n L_n\,\Omega_n(Q) \quad ; \qquad \bar{T}(Q) \equiv \sum_n \bar{L}_n\,\bar{\Omega}_n(Q) \tag{1.10}$$

where $\{\Omega_n\}$ ($\{\bar{\Omega}_n\}$) is a basis for meromorphic quadratic differentials which are holomorphic (antiholomorphic) outside $P_\pm$, and

$$L_r = 1/2 \sum_{n,m} l_r^{nm} :\alpha_n . \alpha_m: , \qquad l_r^{nm} = \frac{1}{2\pi i} \oint_{C_\tau} e_r \, \omega_n \, \omega_m \tag{1.11}$$

(e_n are meromorphic vector fields dual to Ω_n). The operators L_n satisfy the KN algebra, namely (g_0=3g/2)

$$[L_i, L_j] = \sum_{s=-g_0}^{g_0} C^s_{ji} \, L_{i+j-s} + c/12 \, \kappa_{ij} \; ; \qquad c=D \tag{1.12}$$

where C^s_{ij} are the KN structure constants and κ_{ij} is the cocycle (see section 4). This algebra has the property that at g=0 reduces to the usual Virasoro algebra.

The scattering amplitude A^M of M external particles is defined by

$$A^M = \sum_{g=0}^{\infty} k^{2g+M-2} \int d(\text{moduli}) \; A^M[\Sigma] \tag{1.13}$$

with $A^M[\Sigma] = \langle 0| \; T\{V_1 \ldots V_M\} |0\rangle$. d(moduli) is the modular invariant measure and $\{V_i, \; i=1,\ldots,M\}$ are suitable vertex operators representing the external particles. In the next section we will find the vertices which lead to BRST invariant amplitudes, and in section 3 we will see that $A^M[\Sigma]$ is also modular invariant and reduces to the well-known results.

2. Vertex operators for string theory

Let us start remembering some features of genus zero. A vertex operator is an object of the form

$$V = \int_{S^2} i(dz \wedge d\bar{z}/(z\bar{z})) \; w(z,\bar{z}) \tag{2.1}$$

where $z=\exp(\tau+i\sigma)$, τ being the euclidean time parameter and σ the angular coordinate of the string. It is required to have conformal dimension (1,1) in the sense [9]

$$[L^v_n, w(z,\bar{z})] = (z^n \partial_z + n \, z^{n-1}) \; w(z,\bar{z})$$

$$[\bar{L}^v_n, w(z,\bar{z})] = (\bar{z}^n \partial_{\bar{z}} + n \, \bar{z}^{n-1}) \; w(z,\bar{z}) \tag{2.2a}$$

or equivalently, that the (1,1) form $W(z,\bar{z}) = i dz \wedge d\bar{z}/(z\bar{z}) w(z,\bar{z})$ transforms like

$$[L_n^v, W(z,\bar{z})] = L_{e_n} W(z,\bar{z}) \quad , \quad [\bar{L}_n^v, W(z,z)] = L_{\bar{e}_n} W(z,\bar{z}) \tag{2.2b}$$

under the Virasoro algebra, where $e_n = z^{n+1}\partial_z$ is an element of the basis of meromorphic vector fields, holomorphic outside $z=0,\infty$, and L_e means Lie derivative with respect to e. Eqs.(2.2) lead to the guess that an operator vertex at arbitrary genus

$$V = \int_\Sigma W(Q) \tag{2.3}$$

with W(Q) a (1,1) form on Σ, should obey

$$[L_n, W(Q)] = L_{e_n} W(Q) \quad , \quad [\bar{L}_n, W(Q)] = L_{\bar{e}_n} W(Q) \tag{2.4}$$

respect to the KN operators (1.11). The requirements (2.2) are imposed by unitarity; they guarantee that the unphysical degrees of freedom ("ghosts") do not contribute to the amplitudes. By considering the definition of physical space related to the KN algebra [3] the same consecuence follows at arbitrary genus provided that (2.4) holds, as it is explained in ref.[5]. The conditions (2.4) are the basic tool in the study of vertex operators in the operator context.

Let us start with the tachyon vertex operator. It is defined by

$$:W_T(Q): = \Omega(Q) :\exp(ip.X(Q)): \quad , \quad p^2 = 4 \tag{2.5}$$

where Ω(Q) is some integrable (1,1) form on Σ. We would like that (2.4) to hold for $:W_T:$; this requirement will fix (up to a multiplicative constant, of course) Ω(Q). So, let us compute the commutators in (2.4). From the formulas given in section 1 we get

$$[L_n, :\exp(ip.X(Q)):] = -1/(2\sqrt{2}) \sum_{k,l} l_n^{kl} A_l(Q) \times$$

$$\times (p.\alpha_k :\exp(ip.X(Q)): + :\exp(ip.X(Q)): p.\alpha_k) \tag{2.6}$$

In the following calculations the formulas

$$\alpha_k^\mu :f(\alpha,\bar{\alpha}): = :\alpha_k^\mu f(\alpha,\bar{\alpha}): +$$

$$+ \sum_s \tilde{\gamma}_{ks} \frac{:\partial f(\alpha,\bar{\alpha}):}{\partial \alpha_s^\mu} + \sum_s \tilde{s}_{ks} \frac{:\partial f(\alpha,\bar{\alpha}):}{\partial \bar{\alpha}_s^\mu} \tag{2.7a}$$

$$:f(\alpha,\bar{\alpha}):\ \alpha^{\mu}_{k} = :\alpha^{\mu}_{k}\ f(\alpha,\bar{\alpha}): +$$

$$+ \sum_{s} \tilde{\gamma}_{sk}\ \frac{:\partial f(\alpha,\bar{\alpha}):}{\partial \alpha^{\mu}_{s}} + \sum_{s} \tilde{s}_{ks}\ \frac{:\partial f(\alpha,\bar{\alpha}):}{\partial \bar{\alpha}^{\mu}_{s}} \qquad (2.7b)$$

valid for any smooth function f, which follow from eqs.(1.9), will be extremely useful. By using them we obtain

$$[L_{n}, :\exp(ip.X(Q)):] = L_{e_{n}}\ (:\exp(ip.X(Q)):) +$$

$$+ p^{2}/4\ \kappa_{n}(Q)\ :\exp(ip.X(Q)): \qquad (2.8)$$

where the "anomalous" functions $\kappa_{n}(Q)$ are given by

$$\kappa_{r}(Q) = \sum_{n,m} l^{nm}_{r} \sum_{|k|,|l|>g/2} (\tilde{\gamma}_{nk}\ \tilde{\gamma}_{ml} - \tilde{\gamma}_{kn}\ \tilde{\gamma}_{lm})\ \phi_{k}(Q)\ \phi_{l}(Q) +$$

$$+ \sum_{n} l^{n,g/2}_{r} \sum_{|k|>g/2} (\tilde{\gamma}_{nk} + \tilde{\gamma}_{kn})\ \phi_{k}(Q) \qquad (2.9)$$

Now, let us introduce the following Laurent-like expansions obtained by using the methods of refs.[5,10,11]

$$\omega(P,Q;P_{+}) = \sum_{n} \sum_{|k|>g/2} \tilde{\gamma}_{nk}\ \omega^{n}(P)\ \phi_{k}(Q) \quad , \ \tau_{P}>\tau_{Q} \qquad (2.10a)$$

$$\omega(P,Q;P_{-}) = \sum_{n} \sum_{|k|>g/2} \tilde{\gamma}_{kn}\ \omega^{n}(P)\ \phi_{k}(Q) \quad , \ \tau_{Q}>\tau_{P} \qquad (2.10b)$$

where

$$\omega(P,Q;P_{0}) = \partial_{P} \log(E(P,Q)/E(P,P_{0})) - \pi \int_{P_{0}}^{Q} (\eta-\bar{\eta})^{t}\ \tau_{2}^{-1}\ \eta(P) \qquad (2.10c)$$

is a meromorphic 1-form in P. With the help of (2.10) and the definition of the coefficients l^{nm}_{r}, eq.(1.11), we can rewrite (2.9) in the form

$$\kappa_{r}(Q) = 1/(2\pi i) \oint_{C_{Q}} e_{r}(P)\ \omega(P,Q;P_{+})\ \omega(P,Q;P_{-}) \qquad (2.11)$$

where C_{Q} is a contour enclosing Q. The computation of the RHS by using (3.8) is straightforward and we obtain

$$\kappa_{r}(z,\bar{z}) = \partial_{z} e_{r}(z) + \omega_{z}(z,\bar{z})\ e_{r}(z) \qquad (2.12)$$

where

$$\omega_z(z,\bar{z}) = -\partial_z \log(E(z,P_+)E(z,P_-)) - \pi \left(\int_{P_+}^{z} + \int_{P_-}^{z}\right)(\eta-\bar{\eta})^t \, \tau_2^{-1} \, \eta(z) \tag{2.13}$$

Now we impose the conditions (2.4) on $:W_T(Q):$; from eqs.(2.5,8) these are equivalent to

$$\kappa_r(z,\bar{z}) = \nabla_z e_r(z) \qquad , \qquad \bar{\kappa}_r(z,\bar{z}) = \nabla_{\bar{z}} \bar{e}_r(z) \tag{2.14a}$$

where ∇_z, $\nabla_{\bar{z}}$ are the covariant derivatives with respect to the connections

$$\omega_z(z,\bar{z}) = \partial_z \log\Omega_{z\bar{z}}(z,\bar{z}) \qquad , \qquad \omega_{\bar{z}}(z,\bar{z}) = \partial_{\bar{z}} \log\Omega_{z\bar{z}}(z,\bar{z}) \tag{2.14b}$$

From (2.13,14) we read $\Omega(Q)$

$$\Omega(Q) = i \, \nu(Q)\wedge\bar{\nu}(Q)$$

$$\nu(Q) = E(P_+,P_-)/(E(Q,P_+)E(Q,P_-)) \, \exp\left(-\pi/2\int_{P_-}^{Q}(\eta-\bar{\eta})^t \tau_2^{-1}\int_{P_+}^{Q}(\eta-\bar{\eta})\right) \tag{2.15}$$

We remark $\Omega(Q)$ is a well-defined (single-valued) form, though this is not the case for $\nu(Q)$. This integrable form on Σ, $\Omega = i\Omega_{z\bar{z}} dz\wedge d\bar{z}$ with $\Omega_{z\bar{z}}(z,\bar{z}) = |\nu(z,\bar{z})|^2$, defines an Arakelov-type metric [10,11] $h^{\Omega} = \Omega_{z\bar{z}}(dz\otimes d\bar{z} + d\bar{z}\otimes dz)$. Vertex operators for particles belonging to higher massive levels will involve in general terms constructed from covariant derivatives of the fields and the curvature R^{Ω} corresponding to this metric. For example, at massless level we have that

$$:W(Q): \; = i\xi_{\mu\nu}(p) \; :\partial X^{\mu}(Q)\wedge\bar{\partial}X^{\nu}(Q) \; e^{ip.X(Q)}: \tag{2.16}$$

with $\xi_{\mu\nu}(p)$ traceless obeying the transversality conditions

$$\xi_{\mu\nu}(p) \, p^{\nu} = \xi_{\mu\nu}(p) \, p^{\mu} = 0$$

verifies (2.4). The symmetric and antisymmetric parts correspond to the graviton and to the antisymmetric tensor particles respectively. Instead, the dilaton vertex operator involves the curvature; it is given by [10,12]

$$:W_D(Q): = :(i\eta^t_{\mu\nu}\partial X^{\mu}(Q)\wedge\bar{\partial}X^{\nu}(Q) + (D-2)/8 \; \Omega(Q)R^{\Omega}(Q))e^{ip.X(Q)}: \tag{2.17}$$

where $\eta^{t}_{\mu\nu} = \eta_{\mu\nu} - (a_\mu p_\nu + a_\nu p_\mu)$, $a_\mu p^\mu = 1$. It exactly verifies (2.4); however the transverse part involving the arbitrary vector a_μ is a total derivative that decouples from the amplitudes [12].

3. Bosonic amplitudes in string and superstring theories.

Insertion of the vertex operators into $A^M[\Sigma]$ leads us to the calculation of correlation functions of the form [11]

$$<0|:e^{ip_1 \cdot X(Q_1)}:\ldots:e^{ip_M \cdot X(Q_M)}:|0> \tag{3.1}$$

Writing $:e^{ip.X}:=e^{ip.x(\tau)}W_0\bar{W}_0$, with $x^\mu(\tau)=x^\mu - ip^\mu\tau$ and

$$W_0=\exp(-1/2p_\mu \sum_{n<-g/2} \alpha^\mu_n \phi_n(P,P_+)) \exp(-1/2p_\mu \sum_{n>g/2} \alpha^\mu_n \phi_n(P,P_-)) \tag{3.2}$$

One obtains

$$<0|e^{ip_1 \cdot x(\tau_1)}\ldots e^{ip_M \cdot x(\tau_M)}|0> = \prod_{i=1}^{M} e^{p_i^2/2\ \tau_i} \prod_{i<j} e^{p_i \cdot p_j \tau_i} \times \times (2\pi)^{26}\delta(\sum_i p_i^\mu) \tag{3.3}$$

$$<0|W_0(1)\ldots W_0(M)|0> = \prod_{i<j} \exp\ p_i \cdot p_j G(Q_i,Q_j) \tag{3.4}$$

Defining $S(P_i,P_j) \equiv \tau(P_i)+G(P_i,P_j)+\bar{G}(P_i,P_j)$, the amplitude becomes

$$A^M_{tach}[\Sigma] = \int_\Sigma (\prod_{i=1}^{M} \Omega(Q_i)e^{2\tau_i}) \prod_{i\neq j} e^{(p_i \cdot p_j/2)S(Q_i,Q_j)} \times \times (2\pi)^{26}\delta(\sum p_i^\mu) \tag{3.5}$$

(We have used $p_i^2=4$). Now we insert the explicit expressions for $\Omega(Q)$, $\tau(Q)$, and $S(P,Q)$ and arrive to the following result [11]

$$A^M_{tach}[\Sigma] = (2\pi)^{26}\delta(\sum p_i^\mu) \int_\Sigma \prod_{i<j} |E(P_i,P_j)|^{p_i \cdot p_j} \times$$

$$\times \exp[\pi p_i p_j/4 \int_{P_i}^{P_j} (\eta_a - \bar{\eta}_a)\ (\tau_2^{-1})_{ab} \int_{P_i}^{P_j} (\eta_b - \bar{\eta}_b)] \tag{3.6}$$

in agreement with well-known results.

Now, let us consider the case of superstring theories. Bosonic vertices in this framework are defined as follows (see [6] for details)

$$V_{tach} = \int_{\Sigma} d^2\theta\ \Omega^{1/2}\ e^{ip.Y} \qquad (3.7a)$$

$$V_{grav} = \xi_{\mu\nu}(p)\int_{\Sigma} d^2\theta\ DY^{\mu}\wedge\bar{D}Y^{\nu}\ e^{ip.Y} \qquad (3.7b)$$

where $Y^{\mu}=X^{\mu}+i\bar{\theta}\psi^{\mu}-i\theta\bar{\psi}^{\mu}$ and $\bar{D}=\theta\partial+\bar{\partial}/\partial\theta$ .They obey the correct commutation relations with the generalized superVirasoro operators L_n, G_r [6,10,13]. By following the same lines as for the tachyon amplitude in the bosonic string computed above, we find that the M tachyon-scattering amplitude in the RNS superstring is

$$A^M_{tach}[\Sigma]= \int_{\Sigma} \prod_{i=1}^{M} d^2\theta_i \prod_{i<j} |E(P_i,P_j)-2\bar{\theta}_i\bar{\theta}_j \frac{\theta[^{\alpha}_{\beta}](P_i-P_j)}{\theta[^{\alpha}_{\beta}](0)}|^{p_i.p_j} \times$$

$$\times \exp[\pi p_i p_j/4 \int_{P_i}^{P_j} (\eta_a-\bar{\eta}_a)(\tau_2^{-1})_{ab} \int_{P_i}^{P_j} (\eta_b-\bar{\eta}_b)] \qquad (3.8)$$

Since the integrand contains an even number of θ_i , the amplitude vanishes unless M is even, consistent with the notion of a multiplicative conserved "G-parity" quantum number. In the particular case M=2 eq. (3.8) takes the form

$$A^2_{tach}[\Sigma] = \frac{4}{|\theta[^{\alpha}_{\beta}](0)|^2} \int_{\Sigma} |E(P_1,P_2)|^2 |\theta[^{\alpha}_{\beta}](P_1-P_2)|^2 \times$$

$$\times \exp[-\pi \int_{P_1}^{P_2} (\eta_a-\bar{\eta}_a)\ (\tau_2^{-1})_{ab} \int_{P_1}^{P_2} (\eta_b-\bar{\eta}_b)] \qquad (3.9)$$

Similarly one can compute the M-graviton scattering amplitude. We find

$$A^M_{grav}[\Sigma] = \xi^1_{\mu\nu_1}(p_1)\ldots\xi^M_{\mu\nu_M}(p_M)\int_{\Sigma} \prod_{i=1}^{M} d^2\theta_i$$

$$\times \exp[\pi p_i p_j/4 \int_{P_i}^{P_j} (\eta_a-\bar{\eta}_a)\ (\tau_2^{-1})_{ab} \int_{P_i}^{P_j} (\eta_b-\bar{\eta}_b)]\ |F_M|^2 \times$$

$$\times \prod_{i<j} |E(P_i,P_j)-2\bar{\theta}_i\bar{\theta}_j \frac{\theta[^{\alpha}_{\beta}](P_i-P_j)}{\theta[^{\alpha}_{\beta}](0)}|^{p_i.p_j} \qquad (3.10)$$

where

$$F_M=\{\prod_{i<j} \exp[\theta_i \xi_i \cdot p_j \langle \partial X_i X_j \rangle + \theta_j \xi_j \cdot p_i \langle X_i \partial X_j \rangle - \theta_i p_i \cdot \xi_j \langle \psi_i \psi_j \rangle -$$

$$-\theta_j p_j \cdot \xi_i \langle \psi_i \psi_j \rangle + \theta_i \theta_j \xi_i \cdot \xi_j \langle \partial X_i \partial X_j \rangle + \xi_i \cdot \xi_j \langle \psi_i \psi_j \rangle]\}\Big|_{\text{linear in } \xi^\mu_\iota} \tag{3.11}$$

where "<...>" denotes the propagators (see [5,6] for a derivation in this context).

4. The expectation value of the energy-momentum tensor of a CFT

Let us consider a general CFT on Σ, and choose some isothermal coordinates $(z,\bar{z})$ where the metric has the form $h=h_{z\bar{z}}(dz\otimes d\bar{z}+d\bar{z}\otimes dz)$. Its more important feature consists in the existence of a traceless energy-momentum tensor, whose (on-shell) holomorphic and antiholomorphic parts $T(Q)$ and $\bar{T}(Q)$ generate holomorphic and antiholomorphic transformations respectively [13]

$$z \to z + e(z) \qquad , \qquad \bar{z} \to \bar{z} + \bar{e}(\bar{z})$$

In our context T and $\bar{T}$ will be expanded as in eqs.(1.10), where the $L_n, \bar{L}_n$ operators will satisfy the Krichever-Novikov algebra (1.12) with the cocycle of the form [2]

$$\kappa_{ij} = 1/(2\pi i) \oint_C dz \ (\ (1/2 \ \partial_z^3 e_i(z) \ - \ R(z) \ \partial_z e_i(z)) \ e_j(z) \ -$$

$$- \ (i \longleftrightarrow j) \ ) \tag{4.1}$$

where $R(z)$ is the schwartzian connection of the algebra that makes the integrand in (4.1) a well defined 1-form, and C any contour separating P_+ and P_-.

On the other hand, the energy-momentum tensor currently considered in the path integral approach is an object which presents the following operator product expansions [14]

$$T^{PI}(z)T^{PI}(w)=c/2(z-w)^{-4}+2T^{PI}(w)(z-w)^{-2}+\partial_w T^{PI}(w)(z-w)^{-1}+ \text{reg.}$$

$$T^{PI}(z)\phi(w) = \sum_n (z-w)^{-n-2} \ L^V_n\phi(w) \tag{4.2}$$

where ϕ is an arbitrary field. These statements are equivalent to saying that the L^V_n operators in (4.2) satisfy the Virasoro algebra

$$[L^V_i, L^V_j] = (i-j)\, L^V_{i+j} + c/12\, (i^3-i)\, \delta_{i+j,0} \tag{4.3}$$

The KN tensor eqs.(1.10) presents two differences with respect to T^{PI}:
i) it is a well-defined two-differential;
ii) due to the definition via normal ordering of the L_i operators, the vacuum expectation value of T (idem $\bar{T}$) should be zero (and indeed this will be the case in the explicit examples presented below).

In the present formalism we guess that the analoguous object T^s should be one that, in neighborhoods of P_+ where the following expansion holds $(z(P_+)=0)$

$$T^s(z)\, d^2z = \sum_n L^s_n\, z^{-n-2}\, d^2z \quad , \tag{4.4}$$

be so that the L^s_n operators verify (4.3). It is not difficult to show by using the expansions of the Krichever-Novikov bases near P_+ [2,4] that the desired object is given by

$$T^s(z)\, d^2z = T(z)\, d^2z - c/12\, R(z)\, d^2z \tag{4.5}$$

From the remarks exposed above it follows

$$\langle T^s(z)\rangle = \langle 0|T^s(z)|0\rangle = -c/12\, R(z) \tag{4.6}$$

Let us verify this general relation in the case of the scalar field. The KN operators are given by eqs.(1.11). By computing the commutator we get a KN algebra, eq.(1.12), with c=1 and the cocycle given by

$$\kappa_{ij} = 6 \sum_{n,m,k,s} \tilde{\gamma}_{mk}\, \tilde{\gamma}_{ns}\, l^{mn}_i\, l^{ks}_j - (i \longleftrightarrow j) \tag{4.7}$$

From the definition of the l^{nm}_i coefficients eq.(1.11), and by using the expansion of G(P,Q) in eq.(2.13) we can express (4.7) as follows

$$\kappa_{ij} = 6\; 1/(2\pi i)\oint_{C_\tau} e_j(P)\; 1/(2\pi i)\oint_{C_P} e_i(Q)\; G(P,Q)^2 \tag{4.8}$$

By inserting (2.13) into (4.8) we arrive to an expression of the type (4.1) with the schwartzian connection given by [15]

$$R(z) = -12\, (a_2(z) - \pi/2 \sum_{i,j=1}^{g} \eta^i(z)\, (\tau_2^{-1})_{ij}\, \eta^j(z)) \tag{4.9}$$

where $a_2(z)$ is a coefficient defined by the expansion of the prime form $E(P,Q)=E(z,w)dz^{-1/2}dw^{-1/2}$ for w near z [8]

$$E(z,w)^{-1} = (z-w)^{-1} (1 + a_2(z) (z-w)^2 + \ldots) \qquad (4.10)$$

By applying (4.6) we get the well known expectation value of the energy-momentum tensor of a scalar field defined by

$$T^{PI}(z) = - \lim_{w \to z} (\partial_z \phi(z) \partial_w \phi(w) + (z-w)^{-2}) \qquad (4.11)$$

Similar steps for computing the schwartzian connection in the case of a chiral b-c system of weights λ and $(1-\lambda)$ can be followed, finding results also in agreement with the literature (see ref.[15] for details).

References

[1] L.Alvarez-Gaume, C.Gomez and C.Reina, in Proc. of the Trieste Spring School on Superstrings, CERN-TH 4775/87;
L.Alvarez-Gaume, C.Gomez, G.Moore and C.Vafa: "Strings in the operator formalism", BUHEP-87/51.

[2] I.M.Krichever and S.P.Novikov, Funk.Anal.i Pril,21No.2(1987),46.

[3] I.M.Krichever and S.P.Novikov, Funk.Anal.i Pril,21No.4(1987),47.

[4] L.Bonora, A.Lugo, M.Matone and J.Russo: "A global operator formalism on higher genus Riemann surfaces: b-c systems", to appear in Comm.Math.Phys.

[5] A.Lugo and J.Russo: "Hamiltonian formulation and scattering amplitudes in string theory at genus g", to appear in Nucl.Phys.B, and Proc. of the CIME course on "Global Geometry and Mathematical Physics", Montecatini, Italy, preprint SISSA 145/EP/88.

[6] J.Russo: "Hamiltonian, supercharge and scattering amplitudes in superstring theory at genus g", to appear in Nucl.Phys.B.

[7] S.Mandelstam, in Unified String Theories (edited by M.B.Green and D.Gross), World Scientific, Singapore, 1987;
E.D'Hoker and S.Giddings, Nucl.Phys.B291 (1987),90.

[8] J.Fay, "Theta functions on Riemann surfaces", Lectures Notes in Mathematics 352, Springer Verlag (1973).

[9] M.Green, J.Schwarz and E.Witten: "Superstring Theory", vol.I, Cambridge University Press, 1987.

[10] A.Lugo: "Vertex operators for the bosonic closed string theory at arbitrary genus in the operator formalism", preprint SISSA 36/EP/89.

[11] J.Russo: "Multiloop amplitudes for the bosonic closed string theory in the operator formalism", preprint SISSA 151/FM/88, to appear in Phys.Lett.B.

[12] S.Weinberg, Phys.Lett.156B (1985),309;
E.D'Hoker and D.H.Phong, Phys.Rev.D35 (1987),3890;
S.P.De Alwis, Phys.Lett.168B (1986),59;
R.Sasaki and I.Yamanaka, Phys.Lett.165B (1985),283.

[13] J.Russo: "Bosonization and fermion vertices on an arbitrary genus Riemann surface", preprint SISSA 37/EP/89.

[14] A.Belavin, A.Polyakov and A.Zamolodchikov, Nucl.Phys.B241 (1984),333;
D.Friedan, E.Martinec and S.Shenker, Nucl.Phys.B271 (1986),93.

[15] A.Lugo: " The expectation value of the energy-momentum tensor and the Krichever-Novikov algebra in Conformal Field Theories on Riemann surfaces", preprint SISSA 33/EP/89.

DYNAMICS OF OPEN STRINGS IN A BACKGROUND ELECTROMAGNETIC FIELD

V.V.Nesterenko
Laboratory of Theoretical Physics, Joint Institute for Nuclear Research, SU-141980, Dubna, USSR

All our knowledge about strings as physical objects is actually based on the free string theory. It may be said that we have a complete theory of noninteracting strings and superstrings: their mass spectrum is obtained exactly, the corresponding critical dimensions of space-time are explicitly calculated, the algorithm is known for constructing the Hilbert space of the state vectors [1,2]. Our notion of the second-quantised string as an infinite set of local fields with infinitely increasing masses beginning from zero is based also on the mass spectrum of a free string. However, in the old days of Dual Theory [2-4] there were hopes that the interaction of hadronic strings would result in an essential change of their spectrum and the later would become more realistic from the point of view of the hadronic physics (unstable string states would appear, the equidistant property of the mass spectrum would be violated, and so on).

Now the theory of interacting superstrings is considered as a promising candidate for unification of all the fundamental interactions including gravity [5,6]. Unfortunately, here there are no exact results at all. The important moment is the existence of a massless state in the spectrum of a free superstring. Whether this state will remain at the interaction level is not known precisely. In this situation it is worthwhile to explore the exactly solvable models with interactions in the string theory.

The aim of this lecture is to discuss the classical and quantum dynamics of open strings in a background electromagnetic field [7-11]. On the first-quantized level this problem admits an exact solution.

The external electromagnetic field does not affect the spinning (fermionic) variables of the string. Therefore one can confine oneself to a consideration of the bosonic part of the string action

$$S=-\frac{T}{2}\int_{\tau_1}^{\tau_2}d\tau\int_0^{\pi}d\sigma\,|g|^{1/2}g^{\alpha\beta}\partial_\alpha x^\mu\partial_\beta x_\mu-\sum_{a=1}^{2}q_a\int_{C_a}dx_\nu A^\nu(x), \quad (1)$$

$$g=\det(g_{\alpha\beta}),\quad \alpha,\beta=0,1,\quad \mu,\nu=0,1,\dots,D-1.$$

Here T is the string tension, Ca, a=1,2 are trajectories of the string ends on the (τ,σ)-plane, q_1 and q_2 are the electric charges localized at the string ends. Open strings can be of two types: the neutral strings with charges obeying the condition $q_1+q_2=0$ and the charged strings having a net charge $q_1+q_2\neq 0$. Their behaviour in an external electromagnetic field turns out to be very different. The invariance of the string theory under CPT-transformations requires a common consideration of both the neutral and charged open strings.

With the O($1, D-1$)-transformation in the D-dimensional space-time the electromagnetic stress-tensor $F_{\mu\nu}=\partial_\mu A_\nu-\partial_\nu A_\mu$ can be cast into the block diagonal form

$$F_{(A)\mu}^{\cdot\,\nu}=\mathrm{diag}(\overset{(0)}{F},\overset{(1)}{F},\dots,\overset{(D/2-1)}{F}), \quad (2)$$

where F, $A=0,1,\dots,(D/2)-1$ are the (2x2)-matrices

$$\overset{(0)}{F}=\begin{pmatrix}0 & -E\\ -E & 0\end{pmatrix},\quad \overset{(s)}{F}=\begin{pmatrix}0 & -H_s\\ -H_s & 0\end{pmatrix},\quad s=1,2,\dots,(D/2)-1. \quad (3)$$

For definiteness we have put that D is an even number.

The notion of the invariant string mass in a background electromagnetic field can be introduced only for the neutral strings. In the light-like gauge their mass-

-squared spectrum is defined by

$$M^2 = 2\pi T(1-e^2)\sum_{n=1}^{\infty}\sum_{i=2}^{D-1} n\, a_n^{+i} a_n^i - \pi T \frac{D-2}{12}(1-e^2) - M_{tr}^2, \quad (4)$$

where M_{tr}^2 is the tachyonic contribution due to the motion of the string as a whole in transverse directions

$$M_{tr}^2 = \sum_{s=1}^{(D/2)-1} \frac{e^2 + h_s^2}{1+h_s^2} \vec{P}_{s\perp}^{\,2}, \quad e = \frac{q}{T}E, \; h_s = \frac{q}{T}H_s. \quad (5)$$

In eq. (4) a_n^+ and a_n are the ordinary operators of creation and annihilation the transverse string excitations

$$[a_n^i, a_m^{j+}] = \delta_{ij}\delta_{nm}, \quad i,j = 2,3,\ldots,D-1, \quad (6)$$
$$n, m = 1, 2, \ldots .$$

In order that the string oscillations give a positive contribution to M^2, it is necessary to impose the following constraint on the external electric field

$$(qE/T)^2 < 1. \quad (7)$$

This means that the background electromagnetic field should be described by a nonlinear Lagrangian (for example, by the Born - Infeld Lagrangian instead of the Maxwell Lagrangian [10,12]).

The general structure of the string mass spectrum (4) remains the same as in the free case, but the distance between mass levels decreases $(1-e^2)$ times by virtue of the electric field. The basic property of an elementary particle is its mass which is not changed by interaction. In this respect the string is more like a composed object as an atom or a molecule the energy spectrum of which is deformed by external fields.

In the case of charged strings $(q_1 + q_2 \neq 0)$ the situation is more complicated. The mass of such strings is not a conserved quantity. Furthermore, the light-like gauge cannot be introduced for such strings when the external electric field is different from zero. If the charged string is placed in background magnetic fields, the light--like gauge can be employed and the string energy E is conserved. It is convenient to consider this quantity in a special reference frame, where the longitudinal total momentum of the string p^1 is equal to zero. The energy--spectrum of the charged string is given by

$$E^2 = 2\pi T \sum_{s=1}^{(D/2)-1} \Big\{ |\varepsilon^s| a_0^{+s} a_0^s + \qquad (8)$$

$$+ \sum_{n=1}^{\infty} [(n+\varepsilon^s) a_n^{+s} a_n^s + (n-\varepsilon^s) b_n^{+s} b_n^s] \Big\} - E_0^2 .$$

All the operators a_0, a_n and b_n obey the same commutation relations as in eq. (6). The string eigenfrequencies are not integer now but they equal $n \pm \varepsilon^s$, $n = 0, 1, \ldots$, where $tg\, \pi\varepsilon^s = (h_{1s} + h_{2s})/(1 - h_{1s} h_{2s})$, $|\varepsilon^s| < \frac{1}{2}$. Thus, each free eigenfrequency splits into two new ones and there appears new levels in the spectrum connected with zero frequency ε^s. The energy of zero point fluctuations, E_0^2, is

$$E_0^2 / (T\pi) = \frac{D-2}{24} - \sum_{s=1}^{(D/2)-1} |\varepsilon^s| (1 - |\varepsilon^s|) . \qquad (9)$$

It is interesting to note that E_0^2 may vanish for certain values of q_1, q_2 and H_s. If one takes, foe example, all H_s, $s = 1, 2, \ldots, (D/2)-1$ to be equal to H and demands that the charges at the string ends q_1 and q_2 satisfy the condition

$$\frac{H}{T} \frac{q_1 + q_2}{1 - q_1 q_2 (H/T)^2} = ctg \frac{\pi}{\sqrt{6}} ,$$

then one obtains $E_0^2 = 0$, this being true for arbitrary space-time dimension D.

Thus, the interaction of a charged string with an electromagnetic field results in an essential rearrangement of its energy spectrum.

Of course, on the basis of these results one cannot make any definite conclusions about the behaviour of the mass spectrum of strings and superstrings after taking into account their interaction.

From the gauge fields theory we know that the interacting photons remains massless because of the local gauge symmetry. Supersymmetry of the Green- Schwarz action for the superstring is not local but global. Nevertheless, the treatment of superstrings as the gauge theory in the contour space [13] draws it together with the usual local gauge fields.

References

1. Barbashov, B.M. and Nesterenko, V.V. "Introduction to the Relativistic String Theory", World Scientific, Singapore (1990).
2. Scherk, J., Rev.Mod.Phys., 47, 123 (1975).
3. Frampton, P.H., "Dual Resonance Models and Superstrings", World Scientific, Singapore (1986).
4. "Dual Theory", ed. Jacob, M., North-Holland (1974).
5. Green, M.B., Schwarz, J.H. and Witten, E., "Superstrings theory", Vols. 1,2, Cambridge University Press, Cambridge (1987).
6. Barbashov, B.M. and Nesterenko, V.V., Soviet Physics Uspekhi 29, 1077-1096 (1986).
7. Nesterenko, V.V., I.J. Mod.Phys. A (1989).
8. Barbashov, B.M., Koshkarov, A.L. and Nesterenko, V.V. Teor. Mat.Fiz., 32, 176 (1977).
9. Barbashov, B.M., Nesterenko, V.V. and Chervjakov, A.M. Teor. Mat.Fiz., 32, 336 (1977).
10. Aboulsaood, A.M., Callan, C.G., Nappi, C.R. and Yost, S.A. Nucl.Phys., B280 [FS18], 599 (1987).
11. Burgess, C.P., Nucl.Phys., B294, 427 (1987).
12. Fradkin, E.S. and Tseytlin, A.A., Phys.Lett. 158B, 516 (1985).
13. Kaku, M. "Introduction to Superstrings", Springer Verlag, Berlin (1989).

II. STRINGS: FIELD THEORY

Strings, Superstrings and Two-Dimensional Lagrangian Field Theory [1]

S. James Gates, Jr.[2]

Department of Physics and Astronomy
University of Maryland, College Park, MD 20742

Abstract

In these lectures, I will emphasize the use of two dimensional lagrangian field theory techniques to investigate the geometry of the superworld-sheet and manifest realization of symmetries in string theories.

(I.) Introductory Remarks

String and superstring models [1] are among the most interesting but least understood topics in the realms of mathematical physics, classical and quantum field theory, and gauge theories. On the one hand, they seem to hold the promise of providing the only known resolution to the problem of divergences in quantum gravity while simultaneously describing internally consistent "unified field" theories as first envisioned by Einstein. On the other hand, we simply do not have a fundamental understanding of what is the geometrical principle which underlies this powerful new class of theories. Indeed, this last fact is exemplified in the most elementary way by our lack of a "stringy" field strength in any known theory. Field strengths are one of the signet hallmarks of gauge theories and until these appear, I would argue that a fundamental understanding is lacking.

This present situation is extremely unsatisfactory and leaves enormous room for speculation on the behavior of string theories. For example, one such problem concerns "the four dimensional vacua" of string theories. Presently, it is widely believed that a unique vacuum for heterotic string theory should be chosen by the dynamics. I like to think of this as the "Ptolemaic view" of string theory. The belief in a unique vacuum

[1] Invited lectures given at the XXV Winter School of Theoretical Physics, Karpacz, Poland, Feb. 20-March, 4 1989.

[2] Work supported by National Science Foundation Grant PHY 87-46846.

is akin to the belief in a unique place for the earth in earlier cosmology. The alternate "Copernican view" is more to my liking. After all, why should only one vacuum be consistent? In general relativity, the next most complicated gauge theory, the vacuum is *not* determined by the theory. Until a nonperturbative understanding of strings is achieved this question will remain unanswerable. Another nonperturbative problem concerns an interesting high-energy string phenomenon. There is good evidence [2] that at extremely high energies there is a phase transition where the number of degrees of freedom associated with usual spacetime fields are greatly reduced. It seems unlikely, without a geometrical understanding of string field theory, that much of this mostly unimaginable "new physics" can be studied. A geometrical approach would be helpful to study this question. Without this we are in a position analogous to trying to study the physics of black-holes using only weak field approximations! Similarly the lack of a guiding principle in string theory, which would extend the equivalence principle of general relativity or the local gauge principle in Yang-Mills theory, allows side issues to be confused with fundamental ones. Yet after a vast effort on the part of many, many talented physicists, we do not yet have any substantial progress in understanding the geometry of string field theory.

In the following lectures, which are meant to be purely introductory but from a somewhat unorthodox viewpoint, I have tried to concentrate on aspects of two dimensional field theory which may lead us to a deeper understanding of string theory in the long run. Note that using the world-sheet approach we are far removed from attacking the problem of string field theory. Still this has been such a valuable tool in the past that it is perhaps not unreasonable to think that new insights may be gained through further investigation. I have purposely left out numbers of topics which others might consider crucial. My reason for this is two-fold. First, by now, there are a large number of books, conference and school proceedings, and review articles where the interested reader can find excellent introductions. Second, I want to emphasize the use of two dimensional lagrangian field theory techniques as a way of approaching string theory. This is somewhat at variance with most of the present discussions of string theories. By starting from a world-sheet action principle and demanding that <u>all</u> symmetries of the string be manifestly present in an appropriate action which describes the string, we are being much more restrictive than is often the case. A great many discussions of string theories are based only on the existence of conformal (or superconformal) algebras of the Virasoro or Kac-Moody type with little or no attention being paid to whether there exist an appropriate world-sheet action which covariantly encodes a particular algebra. A similar earlier epoch existed in particle physics when it was thought that "current algebra" should supplant the use of lagrangian field theory. Later, it was demonstrated that lagrangian field theory was by far more useful in a wider variety of problems. It may be prudent not to forget this lesson in our pursuit of understanding strings. One other point to recall is the history of the spinning string. After it was written in the form a world sheet action for a two dimensional supergravity theory, it actually spurred the development of supergravity and supersymmetry! So lagrangian approaches seem to me to best allow for possible connections to other parts of quantum field theory.

The following summarizes the five lectures which I gave at the school. I have tried to make this material as self-contained as possible given the limitation on space. Most of the material is introductory but I have also added some less well known facts. One example of this is the $Sp(2)$-covariant "flat space" gauge quantization of 2-d (super)gravity. The final lecture presents highly speculative material and some observations which were spurred by discussions with M. Kaku. I hope these raise in the minds of the students some questions which may lead to possible avenues of future lines of research to help to increase our understanding.

(II.) A Natural Parametrization of Two-Surfaces

Within the setting of two-dimensional field theory, information concerning the topological and geometrical properties of a two-manifold is encoded in the zweibein fields $V_a{}^m(\tau,\sigma)$, their inverse $V_m{}^a$, or the metric $g_{mn} \equiv \frac{1}{2}(V_m^{++}V_n^{--} + V_m^{--}V_n^{++})$. The geometry of a two-surface may also be specified by giving an invariant line element.

$$(ds)^2 = d\sigma^m g_{mn} d\sigma^n = \left(d\sigma^m V_m{}^{++}\right)\left(d\sigma^n V_n{}^{--}\right) . \tag{2.1}$$

However, it is often useful to parametrize the zweibeins in a way where the various symmetries for which the zweibeins provide representations are made more explicit. This is done by using "prepotentials," a concept first introduced for $D = 4, N = 1$ supergravity [3]. We may write the zweibein or frame fields as [4]

$$\begin{aligned} V_{++}{}^m \partial_m &= \psi^{-1} e^{\ell}\left(\partial_{++} + h_{++}{}^{--}\partial_{--}\right) \equiv V_{++} = \psi^{-1}e^{\ell}\hat{V}_{++} \ , \\ V_{--}{}^m \partial_m &= \psi^{-1} e^{-\ell}\left(\partial_{--} + h_{--}{}^{--}\partial_{++}\right) \equiv V_{--} = \psi^{-1}e^{\ell}\hat{V}_{--} \ , \end{aligned} \tag{2.2}$$

where $h_{++}{}^{--}(\tau,\sigma)$, $h_{--}{}^{++}(\tau,\sigma)$, $\psi(\tau,\sigma)$ and $\ell(\tau,\sigma)$ are independent fields. These are the prepotentials. The conformal prepotentials, $h_{--}{}^{++}$ and $h_{++}{}^{--}$, are also known as quadratic differentials or Beltrami differentials. The fields ψ and ℓ are, respectively, the Weyl and Lorentz density-type compensators. The invariant line element can also be written in terms of prepotentials as

$$\begin{aligned} (ds)^2 &= (\Delta^{-1}\psi)^2(d\sigma^{++} - d\sigma^{--}h_{--}{}^{++})(d\sigma^{--} - d\sigma^{++}h_{++}{}^{--}) \ , \\ \Delta &\equiv 1 - h_{++}{}^{--}h_{--}{}^{++} \ , \end{aligned} \tag{2.3}$$

and a quick calculation reveals

$$\sqrt{|det(g_{mn})|} = \psi^2\Delta^{-1} = [det(V_a{}^m)]^{-1} . \tag{2.4}$$

Since the prepotentials are only defined up to gauge transformations, the ambiguity in ψ may be used to define a different function $\psi' \equiv \Delta^{-\frac{1}{2}}\psi$ so that $\sqrt{|g|} = (\psi')^2$. Note that two of the prepotentials (ψ and ℓ) are closely related to local Weyl and Lorentz transformations, respectively. This can be seen by taking variations of these equations. Let $\delta\psi = -k_W\psi$ where k_W is an infinitesimal function. Clearly

$$\delta V_{++}^m = k_W V_{++}{}^m \ , \quad \delta V_{--}{}^m = k_W V_{--}{}^m \ , \tag{2.5}$$

and these are recognizable as infinitesimal Weyl transformations. Similarly if $\delta\ell = k_L$ for infinitesimal function k_L, then

$$\delta V_{++}{}^m = k_L V_{++}{}^m \ , \quad \delta V_{--}{}^m = -k_L V_{--}{}^m \ , \tag{2.6}$$

which are recognizable as infinitesimal Lorentz transformations. Note that $(ds)^2$ is independent of ℓ. This corresponds to the fact that g_{mn} is independent of local Lorentz transformations.

One other local symmetry must be considered; local reparametrization symmetry. Under the action of such an infinitesimal transformation the coordinates have variations given by $\delta\sigma^{++} = k^{++}$ and $\delta\sigma^{--} = k^{--}$ where k^{++} and k^{--} are two more independent local gauge parameters. The prepotentials also transform under local reparametrizations. The transformation laws of the prepotentials under this symmetry have been derived [4] previously with the resulting variations given by

$$\begin{aligned} \delta_G h_{--}{}^{++} &= -\Delta[\ \hat{V}_{--}k^{++} \ - \ k^{++}\hat{c}_{++,--}{}^{++}\ ] \ , \\ \delta_G h_{++}{}^{--} &= -\Delta[\ \hat{V}_{++}k^{--} \ + \ k^{--}\hat{c}_{++,--}{}^{--}\ ] \ , \end{aligned} \tag{2.7}$$

where $\hat{c}_{a,b}{}^c$ is defined by $[\hat{V}_a, \hat{V}_b] \equiv \hat{c}_{a,b}{}^c\hat{V}_c$. For ψ and ℓ the variations under reparametrization transformations are

$$\begin{aligned} \delta_G\psi &= [(\ k^{++}\hat{V}_{++} \ + \ k^{--}\hat{V}_{--})\psi\ ] \ + \ \lambda_S\psi \ , \\ \delta_G\ell &= [(\ k^{++}\hat{V}_{++} \ + \ k^{--}\hat{V}_{--})\ell\ ] \ - \ \lambda_L \ . \end{aligned} \tag{2.8}$$

$$\begin{aligned} \lambda_S &= \tfrac{1}{2}[\ \hat{V}_{++}(k^{++} \ + \ k^{--}h_{--}{}^{++}) \ + \ \hat{V}_{++}(k^{--} \ + \ k^{++}h_{++}{}^{--})\ ] \ , \\ \lambda_L &= \tfrac{1}{2}[\ \hat{V}_{++}(k^{++} \ + \ k^{--}h_{--}{}^{++}) \ - \ \hat{V}_{++}(k^{--} \ + \ k^{++}h_{++}{}^{--})\ ] \ . \end{aligned} \tag{2.9}$$

It is the appearance of λ_S and λ_L in the variations ψ and ℓ above which is characteristic of a density-type compensator. Substituting the variations in (2.7) and (2.8) into (2.2), implies the zweibeins vary as

$$\delta V_{++}{}^n = k^m\partial_m V_{++}{}^n - V_{++}{}^m\partial_m k^n + (k_W + k_L)V_{++}{}^n \ ,$$

$$\delta V_{--}{}^n = k^m\partial_m V_{--}{}^n - V_{--}{}^m\partial_m k^n + (k_W - k_L)V_{--}{}^n \ . \tag{2.10}$$

In general there are four independent gauge parameters k^{++}, k^{--}, k_L and k_W characterizing the symmetries of the theory. Choices of g_{mn} or more correctly $V_a{}^m$ correspond to gauge fixing conditions on the prepotentials. Flat space is described by $(ds)^2 = d\sigma^{++}d\sigma^{--}$. The residual symmetry of this gauge, which preserves the gauge condition, forces restrictions on the *a priori* independent gauge parameters. To maintain this gauge one finds

$$\begin{aligned} &k^{++} = k^{++}(\sigma^{++}) \ , \ \ k^{--} = k^{--}(\sigma^{--}) \ , \ \ k_W = \tfrac{1}{2}[\partial_{++}k^{++} + \partial_{--}k^{--}] \ , \\ &k_L = \tfrac{1}{2}[\partial_{++}k^{++} - \partial_{--}k^{--}] \ . \end{aligned} \tag{2.11}$$

For a covariantly transforming matter field $\varphi^{(0,0)}_{(m,n)}$ of scale weight $w = \frac{1}{2}(m-n)$ (see the appendix) and satisfying $V_{--}\varphi^{(0,0)}_{(m,n)} = 0$, the general transformation law of a covariant field specializes in this gauge to

$$\delta\varphi^{(0,0)}_{(m,n)} = k^{++}\partial_{++}\varphi^{(0,0)}_{(m,n)} + \tfrac{1}{2}(m-n)(\partial_{++}k^{++})\varphi^{(0,0)}_{(m,n)} \quad . \tag{2.12}$$

In the conformal field theory formalism, this is known as the transformation law of a "primary field" of weight $\frac{1}{2}(m-n)$ and this definition is seen to depend on the gauge choice of the zweibein. The commutator of two coordinate transformations takes the form,

$$\left[K^L(k_1^{++}),\, K^L(k_2^{++})\right] = K^L(k_3^{++}) \quad , \quad K^L(k_1^{++}) \equiv k_1^{++}\partial_{++} \quad ,$$

$$k_3^{++} = k_1^{++}\partial_{++}k_2^{++} - k_2^{++}\partial_{++}k_1^{++} \quad , \tag{2.13}$$

with the same result for $K^R(k_1^{--})$ after interchanging all plus signs for minus signs. In general upon quantization the algebra changes by the presence of a Schwinger term,

$$\left[K^L(k_1^{++}),\, K^L(k_2^{++})\right] = K^L(k_3^{++}) - i\frac{\nu_L}{24\pi}\,a^{(L)} \,,$$

$$a^{(L)}(k_1,k_2) = \int d\sigma^{++}\left[k_1^{++}(\partial_{++}^3 k_2^{++}) - k_2^{++}(\partial_{++}^3 k_1^{++})\right] \,, \tag{2.14}$$

The quantity ν_L may be called the left central charge or left anomaly. When expanded over a complete set of modes (2.14) and (2.13) are seen to lead to the usual expressions for the Virasoro algebra with and without central extension, respectively.

Since the zweibein and thus the prepotentials only play the role of lagrange multipliers, the variation of any two-dimensional lagrangian density with respect to them imposes constraints on the remaining fields. That is, given an action $\mathcal{S}$, the equations of motion of the prepotentials take the forms

$$\frac{\delta\mathcal{S}}{\delta h_{++}{}^{--}} \equiv t_{--,--} = 0 \quad , \quad \frac{\delta\mathcal{S}}{\delta h_{--}{}^{++}} \equiv t_{++,++} = 0 \quad ,$$

$$\frac{\delta\mathcal{S}}{\delta\psi} \equiv t_W = 0 \quad , \quad \frac{\delta\mathcal{S}}{\delta\ell} \equiv t_L = 0 \quad . \tag{2.15}$$

In a theory with local Weyl and Lorentz invariance, after an appropriate redefinition, the action will be independent of ψ and ℓ. Thus, the latter two equations will be trivialities. (The redefinitions required are of the forms of local Weyl and Lorentz transformations acting on the fields.)

We may easily proceed to discuss the quantization of theories which involve the zweibein. For concreteness we may choose the bosonic string. The generating functional at zero source value is given by

$$\int [DV_a{}^m][DX^{\underline{m}}]e^{(iS)} \equiv \int [Dh_{++}{}^{--}][Dh_{--}{}^{++}][D\ell][D\psi][DX^{\underline{m}}]e^{(iS)} \quad ,$$

$$S_B = S(X^{\underline{m}}, h_{++}{}^{--}, h_{--}{}^{++}) = \frac{1}{2\pi\alpha'}\int d^2\sigma \; \Delta^{-1}[\, -\tfrac{1}{2}\eta_{\underline{mn}}(\hat{V}_{++}X^{\underline{m}})(\hat{V}_{--}X^{\underline{n}})\,] \quad . \tag{2.16}$$

Note that ψ and ℓ do not appear at all in the action. This is a consequence of the classical Weyl and Lorentz symmetries of the action. The variations of the prepotentials may easily be evaluated on the gauge fixing "surface"

$$\delta h_{++}{}^{--}|_{GF} \equiv \mathcal{P}_1(k^{--}) \quad , \quad \delta h_{--}{}^{++}|_{GF} \equiv \mathcal{P}^\dagger{}_1(k^{++}) \quad ,$$

$$\delta\psi|_{GF} = -k_W + \lambda_S(k^{++}, k^{--}) \;\; , \;\; \delta\ell|_{GF} = k_L - \lambda_L(k^{++}, k^{--}) \quad . \tag{2.17}$$

Using the "standard" field theory ansatz $\delta_G h_{++}{}^{--} \to \mathcal{Q}h_{++}{}^{--}, \delta_G h_{--}{}^{++} \to \mathcal{Q}h_{--}{}^{++}$, $k^{++} \to ic^{++}, k^{--} \to ic^{--}$ etc. yields the BRST variations which imply the following quantum gauge fixed action,

$$\begin{aligned} S_{eff}|_{GF} = \; & \frac{1}{2\pi\alpha'}\int d^2\sigma \; [\, -\tfrac{1}{2}\eta_{\underline{mn}}(\partial_{++}X^{\underline{m}})(\partial_{--}X^{\underline{n}})\;] \\ & + \int d^2\sigma \; [\; ib_{++}{}^{--}\partial_{--}c^{++} + ib_{--}{}^{++}\partial_{++}c^{--} \\ & -ib_W(c_W - \tfrac{1}{2}(\partial_{++}c^{++} + \partial_{--}c^{--})) + ib_L(c_L - \tfrac{1}{2}(\partial_{++}c^{++} - \partial_{--}c^{--}))\;] \;, \end{aligned} \tag{2.18}$$

evaluated on the gauge fixing surface. This is the well-known "flat space" gauge-fixed result for the bosonic string. It may be noted that the functional integrals over the "Weyl" anti-ghost field b_W and the "Lorentz" anti-ghost field b_L are trivial, so the last line in the effective action may be set to zero.

The result above has a BRST symmetry but not an anti-BRST symmetry. It might appear that there is no way to choose the "flat space" gauge and simultaneously introduce an anti-BRST symmetry. This is not so. To show this, first note that the usual BRST generator $\mathcal{Q}$ and anti-BRST generator $\bar{\mathcal{Q}}$ may be assembled into an $Sp(2)$-spinor $\mathcal{Q}^{(\alpha)} \equiv (\mathcal{Q}, \bar{\mathcal{Q}})$. The goal is now to find a gauge-fixed effective action which possesses the $Sp(2)$ symmetry. The next step is to note that the prepotentials $h_{--}{}^{++}$ and $h_{++}{}^{--}$ may be assigned to $Sp(2)$ "vector supermultiplets"

$$(h_{++}{}^{--}, c^{(\alpha)\,--}, d^{--}) \quad , \quad (h_{--}{}^{++}, c^{(\alpha)\,++}, d^{++}) \quad . \tag{2.19}$$

The $Sp(2)$ covariant BRST operator $Q^{(\alpha)}$ acting on the prepotentials is proportional to the partial derivative of the $c^{(\alpha)}$'s in the usual way. Finally the conditions

$$Q^{(\alpha)}c^{(\beta)\,--} = iC^{(\alpha)(\beta)}d^{--} \quad , \quad Q^{(\alpha)}c^{(\beta)\,++} = iC^{(\alpha)(\beta)}d^{++} \quad , \tag{2.20}$$

(where $C^{(\alpha)(\beta)}$ is i times the Levi-Civita tensor) give the "top" components of the multiplets. Additionally, two $Sp(2)$ "scalar supermultiplets" can be defined by

$$(a_{++}{}^{--}, b_{++}^{(\alpha)\,--}, f_{++}{}^{--}) \quad , \quad (a_{--}{}^{++}, b_{--}^{(\alpha)\,++}, f_{--}{}^{++}) \quad . \tag{2.21}$$

The components of these "scalar supermultiplets" satisfy

$$Q^{(\alpha)}a_{--}{}^{++} = ib_{--}^{(\alpha)\,++} \quad , \quad Q^{(\alpha)}a_{++}{}^{--} = ib_{++}^{(\alpha)\,--} \quad ,$$

$$Q^{(\alpha)}b_{--}^{(\beta)\,++} = iC^{(\alpha)(\beta)}f_{--}{}^{++} \quad , \quad Q^{(\alpha)}b_{++}^{(\beta)\,--} = iC^{(\alpha)(\beta)}f_{++}{}^{--} \quad . \tag{2.22}$$

Note also by definition that $Q^{(\alpha)}$ annihilates the "top" component of any multiplet.

Now we can write the $Sp(2)$-covariant "flat space" gauge-fixed effective action on the constraint surface in a concise way

$$\begin{aligned} S_{eff}|_{GF} = {} & \frac{1}{2\pi\alpha'}\int d^2\sigma\,[\,-\tfrac{1}{2}\eta_{\underline{mn}}(\partial_{++}X^{\underline{m}})(\partial_{--}X^{\underline{n}})\,] \\ & + \int d^2\sigma Q^{(\alpha)}Q_{(\alpha)}[\,a_{--}{}^{++}h_{++}{}^{--} + a_{++}{}^{--}h_{--}{}^{++}\,] \;. \end{aligned} \tag{2.23}$$

In this expression, the index on $Q^{(\alpha)}$ is raised and lowered with $C^{(\alpha)(\beta)}$ and its inverse. It is simple to work out the explicit form of this action using the rules stated. It contains an $Sp(2)$ doublet of the usual anticommuting $b-c$ fields and one bosonic term where the fields a and d replace b and c.

A modification of the quantization procedure discussed above is to make use of the background-quantum field technique [5]. One of its useful attributes is that it permits a simple field-theoretical approach to understanding local anomalies in string theories. It leads to a set of propagators and vertices. The renormalized background effective action is calculated by integrating out quantum fields and ghosts. We next use the Adler-Rosenberg method to check to see if the effective action breaks background gauge invariance in a way that can not be removed by local counter terms. In particular, the part of the effective action Γ_{Anom}, that breaks scale and Lorentz invariance takes the form below in terms of background gravity fields $h_{--}^{(B)++}$ and $h_{++}^{(B)\,--}$

$$\Gamma_{Anom.} = -\frac{1}{96\pi}\int d^2\sigma\{\,\nu_L h_{--}^{(B)\,++}[\frac{(\partial_{++})^4}{\Box}h_{--}^{(B)\,++}] + \nu_R h_{++}^{(B)\,--}[\frac{(\partial_{--})^4}{\Box}h_{++}^{(B)\,--}]\,\}\;, \tag{2.24}$$

where ν_R and ν_L are numbers which correspond to the right and left central charges in a Virasoro algebra. As an example, starting from the gauge-fixed effective action in (2.18) one finds $\nu_L = \nu_R = D(1) - 26$. The first factor comes from the D X's in a one loop graph. The factor of -26 comes from the $b - c$ pair in a similar graph. The same calculation can be repeated for the $Sp(2)$-covariant gauge-fixed action (2.23) where $\nu_L = \nu_R = D(1) - 52 + 26$. The factor of -52 arises from the $Sp(2)$-doublet $b - c$ fields and the factor of $+26$ comes from the the $a - d$ fields.

It is often useful to introduce a world gravitationally covariant derivative $\mathscr{D}_A$

$$\mathscr{D}_{++} \equiv V_{++} + \omega_{++}\mathcal{M} \quad , \quad \mathscr{D}_{--} \equiv V_{--} + \omega_{--}\mathcal{M} \quad , \quad \omega_{++} = c_{++,--}{}^{--} \quad ,$$

$$\omega_{--} = c_{++,--}{}^{++} \quad , \quad V_a \equiv V_a{}^m \partial_m \quad , \quad [V_a, V_b] = c_{a,b}{}^c V_c \quad , \qquad (2.25)$$

to include a spin connection ω_a when acting on general matter fields. Due the definitions of the spin connection above we have

$$[\mathscr{D}_{++}, \mathscr{D}_{--}] = -r(V)\mathcal{M} \quad , \qquad (2.26)$$

where $r(V)$ is the curvature scalar of the world sheet.

There are of course many, many other gauge choices possible besides the flat space gauge. For example, the conformal gauge is given by

$$h_{++}{}^{--} = h_{--}{}^{++} = \ell = 0 \, , \quad \psi = e^{\frac{1}{2}\varphi} \, ,$$

$$\rightarrow (ds)^2 = e^{\varphi} \, d\sigma^{++} \, d\sigma^{--} \, . \qquad (2.27)$$

Polyakov's light-cone gauge is obtained from

$$h_{++}{}^{--} = \psi - 1 = \ell = 0 \, , \quad h_{--}{}^{++} = -f \, ,$$

$$\rightarrow (ds)^2 = \left(d\sigma^{++} + d\sigma^{--} f\right) \, d\sigma^{--} \, . \qquad (2.28)$$

There are an infinite number of other gauge choices available. A very special class of gauges corresponds to the use [6] of quadratic differentials for which Strebel's theorem applies. It seems natural to identify the function elements of ref. [6] with $h_{++}{}^{--}$ and $h_{--}{}^{++}$ since by construction these have the proper transformation laws. This suggest choosing gauges of the form

$$h_{++}{}^{--} = \sum_i \frac{a_i}{(\sigma^{++} - p_i{}^{++})^2} \, , \quad h_{--}{}^{++} = \sum_i \frac{a_i}{(\sigma^{--} - p_i{}^{--})^2} \, ,$$

$$\ell = \psi - \Delta = 0 \, ,$$

$$\to (ds)^2 =$$

$$\left[d\sigma^{++} - d\sigma^{--}\left(\sum_i \frac{a_i}{(\sigma^{--} - p_i{}^{--})^2}\right)\right]\left[d\sigma^{--} - d\sigma^{++}\left(\sum_i \frac{a_i}{(\sigma^{++} - p_i{}^{++})^2}\right)\right] , \tag{2.29}$$

where a_i , $p_i{}^{++}$ and $p_i{}^{--}$ are fixed constants. Given (2.29) it is simple to show that the corresponding frame fields are

$$V_{++} = \Delta^{-1}\left[\partial_{++} + \left(\sum_i \frac{a_i}{(\sigma^{++} - p_i^{++})^2}\right)\partial_{--}\right] ,$$

$$V_{--} = \Delta^{-1}\left[\partial_{--} + \left(\sum_i \frac{a_i}{(\sigma^{--} - p_i^{--})^2}\right)\partial_{++}\right] . \tag{2.30}$$

In passing, we note the Weyl gauge choice $\psi = 1$ simplifies the frame fields even further at the expense of introducing an overall factor of Δ^{-2} into the rhs of the expression for $(ds)^2$.

Up until this point, there is an issue concerning the path integral over the conformal prepotentials $h_{--}{}^{++}$ and $h_{++}{}^{--}$ which we have not discussed. Note that in the last gauge choice discussed above, after performing the path integral over the fields the result will depend on a set of parameters $(p_i^{++}, p_i^{--}, a_i)$. These parameters contain information about the *global* or topological structure of the two-surface. Such parameters are referred to as *modular parameters.* This type of information is largely independent of the choice of coordinates with which the two-surface is described. In fact, the various distinct topological classes of the two-surface are characterized by parameters like these. So the final step in performing these path integrals must include a summation over all of the distinct topologies and will in general include ordinary integration over the parameters. This step is necessary to insure that the final results are independent of those coordinate transformations which are not continuously connected to the identity. It is only then that we reach results which are modular invariant.

(III.) Compactification or Non-compactification

When heterotic and spinning strings were discovered, it seemed as though the Minkowski space-time (as opposed to world-sheet) must have more than three space-like directions. Since this occurred at a time when the study of Kaluza-Klein theories was fashionable, this "Kaluza-Klein interpretation" was widely accepted. It was known that ordinary Kaluza-Klein field theories possess field equations whose solutions have the structure of a four-dimensional fiber bundle. That is, the structure contains four-dimensional Minkowski spacetime tensored with an internal fiber space. Fiber bundles have been known to physicists as providing the mathematical framework for the formulation and study of Yang-Mills theories.

However, when viewed from the most recent constructions of four dimensional string theories, it is clear that the "fiber bundle interpretation" itself stands as an independent and logically consistent view. In other words, it is possible to consider *ab initio* four dimensional Minkowski spacetimes for string theories. The price for this is that simultaneous with the introduction of string coordinates, $X^{\underline{m}}(\tau, \sigma)$, sets of "isotopic charge" space coordinates $\phi_L^{\hat{\alpha}}(\tau, \sigma)$ and $\phi_R^{\hat{I}}(\tau, \sigma)$ must be introduced. The notion of isotopic charge space is very familiar from particle theory models outside of the context of string theories. The neutrino and left-handed electron form a "spinor" representation in $SU_W(2)$ isotopic space. Similarly, the pion triplet provides a vector representation. Quarks transform as the fundamental representation of $SU_C(3)$. The implicit use of isotopic charge space occurs throughout the "Standard Model", GUT's, etc. It is thus not surprising to extend it to string theories. So we are not logically bound to the view of relativistic strings as some higher dimensional objects which must undergo "dimensional compactification." These models may be regarded as four-dimensional models whose self-consistency forces the presence of internal spaces to describe color and flavor.

As our notation indicates, the coordinates for isotopic charge space come in two varieties, $\phi_L^{\hat{\alpha}}$ and $\phi_R^{\hat{I}}$. The first of these is associated with a "left-mover" isotopic space and the second with a "right-mover" isotopic space. In general, those associated with $\phi_L^{\hat{\alpha}}$ are not necessarily related to those associated with $\phi_R^{\hat{I}}$. In formulations which make manifest the internal symmetry groups, these isotopic coordinates are valued in either group manifolds or coset spaces. Without manifest realization of the internal symmetry groups, $\phi_L^{\hat{\alpha}}$ and $\phi_R^{\hat{I}}$ may be valued in the root spaces, Cartan subspaces, or maximal compact subgroups associated with the symmetry groups. Alternately, the isotopic charge coordinates may be replaced all together by two-dimensional fermion fields. The fact that so many different ways exist to represent the isotopic space is a consequence of "bosonization" (in the technical sense) of two dimensional field theories.

After the intial consideration of the problem of heterotic string compactification [7], there appeared to be a myriad of different techniques;

(a) *Calabi – Yau*
(b) *Calabi – Yau with torsion*
(c) *Covariant lattices*
(d) *Orbifolds*
(e) *Free fermions*
(f) *Asymmetric orbifolds*
(g) *Chiral WZNW coset models*

with which to accomplish this goal. I will now present a suggestive argument to the effect that the last one of these realizes the philosophy described in the previous paragraphs. The simplest way to give the argument is present it in the form of a pictogram.

"Tree of Compactification Techniques"

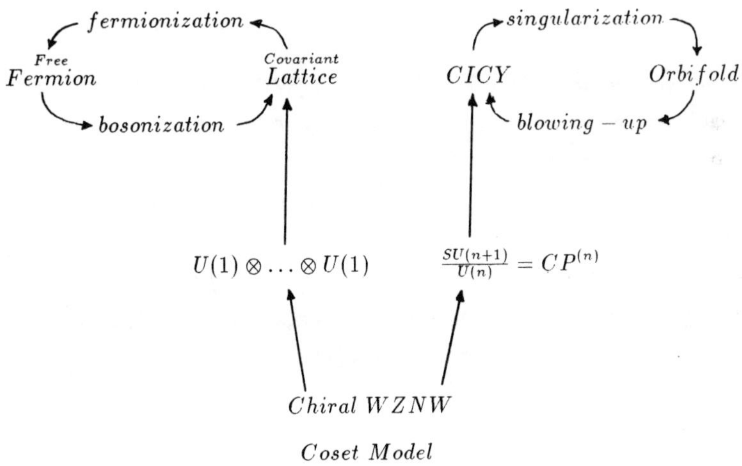

The viewpoint, I am suggesting is that the isotopic charge space coordinates, $\phi_L^{\hat{\alpha}}$ and $\phi_R^{\hat{I}}$, are the basic variables which should appear in a chiral WZNW coset space model. The relations to all other compactification techniques should occur as specializations or through bosonization. In the table above, I am proposing that the relation to Calabi-Yau compactification ought to be begin with the identification between certain coset spaces with the complex projective spaces upon which many of the explicit constructions of Calabi-Yau depend [8].

The basic two-dimensional field required is the chiral boson. This is a two dimensional scalar field which is "chiral" (i.e. only moves in one direction). An action for a single right-moving chiral boson or "righton" was suggested [9] some time ago.

$$S_R = -\tfrac{1}{2}\int d^2\sigma[\ (\partial_{++}\phi_R)(\partial_{--}\phi_R)\ +\ \lambda_{--}{}^{++}(\partial_{++}\phi_R)^2\ ]\ . \qquad (3.1)$$

Variation with respect to $\lambda_{--}{}^{++}$ and ϕ_R produce the equations of motion

$$\frac{\delta \mathcal{L}_R}{\delta \lambda_{--}{}^{++}} = 0\ ,\ \rightarrow (\partial_{++}\phi_R)^2 = 0\ ,\ \rightarrow \partial_{++}\phi_R = 0\ ,$$

$$\frac{\delta \mathcal{L}_R}{\delta \phi_R} = 0\ ,\ \rightarrow \partial_{--}\partial_{++}\phi_R + \partial_{++}(\lambda_{--}{}^{++}\partial_{++}\phi_R) = 0\ . \qquad (3.2)$$

The solution to the first equation is clearly of the form $\phi_R(\sigma,\tau) = \phi_R(\sigma - \tau)$ and as a consequence, the second equation is also satisfied. Since no equation of motion arises for $\lambda_{--}{}^{++}$, it must be a redundant degree of freedom. This is born out by the presence of a "Siegel" symmetry

$$\delta\phi_R = \xi^{++}(\partial_{++}\phi_R)\ \ ,\ \ \delta\lambda_{--}{}^{++} = -\partial_{--}\xi^{++} - \lambda_{--}{}^{++}\overleftrightarrow{\partial}_{++}\xi^{++}\ \ , \qquad (3.3)$$

which leaves the action invariant. Correspondingly there is an action for a left-moving chiral boson or "lefton" obtainable by simply interchanging all plus and minus signs in the discussion above.

The righton action may be quantized by using a path integral generating functional

$$\mathcal{Z}(J) = \int[\mathcal{D}\phi_R][\mathcal{D}\lambda_{--}{}^{++}]exp\{i(\ S_R + \int d^2\sigma\phi_R J\ )\} \qquad (3.4)$$

and since $\lambda_{--}{}^{++}$ is a gauge field, the standard Faddeev-Popov procedure is used. The simplest gauge choice is $\lambda_{--}{}^{++} = 0$ and implementing this leads to an effective action of the form

$$\mathcal{Z}(J) = \int[\mathcal{D}\phi_R][\mathcal{D}b_{++}{}^{--}][\mathcal{D}c^{++}]exp\{i(\ S_{R,eff} + \int d^2\sigma\phi_R J\ )\}\ ,$$

$$S_{R,eff} = \int[-\tfrac{1}{2}(\partial_{++}\phi_R)(\partial_{--}\phi_R) + ib_{++}{}^{--}(\partial_{--}c^{++})]\ . \qquad (3.5)$$

This is now exactly like the quantization of the "holomorphic half" of two-dimensional gravity. Following the result first given by Polyakov, we know that there is an apparent anomaly which can be expressed using a background field formalism as

$$\mathcal{A} = -\tfrac{1}{96\pi}(1-26)\int d^2\sigma \lambda_{--}^{(B)++} \frac{(\partial_{++})^4}{\Box^2} \lambda_{--}^{(B)++} \quad , \tag{3.6}$$

where $\lambda_{--}^{(B)++}$ is a background field. The factor of one comes from a one-loop Feynman diagram with ϕ_R in the loop. The factor of -26 is the contribution from the $b-c$ fields. This anomaly has been referred to as the "Siegel" anomaly to distinguish it from the usual Virasoro anomaly. For a long time this appeared to prevent a consistent quantization of this system.

One point of confusion surrounding "Siegel" symmetry is whether these are really the same as chiral diffeomorphisms. The answer is negative. One way to see this is to note the difference in the transformation laws of the "Siegel" gauge field in (3.3) and the corresponding zweibein prepotential in (2.7). But let us consider the general structure of a theory which possesses both "Siegel" symmetry and local reparametrization symmetry. In general, an action which realizes both of these will be of the form,

$$\mathcal{S} = \int d^2\sigma V^{-1} \mathcal{L}(\lambda_{--}{}^{++}, \lambda_{++}{}^{--}, \psi) \quad , \tag{3.7}$$

where ψ denotes all other fields in the action. Variation with respect to $h_{++}{}^{--}$ and $h_{--}{}^{++}$ produces the usual Virasoro constraints. Variation with respect to $\lambda_{++}{}^{--}$ and $\lambda_{--}{}^{++}$ produce an independent set of Virasoro-like constraints. In fact, if n sets of Siegel gauge fields are introduced, $\lambda^i_{++}{}^{--}$ and $\lambda^i_{--}{}^{++}$, the resulting theory will produce $n+1$ Virasoro-like constraints. It is clear that the coefficient multiplying the action in (3.6) defines a Siegel central charge $\tilde{\nu}_L$ analogous to ν_L the usual central charge in (2.14) or (2.24). So in a theory which involves Siegel gauge fields, not only must the contributions to the pure Virasoro central charge coming from all quantum fields and ghosts vanish, a total of $\frac{1}{2}n(n+1)-1$ additional pure Siegel and mixed Siegel-Virasoro central charges must also vanish!

However, the chiral boson action is not the only example of a system which possesses "Siegel" symmetry. Consider the following action for a fermionic field φ_+,

$$S_{NM} = -i\tfrac{1}{2}\int d^2\sigma [\, \varphi_+ \partial_{--} \varphi_+ + \lambda_{--}{}^{++} \varphi_+ \partial_{++} \varphi_+ \,] \quad . \tag{3.8}$$

This action is invariant under a "Siegel" transformation. The equations of motion are

$$\frac{\delta \mathcal{L}_{NM}}{\delta \lambda_{--}{}^{++}} = 0 \ , \ \rightarrow \varphi_+(\partial_{++}\varphi_+) = 0 \ ,$$

$$\frac{\delta \mathcal{L}_{NM}}{\delta \varphi_+} = 0 \ , \ \to \partial_{--}\varphi_+ + \tfrac{1}{2}(\partial_{++}\lambda_{--}{}^{++})\varphi_+ = 0 \ . \tag{3.9}$$

Multiplying the second by φ_+, we learn $\varphi_+\partial_{--}\varphi_+ = 0$ and these two equations together imply that derivatives of φ_+ must be at most proportional to a constant two-vector times φ_+. But there are no constant two-vector available thus derivatives of φ_+ must vanish! So classically, φ_+ describes a constant which we take to be zero. However, we can now add the action of (3.8) to (3.1) and calculate again the generating functional and the anomaly. The important change in the anomaly is that the coefficient $(1-26)$ becomes $(1+\frac{1}{2}-26)$. Adding fifty of these spinorial no-mover fields to S_R in (3.1) leads to an absence of the anomaly described in (3.6)! This is the Hull mechanism for cancelling the "Siegel anomaly." It is analogous to the anomaly cancellation mechanism between quarks and leptons used in spontaneously broken Weak Interaction models. The no-mover ("noton") in (3.8) was the first example discovered [10]. It is not the only such system. It has been shown by Hull [11] that many such systems exist [12]. These other notons are required for the consistent quantization of the most general actions which contain chiral bosons.

The generating functions derived from two distinct bosonic actions may, in fact, be exactly the same. Similarly, the generating function of a fermionic theory may be exactly the same as that of a bosonic theory. The process of going from one bosonic theory to another which possesses the same generating function is known as "bosonization." Going from a fermionic theory to an equivalent bosonic theory is also known as "bosonization." Finally, going from a bosonic theory to an equivalent fermionic one is known as "fermionization." A lagrangian example of fermionization is to go from the action in (3.1) to the action of a single complex Weyl-spinor or a Majorana-Weyl doublet spinor $\lambda_-^{\hat{I}}$ with $\hat{I} = 1, 2$. Thus under fermionization we have

$$S_R \to S_F \quad , \quad S_F = i\tfrac{1}{2}\int d^2\sigma V^{-1}\lambda_-^{\hat{I}}\mathcal{D}_{++}\lambda_-^{\hat{I}} \ . \tag{3.10}$$

Going the reverse way from the action immediately above to the bosonic action in (3.1) is an example of "abelian bosonization." There is another more general type of bosonization [13] known as "non-abelian bosonization." To see this we first extend the range of the $\hat{I}$ index on $\lambda_-^{\hat{I}}$. Generalizing the result of Witten to the case of "non-abelian chiral bosonization", is the action below

$$\begin{aligned} S_R(u) = -\frac{1}{4\pi}\int d^2\sigma V^{-1} Tr\{ \, (\mathcal{D}_{++}u^{-1})(\mathcal{D}_{--}u) + \lambda_{--}{}^{++}(\mathcal{D}_{--}u^{-1})(\mathcal{D}_{--}u) \\ + \int_0^1 dy(\tilde{u}^{-1}\frac{d}{dy}\tilde{u})[\, (\mathcal{D}_{++}\tilde{u}^{-1})(\mathcal{D}_{--}\tilde{u}) - (\mathcal{D}_{--}\tilde{u}^{-1})(\mathcal{D}_{++}\tilde{u}) \,] \, \} \ . \end{aligned} \tag{3.11}$$

where $u \equiv exp[i\phi^{\hat{a}}(\tau,\sigma)t_{\hat{a}}]$. Here $t_{\hat{a}}$ denotes the generators of a compact Lie algebra where $\hat{a} = 1,\ldots,d_G$, $[t_{\hat{a}},\ t_{\hat{b}}] = i f_{\hat{a}\hat{b}}{}^{\hat{c}} t_{\hat{c}}$, $f_{\hat{a}\hat{b}\hat{c}} f^{\hat{a}\hat{b}}{}_{\hat{d}} = c_2 \delta_{\hat{c}\hat{d}}$, and $Tr(t_{\hat{a}}t_{\hat{b}}) = 2k\delta_{\hat{a}\hat{b}}$. The quantity k is known as the level number. However, to correspond to a bosonization, the action of (3.11) must produce the same anomaly as (3.10). The anomaly associated with (3.11) is known to be [14] given by $\nu_R = \frac{d_G}{1+\frac{c_2}{2k}}$ and for any group such that this is equal to one this represents a bosonization of (3.10). More generally by starting from m fermions a nonabelian bosonization may be achieved for various groups. Had the starting point been a set of left-moving spinors, their non-abelian chiral bosonization leads to

$$S_L(u) = -\frac{1}{4\pi}\int d^2\sigma V^{-1} Tr\{\ (\mathcal{D}_{++}u^{-1})(\mathcal{D}_{--}u) + \lambda_{++}{}^{--}(\mathcal{D}_{--}u^{-1})(\mathcal{D}_{--}u)$$
$$- \int_0^1 dy(\tilde{u}^{-1}\frac{d}{dy}\tilde{u})[\ (\mathcal{D}_{++}\tilde{u}^{-1})(\mathcal{D}_{--}\tilde{u}) - (\mathcal{D}_{--}\tilde{u}^{-1})(\mathcal{D}_{++}\tilde{u})\ ]\ \}\ . \tag{3.12}$$

There is yet one more generalization of the actions in (3.11) and (3.12) which will be important. This corresponds to replacing the Lie algebras by coset spaces and represents the lagrangian field theory equivalent of the GKO construction [15]. The basic idea is that the group generators $t_{\hat{a}}$ which represent a group G are replaced by the elements of a coset space $\frac{G}{H}$ where H is some subgroup of G. The introduction of the subgroup implies the introduction of its own level number k' and second Casimir c'_2. The form of the action is still that in (3.11) and (3.12) but the elements u are only defined over $\frac{G}{H}$. Following Park [16] the extended elements $\tilde{u}$ are still valued in G, but their boundary condition is changed so that $\tilde{\phi}^{\hat{a}}(\tau,\sigma,y=1)$ is only valued in $\frac{G}{H}$. The central charge associated with the resulting righton WZNW coset action is

$$\nu_R = \frac{d_G}{1+\frac{c_2}{2k}} - \frac{d_H}{1+\frac{c'_2}{2k'}} \quad . \tag{3.13}$$

A similar result is true for the lefton version.

We are now in position to write explicit actions which describe a bosonic string with spacetime coordinate $X^{\underline{m}}$ and coset space coordinates $\phi_L^{\hat{\alpha}}$ and $\phi_R^{\hat{I}}$. The appropriate action for this is given by

$$\mathcal{S} = S_B + S_L(u_L) + S_R(u_R) + S_N\ ,$$
$$u_R \equiv exp[i\phi_R^{\hat{I}}t_{\hat{I}}]\quad ,\quad u_R \equiv exp[i\phi_L^{\hat{\alpha}}t_{\hat{\alpha}}]\quad , \tag{3.14}$$

where S_B is given by (2.16), $S_R(u_R)$ is given by (3.11) and $S_L(u_L)$ is given by (3.12). The term S_N represents an appropriately chosen noton action which is required to cancel the "Siegel" anomaly. (See [11,12] for more details.) This action provides a manifest realization of the symmetries of a D-dimensional spacetime manifold $\mathcal{M}_D$, lefton coset space $\frac{G_L^k}{H_L^{k'}}$ and righton coset space $\frac{G_R^k}{H_R^{k'}}$. This is an intrinsic definition of a compactified string. It is totally defined in terms of the world sheet and the maps described by $X^{\underline{m}}$, $\phi_L^{\hat{\alpha}}$ and $\phi_R^{\hat{I}}$. The absence of anomalies simply acts as a veto to either accept or reject as consistent models various choices of the spacetime dimension, lefton coset space, righton coset space and level numbers

The action in (3.14) represents the string propagating freely through configuration and isotopic charge space. The final point concerning this action is that it provides a universal way to introduce the coupling of a compactified string to all of its massless background fields. This is done by using (3.14) as the starting point in the construction of a nonlinear σ-model. For the graviton $e_m{}^{\underline{a}}(X)$, axion $b_{ab}(X)$, and dilaton $\Phi(X)$ background fields, the coupling to the string occurs through the usual modified nonlinear σ-model [17]. Defining $\Pi_{\pm\pm}{}^a \equiv V_{\pm\pm}{}^m(\partial_m X^m)e_{\underline{m}}{}^a(X)$, we have

$$S(e_{\underline{m}}{}^{\underline{a}},\ b_{\underline{ab}},\ \Phi) = \\ \frac{1}{2\pi\alpha'}\int d^2\sigma V^{-1}[-\tfrac{1}{2}(\eta_{\underline{ab}} + b_{\underline{ab}}(X))\Pi_{++}{}^{\underline{a}}\Pi_{--}{}^{\underline{b}} + \alpha'\Phi(X)r(V)\ ] \ . \tag{3.15}$$

The remaining fields of the most general compactified string are a set of left gauge-group vectors $A_{\underline{m}}{}^{\hat{\alpha}}$, right gauge-group vectors $A_{\underline{m}}{}^{\hat{I}}$ and internal scalars $\Phi_{\hat{\alpha}\hat{I}}$. To show the explicit coupling of these to the string and coset space variables requires the introduction of the lefton-righton Thirring model [18]. The essential point concerning this model is that it is the non-abelian chirally bosonized version of the usual fermionic massless Thirring model in the presence of different left and right gauge fields. Before gauge fixing, the form of this model is rather complicated (see ref. [18] for details). However, in the gauge $\lambda_{--}{}^{++} = \lambda_{++}{}^{--} = 0$, the action dramatically simplifies to

$$\begin{aligned} S_{Coupled} =& S_L(u_L) + S_R(u_R) \\ &+ i\frac{1}{\sqrt{2\pi\alpha'}}\int d^2\sigma V^{-1} Tr\{\ 2(u_L^{-1}\mathcal{D}_{++}u_L)\Pi_{--}{}^{\underline{a}}A_{\underline{a}}{}^{\hat{\alpha}}(X)t_{\hat{\alpha}} \\ &\qquad + 2(u_R^{-1}\mathcal{D}_{--}u_R)\Pi_{++}{}^{\underline{a}}A_{\underline{a}}{}^{\hat{I}}(X)t_{\hat{I}} \\ &\qquad + i4\sqrt{2\pi\alpha'}(u_L^{-1}\mathcal{D}_{++}u_L)(u_R^{-1}\mathcal{D}_{--}u_R)\Phi^{\hat{\alpha}\hat{I}}(X)t_{\hat{\alpha}}t_{\hat{I}}\ \} \ . \end{aligned} \tag{3.16}$$

where the first two terms are also evaluated in the same gauge. In taking the traces, it should be noted that $t_{\hat{\alpha}}$ and $t_{\hat{I}}$ are in two different matrix spaces. I should mention that the extension of this action to supergroups has not been investigated.

(IV.) Superspace Geometry for NSR Heterotic Strings

The NSR formulation of string theories basically amounts to adding the requirement of world sheet supersymmetry to the previous discussion. The most efficient way to do this is simply to work directly with a two dimensional superspace and superfields. Let me also point out that the superfield discussion below is carried out for "off-shell" (1,0) superspace. In an "off-shell" superspace, no equations of motion are imposed initially. (This will be made clear in our discussion.) This is distinguished from the "on-shell" approach usually found in discussions of superconformal field theory [19]. The reason this distinction is important is that it allows the effect of higher genus supermodular parameters to be incorporated easily. The usual on-shell approach is best suited to the genus zero theory.

The (1,0) supergravity covariant derivative ∇_A, $(\nabla_+, \nabla_{++}, \nabla_{--})$ is of the form $\nabla_A = E_A{}^M D_M + \omega_A \mathcal{M}$ where $E_A{}^M$ and ω_A are the zweibein and spin connection superfields respectively and $D_M \equiv (D_+, \partial_{++}, \partial_{--})$. Torsion and curvature super tensors are defined by graded commutation $[\nabla_A, \nabla_B\} \equiv T_{AB}{}^C \nabla_C + R_{AB}\mathcal{M}$, subject to constraints [20] which imply,

$$[\nabla_+, \nabla_+\} = i2\nabla_{++} \quad , \quad [\nabla_+, \nabla_{++}\} = 0 \quad ,$$

$$[\nabla_+, \nabla_{--}\} = -i2\Sigma^+\mathcal{M} \quad , \quad [\nabla_{++}, \nabla_{--}\} = -(\ \Sigma^+\nabla_+ + \mathcal{R}\mathcal{M}) \quad . \qquad (4.1)$$

The constraints on ∇_A can be completely solved in terms of five superfield prepotentials $H_+{}^{--}$, $H_{--}{}^{++}$, H^+, L, and Ψ. These are the superfields which generalize the prepotentials in (2.2). The vielbein superfield can be expressed as (c.f. (2.2))

$$E_+ = e^{\frac{1}{2}H^+D_+} e^{\frac{1}{2}L} \Psi^{-\frac{1}{2}} \hat{E}_+ e^{-\frac{1}{2}H^+D_+} \quad ,$$

$$E_{++} = e^{\frac{1}{2}H^+D_+} e^{L} \Psi^{-1} [\ \hat{E}_{++} - i\tfrac{1}{2}(\hat{C}_{+,--}{}^{--} - 2(\hat{E}_+ ln\Psi))\hat{E}_+] e^{-\frac{1}{2}H^+D_+} \quad ,$$

$$E_{--} = e^{\frac{1}{2}H^+D_+} e^{-L} \Psi^{-1} [\ \hat{E}_{--} + i\tfrac{1}{2}\hat{C}_{+,--}{}^{++}\hat{E}_+] e^{-\frac{1}{2}H^+D_+} \quad ,$$

$$\hat{E}_+ \equiv D_+ + H_+{}^{--}\partial_{--} \quad , \quad \hat{E}_{--} \equiv \partial_{--} + H_{--}{}^{++}\partial_{++} \quad ,$$

$$\hat{E}_{++} \equiv \partial_{++} - i[D_+H_+{}^{--} + H_+{}^{--}\partial_{--}H_+{}^{--}]\partial_{--} \equiv \partial_{++} + H_{++}{}^{--}\partial_{--} \quad ,$$

$$E^{-1} \equiv [sdet(E_A{}^M)]^{-1} = \Psi^{-\frac{3}{2}}(1 \cdot e^{\frac{1}{2}H^+\overleftarrow{D}_+})^{-1} e^{\frac{1}{2}L}[1 - H_{--}{}^{++}H_{++}{}^{--}]^{-1} \quad , \qquad (4.2)$$

where $[\hat{E}_A, \hat{E}_B\} = \hat{C}_{A,B}{}^C \hat{E}_C$. The spin connection ω_A can be solved for in terms of $E_A{}^M$ and its first derivative as in ordinary general relativity. The generalizations of (2.7) and (2.8) are also known [4]. The supermetric definition which generalizes (2.3) is expressed as $(ds)^2 = F^{++}F^{--}$ with

$$F^{++} = \Delta^{-1}\Psi e^{-L}[\ d\hat{z}^{++} - d\hat{z}^{--}H_{--}{}^{++} + d\hat{z}^{+}H_+{}^{--}H_{--}{}^{++}\ ] \ ,$$

$$F^{--} \;=\; \Delta^{-1}\Psi e^{L}[\, d\hat{z}^{--} \;-\; d\hat{z}^{++}H_{++}{}^{--} \;-\; d\hat{z}^{+}H_{+}{}^{--}\,] \;. \tag{4.3}$$

with $d\hat{z}^M \equiv (d\zeta^+, d\sigma^{++} - i\zeta^+ d\zeta^+, d\sigma^{--})$. This result for the invariant line element is valid only for the gauge choice $H^+ = 0$.

Before carrying out any gauge fixing the gauge symmetries are superdiffeomorphism (parameters K^+, K^{++} and K^{--}), super-Weyl transformations (parameter K_W) and super-Lorentz transformations (parameter K_L). However, the "flat superspace" gauge relates most of these parameters as was found in (2.11). To see how this happens, it is sufficient to first consider then linearized theory and then to impose the condition of flat superspace.

For $(1,0)$ or heterotic superfield supergravity, the linearized supergravity covariant derivative takes the form $\nabla_A = D_A + H_A{}^B D_B + \omega_A \mathcal{M}$ here $H_A{}^B \equiv$ linear super vielbein, and $\omega_A \equiv$ linearized Lorentz connection (see [4] for the explicit expressions). The gauge transformations (local superdiffeomorphisms) of ∇_A are simply given by $\delta\nabla_A = [K, \nabla_A\}$ where K is the supervector field defined by

$$K = K^+ D_+ + K^{++}\partial_{++} + K^{--}\partial_{--} \;\;, \tag{4.4}$$

with $K^A \equiv (K^+, K^{++}, K^{--})$ being an arbitrary superfunction of the supercoordinates. The variations of $H_A{}^M$ are (with $H^+ \equiv H_+{}^{++}$)

$$\delta H^+ = -D_+K^{++} - i2K^+ \;,\;\; \delta H_+{}^{--} = -D_+K^{--} \;,\;\; \delta H_{--}{}^{++} = -\partial_{--}K^{++} \;, \tag{4.5}$$

Imposing the condition $H^+ = 0$ implies that an algebraic relation between K^+ and K^{++}. Similarly imposing $H_+{}^{--} = 0$ and $H_{--}{}^{++} = 0$ imply that $D_+K^{--} = 0 \rightarrow \partial_{++}K^{--} = 0$ or $K^{--} = K^{--}(\sigma^{--})$ and $\partial_{--}K^{++} = 0$ or $K^{++} = K^{++}(\zeta^+, \sigma^{++})$, respectively. So for the flat space gauge, the gauge group described by (4.4) splits into two commuting pieces

$$\begin{aligned} K^{Tot} &= K^L(K^{++}) + K^R(K^{--})\,, \\ K^L(K^{++}) &= -i\frac{1}{2}(D_+K^{++})D_+ + K^{++}\partial_{++}\,, \quad K^{++} = K^{++}(\zeta^+, \sigma^{++})\,, \\ K^R(K^{--}) &= K^{--}\partial_{--}\,, \quad K^{--} = K^{--}(\sigma^{--})\,. \end{aligned} \tag{4.6}$$

Choosing functions to be periodic in bosonic coordinates implies K^L and K^R are the generators of $sDiff\; S^1$ and $Diff\; S^1$ respectively. For $K^R(X^{--})$ this is obvious but for K^L, it is illuminating to see this in a bit more detail. Calculating a commutator leads to

$$\left[K^L(X^{++}),\, K^L(Y^{++})\right] = K^L(Z^{++})\,, \tag{4.7}$$

$$Z^{++} = X^{++}\partial_{++}Y^{++} - Y^{++}\partial_{++}X^{++} - i\frac{1}{2}(D_+X^{++})(D_+Y^{++}) , \qquad (4.8)$$

Defining components by

$$X^{++} = \xi + i\zeta^+\psi , \quad Y^{++} = \eta + i\zeta^+\chi , \quad Z^{++} = \alpha + i\zeta^+\beta , \qquad (4.9)$$

leads to the results

$$\alpha = \xi\eta' - \eta\xi' + i\frac{1}{2}\psi\chi , \quad \beta = (\xi\chi' - \frac{1}{2}\xi'\chi) - (\eta\psi' - \frac{1}{2}\eta'\psi) , \qquad (4.10)$$

where prime denotes differentiation with respect to the argument of the function. Up to differences in normalizations, this is the algebra of $sDiff\ S^1$ noted by Kirillov [21].

The algebra in (4.7) is a purely classical results. Upon quantization these are modified by Schwinger terms. (These have been calculated in the (1,0) case previously [22].) Related to these are a two-cocycle, $a^{(R)}$, and a super two-cocycle, $a^{(L)}$, defined by

$$a^{(R)}(X,Y) = \int d\sigma^{--} \left[(\partial_{--}X^{--})(\partial^2_{--}Y^{--}) - (\partial_{--}Y^{--})(\partial^2_{--}X^{--})\right] ,$$

$$a^{(L)}(X,Y) = -i\int d\sigma^{++}d\zeta^- \left[(\partial_{++}X^{++})(\partial_{++}D_+Y^{++}) - (\partial_{++}Y^{++})(\partial_{++}D_+X^{++})\right] . \qquad (4.11)$$

The first of these is already explicit since "right-handed" quantities (i.e. depending only on σ^{--}) are non-supersymmetric. The component level expression for $a^{(L)}$ is simply

$$a^{(L)}(X,Y) = \int d\sigma^{++} \left[\xi'\eta'' - \eta'\xi'' + i2\psi'\chi'\right] . \qquad (4.12)$$

These cocycles modify the usual algebra of the K^R and K^L generators at the quantum level,

$$\left[K^R(X) , K^R(Y)\right] = K^R(Z) - i\frac{\nu_R}{24\pi}\, a^{(R)} , \qquad (4.13)$$

and

$$\left[K^L(X), K^L(Y)\right] = K^L(Z) - i\frac{\nu_L}{24\pi}\, a^{(L)} , \qquad (4.14)$$

with $Z^{--} = X^{--}\partial_{--}Y^{--} - Y^{--}\partial_{--}X^{--}$.

Now I can discuss the difference between an off-shell versus an on-shell superspace. Note that K^R and K^L do not depend on all of the coordinates of the (1,0) superspace. This is characteristic of an on-shell superspace. In an off-shell superspace, which is a appropriate for a lagrangian formalism, such restrictions are only allowed to arise as equations of motion or gauge conditions. For many discussions of of superconformal field theory, such restrictions are taken as definitions. We also see how these definitions are especially related to the "flat space" gauge.

Before discussing quantization, it is perhaps appropriate here to discuss instead the reduction of superfield results to the component level. In the appendix on notation and conventions, the "slash" operation, which merely takes the $\zeta^+ = 0$ limit of a superfield, is defined. As examples of this there are definitions,

$$X^{\underline{m}}(\tau,\sigma) \equiv X^{\underline{m}}(\zeta^+,\tau,\sigma)| \quad , \quad \psi^{\underline{m}}(\tau,\sigma) \equiv \nabla_+ X^{\underline{m}}(\zeta^+,\tau,\sigma)|$$

$$\lambda_-{}^{\hat{I}} \equiv \eta_-{}^{\hat{I}}| \quad , \quad F^{\hat{I}} \equiv -i\nabla_+\eta_-{}^{\hat{I}}| \quad . \tag{4.16}$$

This is the basis of the "projection" technique [23] that provides the most efficient way to obtain component results. For example, the (1,0) NSR component-level heterotic superstring action may be obtained using the density projector formula the appendix. The superfield action heterotic string [24] action is simply,

$$S_H = \int d^2\sigma \, d\zeta^- \; E^{-1} i\tfrac{1}{2}[\; \eta_{\underline{mn}}(\nabla_+ X^{\underline{m}})(\nabla_{--}X^{\underline{n}}) + i\delta_{\hat{I}\hat{J}}\eta_-{}^{\hat{I}}\nabla_+\eta_-{}^{\hat{J}} \;] \quad . \tag{4.17}$$

(For convenience, I will neglect factors of $2\pi\alpha'$ in the subsequent discussion.) Next via the Liebnitz rule the derivative in the projection formula is applied, the algebra of the ∇_A operators is used and finally the slash operation is applied. This yields the component-level action

$$\int d^2\sigma \, V^{-1}\{\eta_{\underline{mn}}[\; -\tfrac{1}{2}(V_{++}X^{\underline{m}})(V_{--}X^{\underline{n}}) - i\tfrac{1}{2}\psi_+{}^{\underline{m}}\mathscr{D}_{--}\psi_+{}^{\underline{n}} - \psi_{--}{}^{+}\psi_+{}^{\underline{m}}(V_{++}X^{\underline{n}}) \;]$$

$$+ i\tfrac{1}{2}\lambda_-{}^{\hat{I}}\mathscr{D}_{++}\lambda_-{}^{\hat{I}} + \tfrac{1}{2}F^{\hat{I}}F^{\hat{I}} \;\} \quad . \tag{4.18}$$

The quantization of (1,0) supergravity just mimics the previous discussion for the bosonic theory. For example we may go directly to the Sp(2)-covariant (1,0) "flat space" gauge-fixed heterotic string. The generating functional at zero source value just takes the form,

$$\int [DE_A{}^M] e^{(iS_H)} \equiv \int [DH_+{}^{--}][DH_{--}{}^{++}][DH^+][DL][D\Psi] e^{(iS_H)} \quad . \tag{4.19}$$

Upon substitution of the solution from (4.2), this expression is found to depend on <u>all</u> of the prepotentials unlike the bosonic case. Only after appropriate field redefinitions of the forms of Lorentz, Weyl and K^+-diffeomorphism transformations can the action be bought to a form which depends only the conformal prepotentials. (This is permitted as long as there are 10 X-superfields and 32 $\eta^{\hat{I}}$-superfields.)

$$S_H = \int d^2\sigma \, d\zeta^- \; \Delta^{-1} i\tfrac{1}{2}[\; \eta_{\underline{mn}}(\hat{E}_+ X^{\underline{m}})(\hat{E}_{--}X^{\underline{n}}) + i\delta_{\hat{I}\hat{J}}\eta_-{}^{\hat{I}}\hat{E}_+\eta_-{}^{\hat{J}} \;] \quad , \tag{4.20}$$

where here $\Delta \equiv 1 - H_{--}{}^{++}H_{++}{}^{--}$. Now we impose the flat space gauge conditions following the $Sp(2)$-covariant technique described below (2.19). The action of the

$Sp(2)$-covariant BRST generators on $H_{--}{}^{++}$ and $H_{+}{}^{--}$ is defined through the usual BRST ansatz using the linearized gauge transformations in (4.5). So on the gauge fixing surface we just find,

$$\begin{aligned} S_{eff}|_{GF} \;=\; & \int d^2\sigma d\zeta^- \; i\tfrac{1}{2}[\; \eta_{\underline{mn}}(D_+X^{\underline{m}})(\partial_{--}X^{\underline{n}}) + i\delta_{\hat{I}\hat{J}}\eta_-{}^{\hat{I}}D_+\eta_-{}^{\hat{J}} \;] \\ & + \int d^2\sigma d\zeta^- \mathcal{Q}^{(\alpha)}\mathcal{Q}_{(\alpha)}[\; a_{--}{}^{++}H_+{}^{--} \;+\; a_{++}{}^{-}H_{--}{}^{++} \;] \;. \end{aligned} \tag{4.21}$$

The superfield background-quantum field technique [25] is the simple extension of the nonsupersymmetric one. It leads via the Adler-Rosenberg method to a superfield effective action which contains the supersymmetric version of Γ_{Anom}

$$\begin{aligned} \Gamma_{Anom} \;=\; & \frac{1}{96\pi}\int d^2\sigma\, d\zeta^- \;\{\; i\nu_L H_{--}^{(B)\;++}[\frac{(\partial_{++})^3}{\Box}D_+H_{--}^{(B)\;++}] \\ & + \nu_R H_{+}^{(B)\;--}[\frac{(\partial_{--})^4}{\Box}\;D_+H_{+}^{(B)\;--}]\;\} \;, \end{aligned} \tag{4.22}$$

and with a few simple one-loop calculations using the background-quantum field versions of (4.21) we calculate the values of $\nu_L \;=\; D(\frac{3}{2}) - 30 + 15$ and $\nu_R = D(1) + N_F(\frac{1}{2}) - 52 + 26$. So for $D = 10$ and $N_F = 32$ the anomalies vanish.

For (1,0) supergravity too, many gauge choices are available. The superspace generalizations of the conformal and light-cone gauges are all known. We may also attempt to generalize the Strebel gauge discussed for the bosonic theory in [6]. We first note that for the bosonic string (p_i^{++}, p_i^{--}) has the interpretation of being the location of the i-th puncture on a closed Riemann surface. It is natural to demand that to each such point there should correspond a constant Grassmann parameter π_i^+. In this way, $z_i^M \equiv (\pi_i^+, p_i^{++}, p_i^{--})$ becomes the location of the i-th "superpoint" puncture in (1,0) superspace. Using arguments of dimensionality and 2-d Lorentz covariance, there are two ansatže for the (1,0) super Strebel gauge. We may choose $H^+ = L = \Psi - \Delta = 0$ and

$$\begin{aligned} H_{--}{}^{++} &= \sum_i \Big[\frac{a_i}{(\sigma^{--} - p_i{}^{--})^2} + i\,\zeta^+\frac{\pi_i^+}{(\sigma^{--} - p_i{}^{--})}\Big] \;, \\ H_+{}^{--} &= \sum_i \Big[-i\tfrac{1}{2}\frac{a_i\pi_i^+}{(\sigma^{++} - p_i{}^{++})^2} + i\,\zeta^+\frac{a_i}{(\sigma^{++} - p_i{}^{++})^2}\Big] \;, \end{aligned} \tag{4.23}$$

but since the lowest component of $H_+{}^{--}$ is totally arbitrary there is also the simpler ansatz

$$H_+{}^{--} = i\sum_i \zeta^+\frac{a_i}{(\sigma^{++} - p_i{}^{++})^2} \;. \tag{4.24}$$

The complete coupling of the super modular parameters $(\pi_i^+, p_i^{++}, p_i^{--})$ to matter fields is determined by substituting this gauge choice into the general action formulae discussed in (4.20). Again an indication of an off-shell superspace is the fact that the restricted coordinate dependence of the conformal prepotentials in (4.23) or (4.24) is just the result of a gauge choice.

As stated in the compactification or non-compactification section, a fermionic representation (as in (4.17)) of the fiber degrees of freedom does not yield a manifest realization of all internal symmetry groups. To find this requires a non-abelian bosonization of the fermionic action. Before considering the $D = 4$ case, it is useful to carry this out for the $D = 10$ theory especially in the case of the $E_8 \otimes E_8$ string. The 32 spinors in the heterotic string action are capable of providing a manifest realization of at most a $SO(16) \otimes SO(16)$ subgroup of the the full $E_8 \otimes E_8$. At this point in most discussions (see for instance [24]) the Frenkel-Kac construction and solitonic modes are invoked to realize the remainder of the group. This is not sufficient if we want to realize the symmetry group directly on the world sheet action of the string. However, there is another way to proceed by using the supersymmetric righton coordinates $\Phi_R^{\hat{I}}$ for the group manifold of $E_8 \otimes E_8$. But is the world sheet supersymmetry of the NSR formulation compatible with the "Siegel" symmetry? We need to first investigate this issue.

A superfield proposal for the (1,0) action of the righton multiplet was made sometime ago [26]. It takes the form,

$$S_R = i\tfrac{1}{2} \int d^2\sigma \, d\zeta^- \; [\, (D_+ \Phi_R)(\partial_{--} \Phi_R) \; + \; \Lambda_{--}{}^{++} (D_+ \Phi_R)(\partial_{++} \Phi_R) \,] \;\; , \qquad (4.25)$$

which at the component level leads to [10]

$$\begin{aligned} S_R \;=\; - \int d^2\sigma \tfrac{1}{2} [\, & (\partial_{++} \phi_R)(\partial_{--} \phi_R) \; + \; \lambda_{--}{}^{++} (\partial_{++} \phi_R)^2 \\ & + i\rho_+ (\partial_{--} \rho_+) \; + \; i\lambda_{--}{}^{++} \rho_+ (\partial_{++} \rho_+) \\ & - \lambda_{--}{}^{+} \rho_+ (\partial_{++} \phi_R) \,] \;\; , \end{aligned} \qquad (4.26)$$

after defining component fields by $\phi_R \equiv \Phi_R|$, $\rho_+ \equiv \nabla_+ \Phi_R|$, $\lambda_{--}{}^{+} \equiv i \nabla_+ \Lambda_{--}{}^{++}|$, $\lambda_{--}{}^{++} \equiv \Lambda_{--}{}^{++}|$. Setting all of the fermionic fields to zero clearly yields (3.1). Another interesting limit is provided by keeping only ρ_+ and $\lambda_{--}{}^{++}$ nonzero. This yields the action for the fermionic noton in (3.8). Here it is just the manifest realization of (1,0) supersymmetry which requires the noton! (We will consider the situation with super "Siegel" anomalies [27] shortly.) The final point of concern is whether (4.25) has a super version of the "Siegel" gauge transformation which leaves the action invariant? It can be proven that the action remains invariant up to surface terms under the variations

$$\delta\Phi_R = -i\xi^{++}D_+R_+ - i\tfrac{1}{2}(D_+\xi^{++})R_+ \quad , \quad R_+ \equiv D_+\Phi_R \ ,$$

$$\delta\Lambda_{--}{}^{++} = -\partial_{--}\xi^{++} + \xi^{++}\overleftrightarrow{\partial}_{++}\Lambda_{--}{}^{++} + i\tfrac{1}{2}(D_+\Lambda_{--}{}^{++})(D_+\xi^{++}) \ . \tag{4.27}$$

Now to address the issue of the super "Siegel" anomaly, the steps which led from (3.4) to (3.6) can be repeated here. The expression for the super "Siegel" anomaly is not surprisingly given by

$$\Gamma^{(L)}_{Siegel-Anom} = \tilde{\nu}_L(\frac{1}{96\pi})\int d^2\sigma\, d\zeta^- \ \{ \ i\Lambda^{(B)}_{--}{}^{++}[\frac{(\partial_{++})^3}{\Box}D_+\Lambda^{(B)}_{--}{}^{++} \ ] \ \} \ . \tag{4.28}$$

where $\tilde{\nu}_L = \frac{3}{2} - 15$. The factor of $\frac{3}{2}$ arises from the righton and the factor of -15 comes from the b-c ghost system arising from the gauge condition $\Lambda_{--}{}^{++} = 0$. Following the Hull mechanism for cancelling this anomaly requires that a noton sector should be added to (4.25). Since there is some confusion [28] in the literature concerning this matter, I will discuss it in some detail.

Hull has considered general systems of (1,0) supersymmetric noton multiplets [12]. As an independent basis, the most general action for these seems to contain some number N_W of "scalar" notons W, $N_{\tilde{\eta}}$ noton minus spinors $\tilde{\eta}_-$, $N_{\hat{B}}$ pair superfields $\hat{B}_+^{(s',r')} - \hat{C}^{(s',r')}$ and $N_{\tilde{B}}$ pairs $\tilde{B}_{--}^{(r,s)} - \tilde{C}^{(s,r)}$. The action is given by

$$\begin{aligned} S_N = \ & i\tfrac{1}{2}\int d^2\sigma d\zeta^-\Big\{ \ \tilde{\eta}_-(D_+ + \Lambda_+{}^{--}\partial_{--})\tilde{\eta}_- + \tilde{B}_{--}^{(r,s)}(D_+ + \Lambda_+{}^{--}\partial_{--})\tilde{C}^{(s,r)} \\ & + \hat{B}_+^{(r',s')}[\partial_{--} + \Lambda_{--}{}^{++}\partial_{++} - i\tfrac{1}{2}(D_+\Lambda_{--}{}^{++})D_+]\hat{C}^{(s',r')} \\ & - \tfrac{1}{2}[(s'-r')(\partial_{++}\Lambda_{--}{}^{++})\hat{B}_+{}^{(r',s')}\hat{C}^{(s',r')} - (s-r)\tilde{B}_{--}^{(r,s)}(\partial_{--}\Lambda_+{}^{--})\tilde{C}^{(s,r)}] \\ & + \frac{1}{1-i\Lambda^2}[\partial_{--}W + i\tfrac{1}{2}(D_+\Lambda_{--}{}^{++})D_+W][D_+W + iD_+(\Lambda_{--}{}^{++}\Lambda_+{}^{--}D_+W)]\Big\} \ . \end{aligned} \tag{4.29}$$

I have used a convention where the notation $\tilde{C}^{(s,r)}$ implies that $\tilde{C}^{(s,r)}$possesses s upper (+) indices and r upper (−) indices and similarly for the other superfields. The factor Λ^2 is defined by $\Lambda^2 \equiv \Lambda_{--}{}^{++}D_+\Lambda_+{}^{--}$. This action possesses both righton and lefton Siegel symmetries (see [29]).

To cancel the anomaly in (4.28) I choose $N_W = 9$. However, the scalar noton also has the 'other' anomaly due to the presence of the lefton Siegel gauge superfield $\Lambda_+{}^{--}$. For convenience sake I can choose $N_{\hat{B}} = 0$ and $N_{\tilde{B}} = 0$, then the other Siegel anomaly is given by

$$\Gamma^{(R)}_{Siegel-Anom} = \tilde{\nu}_R(\frac{1}{96\pi})\int d^2\sigma\, d\zeta^- \ \{\Lambda_+^{(B)}{}^{--}[\frac{(\partial_{--})^4}{\Box}D_+\Lambda_+^{(B)}{}^{--}] \ \} \ , \tag{4.30}$$

where $\tilde{\nu}_R = (9 + N_{\tilde{\eta}}(\frac{1}{2}) - 26)$. Here the factor of 9 comes from the scalar notons, the factor of $N_{\tilde{\eta}}(\frac{1}{2})$ comes from the minus spinor notons and the factor of -26 arises from the ghosts due to the gauge fixing condition $\Lambda_+{}^{--} = 0$. Picking $N_{\tilde{\eta}} = 34$ cancels the anomaly. This is just one solution to the problem of the "Siegel" anomaly. There are an infinite number of others.

The situation for (1,0) lefton multiplets is even simpler. The action for such a multiplet is given by [9]

$$S_L \;=\; i\tfrac{1}{2} \int d^2\sigma \, d\zeta^- \; [\; (D_+\Phi_L)(\partial_{--}\Phi_L) \;+\; \Lambda_+{}^{--}(\partial_{--}\Phi_L)^2 \;] \;, \tag{4.31}$$

which at the component level leads to [10]

$$S_L \;=\; -\int d^2\sigma \tfrac{1}{2}[\; (\partial_{++}\phi_L)(\partial_{--}\phi_L) \;+\; \lambda_{++}{}^{--}(\partial_{--}\phi_L)^2 + i\beta_+\partial_{--}\beta_+ \;] \;, \tag{4.32}$$

after using definitions similar to those below (4.26). (I have also used the Wess-Zumino gauge condition $\Lambda_+{}^{--}| = 0$ since the invariance of the action under (4.33) allows this.) The superfields in the action possess lefton Siegel-gauge variations

$$\delta\Phi_L = \xi^{--}\partial_{--}\Phi_L \;,$$

$$\delta\Lambda_+{}^{--} = -D_+\xi^{--} - \Lambda_+{}^{--}\partial_{--}\xi^{--} + \xi^{--}\partial_{--}\Lambda_+{}^{--} \;. \tag{4.33}$$

that leave the action invariant up to surface terms. Upon quantization of S_L it leads to the Siegel anomaly described by (4.28) but with $\tilde{\nu}_R = 1 - 26$. To cancel this, I simply add S_N to S_L and for the simplest case choose $N_W = N_{\hat{B}} = N_{\tilde{B}} = 0$ and $N_{\tilde{\eta}} = 50$. (This case is just like the corresponding bosonic theory in (3.8).) In general, with arbitrary systems of leftons and rightons, by appropriate choices of N_W, $N_{\hat{B}}$, $N_{\tilde{B}}$, $N_{\tilde{\eta}}$, s, r, s' and r' it is trivial to eliminate the Siegel anomaly.

In the above discussion, I have neglected the presence of (1,0) supergravity. This may cause the astute reader to wonder if this changes anything? It does not. But there is one point which is somewhat subtle. I have shown that the pure "Siegel" anomaly can be made to cancel by the addition of notons. There are also the mixed "Siegel-Virasoro" anomalies. The use of the background-quantum field formalism in the presence of (1,0) supergravity, implies a beautifully simple method [30,31] to see that they vanish. In the presence of a nontrivial supergravity background we are free to choose the gauge fixing conditions on the quantum Siegel gauge fields as

$$\Lambda_+^{(Q)\;--} = -H_+^{(B)\;--} \;\;, \;\; \Lambda_{--}^{(Q)\;++} = -H_{--}^{(B)\;++} \;. \tag{4.34}$$

This will insure the cancellation of all mixed "Siegel-Virasoro" anomalies because all possible vertices which could contribute to such diagrams vanish identically! This gauge choice also has another bonus. Using it shows that the ghosts which arise from the gauge fixing of Siegel gauge fields do not contribute to the usual Virasoro currents. An alternate way to view all of this is to note that if the gauge choice $\Lambda_{+}^{(Q)\,--} = \Lambda_{--}^{(Q)\,++} = 0$ is made, there are always counter terms whose forms are those determined as if we had used (4.34). The counterterms will eliminate all mixed anomalies and further remove any dependence that the usual Virasoro charges may have on Siegel-gauge ghosts. Thus, contrary to a number of statements in the literature[28,32], the ghosts coming from Siegel gauge fields do not effect the usual central charges of the super Virasoro algebra (4.13,4.14). Everything is consistent and the frequently reported demise of chiral bosons is seen to be greatly exaggerated.

Having seen that the "Siegel" symmetry of the righton admits a (1,0) extension implies that an abelian bosonization of two η-superfields into a single Φ_R superfield is consistent with the world sheet supersymmetry. We are thus able to bosonize the 32 spinor superfields into 16 abelian righton plus an appropriately chosen noton sector. But our goal of manifest realization of the $E_8 \otimes E_8$ symmetry requires a non-abelian bosonization. This is provided by (1,0) supersymmetric nonabelian rightons [33]. To describe this model,I introduce the generators, $t_{\hat{I}}$, of some compact Lie algebra satisfying

$$[t^{\hat{I}}, t^{\hat{J}}] = i f^{\hat{I}\hat{J}\hat{K}} t^{\hat{K}} \quad , \quad tr(t^{\hat{I}} t^{\hat{J}}) = 2k_R \delta^{\hat{I}\hat{J}} \quad . \tag{4.35}$$

Next I define the currents associated with the group elements constructed from $t_{\hat{I}}$,

$$(D_A R) R^{-1} = i R_A^{\hat{I}} t_{\hat{I}} \quad , \quad R \equiv exp(i \Phi_R^{\hat{I}} t_{\hat{I}}) \quad . \tag{4.36}$$

The action for the nonabelian (1,0) supersymmetric righton is given by [33]

$$\begin{aligned} S_R = -\frac{1}{4\pi} \int d^2\sigma d\zeta^- \{ & -2k_R R_+^{\hat{I}} R_{--}^{\hat{I}} + i2k_R \Lambda_{--}{}^{++} R_+^{\hat{I}} D_+ R_+^{\hat{I}} \\ & + i\frac{4}{3} k_R f_{\hat{I}\hat{J}\hat{K}} \Lambda_{--}{}^{++} R_+^{\hat{I}} R_+^{\hat{J}} R_+^{\hat{K}} \\ & + \int_0^1 dy [(\frac{d\tilde{R}}{dy} \tilde{R}^{-1})[(\partial_{--}(D_+ \tilde{R}\tilde{R}^{-1}) - (D_+(\partial_{--}\tilde{R}\tilde{R}^{-1})]] \} \quad . \end{aligned} \tag{4.37}$$

The final line in this action is the usual extension of the WZNW terms including the addition of the redundant coordinate y required to write such terms in closed form. On choosing the generators $t_{\hat{I}}$ to correspond to $E_8 \otimes E_8$, this corresponds to the manifest realization required. The 32 fermions in (4.17) can now be replaced by the 496 rightons.

The heterotic string action discussed previously describes a $D = 10$ heterotic string. As I argued for the bosonic string, the question of four dimensional heterotic strings can be thought of as finding world sheet actions for a fiber bundle of the structure $\mathcal{M}_{(4,N)} \otimes \mathcal{G}_L \otimes \mathcal{G}_R$. Here $\mathcal{M}_{(4,N)}$ refers to a four dimensional spacetime with N supersymmetries but the appearance of the spacetime supersymmetries is not manifest. The manifest realization of the internal symmetries can once again be accomplished by introducing coordinates $\Phi_L^{\hat{\alpha}}$ and $\Phi_R^{\hat{I}}$ on the groups or coset spaces $\mathcal{G}_L$ and $\mathcal{G}_R$, respectively. In the NSR approach we need only incorporate the world-sheet supersymmetry into our previous bosonic discussion. A far superior approach would consists of carrying out this type of construction within the context of a Green-Schwarz type of approach [34]. (I will make some comments about this in my final lecture.)

The obvious key ingredient needed is the nonabelian (1,0) lefton multiplet [33]. Its action is just

$$S_L = -\frac{1}{4\pi}\int d^2\sigma d\zeta^- \{ \; -2k_L L_+^{\hat{\alpha}} L_{--}^{\hat{\alpha}} - 2k_L \Lambda_+{}^{--} L_{--}^{\hat{\alpha}} L_{--}^{\hat{\alpha}}$$
$$- \int_0^1 dy [(\frac{d\tilde{L}}{dy}\tilde{L}^{-1})[(\partial_{--}(D_+\tilde{L}\tilde{L}^{-1}) - (D_+(\partial_{--}\tilde{L}\tilde{L}^{-1})]] \} \; , \tag{4.38}$$

where I have used the properties and definitions,

$$[t^{\hat{\alpha}}, t^{\hat{\beta}}] = i f^{\hat{\alpha}\hat{\beta}\hat{\gamma}} t^{\hat{\gamma}} \;\; , \;\; tr(t^{\hat{\alpha}} t^{\hat{\beta}}) = 2k_L \delta^{\hat{\alpha}\hat{\beta}} \;\; ,$$
$$(D_A L)L^{-1} = i L_A^{\hat{\alpha}} t_{\hat{\alpha}} \;\; , \;\; L \equiv exp(i\Phi_L^{\hat{\alpha}} t_{\hat{\alpha}}) \;\; . \tag{4.39}$$

Here $t_{\hat{\alpha}}$ are the generators of a compact Lie algebra which is not necessarily related to $t_{\hat{I}}$.

Thus the form of the most general four dimensional heterotic string action in the NSR formalism is

$$S_{Tot} = S_{SM} + S_L + S_R + S_N \;\; . \tag{4.40}$$

One of the simplest four dimensional models is the $N = 4, SO(44)$ heterotic string theory [35]. Here it is described by choosing $\mathcal{G}_R = SO(44)$ and as in all $N = 4$ models $\mathcal{G}_L = U(1)^6$. Another is the $E_8 \otimes E_8 \otimes E_6$ model which obviously corresponds to the different choice $\mathcal{G}_R = E_8 \otimes E_8 \otimes E_6$. Note the manifest symmetry requirement on these theories at the level of the world sheet action <u>absolutely requires</u> the existence of lefton multiplets. The presence of $N < 4$ space-time supersymmetry in these σ-models seems connected to the <u>consistent</u> appearance of coset spaces G/H in the righton sector of the (1,0) supersymmetric action. The "off-diagonal" elements of a coset space are associated with the appearance of spin-0 bosons that are not in the adjoint representation of a group. Spin-1 gauge bosons must always be in the adjoint representation. So the consistent inclusion of a coset space signals, at least, a number of $N = 2$ hypermultiplets. For these $N < 4$ theories, also the left coset $\mathcal{G}_L$ may be chosen to be a group other than $U(1)^6$. Although the discussion only considered group manifolds so far, the extension to coset spaces can be obtained [36]. This provides the (1,0) supersymmetric lagrangian realization of the GKO [29] construction. The actions are still those in (4.37,4.38) but the terms involving L and R are now valued in G/H. (This, of course, implies the freedom of choice of a set of level numbers.) The terms involving $\tilde{L}$ and $\tilde{R}$ are valued in all of G but at the boundary of y-space they are only nontrivial in G/H. At our present level of understanding, the appearance of $N = 1$ models seems to be equivalent to the requirement that a subsector of the lefton and righton action should provide a representation of world sheet $N = 2$ supersymmetry [3]. This is presently being studied in more detail [38].

The most general coupling of the massless background fields of four dimensional heterotic strings has only recently been obtained [29]. The form of this action (in the gauge $\Lambda_+{}^{--} = \Lambda_{--}{}^{++} = 0$) is the expected generalization of (4.40)

$$S_{Tot} = S_{SM} + S_L + S_R + S_N$$

$$+i2\int d^2\sigma\, d\zeta^-\ E^{-1}\Big[\ Tr\{(\nabla_+ L)L^{-1}\Pi_{--}{}^{\underline{a}}A_{\underline{a}}{}^{\hat{\alpha}}(X)t_{\hat{\alpha}} + (\nabla_{--}R)R^{-1}\Pi_+{}^{\underline{a}}A_{\underline{a}}{}^{\hat{I}}(X)t_{\hat{I}}$$

$$+i8(\nabla_{--}R)R^{-1}(\nabla_+ L)L^{-1}\Phi^{\hat{\alpha}\hat{I}}(X)t_{\hat{I}}t_{\hat{\alpha}}\ \}\Big]\ . \qquad (4.41)$$

where S_L and S_R are taken from (4.37) and (4.38) respectively in the limit of vanishing Siegel gauge fields. The base space action S_{SM} is just

$$S_{SM} = \int d^2\sigma d\zeta^- E^{-1}[i\tfrac{1}{2}(\eta_{\underline{ab}} + b_{\underline{ab}}(X))\Pi_+{}^{\underline{a}}\Pi_{--}{}^{\underline{b}}\ +\ \Phi(X)\Sigma^+\ ]\ ,$$

$$\Pi_+{}^{\underline{a}} \equiv (\nabla_+ X^{\underline{m}})e_m{}^{\underline{a}}(X)\ \ ,\ \ \Pi_{--}{}^{\underline{a}} \equiv (\nabla_{--}X^{\underline{m}})e_m{}^{\underline{a}}(X)\ . \qquad (4.42)$$

This model provides a starting point for the β-function method [17,39] of the derivation of the string effective action.

I hope to have convinced the reader that four dimensional strings and superstrings are not so very different from four dimensional field theory when symmetries are manifestly realized on the world sheet action of the string. In building the standard model, choices of the dimension of space, gauge groups, particle spectrum and coupling constants had to be made and matched by us to nature. (These are the "knobs" with which we tune the model into reality.) So it is similarly for strings that choices of dimension, lefton coset, righton coset, and level numbers must be made. In the standard model not all choices of gauge groups and particle spectrum are mathematically consistent (i.e. free of ABJ anomalies). For a four dimensional gauge theory with gauge group G and generators $t_{\hat{a}}$ this is summarized by the group-theoretical statement

$$\Sigma Tr(t_{\hat{a}}\{t_{\hat{b}}, t_{\hat{c}}\}) = 0 \quad . \tag{4.43}$$

In string theory, mathematical consistency in the form of anomaly freedom and its extension, modular invariance, act as a veto to some choices in the same way. The conditions for world-sheet ABJ anomaly freedom in the string are

$$\Sigma\nu_L = 0 \quad , \quad \Sigma\nu_L = 0 \quad . \tag{4.44}$$

But what we presently lack is a similar group-theoretical statement about modular invariance directly formulated in terms of $\mathcal{M}_{(4,N)} \otimes \mathcal{G}_L \otimes \mathcal{G}_R$. There should exist some function $\mathcal{F}$ such that when

$$\mathcal{F}(\mathcal{M}_{(4,N)}, \mathcal{G}_L, \mathcal{G}_R) = 0 \quad , \tag{4.45}$$

then the resulting string theory will be free of modular anomalies. Once this statement is discovered, then four dimensional model-building with string theories will be as simple as with ordinary gauge theories. Statements of this type are known[3], however, in the non-manifest symmetry realizations provided by Calabi-Yau manifolds, free fermions or covariant lattices [7]. Turning these into manifestly covariant statements is an unsolved problem.

[3] This is analogous to stating the anomaly-freedom condition of ordinary gauge theories in terms of the Dynkin diagrams, weights and roots of Lie groups. Imagine how complicated it would be to express (4.43) this way!

(V.) "Crazy Ideas" about Strings

During the course of this school, Prof. Kaku has deliver some thought provoking lectures concerning a geometric approach [40] to string field theory. Although some of the proposals he made are controversial, the essence of his lectures dealt with trying to resolve the central question in string theory; "What is the stringy equivalent of the Equivalence Principle?"

As I stated in my opening remarks, no satisfactory understanding of string theory will exist until this question is answered! It seems plausible to believe that should a geometrical formalism exists, much of the intricate and arcane methods presently used to study string theory would become obsolete. I suspect that the way we enforce modular invariance will be one of these. Consider how for the present, identities involving theta functions are an essential tool for this purpose. If we ever find the truly covariant geometrical approach to string theory, I think that this will be seen to be an analogy to the following construction in ordinary field theory. Recall that the propagators in field theory may be expressed as Hankel functions. In the ordinary course of evaluating Feynman diagrams this is almost never done. But if it is done, then in evaluating multi-loop diagrams, one would certainly encounter integrals over products of these functions. To simplify these, certain identities involving Bessel functions would play an important role. However, since we have a covariant formalism for treating field theory this is unnecessary. Furthermore, by studying identities involving Bessel functions, I do not easily see how we could work our way backward to discover covariant quantum field theory! Yet this is roughly where we are in our study of string theory. This is another example of what Prof. Kaku means by the field has been evolving backward since its discovery.

With the keen awareness of just how meager is our present understanding of string theory, I would like in this talk to make some observations. These observations are the 'crazy ideas' to which the title refers because to prove that they are more than that will require far more insights and hard work. I hope these 'ideas' will raise some questions in the mind of the some of the school's students. Perhaps through answering some of these questions we will be able to make further progress in string theory.

Idea 1: Treat the String Field as a Prepotential!

Why does the string field look analogous to a superfield prepotential? When I look at string theory with our present level of understanding, I am struck by how similar it is to superfield theory. In the latter we have an extension of spacetime by the addition of Grassmann coordinates. In the former, the spacetime is essentially extended by the mode greater than one bosonic oscillators and in covariant approaches, the BRST ghosts oscillators. Expansions of superfields over the extra Grassmann coordinates yield ordinary component fields. Expansions of string fields over the extra oscillators yield ordinary fields. So trying to use superfield theory as a model, I am led to suspect that the string field functional has a direct analogy in superfield theory. I do not have to look very far to find it...N $=1$ four dimensional superspace Yang-Mills or supergravity theory. Let me choose super Yang-Mills for simplicity.

It is well known that the component fields of the Yang-Mills multiplet consist of a gauge boson $A_{\underline{m}}^{\hat{I}}$, gaugino $\lambda^{\hat{I}}$, and pseudo-scalar auxiliary field $d^{\hat{I}}$ which occur in the expansion of the 'vector' superfield $V = V^{\hat{I}} t_{\hat{I}}$

$$V^{\hat{I}} = (\bar{\theta}\sigma^{\underline{a}}\theta) A_{\underline{a}}^{\hat{I}} + [i(\bar{\theta}\bar{\theta})(\theta\lambda^{\hat{I}}) + h.c] + (\theta\theta)(\bar{\theta}\bar{\theta}) d^{\hat{I}} \quad . \tag{5.1}$$

The action for this superfield takes the form of

$$S = \int d^4x d^2\theta d^2\bar{\theta} \; Tr\{ \; V\mathcal{O}_1 V + V\mathcal{O}_2 V\mathcal{O}_3 V + V\mathcal{O}_4 V\mathcal{O}_5 V\mathcal{O}_6 V \; \} \tag{5.2}$$

for some operators $\mathcal{O}_i$ whose explicit forms are known. When expressed in terms of component fields, the usual gauge invariance of a spin one field is present. At the linearized level this gauge transformation can be cast in the form $\delta V = i(D^2\Lambda - \bar{D}^2\bar{\Lambda})$. We may also write this as $\delta V = \mathcal{O}_7\Lambda$ where $\Lambda = exp(i(\bar{\theta}\sigma^{\underline{a}}\theta)\partial_{\underline{a}})\lambda(x)$. It appears that this theory is described by a polynomial action. However, this is not the case! The action above is a Wess-Zumino gauge-fixed action even though it possesses the usual component level gauge invariance. In general gauges, the complete theory is nonpolynomial. I may be a bit more explicit by noting that $\mathcal{O}_1 = D^\alpha \bar{D}^2 D_\alpha$. So the operator which generates gauge transformations is not the same as that which gives the linearized action. This is the key point which allows for the existence of a field strength superfield $W_\alpha = i\bar{D}^2 D_\alpha V = i\mathcal{O}_3 V$. The linearized equation of motion is expressible in terms of this field strength $D^\alpha W_\alpha - \bar{D}^{\dot{\alpha}} \bar{W}_{\dot{\alpha}} = 0$.

This situation may be compared with the present situation in string field theory. In the case of the Witten action [41] for the open string, the lagrangian takes the form

$$\mathcal{L} = \Phi Q \Phi + \tfrac{2}{3}\Phi * \Phi * \Phi \quad . \tag{5.3}$$

Here Q is the first quantized BRST operator. Attempts have been made to extend this to closed string theories and superstring [42,43] theories also. The linearized gauge invariance of the action is $\delta\Phi = Q\Lambda$. Thus by use of the analogy, the first thing which seems "unnatural" about our present BRST formulations of string field theory is that it is the BRST operator which appears both in the linearized action and as the generator of gauge transformations. This suggests that a geometrical formulation can only be achieved if we are able to factorize [3] the usual BRST operator in (5.3). As the Klein-Gordon operator had to be factorized to obtain the Dirac operator, it seems necessary to factorize the BRST operator to obtain a geometrical string field theory. Until this is done, I see no way to define a field strength in string field theories. Such a 'stringy' field strength would be the direct analog of W_α. A related question which arises is where is the stringy extension of conformal gravity? Our previous experience with superspace supergravity has taught us that when it is possible to isolate a gravitational prepotential, then it is also possible to consider higher derivative theories. I believe we should consider this possibility for strings.

A second point about the present formulation of string field theory which I find very improbable is that string actions take a polynomial form. I think it is much more likely that our present generation of string field theory actions will ultimately be seen to be similar to the Wess-Zumino gauge-fixed action for the supersymmetric Yang-Mills theory above. In fact, there is a heuristic argument which indicates why this is so. It is well known that closed string theories contain ordinary gravity and that both closed and open superstring theories contain supergravity and super Yang-Mills theories. All of these 'non-stringy' theories possess highly non-polynomial actions. It would therefore be quite natural if the corresponding string theories which contain these should also be highly non-polynomial. This is why I think the recent suggestions [6] concerning non-polynomial approaches to string field theory are so crucial. In this regard, they pass the test of Occam's razor.

Finally, on the basis of this analogy, I can make a prediction concerning the geometrization of string field theory. In (5.3), it is seen that $\mathcal{O}_3$ is proportional to the linearized field strength superfield. Similarly in every other known non-abelian gauge theory (ordinary Yang-Mills , gravitation, etc.) the linearized field strength also occurs in the 3-point coupling of the gauge field. Thus, on the basis of this analogy, in a non-polynomial and truly geometrical approach to string field theory, the linearized stringy field strength must be contained in the 'pants' diagram. So the further study of non-polynomial string field theory is one, *if not the*, most crucial topic currently underway.

---

[3]The possibility of factorization of linearized string field equation of motion has also been discussed by Gervais [44] from another point of view.

Idea 2: Look for Consistent $D = 4, SO(N_L) \otimes SO(N_R)$ GS-type Actions.

In the discussion of four dimensional strings, both in the bosonic and NSR cases, the structure that emerged could be interpreted as a four dimensional fiber bundle of the form $\mathcal{M}_4 \otimes \mathcal{G}_L \otimes \mathcal{G}_R$. This is highly suggestive that the most efficient way to describe superstrings is to look for a Green-Schwarz [34] type of action which can be interpreted as the theory of a fiber bundle of the form $s\mathcal{M}_{(4,N)} \otimes \mathcal{G}_L \otimes \mathcal{G}_R$. Here $s\mathcal{M}_{(4,N)}$ refers to a $D = 4$ superspace with $N = N_L + N_R$ manifest rigid supersymmetries. At present there is not much known concerning the explicit action for this theory. I expect its total action to follow the general form of (3.14) and (4.40)

$$S_{Tot} = S_{GS'} + S_L(u_L) + S_R(u_R) + S_N \quad , \tag{5.4}$$

with $S_L(u_L)$, $S_R(u_R)$ and S_N to be some appropriate generalizations of the previous found examples.

The action $S_{GS'}$ for the base superspace is, however, easy to construct. First I introduce a four dimensional string coordinate $X^{\mu\dot{\mu}}(\tau, \sigma)$ and four dimensional Weyl-spinor Grassmann coordinates $\Theta^{\mu\ i}(\tau, \sigma)$ and $\Theta^{\mu\ i'}(\tau, \sigma)$ and their complex conjugates. The indices i and i' take on values $1, ..., N_L$ and $1, ..., N_R$ respectively. Collectively the symbol $Z^{\underline{M}} \equiv (\Theta^{\mu\ i}, \Theta^{\mu\ i'}, \bar{\Theta}^{\dot{\mu}}{}_i, \bar{\Theta}^{\dot{\mu}}{}_{i'}, X^{\mu\dot{\mu}})$ denotes the supercoordinate position vector. Next I introduce the symbol $\hat{G}_{\underline{ABC}}$ defined by

$$\hat{G}_{\underline{ABC}} = i\tfrac{1}{2} C_{\alpha\gamma} C_{\dot{\beta}\dot{\gamma}} \left\{ \begin{array}{ll} \delta_i{}^j & : if\ \underline{A} = \alpha\ i\ ,\ \underline{B} = \dot{\beta}\ j\ ,\ \underline{C} = \gamma\dot{\gamma}\ or \\ & \quad any\ even\ permutation, \\ -\delta_i{}^j & : for\ any\ odd\ permutation, \\ -\delta_{i'}{}^{j'} & : if\ \underline{A} = \alpha\ i'\ ,\ \underline{B} = \dot{\beta}\ j'\ ,\ \underline{C} = \gamma\dot{\gamma}\ or \\ & \quad any\ even\ permutation, \\ \delta_{i'}{}^{j'} & : for\ any\ odd\ permutation, \\ 0 & : \ otherwise. \end{array} \right\} \tag{5.5}$$

Also to write the action concisely, I introduce the notations,

$$\Pi_{\pm\pm}{}^{\underline{A}} \equiv V_{\pm\pm}{}^m (\partial_m Z^{\underline{M}}) E_{\underline{M}}{}^{\underline{A}} \quad , \quad \hat{\Pi}_y{}^{\underline{A}} \equiv (\partial_y \hat{Z}^{\underline{M}}) \hat{E}_{\underline{M}}{}^{\underline{A}} \quad , \quad \hat{E}_{\underline{M}}{}^{\underline{A}} \equiv E_{\underline{M}}{}^{\underline{A}}(\hat{Z}) \ . \tag{5.6}$$

The "hatted" coordinates are the usual extensions used in the Vainberg construction [45]. The spacetime vielbein $E_{\underline{M}}{}^{\underline{A}}$ is fixed at its flat superspace value. Now the action $S_{GS'}$ takes the familiar geometrical form,

$$S_{GS'} = \frac{1}{2\pi\alpha'} \int d^2\sigma\, V^{-1} [\, -\tfrac{1}{2}\Pi_{++}{}^{\alpha\dot{\alpha}} \Pi_{--\ \alpha\dot{\alpha}} + \int_0^1 dy\, \hat{\Pi}_y{}^{\underline{C}} \hat{\Pi}_{++}{}^{\underline{B}} \hat{\Pi}_{--}{}^{\underline{A}} \hat{G}_{\underline{ABC}}\,] \ . \tag{5.7}$$

This action is κ-invariant under the variations

$$\begin{aligned}
\delta_\kappa Z^{\underline{N}} = & \; i\Pi_{++}{}^{\alpha\dot{\alpha}}(\bar{\kappa}_{--\;\dot{\alpha}}{}^{i} E_{\alpha\; i}{}^{\underline{N}} + \kappa_{--\;\alpha\; i} E_{\dot{\alpha}}^{i\;\underline{N}}) \quad , \\
& + i\Pi_{--}{}^{\alpha\dot{\alpha}}(\bar{\kappa}_{++\;\dot{\alpha}}{}^{i'} E_{\alpha i'}{}^{\underline{N}} + \kappa_{++\;\alpha\; i'} E_{\dot{\alpha}}^{i'\underline{N}}) \quad , \\
\delta_\kappa V_{--}{}^{m} = & \; -(\kappa_{--\; i}{}^{\alpha}\Pi_{--\;\alpha}{}^{i} + \bar{\kappa}_{--}{}^{\dot{\alpha}\; i}\bar{\Pi}_{--\;\dot{\alpha}\; i})V_{++}{}^{m} \quad , \\
\delta_\kappa V_{++}{}^{m} = & \; -(\kappa_{++\; i'}{}^{\alpha}\Pi_{++\;\alpha}{}^{i'} + \bar{\kappa}_{++}{}^{\dot{\alpha}\; i'}\bar{\Pi}_{++\;\dot{\alpha}\; i'})V_{--}{}^{m} \quad .
\end{aligned} \tag{5.8}$$

The unfortunate fact is that at present I do not know how to explicitly write the forms of the remaining terms in $\mathcal{S}_{Tot}$ in complete generality. However, in the most interesting case of $N_L = 1, N_R = 0$, things do simplify somewhat. Here all quantities with i' indices vanish and there is only one spacetime supersymmetry! Under these conditions, it turns out remarkably that $S_R(u_R)$ given in (3.11) is also κ-invariant under the variations

$$\delta_\kappa \phi_R^{\hat{I}} = 0 \quad , \quad \delta_\kappa \lambda_{--}{}^{++} = (\kappa_{--}{}^{\alpha}\Pi_{--\;\alpha} + \bar{\kappa}_{--}{}^{\dot{\alpha}}\bar{\Pi}_{--\;\dot{\alpha}}) \quad . \tag{5.9}$$

So the real remaining problem is to construct an action $S_L(u_L)$ that is simultaneously invariant under a κ-symmetry and a generalized lefton "Siegel" symmetry! Once this problem is solved it will then be possible to treat the $N_L > 1, N_R \neq 0$ theories. One of the amusing implications of this program is that it will imply from anomaly-freedom that the $N_L = 4, N_R = 4$ case (i.e. a version dual to the old "$N = 8$" theory [46]) is the maximal supergravity model. In this way for each value of N there will emerge a supergravity theory from the low energy limit of a superstring and which can be coupled to the corresponding GS action.

Idea 3: Study Models with $(Supersymmetry)^2$.

Why do both NSR and Green-Schwarz formulations of superstrings exist? This is a very strange situation to find. It is so striking that it suggests that we have found two parts of a larger structure. For instance, these two different formulations might correspond to two different gauge choices in a larger theory. In an effort to obtain this larger structure, sometime ago Hitoshi Nishino and I [47] proposed a new class of models which possess both spacetime and world-sheet supersymmetries. We called such models, $(supersymmetry)^2$ models. The first such model constructed was an extension of the NSR string. Later when the requirement of κ-supersymmetry was added [48] it was found that only world-sheet (1,0) supersymmetry was consistent. These models were later called spinning superstrings and [49] their particle limits were referred to as spinning superparticles. This class of models is still not very well understood with even their consistency being a matter of some debate. However, one of their interesting properties is that they necessarily contain NSR fermions and spin-fields!

To show an example of such a model, it will suffice to construct the first-class superparticle which possesses $(supersymmetry)^2$ [50]. Since this model possesses a supersymmetry along the world line of the as well as a spacetime supersymmetry, it is convenient to first discuss the world sheet supergravity. We introduce coordinates (τ, ζ) with τ being the time-like bosonic coordinate and ζ being the one-component real Grassmann coordinate. The spinorial supercovariant derivative may be chosen to take the form $\mathbf{D} = \partial/\partial\zeta + i\zeta\partial$, $\partial \equiv \partial/\partial\tau$ so as to satisfy the equation

$$\{\mathbf{D}, \mathbf{D}\} = i2\partial \quad . \tag{5.10}$$

The introduction of a supergravity theory on the world-line requires a vielbein with one fermionic component E_F and one bosonic component E_B which satisfy

$$\{E_F, E_F\} = i2E_B \quad . \tag{5.11}$$

As in our discussions in near (4.2), there is a solution to this equation which can be cast in terms of a real fermionic superfield H_F and a real scalar superfield Ψ. Explicitly this solution is

$$E_F = \Psi^{-\frac{1}{2}}\hat{E}_F \quad , \quad E_B = \Psi^{-1}[\ \hat{E}_B + i\tfrac{1}{2}(\hat{E}_F ln\Psi)\hat{E}_F\ ] \quad ,$$

$$\hat{E}_F \equiv \mathbf{D} + iH_F\partial \quad , \quad \hat{E}_B \equiv [\ 1 + H_B\ ]\partial \quad , \quad H_B \equiv [\ \mathbf{D}H_F + iH_F\partial H_F\ ] \quad ,$$

$$E^{-1} \equiv sdetE^{-1} = \Psi^{\frac{1}{2}}[\ 1 + H_B\ ]^{-1} \quad . \tag{5.12}$$

Now remarkably enough there exists a world-line supersymmetric extension of the "alphabet" algebra [51,52]. For the simplicity of the presentation, the $D = 4$, $N = 1$ superspace may be chosen for the spacetime[50] and we consider the limit where $\Psi = 1$ and $H_F = 0$. The appropriate spinning superparticle coordinates are given by the doubly supersymmetric coordinates $\mathbf{Z}^{\underline{A}} = (\mathbf{X}^{\underline{a}}, \mathbf{\Theta}^{\alpha}, \bar{\mathbf{\Theta}}^{\dot{\alpha}})$. Due to the dual requirements of spacetime and world sheet supersymmetry, particle coordinates, Grassmann coordinates, NSR fermions and spin-field like quantities are all simultaneously present.

$$\mathbf{X}^{\underline{m}}(\zeta, \tau) = \psi^{\underline{a}}(\tau) + \zeta x^{\underline{a}}(\tau) \quad , \quad \mathbf{\Theta}^{\mu}(\zeta, \tau) = \theta^{\alpha}(\tau) + \zeta S^{\alpha}(\tau) \tag{5.13}$$

Similarly a set of canonically conjugate momenta may be introduced $\mathcal{P}_{\underline{A}} = (\mathbf{p}_{\underline{a}}, d_{\alpha}, \bar{d}_{\dot{\alpha}})$ can be introduced with the only nonvanishing Poisson brackets given by

$$\{\mathbf{p}^a, \mathbf{p}^b\} = 2\eta^{ab} \quad , \quad \{d_{\alpha}, \bar{d}_{\dot{\alpha}}\} = 2\mathbf{D}\mathbf{p}_{\alpha\dot{\alpha}} \tag{5.14}$$

The generators of the "alphabet" algebra are now given by

$$\mathcal{A} = \tfrac{1}{2}\mathbf{pDp} \quad , \quad \mathcal{B}^\alpha = (\mathbf{Dp}^{\alpha\dot\beta})\bar{d}_{\dot\beta} \quad ,$$

$$\mathcal{C}_{\underline{a}} = \tfrac{1}{4}d_{[\alpha}\bar{d}_{\dot\alpha]} + \tfrac{1}{2}[\ \mathcal{M}_{\alpha\gamma}(\mathbf{Dp}^\gamma{}_{\dot\alpha}) - \bar{\mathcal{M}}_{\dot\alpha\dot\gamma}(\mathbf{Dp}_\alpha{}^{\dot\gamma})\ ]$$

$$\mathcal{D}_{\underline{a}} = \mathcal{M}_{\underline{ab}}(\mathbf{Dp}^{\underline{b}}) + k(\mathbf{Dp}_{\underline{a}}) \quad , \quad \mathcal{E}_{\underline{a}} = d_\alpha(\mathbf{D}\bar{d}_{\dot\alpha}) \quad , \quad \mathcal{F}_{\underline{a}} = (\mathbf{D}d_\alpha)\bar{d}_{\dot\alpha} \quad . \tag{5.15}$$

These generators form a generalized $(supersymmetry)^2$ algebra.

$$\{\mathcal{A},\mathcal{A}\} = \tfrac{1}{2}\mathbf{D}\mathcal{A} \quad , \quad \{\mathcal{B}^\alpha,\mathcal{B}^{\dot\alpha}\} = 4(\mathbf{D}\mathcal{A})(\mathbf{Dp}^{\alpha\dot\alpha}) \quad , \quad [\mathcal{B}^\alpha,\mathcal{C}_{\beta\dot\gamma}] = 2\delta^\alpha_\beta\bar{d}_{\dot\gamma}(\mathbf{D}\mathcal{A}) \quad .$$

$$[\mathcal{C}_{\underline{a}},\mathcal{C}_{\underline{b}}] = \tfrac{1}{2}[\ \mathcal{C}_{\alpha\dot\beta}(\mathbf{Dp}_{\beta\dot\alpha}) - \mathcal{C}_{\beta\dot\alpha}(\mathbf{Dp}_{\alpha\dot\beta})\ ] + \mathcal{M}_{\underline{ab}}\mathbf{D}\mathcal{A} \quad ,$$

$$[\mathcal{C}_{\underline{a}},\mathcal{D}_{\underline{b}}] = \tfrac{1}{2}\epsilon_{\underline{abcd}}(\mathbf{Dp}^{\underline{c}})\mathcal{D}^{\underline{d}} \quad , \quad [\mathcal{D}_{\underline{a}},\mathcal{D}_{\underline{b}}] = -2\mathcal{M}_{\underline{ab}}\mathbf{D}\mathcal{A} - (\mathbf{Dp}_{[\underline{a}})\mathcal{D}_{\underline{b}]} \quad ,$$

$$[\mathcal{B}^\alpha,\mathcal{E}_{\beta\dot\gamma}] = 2\delta^\alpha_\beta(\mathbf{D}\mathcal{A})\mathbf{D}\bar{d}_{\dot\gamma} \quad , \quad [\mathcal{B}^\alpha,\mathcal{F}_{\beta\dot\gamma}] = 2\delta^{\dot\alpha}_{\dot\gamma}(\mathbf{D}\mathcal{A})\mathbf{D}d_\beta \quad ,$$

$$[\mathcal{C}_{\underline{a}},\mathcal{E}_{\underline{b}}] = 2(\mathbf{Dp}_{\beta\dot\alpha})\mathcal{E}_{\alpha\dot\beta} \quad , \quad [\mathcal{C}_{\underline{a}},\mathcal{F}_{\underline{b}}] = 2(\mathbf{Dp}_{\alpha\dot\beta})\mathcal{F}_{\beta\dot\alpha} \quad ,$$

$$[\mathcal{E}_{\underline{a}},\mathcal{F}_{\underline{b}}] = 4(\mathbf{Dp}_{\alpha\dot\beta})\mathbf{D}\mathcal{E}_{\beta\dot\alpha} \quad . \tag{5.16}$$

In the algebra above $\mathcal{M}_{\underline{ab}}$ denotes the generator of spacetime Lorentz transformations and k is the eigenvalue of the spacetime dilatation generator. It is simple to write the form of the action which is invariant under the symmetries generated by $G_i \equiv (\mathcal{A},\mathcal{B}$, etc.). It is given by

$$S = \int d\tau d\zeta E^{-1}[\ (E_B Z^M)E_M{}^A\mathcal{P}_A - i\Lambda^i G_i\ ] \quad , \tag{5.17}$$

where we have introduced the spacetime supervielbein $E_{\underline{M}}{}^{\underline{A}}$ and restored the worldline vielbein. It is possible to construct many such models.

An entirely different example is provided by the particle limit of the original $(supersymmetry)^2$ model [46,47]

$$S_2 = \int d\tau d\zeta\ E^{-1}[\ -\tfrac{1}{2}\eta_{\underline{ab}}\Pi_F{}^{\underline{a}}\Pi_B{}^{\underline{b}}\ +\ \int_0^1 dy\ \hat\Pi_y{}^{\underline{C}}\hat\Pi_F{}^{\underline{B}}\hat\Pi_B{}^{\underline{A}}\hat H_{\underline{ABC}}\ ] \quad , \tag{5.18}$$

where here we define

$$\Pi_F{}^{\underline{A}} \equiv (E_F\tilde{\mathbf{Z}}^{\underline{M}})E_{\underline{M}}{}^{\underline{A}} \quad , \quad \Pi_B{}^{\underline{A}} \equiv (E_B\tilde{\mathbf{Z}}^{\underline{M}})E_{\underline{M}}{}^{\underline{A}} \quad , \quad \tilde{\mathbf{Z}}^{\underline{m}} \equiv x^{\underline{m}} + i\zeta\psi^{\underline{m}} \quad , \tag{5.19}$$

and $\tilde{\mathbf{\Theta}}^\mu$ defined in (5.12). This action is invariant under a κ-variation (found by truncating the results in [51])

$$\delta_\kappa\tilde{\mathbf{Z}}^{\underline{N}} = i\tfrac{1}{2}\sigma_{\underline{a}}{}^{\alpha\beta}\Pi_B{}^{\underline{a}}\kappa_\alpha E_\beta{}^{\underline{N}} \quad , \quad \delta_\kappa H_F = -i\Psi^{-\frac{1}{2}}(\kappa_\alpha\Pi^\alpha_F)(1+H_B) \quad ,$$

$$\delta_\kappa\Psi = -i(\kappa_\alpha\Pi^\alpha_F)(E_F\Psi) \to \delta_\kappa S_2 = 0 \tag{5.20}$$

but not under variations corresponding to the generators of the entire alphabet algebra.

Notation and Conventions

Usually the coordinates of the two-dimensional world sheet are denoted by τ (time-like) and σ (space-like) or collectively by $\sigma^m = (\tau, \sigma)$ ($\sigma^o = \tau$, $\sigma^1 = \sigma$). However, it is much simpler to work with the "light-cone" coordinates $\sigma^{++} \equiv \sigma + \tau$ and $\sigma^{--} \equiv \sigma - \tau$. In this case $\sigma^m \equiv (\sigma^{++}, \sigma^{--})$. A similar type of basis may be chosen for the partial derivative $\partial_m \equiv (\partial_{++}, \partial_{--})$.

$$\partial_{++} \equiv \frac{\partial}{\partial\sigma} + \frac{\partial}{\partial\tau} , \quad \partial_{--} \equiv \frac{\partial}{\partial\sigma} - \frac{\partial}{\partial\tau} .$$

note that as a consequence $\partial_m \sigma^n = 2\delta_m{}^n$! The metric on the world sheet is denoted by $g_{mn}(\tau, \sigma)$ or just g_{mn}. Due to the presence spinors inherent in all supersymmetry theories, it is convenient to introduce a zweibein field $V_m{}^a$ via

$$g_{mn} \equiv \tfrac{1}{2}(V_m{}^{++} V_n{}^{--} + V_m{}^{--} V_n{}^{++})$$

where an early-late convention on the indices indicates either holonomic $(m, n, \ldots)$ or anholonomic $(a, b \ldots, \ell)$ varieties. It is useful to note the two-dimension Levi-Civita tensor density ϵ_{mn} is given by

$$\epsilon_{mn} \equiv \tfrac{1}{2}(V_m{}^{++} V_n^{--} - V_m^{--} V_n^{++}) .$$

The plus and minus indices are used as spinorial indices, denoting spins of plus and minus one-half, respectively. Within use of this convention, the Grassmann coordinates required to describe a superspace extension of the two-manifold can be expressed as ζ^+ and ζ^-. Thus, the coordinates of a world sheet "superpoint" z^M in (1,0) superspace are $(\zeta^+, \sigma^{++}, \sigma^{--})$. A similar superpoint in (1,1) superspace has the form $z^M = (\zeta^+, \zeta^-, \sigma^{++}, \sigma^{--})$.

String theories also contain quantities which are valued in space of coordinates of the spacetime manifold. For example, the string coordinate itself carries the index of a D-vector. For this reason, it is required that a convention for such indices should exist. The convention used here is that underlined Latin letters are used for this purpose. Thus, for example, the bosonic string coordinate is denoted by $X^{\underline{m}}(\tau, \sigma)$ with $\underline{m} = 0, \ldots D-1$. Here too an early-late convention is used to distinguish between indices which refer to anholonomic or holonomic bases. This extension is also made to spacetime superspace coordinates $z^{\underline{M}} \equiv (\Theta^\alpha, X^{\underline{m}})$ etc.

The components of two dimensional tensorial or multi-spinorial tensors are always referred to flat (anholonomic bases). Thus our most general two dimensional fields φ have components denoted by

$$\varphi\,{}^{(p,q)}_{(m,n)} \equiv \varphi^{+_1 \cdots +_p, -_1 \cdots -_q}_{+_1 \cdots +_m, -_1 \cdots -_n}$$

with m (p) lower (upper) plus signs and n (q) lower (upper) minus signs.

Under a local Lorentz transformation with infinitesimal parameter k_L, any such covariant field varies according to the rule

$$\delta_L \varphi_{(m,n)}^{(p,q)} = \frac{1}{2} k_L \left(m - n - p + q \right) \varphi_{(m,n)}^{(p,q)} \ .$$

Thus every "up" plus index is equivalent to a "down" minus index. This equation permits the definition of the action of the Lorentz generator, $\mathcal{M}$, by writing $\delta_L \varphi = k_L [\mathcal{M}, \varphi]$ which implies

$$\left[\mathcal{M}, \varphi_{(m,n)}^{(p,q)} \right] = \frac{1}{2} \left(m - n - p + q \right) \varphi_{(m,n)}^{(p,q)} \ .$$

Typically if $m + n + p + q$ is even, then φ is bosonic. If $m + n + p + q$ is odd, then φ is usually fermionic.

Under a local Weyl transformation with infinitesimal parameter k_W the Weyl variation of a field of scale weight w takes the form

$$\delta_W \varphi_{(m,n)}^{(p,q)} = \frac{1}{2} k_W w \, \varphi_{(m,n)}^{(p,q)} \ .$$

Alternately, this equation can be used to define the action of the Weyl generator d on the field by writing $\delta_W \varphi = k_W [d, \varphi]$ which implies

$$\left[d, \varphi_{(m,n)}^{(p,q)} \right] = \frac{1}{2} \left(w \right) \varphi_{(m,n)}^{(p,q)} \ .$$

Finally, under a general coordinate transformation the covariant field transforms as

$$\delta_G \varphi_{(m,n)}^{(p,q)} = k^{++} V_{++}{}^m \partial_m \varphi_{(m,n)}^{(p,q)} + k^{--} V_{--}{}^m \partial_m \varphi_{(m,n)}^{(p,q)} \ .$$

So that the total variation of a covariant field transforming under all of the local world sheet symmetries is

$$\begin{aligned} \delta_G \varphi_{(m,n)}^{(p,q)} = {} & k^{++} V_{++}{}^m \partial_m \varphi_{(m,n)}^{(p,q)} + k^{--} V_{--}{}^m \partial_m \varphi_{(m,n)}^{(p,q)} \\ & + \tfrac{1}{2} (k_W \left(m - n - p + q \right) + k_L w) \varphi_{(m,n)}^{(p,q)} \end{aligned}$$

A superfield is just a function of z^M. Assuming analyticity in the Grassmann direction, we have functions such as $X(z)$ given by

$$X(z) = X(\sigma^{++} , \sigma^{--}) + \zeta^+ \psi_+(\sigma^{++} , \sigma^{--}) \ .$$

The component field content of this multiplet may also be defined by $X(\tau,\sigma) \equiv X(\zeta^+,\tau,\sigma)\,|$ and $\psi_+(\tau,\sigma) \equiv \nabla_+ X(\zeta^+,\tau,\sigma)\,|$ where $|$ indicates the limit as ζ^+ approaches zero. This permits a simple way to obtain component results from superspace via the identity

$$\int d\zeta^- E^{-1}\mathcal{L}_- = [det(V_a{}^m)]^{-1}((\nabla_+ - i\psi_{++}{}^+)\mathcal{L}_-)| \equiv V^{-1}((\nabla_+ - i\psi_{++}{}^+)\mathcal{L}_-)|$$

for any lagrangian $\mathcal{L}_-$. The actual component fields of (1,0) supergravity are the zweibein $V_a{}^m(\sigma^{++},\sigma^{--})$ and the gravitino $\psi_a{}^+(\sigma^{++},\sigma^{--})$. As in most superspace theories, it is convenient to introduce the spinorial covariant derivative

$$D_+ \equiv \partial/\partial\zeta^+ + i\zeta^+\partial_{++}$$

Euclideanization

It is almost universal now to couch the formulation of string and string field theofy in terms of Riemann surfaces. There are a number of reasons for doing this. Two dimensional Minkowski space, like its higher dimensional analogs, is noncompact. In defining path integrals, for example, it is much more convenient to work with a Wick-rotated time coordinate $\tau \to -i\tau$ just as in conventional four dimensional field theory,

$$\sigma^{++} \to \sigma - i\tau \equiv \bar{z}\ , \quad \sigma^{--} \to \sigma + i\tau \equiv z\ .$$

As an illustration, consider a real scalar field ϕ with Minkowski space action

$$S = \int d\tau d\sigma \left[-\frac{1}{2}(\partial_{++}\phi)(\partial_{--}\phi)\right]\ , \quad = \int d\tau d\sigma\{-\frac{1}{2}\left[\left(\frac{\partial\phi}{\partial\sigma}\right)^2 - \left(\frac{\partial\phi}{\partial\tau}\right)^2\right]\}\ ,$$

After Euclideanization this becomes

$$S \to i\int dz d\bar{z}\frac{1}{2}(\partial_{\bar{z}}\phi)(\partial_z\phi) = iS_E\ ,$$

so $e^{iS} \to e^{-S_E}$. The equation of motion is clearly $\partial_z\partial_{\bar{z}}\phi = 0$ implying $\phi = \psi(z) + \bar{\psi}(\bar{z})$ in terms the complex field $\psi(z)$. But one of the features of Euclideanization is that not all Minkowski space field theories admit Euclideanization or vice-versa. This is a well-known phenomenon from four dimensional gauge theories. Real self-dual Maxwell field strengths exist for four dimensional euclidean spaces but not for four-dimensional Minkowskian spaces! In two-dimensional field theories, Majorana-Weyl spinors exist in Minkowskian space but not in Euclidean space. This leads to an interesting situation for strings defined in Minkowski space and the corresponding theories on Riemann surfaces. In the table below we list the usual transcription of scalar and spinor fields from Minkowskian to Euclidean 2-spaces

	Minkowski Space Field	Euclidean Space Field	
Real Scalar	$\phi = \bar{\phi}$	$\phi = \bar{\phi}$	Real Scalar
Majorana-Weyl Spinor	$\psi_+ = \bar{\psi}_+$	$\psi_+ \neq \bar{\psi}_+$	Weyl spinor

It is seen that the Majorana-Weyl spinor of the Minkowski formulation is replaced by a Weyl spinor in the Euclidean formulation.

A theory where the issue of Euclideanization has not been explored thoroughly is in the theory of chiral bosons. The Siegel action for single such field after Euclideanization becomes

$$S_E = \frac{1}{2} \int dz d\bar{z} \left[(\partial_{\bar{z}} \phi)(\partial_z \phi) + \lambda (\partial_{\bar{z}} \phi)^2 \right] \ .$$

But it is clear that this action is *not* even real! So the chiral boson (just like the Majorana-Weyl spinor) which possesses a perfectly well-behaved Minkowski space action does not admit any simple extension to Euclidean space. In the conventional conformal algebra formalism, the use of actions is avoided altogether. A chiral boson in the conformal field theory approach is defined to be the holomorphic (or anti-holomorphic) half of the solution to the scalar equation of motion (i.e. $\psi(z)$ or $\psi(\bar{z})$). So to actually study the problem of modular invariant partition functions for chiral bosons á la Polyakov requires developing techniques outside of those presently used in conformal field theory.

Acknowledgement

I wish to recognize the collaborative efforts of R. Brooks, D. Depireux, F. Gieres, T. Hübsch, F. Muhammad, Q-Han Park and W. Siegel in carrying out the research discussed in my lectures. I further wish to thank the organizers of the school for the opportunity to participate and to acknowledge the editor, Dr. Zbigniew Haba, for his consideration during the preparation of this manuscript. Finally, I thank W.Siegel at ITP,SUNY, Stony Brook along with G.Semenoff and N.Weiss at the Univ.of British Columbia for hospitality and assistance during the preparation of this manuscript.

REFERENCES

[1] M. B. Green, J. H. Schwarz, and E. Witten, *Superstring Theory*, Camb. Press. (1987) NY, NY; M. Kaku, *Introduction to Superstrings* Springer-Verlag (1988) NY, NY.

[2] D. J. Gross and P. Mende, Princeton Univ. Preprint # PUPT-1067, D. Amati, M. Ciafaloni and G. Veneziano, Int.J.of Mod.Phys. 7 (1988) 1615;D. J. Gross, Phys.Rev.Lett. **60** (1988) 1229.

[3] S. J. Gates, Jr., M. T. Grisaru, M. Roček,and W. Siegel, *Superspace or One Thousand and One Lessons in Supersymmetry* (Benjamin/Cummings, Reading, 1983) p. 246.

[4] S. J. Gates, Jr. and F. Gieres, Nucl. Phys. **B320** (1989) 310.

[5] B. DeWitt, Phys. Rev. 162 (1967) 1195; ibid. 162 (1967) 1239; J. Honerkamp, Nucl. Phys. **B36** (1971) 130; ibid. **B48** (1972) 269; G. t'Hooft, Nucl. Phys. **B62** (1973) 444; L. F. Abbot, Nucl. Phys. **B185** (1981) 189; L. Alvarez-Gaume, D. Z. Freedman and S. Mukhi, Ann.Phys. **163** (1985) 318; S. Mukhi, Nucl. Phys. **B264** (1986) 640, E. Braaten, T. Curtright and C. Zachos, Nucl. Phys. **B260** (1985)

[6] M. Saadi and B. Zweibach, M.I.T. preprint MIT-CTP # 1704; H. Sonoda and B. Zweibach, M.I.T. preprint MIT-CTP # 1758; T. Kugo, H. Kunitomo and K. Suehiro, Kyoto Univ. Preprint # KUNS 965, HE(TH)89/01.

[7] P. Candelas, G. Horowitz, A. Strominger, and E. Witten; Nucl. Phys. **B258** (1985) 46, L. Dixon, J. Harvey, C. Vafa and E. Witten, Nucl. Phys. **B261** (1985) 678; K. S. Narain, Phys. Lett. **169B** (1986) 41; I. Antoniadis, C. Bachas, C. Kounnas

and P. Windey, Phys. Lett. **171B** (1986) 51; H. Kawai, D. C. Lewellyn, and S.-H.-H. Tye, Phys. Rev. Lett. 57 (1986) 1832.

[8] S. T. Yau, *Proc. of Symposium on Anomalies, Geometry and Topology*, eds. W. A. Bardeen and A. R. White, (World Scientific, Singapore, 1985); A. Strominger and E. Witten, Commun. Math. Phys. 101 (1986) 341; B. R. Greene, K. H. Kirklin, P. J. Miron and G. G. Ross, Nucl. Phys. **B278** (1986) 341; T. Hübsch, Commun. Math. Phys. 108 (1987) 291; P. S. Green and T. Hübsch, Commun.Math.Phys. 109 (1987) 99; ibid. 115 (1988) 231; R. Schimmrigk, Phys. Lett. **193B** (1987) 175; P. S. Aspinwall, B. R. Greene, K. H. Kirklin and P. J. Miron, Nucl. Phys. **B294** (1987) 193; P. Candelas, A. M. Dale, C. A. Lütken and R. Schimmrigk, Nucl.Phys. **B298** (1988) 493; P. Candelas, C. A. Lütken and R. Schimmrigk, Nucl. Phys. **B306** (1988) 113.

[9] W. Siegel, Nucl. Phys. **B238** (1984) 307.

[10] S. J. Gates, Jr., F. Muhammad and R. Brooks, Phys. Lett. **194B** (1987) 35.

[11] C. M. Hull, Phys. Lett. **206B** (1988) 234; W. Siegel, *Introduction to String Field Theory* (World Scientific, Hong Kong, 1988) p. 124.

[12] C. M. Hull, Phys. Lett. **212B** (1988) 437.

[13] E. Witten, Commun. Math. Phys. 92 (1984) 455.

[14] D. Nemeschansky and S. Yankelowicz, Phys. Rev.Lett.54 (1985) 620; E.Bergshoeff, S. Randjbar-Daemi, A. Salam, H. Sarmadi, and E. Sezgin, Nucl. Phys. **B269** (1986) 77.

[15] P. Goddard, A. Kent, and D. Olive, Comm. Math. Phys. 103 (1986) 105.

[16] Q-Han Park, Phys. Lett. **223B** (1989) 175.

[17] D. Friedan, Phys. Rev. Lett. 45 (1980) 1057; D. Friedan, Ann. Phys. 163 (1985) 318, E. S. Fradkin and A. A. Tseytlin, Phys. Lett **158B** (1985) 316; idem. Nucl. Phys. **261** (1985) 1; S. J. Gates, Jr., C. M. Hull, and M. Roček, Nucl. Phys. **248** (1984) 157; T. Curtright and C. Zachos, Phys. Rev. Lett. 53 (1984) 1799; P. Howe and H. Sierra, Phys. Lett. **148B** (1984) 451.

[18] D. A. Depireux, S. J. Gates, Jr. and Q-Han Park, Phys. Lett. **224B** (1989) 364.

[19] V. G. Knizhnik and A. B. Zamolodchikov, Nucl. Phys. **B241** (1984) 333, D. Friedan, E. Martinec and S. Shanker, Nucl. Phys. **B227** (1986) 93, idem. D. Friedan,

E. Martinec and S. Shanker, Phys. Lett. **160B** (1985) 55; V. G. Knizhnik, Phys. Lett. **160B** (1985) 403.

[20] R. Brooks, F. Muhammad and S. J. Gates, Jr. Nucl. Phys. **B268** (1986) 599, M. Evans and B. Ovrut, Phys. Lett. **171B** (1986) 177; ibid. **175B** (1986) 145; G. Moore, P. Nelson and J. Polchinski, Phys. Lett.**169B** (1986) 47; G. Moore and P. Nelson, Nucl. Phys.**B274** (1986) 177.

[21] A. A. Kirillov, *Twistor Geometry and Nonlinear Systems*, eds. H. Doebner and T. Palev (Springer-Verlag, 1982), p. 101.

[22] J. Louis,B. Ovrut, R. Garries and J. Wess, Univ.of Penn.Preprint # UPR-0342T.

[23] J. Wess and B. Zumino, Phys. Lett. **74B** (1978) 51.

[24] D. J. Gross, J. A. Harvey, E. Martinec and R. Rohm, Phys.Rev.Lett 54 (1985) 502; Nucl. Phys. **B256** (1986) 253; ibid. **B267** (1986) 75.

[25] S. J. Gates, Jr., M. T. Grisaru, L. Mezincescu and P. K. Townsend, Nucl. Phys. **B286** (1987) 1.

[26] M. T. Grisaru, L. Mezincescu and P. K. Townsend, Phys. Lett. **179B** (1986) 247.

[27] Y. Frischman and J. Sonnenschein, Nucl. Phys. **B301** (1988) 346; C. Imbimbo and A. Schwimmer, Phys. Lett. **193B** (1987) 455; J. M. F. Labastida and M. Pernici, Class. Quant. Grav. 4 (1987) L197; Nucl. Phys. **B297** (1988) 3067; R. Floreanini and R. Jackiw, Phys. Rev. Lett. 59 (1987) 1873; M. Sanielevici, G. Semenoff and Y. S. Wu, Nucl. Phys. **B312** (1989) 197; R. Link, Phys. Lett. **216B** (1989) 145; F. Yu and Y. S. Wu, Int. J. Mod. Phys. **A4** (1989) 701; B. McClain and Y. S. Wu, Univ. of Utah preprint, June 1989; F. Schaposnik and J. Solomin, Int. J. of Mod. Phys. **A4** (1989) 249.

[28] P. Fré, F. Gliozzi and M. Rego Monteiro, Phys. Lett. **219B** (1989) 419.

[29] S. Bellucci, D. A. Depireux and S. J. Gates, Jr., Univ. of Maryland Preprint # UMDEPP 89-123 (July, 1989).

[30] S. J. Gates, Jr. and W. Siegel, unpublished.

[31] S. Bellucci, R. Brooks and J. Sonnenschein, Mod. Phys. Lett. **A3** (1988) 1537.

[32] F. Yu and Y. S. Wu, see ref. 27.

[33] S. J. Gates, Jr. and W. Siegel, Phys. Lett. **206B** (1988) 631.

[34] M. B. Green and J. H. Schwarz, Phys. Lett. **136B** (1984) 367; idem. Nucl. Phys. **B198** (1982) 252, 441.

[35] K. S. Narain, see ref. 7.

[36] K. Bardakci, E. Rabinovici and B. Säring, Nucl. Phys. **B299** (1988) 151; K. Gawedzki and A. Kupianen, Phys. Lett. **215B** (1988) 119; I.H.E.S. C.N.R.S. Bures-sur-Yette, Preprint # IHES/P/89/06; D. Karabali, Q-Han Park and H. Schnitzer, Brandeis Univ. Preprint # BRX TH-247; H. Schnitzer, Brandeis Univ. Preprint # BRX TH-254; D. Karabali, Q-Han Park, H. Schnitzer and Z. Yang, Phys. Lett. **216B** (1989) 307.

[37] W. Boucher, D. Friedan, and A. Kent, Phys. Lett. **172B** (1986) 316; A. B. Zamolodchikov and V. A. Fateev, Zh. Eksp.Theor.Fiz. 90 (1986) 1553; S. Nam, Phys. Lett. **172B** (1986) 323; D. Gepner, Nucl. Phys. **B296** (1982)575; Phys. Lett. **199B** (1987) 380.

[38] S. J. Gates, Jr. and T. Hübsch, in preparation.

[39] C. M. Hull and E. Witten, Phys. Lett. **160B** (1985) 398; C. G. Callan, D. Friedan, E. Martinec and M. Perry, Nucl. Phys. **B262** (1985) 593; M. T. Grisaru, A. Van de Ven and D. Zanon; Phys. Lett. **173B** (1986) 423; Nucl. Phys. **B277** (1986) 388, 409.

[40] The first suggestions of nonpolynomiality for string field theory were due to M. Kaku. See chapter eight of the second work of reference [1].

[41] E. Witten, Nucl. Phys. **B268** (1986) 253; idem. **B276** (1986) 291.

[42] H. Terao and S. Uehara,Phys. Lett. **173B** (1986) 134; T. Banks, M. E. Peskin, C. R. Preitschopf, D. Friedan and E. Martinec, Nucl. Phys. **B274** (1986) 71; A. Leclair and J. Distler, Nucl. Phys. **B273** (1986) 552.

[43] J. P. Yamron, Phys. Lett. **174B** (1986) 69; G. D. Date, M. Günaydin, M. Pernici, K. Pilch and P. van Nieuwehhuizen, Phys. Lett. **171B** (1986) 182;

[44] J. L. Gervais, Nucl. Phys. **B276** (1986) 339, 349.

[45] M. M. Vainberg, *Variational Methods for the Study of Nonlinear Operators* (Holden Day, San Francisco, 1964), p. 135; E. Witten, Nucl. Phys. **B223** (1983) 422; M. Henneaux and Phys. Lett. **152B** (1985) 340.

[46] E. Cremmer and B. Julia, Nucl. Phys. **B159** (1979) 141.

[47] S. J. Gates, Jr.and H. Nishino, Class.Quant.Grav. 3 (1986) 391.

[48] R. Brooks, F. Muhammad and S. J. Gates, Jr., Class.Quant.Grav.3 (1986) 745; R. Brooks, Phys. Lett. **186B** (1987) 313.

[49] S. Aoyama, J. Kolwalski-Glikman, J. W. van Holten and J. Lukierski, Phys.Lett. **201B** (1988) 487; idem. **216B** (1989) 133.

[50] S. J. Gates, Jr.and P. Majumdar, Mod.Phy.Lett. 4 (1989) 339.

[51] W. Siegel, Nucl. Phys. **B263** (1985) 93.

[52] R. Kallosh, W. Troost and A. van Proeyen, Phys. Lett. **212B** (1988) 428.

RENORMALIZATION GROUP AND STRING DYNAMICS

ANTAL JEVICKI
Physics Department
Brown University
Providence, R.I., 02912
U.S.A.

ABSTRACT

We review approaches to generate dynamical string equations from 2–dimensional renormalization theory. A string moving in a general background defines a 2d nonlinear σ–model. A requirement of finiteness or Weyl invariance that leads to consistency conditions giving dynamical equations for the fields. We summarize the recent progress and outline the open questions.

1. INTRODUCTION

Conformal invariance has appeared as a central feature of string theory. The nontrivial vacua of the theory all correspond to conformal invariant 2d field theories. Even more the requirement of conformal invariance can be elevated to a dynamical principle. If one considers a string moving in a background field and requires conformal symmetry this requirement leads to equations of motion for the fields. This has been completely demonstrated for massless background fields [1–5] where the couplings lead to renormalizable 2d σ–models.

The vanishing of β–functions assures scale (and conformal) invariance and produces the dynamical equations for the massless background fields. There is a large literature on this subject with detailed investigation of various theories and higher order loop and α' corrections.

The more general question of massive backgrounds on the other hand has only been investigated recently with some encouraging results [7–17]. A string having an infinite tower of excited states leads to the idea of coupling all higher background fields and imposing the demand of conformal invariance. This would then give the most general 2d conformal field theory and the most general string vacua. The problem of incorporating all the higher massive degrees of freedom is not a trivial one. First of all the higher couplings in the 2d σ model are all nonrenormalizable. Coupling the first massive background field requires the coupling of all of the others (an infinite sequence), only then can one hope to make some sense at the quantum level (renormalizability). The question of what the higher order couplings are is itself a nontrivial one. It is usually phrased as the question of what the most general 2d field theory is. Interesting progress has been achieved during the last two years and we describe the results in these lectures. While this is so the subject still awaits the final most general formulation with clear open questions remaining.

The content of this review goes as follows. The basic ideas and the tachyon background coupling are discussed in sect. II. It is explained how for the first massive background field (the tachyon) one evaluates the ultraviolet divergences. Here we describe the results of Ooguri and Sakai.

In general the field equations must be such that they generate for the correct string S–matrix. In sect. III we summarize some properties of the S- -matrix generating functional [6]. We address the issue of reconstructing the field equations from the S–matrix and derive an equation obeyed by the Lagrangian. Here one sees a problem, the renormalization group equations differ from the S–matrix equations. This indicates a disagreement at the five point level, this problem is still to be understood.

In sect. IV we consider the issue of what the most general 2d couplings should be. An analogy with string field theory is used, it leads to a concrete specification of all higher background fields. We describe an investigation of this general σ–model. A requirement of Weyl (local scale) invariance is shown to lead to the correct string field theory equations of motion. This is seen in the first linearized order where the equations obtained are the free BRST field theory equations. This summarizes the results of a joint work with S. Jain.

In section IV we describe a general Wilson renormalization group approach to the problem. An explicit equation was suggested some time ago by Banks and Martinec, it has been investigated in more detail recently by Polchinski and collaborators with rather interesting results.

In summary it will be seen that the general renormalization group approach to the dynamics of string theory has some concrete successes. On the other hand some open questions and possibly the most exciting developments still remain. In this sense the situation is not dissimilar to the present status of string field theory. One hopes that with fuller understanding both of these will converge to a new powerful dynamical principle.

2. β–Functions and the C-theorem

Consider a string moving in an external massless background field. For closed strings one has the graviton $G_{\mu\nu}(X)$, the antisymmetric tensor field $H_{\mu\nu}(X)$ and the dilator $D(X)$. The action

$$
\begin{aligned}
I(X^\mu) = \frac{1}{4\pi\alpha'} \int d^2z\sqrt{\gamma}\Big[&\gamma^{ab}G_{\mu\nu}(X)\partial_a X^\mu \partial_b X^\nu_* + \\
&+ \epsilon^{ab}H_{\mu\nu}(X)\partial_a X^\mu \partial_b X^\nu \\
&+ R^{(2)}D(X)\,\Big]
\end{aligned}
\tag{2.1}
$$

represents a 2–dimensional nonlinear σ–model. The ultraviolet divergences of generalized nonlinear σ–models have been well understood, it is renormalizable and the divergences are removed by redefining the coupling constants. Here the "coupling constants" are given by arbitrary background functions $G_{\mu\nu}, H_{\mu\nu}, D$.

Their renormalization defines the β–functions,

$$\begin{aligned} G_{\mu\nu} &\longrightarrow G_{\mu\nu} + \beta^G_{\mu\nu} \log \Lambda \\ H_{\mu\nu} &\longrightarrow H_{\mu\nu} + \beta^H_{\mu\nu} \log \Lambda \end{aligned} \tag{2.2}$$

A simple (background field method) calculation gave

$$\begin{aligned} \beta^G_{\mu\nu} &= R_{\mu\nu} - \frac{1}{4} H_\mu^{\lambda\sigma} H_{\nu\lambda\sigma} \\ \beta^H_{\mu\nu} &= \nabla_\lambda H^\lambda_{\mu\nu} - \frac{1}{2} \nabla^\lambda H_{\mu\nu\lambda} \end{aligned} \tag{2.3}$$

A demand of finiteness (vanishing of U.V. divergences) is expressed as

$$\beta_{\mu\nu} = 0 \tag{2.4}$$

This produces constraints on the background fields. The remarkable observation of [1–5] is that the resulting equations are precisely the dynamical equations for the massless fields $G_{\mu\nu}, H_{\mu\nu}$ and D. One recognizes the Einstein equation and the equation for an antisymmetric tensor field.

It is in some sense reasonable that the above phenomenon happens. One could have expected that effective system of a string in a background is only consistent if we have conformal invariance. In two dimensions conformal invariance follows from scale invariance. To see that the vanishing of β–functions implies scale invariance one simply recalls the following facts about the variation of the effective action $\Gamma(\mu, g)$ in terms of scale μ:

$$\mu \frac{\partial}{\partial \mu} \Gamma \sim < \int d^2 z T^a_a > \tag{2.5}$$

Here T_a^a is the trace of the energy–momentum tensor and one has

$$\int < T_a^a >= \int d^2z < \beta^G_{\mu\nu} \partial X^a \partial X^v + \beta^H_{\mu\nu} \partial X \partial X \epsilon >$$

Consequently the vanishing of β–function leads to scale (and conformal) invariance.

There is a very simple argument which illuminates directly why the β–functions equal them massless field equations of motion. Consider the graviton case and write $G_{\mu\nu} = \delta_{\mu\nu} + h_{\mu\nu}$. Expanding in terms of the field $\tilde{h}_{\mu\nu}(k)$ one has

$$Z = \sum \frac{1}{n!} h(1) \ldots h(n) < v(1) v(2) \ldots v(n) > \tag{2.6}$$

where $v_{\mu\nu} = \partial X^\mu \partial X^\nu e^{ik\cdot X}$ is the graviton vertex operator. We see that when the h sources are on shell the σ–model generates the graviton S–matrix elements. There are some general features obeyed by an S–matrix generating functional $S(h)$. For a theory whose action is $I(\phi) = \int d^D X(\mathcal{L}_\circ + \mathcal{L}_{\text{int}})$ one can show (based on general principles, see sect. III) that

$$S(h) = I\left(h + \frac{1}{\Box} \mathcal{L}^{(1)}_{\text{int}}(h)\right) \tag{2.7}$$

where $\mathcal{L}^{(1)}_{\text{int}}$ is the first derivative. In the σ– model perturbation theory one has a power series in $\partial_\mu = \frac{\partial}{\partial X_\mu}$ (the two–dimensional derivative). The way in which the nonanalytic terms $\frac{1}{\Box}$ are reproduced is by U.V. divergences. Namely

$$\frac{1}{\Box} \simeq \int_a^1 dt^{-\Box - 1} \approx \log a + \Box (\log a)^2 + \cdots \tag{2.8}$$

and we have the fundamental correspondence [2,3,4]

massless poles $\Leftrightarrow$ 2d U.V. divergences .

For the σ–model this implies that

$$Z(h) = S(h) = I\left(h + \log a\mathcal{L}^{(1)}_{\text{int}}(h) + \cdots\right) \tag{2.9}$$

This implies the above mentioned renormalization of the coupling h . Furthermore, we have demonstrated that the β–function (the coefficient of U.V. divergences) is the derivative of the Lagrangian and consequently the dynamical equation of motion (at least for test fields $h(k)$).

The Zamolodchikov's C–theorem [5] defines a functional that is decreasing along the RG flow. Consider the components of the energy momentum tensor in a renormalizable 2d field theory. Denote $T_{zz} = T \quad T_{z\bar{z}} = D$ and the correlation functions

$$\begin{aligned} C(z) &= z^4 < T(z)T(o) > \\ E(z) &= z^2\bar{z}z < T(z)D(o) > \\ H(z) &= (\bar{z}z)^2 < D(z)D(o) > \end{aligned} \tag{2.10}$$

One has the conservation law

$$(z\partial_z + \bar{z}\partial_{\bar{z}})(C - 6H - 4E) = -12H \tag{2.11}$$

The dependence on the 2d coordinates is through $\frac{r^2}{a^2}$ where $r^2 = \bar{z}z$ and a is the 2d cutoff. The conservation law then becomes $a\frac{\partial}{\partial a}$ and with the definition of a β–function one obtains

$$\beta^a \frac{\partial}{\partial g^a}(C - 6H - 4E) = -24H \tag{2.12}$$

When $r^2 = a^2$ these are functions of the coupling constants only $C = C(g) \quad H = H(g)$ etc. For a positive definite theory $H \geq 0$ and we have a functional $I(g) =$

$C(g) - 6H - 4E$ which always decreases along a RG trajectory. This exhibits the action principle behind the 2d renormalization theory (it confirms that the β–function is a derivative of an action). At the fixed point $\beta(g^*) = 0$ so $H(g^*) = 0$, it also follows that $E(g^*) = 0$ and the functional $I(g^*) = C(g^*) = c$ is the central change of the Virasorso algebra.

3. Tachyon Background

The massive background field which one would like to study first is the tachyon. Indeed since the early days of dual models the fact that the tachyon had a negative mass squared led to conjectures that the tachyon field should develop a nonzero expectation value inducing a new phase for string theory. In the present context one would like to generalize the β–function method, couple massive background fields [7–19] in the σ–model and derive the effective dynamical equations for them. This was first considered for the tachyon in an interesting paper by Das and Sathiapalan [7]. We follow the discussion given by Ooguri and Sakai [10].

The tachyon vertex operator is $V_k = e^{ik\cdot X(z)}$ and the background coupling is

$$I_\phi = \int dk\tilde{\phi}(k) \int \frac{d^2z}{a^2}\sqrt{\gamma}e^{ik\cdot X(z)} = \tag{3.1}$$

there a^2 represents the dimensional short distance cutoff.

The two dimensional "σ–model"

$$I_\phi = \int d^2z\sqrt{\gamma}\left(\partial X\partial X + \frac{1}{a^2}\phi(X(z))\right) \tag{3.2}$$

represents a scalar field theory with an arbitrary interaction term. There is a phase where the theory is superrenormalizable but Das and Sathiapalan pointed out the analogy with the Sine–Gordon theory where nontrivial U.V. divergences appear. The divergences can be extracted by expanding in ϕ_k.

$$Z = \sum_n \frac{1}{n!} \prod_i \int dz_i (a^2)^{\frac{k_i^2}{2}-1} < \prod_i : e^{ik_i X(z_i)} > \cdot \\ \cdot \phi_{k_1} \phi_{k_2} \cdots \phi_{k_n} \tag{3.3}$$

The normal ordering of $e^{ik\cdot X(z)}$ is performed, it is responsible for the additional $(a^2)^{\frac{k^2}{2}}$ cutoff dependence. We see in this factor the reason for the mass shell condition $\frac{k^2}{2} - 1$, we proceed however by being off mass shell. There are further U.V. divergences, which are isolated by Ooguri and Sakai [10] by the operator product method. The two–point correlation function is

$$< X(z)X(z') >= \log\left(|z|^2 + a^2\right) \tag{3.4}$$

and for two vertex operators

$$V_k(z_1)V_k(z_2) \approx (|z_1 - z_2|^2 + a^2)^{k_1 k_2} V_{k_1+k_2} + \cdots \tag{3.5}$$

For n vertex operators one extracts the singularity by taking $z_1 - z_2 = u$, and $z_1 - z_j = u z_j'$ for $j = 3, \ldots N$ and taking $u \to 0$. This leads to the following result for the partition function

$$Z = \sum_N \frac{1}{N!} \int dk_i \phi_{k_1} \phi_{k_2} \cdots \phi_{k_N} \frac{-2\pi}{2 \sum_{i<j} k_i k_j + 2N - 2} T_{N+1} \cdot \\ \cdot \int d^2 z (a^2)^{\frac{k^2}{2}-1} e^{ik\cdot X(z)} \tag{3.6}$$

with $k = \sum k_i$ and

$$T_{N+1}(-k, k_1, k_2, \ldots k_N) = \int \prod_{i=3}^{N} dz'_i |z'_i - z'_j|^{2k_i \cdot k_j} \tag{3.7}$$

In this we recognize the Koba–Nielsen form, if the momenta are on shell one has the $N+1$ point scattering amplitude

$$T_{N+1} = S_{N+1}(-k, k_1, \ldots k_N) \quad . \tag{3.8}$$

Considering the generating function off shell we see that all the cutoff dependence is given in the factor $(a^2)^{\frac{k^2}{2}-1}$. To define the β–function and renormalize the theory one treats the field ϕ as coupling constants with an implicit cutoff dependence $\phi_k = \phi(a)$ and then

$$\beta_k = \frac{\partial \phi_k}{\partial \ell n a^2} \tag{3.9}$$

The renormalization $\phi_k(a)$ is such that the total partition function is cutoff independent.

$$\frac{dZ}{d\ell n a^2} = \frac{\partial Z}{\partial \ell n a^2} + \frac{\partial Z}{\partial \phi_k} \frac{\partial \phi_k}{\partial \ell n a^2} = 0 \tag{3.10}$$

This leads to the equation

$$(k^2 - 2) W(\phi, k) + \int dk' \frac{\delta W(\phi, k)}{\delta \phi_{k'}} \beta_{k'} = 0 \tag{3.11}$$

with

$$W(\phi, k) = \phi_k + \sum_{N=2} \frac{1}{(N-1)!} \frac{-2\pi}{2 \sum_{i<j} k_i k_j + 2N - 2} T_{N+1}(-k, k_1, \ldots k_N) \cdot \\ \cdot \phi_{k_1} \cdots \phi_{k_N} \tag{3.12}$$

The equation relating β and W derived by Ooguri and Sakai allows one to solve for the β–function (and the effective tachyon field equations of motion).

The solution is given iterativily with a power series in ϕ with the first few terms easily seen to equal

$$\begin{aligned}\beta(\phi) = &(k^2-2)\phi_k + \frac{\pi}{2}T_3(-k,k_1,k_2)\phi_{k_1}\phi_{k_2}+\\ &+\frac{\pi}{3!}\left(T_4 - \frac{2\pi}{s-2} - \frac{2\pi}{t-2} - \frac{2\pi}{u-2}\right)\phi_{k_1}\phi_{k_2}\phi_{k_3}+ \\ &+O(\phi^4)\end{aligned} \tag{3.13}$$

These terms correspond to the quadratic term and the cubic and quartic interaction terms in the Lagrangian. It is also clear that the result of the renormalization group approach is satisfactory to this order, the S–matrix computed from this Lagrangian would be manifestly correct (at least for the three and four point case). This result was also derived by several other authors.

At higher orders there is a problem. The iteration of the β–function equation gives forth (and higher) order terms in the equation of motion. That would correspond to a nonpolynomial Lagrangian which is expected. The specific form for the fourth order term however is such that if integrated (naively) to a Lagragian one would not obtain the correct scattering amplitude. We shall give a general analysis of this in the next section and later in the discussion of the Banks–Martinec approach.

The reason for the failure of the renormalization group approach is not clear. It might be that there are some additional divergences. It might also be that the renormalization group should only be valid on shell [11]. One could also have ambiguities when going from the eq. of motion to a Lagrangian.*

⋆ R. Woodard (private communication)

4. S-Matrix Methods

The background σ–model can be interpreted as an S–matrix generating functional [2]. It is clear that the problem which one is faced with is the one of reconstructing a Lagrangian from a known string theory S–matrix. For this reason it is useful to summarize some general features [6] of the S–matrix functional. What follows will be general, the only assumption is that a Lagrangian for a theory exists.

Consider then a theory with an action $I(\phi) = I_\circ(\phi) + I_{\rm int}(\phi)$ with a set of fields ϕ_a. Introducing linearized on–shell fields φ_a

$$(k^2 + m_a^2)\varphi_a = 0 \tag{4.1}$$

we define the S–matrix generating functional

$$S[\varphi] = \sum_{n=3}^{\infty} \frac{1}{N!} S^{(N)}_{a_1 a_2 \ldots a_N} \varphi_1 \varphi_2 \ldots \varphi_N \tag{4.2}$$

with $S^{(N)}(1, 2, \ldots N)$ being the N–particle scattering amplitudes.

In general the S–matrix is obtained from the Greens function functional

$$Z(j) = e^{W(j)} = \int d\phi e^{i(I(\phi) + j\phi)} \tag{4.3}$$

by a reduction formula

$$S[\varphi] = W\left[j = (\overleftarrow{\partial}^2 + m^2)\varphi\right] \tag{4.4}$$

The path integral can be used to derive an expression for the S–matrix functional in terms of the classical field. One defines $\phi_{c\ell}$ to obey the equations

$$\frac{\delta I}{\delta \phi} = (k^2 + m^2)\phi - I^{(1)}_{\text{int}}(\phi) = 0 \tag{4.5}$$

It can be iteratively built in terms of the linearized fields φ and

$$\phi_{c\ell} = \varphi + \frac{1}{k^2 + m^2} I^{(1)}_{\text{int}}(\phi) \tag{4.6}$$

One then proves that

$$S[\varphi] = I(\phi_{c\ell}(\varphi)) \tag{4.7}$$

with some careful redefinition of the free (quadratic) part. This is the general representation used in Sect. II.

We can also derive an equation relating the S–matrix functional and the derivative of the Lagrangian. The later we denote as the "β"– function:

$$\tilde{\beta}(\phi) \equiv I^{(1)}_{\text{int}}(\phi) \tag{4.8}$$

Now from the generating functional

$$\frac{\delta W(j)}{\delta j} = \phi \tag{4.9}$$

which implies

$$\phi_{c\ell} = \frac{1}{k^2 + m^2} \frac{\delta S[\varphi]}{\delta \varphi} \tag{4.10}$$

Here we used the fact that $j = (k^2 + m^2)\varphi$ and that for φ on–shell $j \to 0$ with ϕ obeying the classical equation of motion (in tree aproximation). This follows form the Legendra transform formula $W(j) = I(\phi) + j\phi, \quad I^{(1)}(\phi) + j = 0$.

We have obtained in general a representation for the nonlinear classical field ϕ in terms of the S–matrix

$$\phi_{c\ell} = \frac{1}{k^2+m^2}\frac{\delta S}{\delta\varphi} = \sum_{N=1}^{\infty}\frac{1}{N!}\frac{1}{k^2+m^2}S_{N+1}\,\varphi_{(1)}\cdots\varphi_{(N)} \tag{4.11}$$

Note that the right hand side is just the functional $W(\phi,k)$ of the previous section.

We now proceed to derive the equation determining the nonlinear equation of motion in terms of the S–matrix. Write the classical equation $I^{(1)}(\phi)=0$ as

$$(k^2+m^2)\phi = \tilde{\beta}(\phi) \tag{4.12}$$

Then denoting $\frac{\delta S(\varphi)}{\delta\varphi} = \tilde{W}(\varphi)$ we have

$$\tilde{W}(\varphi) = \tilde{\beta}(\phi) \tag{4.13}$$

in which we eliminate φ through

$$\varphi = \phi - \frac{1}{k^2+m^2}\tilde{\beta}(\phi) \tag{4.14}$$

to obtain the final fuctional relation

$$\tilde{W}\left(\phi - \frac{1}{k^2+m^2}\tilde{\beta}(\phi)\right) = \tilde{\beta}(\phi) \tag{4.15}$$

This equation relates the equation of motion (β–function) to the S–matrix. Obviously we can solve for $\tilde{\beta}(\phi) = I^{(1)}(\phi)$ iteratively once the $S[\varphi]$ functional is known. In this procedure one needs to define an off shell extension since the field ϕ is the full nonlinear field it does not obey the linearized mass shell condition.

Comparing the above equation with the tachyon equation presented in the previous section we note the following. The correspondence is $\beta = \tilde{\beta} + (k^2 + m^2)\phi$ with the free term added but our equation is nonlinear in $\tilde{\beta}$ while the renormalization group leads to a linear equation for $\beta(\phi)$. This explains the statement that the tachyon eq. fails in higher orders. More explicitly we now give the iterative expansion for $\tilde{\beta}(\phi) = \beta_3(\phi) + \beta_4(\phi) + \beta_5(\phi) + \cdots$ to the fourth order and exhibit the missing terms. A straightforward expansion gives

$$\beta_3 = W_3(\phi) = \frac{1}{2!} S_3(k, k_1, k_2)\phi_k, \phi_{k_2} \tag{4.16}$$

Then

$$\beta_4 = W_4(\phi) - \frac{\delta W_3}{\delta \phi_{k'}} \frac{1}{k'^2 + m^2} \beta_3(\phi, k') \tag{4.17}$$

which leads to

$$\begin{aligned} \beta_4 = \frac{1}{3!} [S_4(k, k_1, k_2, k_3) - \triangle(k_1 + k_2) - \triangle(k_2 + k_3) - \\ - \triangle(k_1 + k_3)] \phi_{k_1}\phi_{k_2}\phi_{k_3} \end{aligned} \tag{4.18}$$

For β_5 the nonlinear terms in our equation begin to contribute

$$\begin{aligned} \beta_5(\phi) = W_5(\phi, k) - \frac{\delta W_3}{\delta \phi_{k'}} \triangle(k')\beta_4 - \frac{\delta W_4}{\delta \phi_{k'}} \triangle(k')\beta_3 \\ + \frac{1}{2} \frac{\delta^2 W_3}{\delta \phi_{k'} \delta \phi_{k''}} (\triangle(k')\beta_3)(\triangle(k'')\beta_3) \end{aligned} \tag{4.19}$$

It is the last term which is absent in the renormalization group approach. Reducing the W_4 contribution we find another term of the same form and the result

$$
\begin{aligned}
\beta_5(\phi, k) = W_5(\phi, k) &- \frac{\delta\beta_3}{\partial\phi}\triangle\beta_4 - \beta_3\triangle\frac{\delta\beta_4}{\partial\phi} \\
&- \frac{\partial\beta_3}{\partial\phi}\triangle\frac{\partial\beta_3}{\partial\phi}\triangle \\
&- \frac{1}{2}\frac{\partial^2\beta_3}{\partial\phi\partial\phi}(\triangle\beta_3)(\triangle\beta_3)
\end{aligned}
\tag{4.20}
$$

All the different contributions have a clear graphical interpretration. From the five–point S–matrix element one is subtracting the Feynman diagrams with poles. The last term corresponds to the diagram

ϕ k ϕ

ϕ ϕ

In the renormalization group approach its coefficient would be (-1) rather than the correct $-\frac{1}{2}$. Consequently the renormalization group does not give the same equation of motion as the present general discussion. This implies that the action obtained by integrating the renormalization group equation fails at fifth order. The five–point scattering amplitude would not be obtained correctly from it, the weight of the above mentioned diagram would be erroneous. Clearly this problem persists in higher orders. It is interesting to note the work of Brunstein, Nemeschansky and Yankielewocz. These authors have argued that the S–matrix ansatz $\phi_{c\ell} = \phi(\varphi)$ (our eq. 4.11) does obey the renormalization group equation and one is left with the problem of reconciling these two results.

5. General Background Couplings and String Field Equations

The most general formulation of the renormalization group method should involve most general vertex operator couplings. The idea is that a renormalization group flow of a most general background coupled $2d$ field theory would give the complete field equations of motion. In the case of open strings these should ultimately be the known equations of Witten's or the CSV string field theory while for closed strings one has a promise to derive a long searched for theory of closed strings. We shall now summarize some work done on this subject in collaboration with S. Jain [14]. Introducing general couplings (involving both the coordinates X_μ and the ghosts) we have formulated equations that follow from demanding Weyl invariance of the theory. A demand of Weyl invariance should be equivalent to the vanishing of β–functions, for massless fields the corresponding equations have been studied in detail. When all (both massless and massive vertex operators) are considered we would like to obtain a correspondence with (open) string field theory.

Consider the first nontrivial level which corresponds to the massive tensor multiplet. The couplings corresponding to the bosonic oscillators are given by

$$I_E + I_B = \int dt\, e^{-1}\left[\nabla X^\mu \nabla X^\nu E_{\mu\nu}(x) + \nabla^2 X^\mu B_\mu(x)\right] \tag{5.1}$$

where $e \equiv e^{\phi/2}$ is the Liouville field and $\nabla = \frac{1}{e}\frac{d}{dt}$ is the covariant derivative along the real axis t (open string). The analogous couplings for the Stuckelberg field ψ and the multiplier S_μ would read

$$I_\psi + I_{S_\mu} = \int dt\, e^{-1}\left[(\nabla^2 y + b\nabla y \nabla y)\psi(x) + \nabla y \nabla x^\mu S_\mu\right] \tag{5.2}$$

Here $y(z)$ are the bosonized ghost coordinates. The variation with respect to $\delta\phi$ comes first from the explicit dependence on ϕ and then from the Weyl anomaly of the correlation functions. To this order one needs

$$\begin{aligned} < xx >&= \frac{1}{2\epsilon} + \phi \\ < \dot{x}x >&= \frac{1}{2}\dot{\phi} \\ < \dot{x}\dot{x} >&= \frac{\rho}{4}\ddot{\phi} + \frac{\mu}{12}\dot{\phi}\dot{\phi} + \frac{\nu}{24}\dot{\phi}\phi e^{\phi} \end{aligned} \tag{5.3}$$

where the coefficients ($\rho = \frac{2}{3}$, $\mu = -\frac{1}{2}$, $\nu = 0$) could be determined using the heat kernel method. In the present computation, they are specified by self consistency. Using the correlation functions we find the ϕ dependence

$$\langle I_{B+E} \rangle = \int dt\, e^{-\frac{1}{2}(1+k^2)\phi} e^{ikX[z]} \left[B' + C'\dot{\phi} + D'\ddot{\phi} + E'\ddot{\phi}\ddot{\phi}\right] \tag{5.4}$$

where for example $B' = \dot{x}^\mu \dot{x}^\nu E_{\mu\nu} + 2\ddot{x}^\mu B_\mu$ and similarly for the other coefficients. One can now perform a variation $\delta\phi$ which gives the condition for Weyl invariance. To the above one adds the following variation for the ghost sector:

$$\langle \delta I_{\psi+S} \rangle = \int dt\, \frac{d}{dt} \left\{ \frac{1}{e} \delta\dot{\phi}\psi(x) + \delta\phi \nabla x^\mu S_\mu \right\} \tag{5.5}$$

In the expression for the total variation $\delta < I_{E+B} > + < \delta I_{\psi+S} >$ one obtains terms proportional $\delta\phi$, $\delta\dot{\phi}$, $\delta\ddot{\phi}$, $\dot{\phi}\delta\phi$, $\ddot{\phi}\delta\phi$, $\dot{\phi}\dot{\phi}\delta\phi$, $\dot{\phi}\delta\dot{\phi}$. Demanding the coefficients of these terms to vanish leads to the equations of motion. The $\delta\phi$ term gives

$$\begin{aligned} (\Box - 1)E_{\mu\nu} - (\partial_\mu S_\nu) &= 0 \\ (\Box - 1)(B_\mu - S_\mu) &= 0 \end{aligned} \tag{5.6}$$

which are the correct field equations for E, B. The $\delta\dot{\phi}$ terms gives the equation $S_\mu = \partial \cdot E - B_\mu + \partial_\mu \psi$ and the $\delta\ddot{\phi}$ gives the constraint $\frac{1}{4}E + \partial B + \frac{3}{2}\psi = 0$.

The other coefficients lead to further equations (with higher derivatives), these turn out to be consistency conditions which vanish for the specific value of the anomaly given.

Compare the result with the open string field equations following from the BRST approach

$$QA = 0 \tag{5.7}$$

The string field $|A>$ is expanded in terms of Hilbert space component fields. Here one distinguishes the terms built on the $|->$ and $|+>$ vacuum states

$$|A>= \Phi|-> +S|+> \tag{5.8}$$

In terms of the ghost oscillators $(\bar{b}_n, \bar{c}_m)$ we have furthermore

$$\begin{aligned} \Phi &= \phi + \phi_1^1 bc + \phi_2^2 (bc)(bc) + \cdots \\ S &= S_0^1 b + S_1^2 b(bc) + S_2^3 b(bc)(bc) + \cdots \end{aligned} \tag{5.9}$$

The fields ϕ_n^n and S_n^{n+1} are expanded in terms of the coordinate oscillators. For the physical field ϕ

$$\phi = \varphi(X_0) + iA_\mu \alpha_{-1}^\mu + B_\mu \alpha_{-2}^\mu + E_{\mu\nu} \alpha_{-1}^\mu \alpha_{-1}^\nu + \cdots \tag{5.10}$$

with φ being the tachyon $(m^2 = -1)$, A_μ the photon $(m^2 = 0)$, B_μ and $E_{\mu\nu}$ the components of a massive $s = 2$ tensor field $(m^2 = +1)$ and so on. The fields $\phi_n^n (n \geq 1)$ which come with ghost type oscillators correspond to Stuckelberg fields and their presence is crucial for complete gauge invariance. The fields

S_n^{n+1} appearing in the $|+>$ vacuum sector are redundant (Nakanishi–Lautrup fields). The BRST charge in this decomposition reads

$$
\begin{gathered}
Q = c_0 K + b_0 M + d + \partial \quad \text{with} \quad K = L_0 - 1 \\
d = \sum_{n>0} c_{-n} L_n \quad \text{and} \quad \partial = \sum_n c_n L_{-n} \quad \text{and}
\end{gathered}
\tag{5.11}
$$

The free Lagrangian is

$$
L = < A|Q|A > = \frac{1}{2}\phi_n^n K \phi_n^n + \phi_n^n \partial S_n^{n+1} - \phi_n^n d S_{n-1}^n + \frac{1}{2} S_{n-1}^n M S_{n-1}^n \tag{5.12}
$$

giving the equations of motion

$$
\begin{aligned}
K\phi_0^0 + \partial S_0^1 &= 0 \\
K\phi_1^1 + \partial S_1^2 + d S_0^1 &= 0
\end{aligned}
\tag{5.13}
$$

and $(d+\partial)\phi + MS = 0$ for the redundant multiplier fields. For example, at the tensor level one has

$$
\begin{aligned}
(\Box - 1)E_{\mu\nu} - (\partial_\mu S_\nu) + \frac{1}{2}\delta_{\mu\nu} S &= 0 \\
(\Box - 1)B_\mu - S_\mu - \partial_\mu S &= 0 \\
-(\Box - 1)\psi - \partial_\mu S + \frac{3}{2} S &= 0
\end{aligned}
\tag{5.14}
$$

and:

$$
\begin{aligned}
S_\mu &= \partial_\nu E_{\mu\nu} + \partial_\mu \psi - B_\mu \\
S &= \frac{1}{4} E + \partial \cdot \beta + \frac{3}{2}\psi
\end{aligned}
\tag{5.15}
$$

where ψ is a Stuckelberg field, and S , S_μ are multipliers. They come from

$$
\begin{aligned}
\phi_1^1 &= \psi b_{-1} c_{-1} \\
S_0^1 &= S b_{-2} + S_\mu b_{-1} \alpha_{-1}^\mu
\end{aligned}
\tag{5.16}
$$

We see an agreement with the background field equations (the background field equations correspond to the partially gauge fixed case $S = 0$).

The complete problem with all the couplings included can be discussed in general. To every Hilbert space state one associates a vertex operator. The formula [8]:

$$\int dt\, V\left[X(t)\right] = \int dt\, A\left[X(t+e^{i\sigma})\right] \tag{5.17}$$

gives the general σ–model coupling. Here $A[x]$ is the general string field functional.

Explicitly a general string field (in the coherent space representation) reads

$$A[X(e^{i\sigma})] = \sum \alpha_{-n_1}\alpha_{-n_2}\dots\alpha_{-n_k}\phi_{\{n_1,n_2,\dots n_3\}} \tag{5.18}$$

Since $\frac{\alpha_{-n}}{n} = \int d\sigma e^{in\sigma}X(e^{i\sigma})$ we have that

$$\frac{\alpha_{-n}(z)}{n} = \int d\sigma e^{-in\sigma}X(z+e^{i\sigma}) = \frac{1}{n!}X^{(n)}(z) \tag{5.19}$$

where $X^{(n)}(z) = \partial_z^n X(z)$. The associated general background coupling in the 2d field theory is then

$$V(X(z)) = \sum \partial_z^{n_1}X(z)\partial_z^{n_2}X(z)\dots\partial_z^{n_k}X(z)e^{ikX(z)}\phi_{n_1,n_2\dots n_k} \tag{5.20}$$

Here one obtains the generalized vertex operators with the corresponding "coupling constants" $\phi_{\{n\}}$. The Liouville field is coupled through a covariant derivative

$$\partial_z \to \nabla_z \equiv e^{-\phi}\partial_z \tag{5.21}$$

The couplings corresponding to Stuckelberg and Nakanishi-Lautrup fields are given by ghost terms using a bosonized coordinate $y(z)$ which plays the role of a 27-coordinate. The bosonic ghost couples to the Liouville field lineary.

For the purpose of obtaining Weyl transformation properties, it is useful to have a momentum variable representation $\tilde{A}(p_\circ, p_1, p_2 \ldots)$ for the wave functional $A(x(t), x_1(t), x_2(t) \ldots)$. Then

$$I_B = \int dt\, e \int \mathcal{D}p\, e^{i \sum_{n=0}^{\infty} p_n x_n(t)} \tilde{A}(p) \tag{5.22}$$

with $x_\circ(t) \equiv x(t)$ and

$$\begin{aligned} X_n &= K_n X(t) \\ K_n &= \frac{1}{n!}(\nabla)^n \end{aligned} \tag{5.23}$$

In this representation the total derivative term for ghosts will read

$$\delta I_{\text{ghost}} = \int dt\, e \frac{1}{e}\frac{d}{dt}\left(\int \mathcal{D}P\, e^{iP\cdot X} \xi(P)\right) \quad . \tag{5.24}$$

Here we have introduced a 27th coordinate, denoted $v(t)$, with $v_1(t) \equiv K_1 v(t) = \delta\phi$ and

$$v_n(t) \equiv K_n v(t) = \frac{1}{n} K_{n-1}(\delta\phi) \quad . \tag{5.25}$$

Thus $X_n^M \equiv (x_n^\mu, v_n) P_n^M \equiv (p_n^\mu, q_n)$ and

$$P \cdot X \equiv \sum_{n=0}^{\infty} (p_n x_n + q_n v_n) \quad . \tag{5.26}$$

$\xi(P)$ is given by

$$\xi(P) = \delta(q_\circ) \sum_{n=1}^{\infty} \xi_n(p) \xi_n'(q) \quad , \quad \xi_n'(q) = i \frac{\partial}{\partial q_n} \delta(q) \quad . \tag{5.27}$$

$\delta(q_\circ)$ ensures that there are no terms $\propto v(t) \sim \int \delta\phi$, and the fact that ξ_n' has only one derivative ensures that we have only terms linear in $\delta\phi$. Performing the

q integration yields

$$\delta I_y = \int dt\, e \int \mathcal{D}p\, e^{ip\cdot x} \sum_{n=1}^{\infty} \left(-v_n \sum_{m=0}^{\infty} (m+1) p_m \frac{\partial}{\partial p_{m+1}} + (n+1) v_{n+1} \xi_n(p) \right) \tag{5.28}$$

We now perform the computation of Weyl anomalies and derive the condition that these vanish. The bare vertex can be written as the product of the normal ordered vertex and a factor containing the coincident 2–point functions of the x_n:

$$I_B = \int dt\, e \int \mathcal{D}p\, :e^{ip\cdot x}: e^{-\frac{1}{2}\sum_{n,m=0}^{\infty} p_n p_m <x_n x_m>} \tilde{A}(p) \quad . \tag{5.29}$$

A Weyl variation of this then consists of 2 parts, one coming from the explicit dependence of the x_n on ϕ, and the other from the dependence of coincident propagators $\left\langle \frac{d^m}{dt^m} x(t) \frac{d^n}{dt^n} x(t) \right\rangle$ on ϕ. Consider first the explicit dependence on ϕ which is present through the covariant derivative $\nabla = \frac{1}{e}\frac{d}{dt}$. It is easy to show from the definition $x_n(t) \equiv K_n\left(\phi, \frac{d}{dt}\right) x(t)$, that

$$\begin{aligned} &\delta x_n(t) = \sum\nolimits_{m=1}^{n} A_{nm} x_m(t) \quad ; \\ &A_{nm} = -\frac{n}{n-m+1} K_{n-m}(\delta\phi), \\ &m = 1, 2, \ldots, n; \quad A_{nm} = 0 \quad , \quad m > n \quad . \end{aligned} \tag{5.30}$$

This results in the variation (using a change of variables $p_m \to \sum_{n+1}^{\infty} p_n(\delta_{nm} + A_{nm})$)

$$\delta_1 I_B = \int dt\, e \int \mathcal{D}p\, e^{ip\cdot x} \left(\delta\phi - DTrA - (pA)_m \frac{\partial}{\partial p_m} \right) \Psi(p) \quad . \tag{5.31}$$

The first term in the bracket comes from the variation of $dt\, e$, the second term $-DTrA$ from the Jacobian of the change of variables, and the last term from the shift in the argument of $\tilde{\Psi}$ due to the change in variables. Using the expression for A_{nm} above,

$$\delta_1 I_B = \int dt\, e \int \mathcal{D}p\, e^{ip\cdot x}\Big\{\delta\phi\left(1 + D\sum_{m=1}^{\infty} m + \sum_{m=1}^{\infty} m p_m \frac{\partial}{\partial p_m}\right) + \sum_{n=1}^{\infty}\frac{1}{n+1}K_n(\delta\phi)\sum_{m=1}^{\infty} m p_{n+m}\frac{\partial}{\partial p_m}\Big\}A \tag{5.32}$$

Making the transcription

$$p_m = \frac{1}{\sqrt{2\alpha'}}\alpha_m \; m = 0,1,2,\ldots ; \quad \frac{\partial}{\partial p_m} = -\frac{\sqrt{2\alpha'}}{m}\alpha_{-m} \; , m = 1,2,3,\ldots \tag{5.33}$$

and noting that the Virasoro generators L_n are given by

$$L_n = \frac{1}{2}\sum_{m=-\infty}^{\infty} :\alpha_{-m}\alpha_{n+m}: - \sum_{m=1}^{\infty} m\frac{\partial}{\partial p_m}\cdot p_{n+m} + \alpha'\sum_{m=0}^{n} p_m p_{n-m} \; , \tag{5.34}$$

we can recognize in the appearance of L_0 in the coefficient of $\delta\phi$ and L_n in the coefficient of $K_n(\delta\phi)$. However L_0, L_n do not completely appear; p_o^2 and $p_m p_{n-m}$ type terms are absent. We find that this contribution comes from the anomaly present in the correlation functions $\left\langle \frac{d^n}{dt^n}\frac{d^n}{dt^n}{}^m x(t)\right\rangle$. This anomalous contribution indeed comes in the form

$$\delta_2 I_B = -\frac{1}{2}\int dt\, e \int \mathcal{D}p\, e^{ip\cdot x}\sum_{n,m=0}^{\infty}\delta_2\langle x_n(t)x_m(t)\rangle\, p_n p_m \tilde{\Psi}(p) \tag{5.35}$$

and the final total Weyl variation $\delta I_B = \delta_1 I_B + \delta_2 I_B$ is given by

$$\delta I_B = \int dt\, e^{\phi}\int \mathcal{D}p\, e^{ip\cdot x}\left(\delta\phi(1 - L_0)\sum_{n=1}^{\infty}\left[\frac{1}{n+1}K_n(\delta\phi)L_n + Y_n\right]\right)\tilde{A}(p) \quad , \tag{5.36}$$

where Y_n are extra terms which are shown to vanish. The equation of motion following from Weyl invariance $\delta I_B + \delta I_g = 0$ is then

$$\begin{aligned} &\delta\phi \left[(1 - L_0)\tilde{A} + L_{-1}\xi_1\right] + \\ &\sum_{n=1}^{\infty} \frac{1}{n+1} K_n(\delta\phi) \left\{-L_n\tilde{A} + L_{-1}\xi_{n+1} + (n+1)\xi_n\right\} = 0 \end{aligned} \tag{5.37}$$

Compare this to Witten's string field theory with the following truncated set of fields

$$|\Phi >= |\tilde{A} > + \sum_{n=1}^{\infty} c_{-n+1} b_{-1} |\xi_n \rangle \tag{5.38}$$

where $|\tilde{A} >$ and $|\xi_n >$ are built out of the α oscillators only acting on a vacuum $|- >= c_1|0 >$, and demand $Q|\Psi >= 0$ we obtain the equations of motion

$$\begin{aligned} &(1 - L_0)\tilde{A} + L_{{}_1}\xi_1 = 0 \\ &- L_n\tilde{A} + L_{-1}\xi_{n+1} + (n+1)\xi_n = 0\,, \quad n = 1, 2, \ldots \quad . \end{aligned} \tag{5.39}$$

Thus with the proviso $Y_n = 0$ we have derived the SFT equations from Weyl invariance.

Several comments are relevant concerning these equations and their derivation. First, the Weyl coefficients are fixed by requiring the vanishing of Y_n terms. The L_n structure – which is present in the SFT equations – then emerges from the Weyl transformation. We have seen that this completely fixes the coefficients for all the 2–point functions. The coefficients obtained are such that the variation of the part linear in ϕ itself produces the SFT equations. The nonlinear terms in ϕ combine in such a way that the consistency of the approach can be checked. The coefficients obtained in this manner satisfy the Leibnitz rule for the 2–point functions, i.e.

$$\frac{d}{dt}\left\langle x^{(p)}x^{(q)}\right\rangle_{\phi}=\left\langle x^{(p+1)}x^{(q)}\right\rangle_{\phi}+\left\langle x^{(p)}x^{(q+1)}\right\rangle_{\phi}\quad . \tag{5.40}$$

In the 2d field theory the anomalies in general depend on the renormalization scheme and it is of interest to find the scheme which yields the above coefficients.

Secondly, our procedure yields partial gauge fixed equations. The partial gauge condition is evident from the truncated field expansion. In the ghost sector it omits states containing $b_{-2}, b_{-3 \ldots}$ oscillators, (e.g. at the first massive level it omits $Sc_{\circ}b_{-2}|- >$, thus working in the $S = 0$ gauge). In SFT, these states are responsible for gauge invariance associated with L_{-2}, L_{-3}, etc. (e.g. at the first massive level the equations $(1 - L_{\circ})\tilde{\Psi} + L_{-1}\xi_1 + L_{-2}S = 0\ ,\ \ -L_1\tilde{\Psi} + L_{-1}\xi_2 + 2\xi_1 = 0\ ,\ \ -L_2\tilde{\Psi} + 3\xi_2 + 4S = 0$, posses the full invariance. It would be of interest to derive the most general gauge invariant equations. They are likely to be associated with a variation more general than Weyl and a candidate is a generalized BRST transformation. This idea was pursued recently by Redlich [16].

6. Renormalization Group and the Banks–Martinec Equation

We have described in the previous section the coupling of general massive background fields to the string. The vertex function describing the couplings were in a one to one relation with states in string field theory. Indeed studying Weyl invariance we have seen that the string field equations emerge. This was done in the leading order and the equation recovered was the free string field equation. One could proceed further and it is rather interesting to deduce the interaction to all orders.

Banks and Martinec [9] have attempted to discuss the renormalization group flows in general and have derived a closed (functional) form for an equation of motion. Recently this approach was studied further by Polchinski and collaborators [12]. We now describe these results.

The idea of Wilsons renormalization group (block spin) method is to consider the theory at some fixed cutoff $\Lambda_\circ$ with a propagator

$$D_\Lambda(p^2) = \frac{1}{p^2+m^2} K(p^2/\Lambda_\circ^2) \quad , \quad K = \begin{cases} 1, & p^2 < \Lambda_\circ^2 \\ 0, & p^2 > \Lambda_\circ^2 \end{cases} \tag{6.1}$$

and derive the effective theory with some lower cutoff Λ_R by integrating out the modes from Λ to Λ_R. This leads to cutoff dependent couplings and new iteration terms in the Lagrangian. The generating functional with a source $j(p)$ having support below the cutoff Λ_R should be cutoff independent

$$\Lambda \frac{d}{d\Lambda} Z(j) = 0 \tag{6.2}$$

This is assured if in the effective Lagrangian the dependence on the cutoff follows

$$\begin{aligned} \Lambda \frac{dL}{d\Lambda} = & -\frac{1}{2} \int d^4p \frac{1}{p^2+m^2} \Lambda \frac{\partial}{\partial \Lambda} K \times \\ & \times \left\{ \frac{\partial L}{\partial \phi} \frac{\partial L}{\partial \phi} + \frac{\partial^2 L}{\partial \phi \partial \phi} \right\} \end{aligned} \tag{6.3}$$

For general string theory one likes to perform local variations to obtain a more general equation. Consider local scale transformations

$$\delta X^\mu(\sigma) = \epsilon(v^a(\sigma)\cdot\partial_a + \gamma(\sigma))X^\mu \tag{6.4}$$

where $\gamma(\sigma)$ is the anomalous dimension.

Scale invariance of correlation functions

$$< \prod_i F(\sigma_i) >= \int [dx^\mu] e^{-I[x]} F \tag{6.5}$$

follows if

$$\int d^2\sigma \left[(v^a\partial_a X^\mu + \gamma^\mu)\frac{\delta I}{\delta X^\mu} - \frac{\delta\gamma^\mu(\sigma)}{\delta X_\mu(\sigma)} \right] = 0 \tag{6.6}$$

An appropriate guess for the anomalous dimension operator $\gamma^\mu(\sigma, X^\mu)$ gives

$$\begin{aligned} &\int d\sigma(v^a\partial_a X^\mu + \psi^\mu)\frac{\delta I}{\delta X_\mu(\sigma)} - \frac{\delta\psi^\mu(\sigma)}{\delta X_\mu(\sigma)} \\ &+ \int d\sigma_1 d\sigma_2 \frac{1}{2}[\delta_v G(\sigma_1\sigma_2))] \left(\frac{\delta I}{\delta X_\mu(1)}\frac{\delta I}{\delta X_\mu(2)} - \frac{\delta^2 I}{\delta X(1)\delta X(2)} \right) \\ &+ X^\mu(\sigma)(G^{(-1)}\delta_v G)(\sigma_1,\sigma_2)\frac{\delta I}{\delta X_\mu(\sigma_2)} = 0 \end{aligned} \tag{6.7}$$

where $G(\sigma_1, \sigma_2)$ is the two–point correlation function.

For conformal transformations $v_a(\sigma)$ obeys

$$\delta_a v_b + \partial_b v_a = \delta_{ab}\partial\cdot v \tag{6.8}$$

The final form for the equation obeyed by the background Lagrangian given by

Polchinski and collaborators is

$$\int dt_1 dt_2 \hat{\theta}(1-(t_1-t_2)^2)\frac{v(t_1)-v(t_2)}{t_1-t_2}\left(\frac{\delta I_B}{\delta X(1)}\frac{\delta I_B}{\delta X(2)}-\frac{\delta^2 I}{\delta X(1)\delta X(2)}\right)$$
$$+\int dt v(t)\partial_t X^\mu \frac{\delta I_B}{\delta X_\mu}=0 \tag{6.9}$$

The equation looks quadratic in the background field since $I_B = \phi_i V_i(X)$ where ϕ_i are the vertex operators. But its meaning might be more subtle than the naive appearance. At the linearized level it agrees with string field theory equations.

The linear term in the background couplings I_B reads

$$\int d_{\tau_1} d_{\tau_2}\theta\frac{v(\tau_1)-v(\tau_2)}{\tau_1-\tau_2}\frac{\partial^2 I_B}{\partial X(\tau_1)\partial X(\tau_2)}+\int d\tau v(\tau)\dot{X}^\mu\frac{\partial I_B}{\partial X(\tau)}$$
$$+\int d\tau\frac{\partial}{\partial\tau}\sum_n\left(\frac{v^{(n)}(\tau)}{n!}\xi_n(\tau)\right)=0 \tag{6.10}$$

Here one has added a total derivative term by hand: it will correspond to the contribution of ghosts.

Expanding in terms of Taylor series

$$v(\tau)=\sum\frac{1}{n!}v^{(n)}\tau^n$$
$$X_\mu(\tau)=\sum\frac{1}{n!}X^{(n)}\tau^n \tag{6.11}$$

and doing the integral results in the following:

$$\sum_{n,m}\left(v^{(m)}X^{(n+1)}\frac{\partial I_B}{\partial X^{(m+n)}}+v^{(m+n+1)}\frac{\partial^2 I_B}{\partial X^{(n)}\partial X^{(m)}}+\right.$$
$$\left.+\sum_k\left(v^{(k+1)}\xi_k+v^{(k)}\dot{\xi}_k\right)=0\right. \tag{6.12}$$

We recognize in the first term the Virasoro generators

$$L_n = \sum_m X^{(m)} \frac{\partial}{\partial X^{(n+m)}} - 2 \sum_{\ell+m=n} \frac{\partial^2}{\partial X^{(\ell)} \partial X^{(m)}} \tag{6.13}$$

and identifying terms in $v^{(n)}$ obtain

$$\begin{aligned} L_\circ \mathcal{L}_B &= L_{-1}\xi_1 + \xi_\circ \\ L_n \mathcal{L}_B &= L_{-1}\xi_{n+1} + (n+1)\xi_n \end{aligned} \tag{6.14}$$

where one has used

$$\frac{\partial}{\partial \tau} = \frac{\partial X}{\partial \tau}\frac{\partial}{\partial X} = \dot{X}\frac{\partial}{\partial X} = \hat{L}_{-1} \quad . \tag{6.15}$$

The derived equations agree with string field theory and also the equations following from the demanding Weyl invariance [14].

A most interesting question concerns the full interacting quadratic equation for the fields. Polchinski, Liu and Hughes have tried to give an (indirect) argument that this theory reproduces closed string theory amplitudes. They have shown that the classical field with teh S–matrix ansatz obeys the Banks–Martinec equation. The integration of it would correspond to a cubic interesting Lagrangian for closed (and open) strings. There is some question about whether this is true. For open strings one has the cubic Witten field theory interaction, the above renormalization group approach gives something different and in that case one expects higher order terms also. For closed strings the known candidate field theories all correspond to nonpolynomial coupling.

ACKNOWLEDGEMENTS

The author acknowledges the pleasant atmosphere of the Karpacz Winter School. This work was supported by the DOE grant DE–AC02–76 ER03130.A026–Task A.

REFERENCES

1. C. Lovelace, *Phys. Lett.* **B135** (1984) 75;
   E. S. Fradkin and A. A. Tseytlin, *Phys. Lett.* **B158** (1985) 315 and *Nucl. Phys.* **B261** (1985) 1;
   C. G. Callan, D. Friedan, E. Martinec and M. Perry, *Nucl. Phys.* **B262** (1985) 593;
   A. Sen, *Phys. Rev.* **D32** (1985) 2102;
   Y. Watabiki, *Phys. Lett.* **210B** (1988) 113;
   C. Callan, C. Lovelace, C. Nappi and R. Yost, *Nucl. Phys.* **290** (1987);
   S. R. Das and S. J. Rey, *Phys. Lett.* **B186** (1987) 328.
2. B. Fridling and A. Jevicki, *Phys. Lett.* **174B** (1986) 75.
3. A. M. Polyakov, *Physica Scripta* **T15** (1987) 191.
4. C. Lovelace, *Nucl. Phys.* **B273** (1986) 413.
5. A. B. Zamolodchikov, *JETP Lett.* **43** (1986) 731.
   S. R. Das, G. Mandal, and S. R. Wadia, TATA preprint TIFR–TH–88/33;
   C. Vafa, Harvard preprint HUTP–88/A034;
   N. E. Mavromatos and J. Miramontes, Oxford preprint April 1988.
6. C. K. Lee and A. Jevicki, *Phys. Rev.* **D37** (1988) 1485.

7. S. R. Das and B. Sathiapalan,, *Phys. Rev. Lett.* **56** (1986) 2664.

8. I. Ichinose and B. Sakita, *Phys. Lett.* **175B** (1986) 423.

9. T. Banks and E. Martinec, *Nucl. Phys.* **B294** (1987) 733.

10. H. Ooguri and N. Sakai, *Phys. Lett.* **197B** (1987) 109.

11. R. Brustein, D. Nemeschansky and S. Yankielowicz, *Nucl. Phys.* **B301** (1988) 224;

    I. Klebanov and L. Susskind, *Phys. Lett.* **200B** (1988) 446;

    T. Inami and H. Nishino, *Phys. Lett.* **196B** (1987) 154;

    N. Nakazawa, K. Sakai, and J. Soda, Hiroshima preprint RKK 88–16;

    U. Ellwanger and J. Fuchs, Heidelberg preprint HD–THEP–88–10.

12. J. Hughes, J. Liu and J. Polchinski, Texas preprint UTTG–13–88.

13. J. Labastida and Maria Vozmediano, IAS preprint HEP–88/11.

14. S. Jain and A. Jevicki, *Phys. Lett.* **B220** (1989) 379.

15. P. C. Argyres and C. Nappi, Princeton preprint, PUPT–1118.

16. A. N. Redlich, IAS preprint, IASSNS–HEP–88/24.

17. D. Sheviz and A. Strominger, Santa Barbara preprint, March 1989.

GEOMETRY OF THE MASSIVE STRINGS

B.M. Barbashov, A.M. Chervyakov
Joint Institute for Nuclear Research, Dubna, USSR

There are many approaches to investigation of the dynamics of the relativistic string with massive ends/1-6/.

In papers/8-10 /the dynamics of a free relativistic string has been studied within the intrinsic geometry of minimal surfaces in Minkowski space. It is therefore natural to apply the geometrical methods not only to the surface but also to the curves restricting it.

Let x^μ (μ =0,1,2) be coordinates of Minkowski space E_2^1 and $u^1=\tau$, $u^2=\sigma$ are curvilinear coordinates of the pseudo-Riemannian manifold $M_1^1 \subset E_2^1$ with the following range of definition $0 \le \sigma \le \pi$, $\tau_1 \le \tau \le \tau_2$ where τ is considered as the evolution parameter. The surface covered by the relativistic string with masses m_1 and m_2 at ends is an extremal of the functional

$$S = -\gamma \iint d^2u \sqrt{-g} - \sum_{i=1}^{2} m_i \int_{\tau_1}^{\tau_2} d\tau \sqrt{\dot{x}^2(\tau,\sigma_i)}\,, \tag{I}$$

$$i=1,2\,,\ \sigma_1=0\,,\ \sigma_2=\pi\,.$$

The constant γ is of dimensionality of the mass squared,

$$g=\det\|g_{ij}\|\,,\ g_{ij}=(\partial x^\mu/\partial u^i)(\partial x_\mu/\partial u^j),\ i,j=1,2\,,$$

is the metric induced on M_1^1 . In what follows use will be made of the notation $\dot{x}^\mu = \partial x^\mu/\partial\tau$, $x'^\mu=\partial x^\mu/\partial\sigma$.

On the surface M_1^1 one can always locality introduce conformal coordinates in terms of which the matrix $g_{ij}(\tau,\sigma)$ gets diagonal

$$g_{11}=\dot{x}^2=-x'^2=-g_{22}=g\,;\ g_{12}=g_{21}=(\dot{x}x')=0\,. \tag{2}$$

The Euler equations for functional (I) in the coordinates (2) look as follows

$$\ddot{x}^{\mu}(\tau,\sigma) - x''^{\mu}(\tau,\sigma) = 0 \tag{3}$$

and the world trajectories of boundaries $x^{\mu}(\tau,0)$ and $x^{\mu}(\tau,\pi)$ of the surface M_1^1 of massive ends of the string are given by the equations

$$\begin{cases} \dfrac{d}{d\tau}\left(\dfrac{\dot{x}^{\mu}(\tau,0)}{\sqrt{\dot{x}^2(\tau,0)}}\right) = \dfrac{\gamma}{m_1} x'^{\mu}(\tau,0), \\ \dfrac{d}{d\tau}\left(\dfrac{\dot{x}^{\mu}(\tau,\pi)}{\sqrt{\dot{x}^2(\tau,\pi)}}\right) = -\dfrac{\gamma}{m_2} x'^{\mu}(\tau,\pi). \end{cases} \tag{4}$$

In addition to the metric tensor (2) we will introduce the second quadratic form of the world surface of the relativistic string with the coefficients

$$b_{ij} = n_{\mu} \frac{\partial^2 x^{\mu}}{\partial u^i \partial u^j}, \quad i,j = 1,2, \tag{5}$$

where n^{μ} is the vector of the unit space-like normal to the surface

$$n^2 = -1, \quad n_{\mu} \frac{\partial x^{\mu}}{\partial u^i} = 0, \quad i = 1,2. \tag{6}$$

In this way, at every point of M_1^1 there naturally arises a moving frame of reference formed by two tangent vectors $\partial x^{\mu}/\partial u^i$ (i=1,2) and the normal n^{μ}.

In the three-dimensional pseudo-Euclidean space the fundamental forms (2) and (5) completely define the world surface of the relativistic string if there hold the following relations / 8 -I0/: the coefficients of the second quadratic form (5) are of the form

$$b_{11} = b_{22} = \frac{A_+(u^+) - A_-(u^-)}{2}, \tag{7}$$
$$b_{12} = b_{21} = \frac{A_+(u^+) + A_-(u^-)}{2}$$

and the metric tensor (2) of the surface M_1^1 is given by the representation

$$g(u^+,u^-) = \frac{A_+(u^+)A_-(u^-)\left[f(u^+) - g(u^-)\right]^2}{4f'(u^+)g'(u^-)}, \qquad (8)$$

with $u^{\pm} = \tau \pm \sigma$ and arbitrary functions $A_{\pm}$ and f, g. The representation (8) allows a common fractional-linear transformation of $f(u^+)$ and $g(u^-)$ / 11 /

$$f(u^+) \Rightarrow \frac{af(u^+)+b}{cf(u^+)+d}, \quad g(u^-) \Rightarrow \frac{ag(u^-)+b}{cg(u^-)+d}. \qquad (9)$$

In what follows to fullfill the condition $g = \dot{x}^2 > 0$ we will assume that $\varepsilon(f') = \varepsilon(g')$ and $\varepsilon(A_+) = \varepsilon(A_-)$ where $\varepsilon(z) = \frac{z}{|z|}$, $z \neq 0$ is the sign function.

Now consider the world trajectories (4) limiting the string surface: Introduce the curvature $K_i(\tau)$ and torsion $\varkappa_i(\tau)$, (i=1,2) of the boundary curves $x^{\mu}(\tau, \sigma_i)$ satisfy the Frenet-Serret equations / 12 /

$$\frac{d}{d\tau}(\dot{x}^{\mu}/\sqrt{g}) = (-1)^{i+1} k_i(\tau)\, x'^{\mu}(\tau, \sigma_i), \qquad (10)$$

$$\frac{d}{d\tau}\eta^{\mu}(\tau, \sigma_i) = \varkappa_i(\tau)\, x'^{\mu}(\tau, \sigma_i), \quad i = 1, 2.$$

Here the vectors $\dot{x}^{\mu}/\sqrt{g}$, $(-1)^{i+1}x'^{\mu}/\sqrt{g}$ and η^{μ} form a moving reference frame of the surface M_1^1 at fixed $\sigma = \sigma_i$ (i=1,2), and are respectively tangents, normals, and binormals to the trajectories (4).

Comparing the boundary conditions (4) with the first relation of (10) at i=1,2 we get

$$K_i(\tau) = \gamma / m_i, \quad i = 1, 2. \qquad (11)$$

Thus, the radii of curvature of the trajectories (4) are constant and equal to m_i/γ, (i=1,2). Torsion of those trajectories can be found by projecting the second relation of (10) at i=1,2 onto the vector $x'^{\mu}(\tau, \sigma_i)$:

$$\varkappa_i(\tau) = \frac{b_{12}(\tau, \sigma_i)}{g} = \frac{A_+ A_- (f-g)^2}{2(A_+ + A_-)f'g'}, \qquad (12)$$
$$i = 1, 2.$$

Projection of the same equation (IO) onto the vector $\dot{x}^\mu(\tau,\sigma_i)$ gives the condition for the trajectories of the string massive ends to be asymptotic lines on the surface

$$b_{11}(\tau,\sigma_i)=0. \tag{13}$$

According to the differential geometry / 4 /, curves of that sort have a zeroth normal curvature at every point. Considering also that their binormals are directed along the vectors $\eta^\mu(\tau,\sigma_1)$ and $-\eta^\mu(\tau,\sigma_2)$, we respectively obtain

$$K_i(\tau)=(-1)^{i+1}\,K_{g_i}(\tau), \tag{14}$$

where K_{g_i}, (i=1,2) are geodesic curvature defined in the metric (2) by the expression / 12 /

$$K_{g_i}(\tau)=\frac{1}{2\sqrt{g(\tau,\sigma_i)}}\frac{\partial}{\partial\sigma}\ln g(\tau,\sigma_i),\ i=1,2. \tag{15}$$

Equations (14) and (15) demonstrate invariance of the curvature $K_i(\tau)$, under isometric deformation of the string surface in E_2^1. From (11) and (14)-(15) it follows that the asymptotic lines (4) are at the same time the lines of constant geodesic curvature. This gives the following boundary conditions for the metric $g(\tau,\sigma_i)$:

$$\frac{\partial}{\partial\sigma}\ln g(\tau,\sigma)\Big|_{\sigma=\sigma_i}=2\frac{\gamma}{m_i}(-1)^{i+1}\sqrt{g(\tau,\sigma_i)},\quad i=1,2. \tag{16}$$

The condition (13) allows us to connect the torsions $\mathcal{X}_i(\tau)$ (i=1,2), with the Gaussian curvature of the world surface of the relativistic string at $\sigma=\sigma_i$ (i=1,2). Inserting (7) into (13) we obtain, instead of the two functions $A_\pm(u^\pm)$, one periodic function

$$A_-(\tau)=A_+(\tau)=A(\tau),\ A(\tau-\pi)=A(\tau+\pi). \tag{17}$$

Using (7) and (17) we rewrite the relations (12) in the form

$$\mathcal{X}_i(\tau)=A(\tau+\sigma_i)/g(\tau,\sigma_i),\ i=1,2. \tag{18}$$

Thus, the world trajectories of the string massive ends are given by equations (4) as the asymptotic lines of constant geodesic curvature.

Substituting (8) into the boundary conditions (16) we get

$$\frac{d}{d\tau}\ln\frac{g'}{f'}+2\,\frac{f'+g'}{f-g}=K_1|A|\,\frac{|f-g|}{\sqrt{f'g'}},$$

$$f=f(\tau),\ g=g(\tau), \qquad (19)$$

$$\frac{d}{d\tau}\ln\frac{g'(-)}{f'(+)}+2\,\frac{f'(+)+g'(-)}{f(+)-g(-)}=-K_2|A(+)|\,\frac{|f(+)-g(-)|}{\sqrt{f'(+)\,g'(-)}},$$

$$f(+)=f(\tau+\pi),\ g(-)=g(\tau-\pi). \qquad (20)$$

A periodic function $A(\tau)$ in (19)-(20) represents a tensor of the second quadratic form of the world sheet M_1^1 under the transformation $\tau\to\bar{\tau}=E(\tau)$, $E'(\tau+\pi)=E'(\tau-\pi)$ and ensures, thereby, the invariance of eqs.(19)-(20). Depending on the values of the gauge fixing function $A(\tau)$, the solution of these equations defines local equivalent world sheets swept by the relativistic string with massive ends.

Using the representation (8) we may write the torsion (18) as follows

$$\varkappa_1(\tau)=\frac{4f'g'}{A(f-g)^2},\quad \varkappa_2(\tau)=\frac{4f'(+)\,g'(-)}{A(+)\big(f(+)-g(-)\big)^2}. \qquad (21)$$

Thus, as it follows from (19)-(20) and (21) the string world surface $M_1^1\subset E_2^1$ is determined completely by the curvature K_i and the torsion $\varkappa_i(\tau)$, i=1,2 of the string end trajectories.

Making use of the relations (21) and (22) we express $A(\tau)$ in terms of the torsion $\varkappa_i$ and substituting it into the boundary conditions (19) and (20) we get

$$\frac{d}{d\tau}\ln\frac{g'}{f'}+2\frac{f'+g'}{f-g}=\frac{4K_1}{|\varkappa_1(\tau)|}\frac{\sqrt{f'g'}}{|f-g|},$$

$$\frac{d}{d\tau}\ln\frac{g'(-)}{f'(+)}+2\frac{f'(+)+g'(-)}{f(+)-g(-)}=-\frac{4K_2}{|\varkappa_2(\tau)|}\frac{\sqrt{f'(+)g'(-)}}{|f(+)-g(-)|}. \tag{22}$$

References

1. Barbashov B.M. (1977) Nucl.Phys. B129, 175-188
2. Barbashov B.M., Nesterenko V.V. (1978) Particles and Nucleus, 9, 709-758 (in Russian)
3. Chernikov N.A., Shavokhina N.S. (1980) Teor.Mat.Fiz. 42, N 1, 59-70; 43, N 3, 356-366 (in Russian)
4. Frampton P.H. (1975) Phys.Rev. D12, 538-545
5. Chodos A., Thorn C.B. (1974) Nucl.Phys. B72, 509-522
6. Shavokhina N.S. (1982) Dokl.Akad.Nauk SSSR, 265, 852-856 (in Russian)
7. Eisenhart L.P. (1964) Riemannian Geometry. Princeton: Princeton University Press.
8. Omnes R. (1979) Nucl.Phys. B149, 269-284
9. Barbashov B.M., Koshkarov A.L. (1979) Teor.Mat.Fiz., 39, No 1, 27-34 (in Russian)
10. Barbashov B.M., Nesterenko V.V., Chervyakov A.M. (1979) Teor. Mat. Fiz. 40, N 1, 15-27 (in Russian)
11. Bianchi L., (1903) Lezioni di geometria differenziale. 2, Pisa: Spoerri.
12. Barbashov V.M. and Chervyakov A.M. (1986) Geometrical Method for Solving the Boundary Value Problem in the Theory of the Relativistic String with Massive Ends. JINR Prepr. P2-86-521, Dubna (in Russian)

HIGH-ENERGY UPPER BOUNDS ON SCATTERING AMPLITUDE IN A SPACE-TIME OF GENERAL DIMENSION

Masud Chaichian

Department of High Energy Physics, University of Helsinki, Siltavuorenpenger 20 C, SF-00170 Helsinki, Finland

and

Jan Fischer

Institute of Physics, Czechoslovak Academy of Sciences, Na Slovance 2, CS-18040 Prague 8, Czechoslovakia

1. Introduction

The high-energy behaviour of string scattering amplitudes has been intensively studied during the past 1-2 years. Remarkable progress has been made in correlating the string theory with the general properties of scattering amplitude. Soldate [1] investigated the high-energy unitarity of the partial-wave expansion of a closed string amplitude in flat space-time of a general dimension $D \geq 6$, and gave general argument that such partial-wave amplitudes in tree approximation would violate unitarity at sufficiently high energy. Gross and Mende [2] and Amati, Ciafaloni and Veneziano [3] studied the high-energy behaviour of the string and the superstring amplitude respectively and found dominating contributions in different kinematic regions. Muzinich and Soldate [4] looked into the behaviour of string amplitude by summing multiple Regge ized graviton exchange in the eikonal approximation.

I will report on our contribution to this development; details is published elsewhere [5]. We generalize the methods of Froissart and Martin [6] and use analyticity and unitarity to obtain high-energy upper bounds on the forward and non-forward

Talk given by M. Chaichian

scattering amplitude in flat space-time of a general dimension $D \geq 3$. Assumptions are analogous; let us mention that analyticity of the elastic scattering amplitude in $\cos\theta$ in an ellipse with foci at $\cos\theta = \pm 1$ is essential.

If N is the number of on-mass-shell particles involved, the number of independent variables is $N(N-3)/2$ and $(D-1)N - D(D+1)/2$ for $N \leq D$ and $N \geq D$ respectively. If $N = 4$, this number equals 2 for any $D \geq 3$; we therefore define s, t and u in the usual way.

2. Explicit form of the bounds

The partial-wave expansion of the four-scalar elastic scattering amplitude in D dimensions has the form

$$F_\lambda(s,t) = \phi_\lambda(s) \sum_{\ell=0}^{\infty} (N_\ell^\lambda)^{-1} C_\ell^\lambda(1) C_\ell^\lambda\left(1 + \frac{t}{2k^2}\right) f_\ell(s) \ , \tag{1}$$

where

$$\lambda = (D-3)/2 \ ,$$

$$\phi_\lambda(s) = 2\Gamma\left(\frac{D}{2} - 1\right)(16\pi)^{D/2-1} s^{2-D/2} \ ,$$

$$N_\ell^\lambda = \frac{2^{1-2\lambda} \ \pi \ \Gamma(\ell + 2\lambda)}{\ell! \ (\ell+\lambda) \ \Gamma^2(\lambda)} \ ,$$

$$C_\ell^\lambda(1) = \frac{\Gamma(\ell + 2\lambda)}{\ell! \ \Gamma(2\lambda)}$$

and the Gegenbauer polynomials $C_\ell^\lambda(x)$ can be represented in the following form :

$$C_\ell^\lambda(x) = M(\ell,\lambda) \int_0^\pi (x + \sqrt{x^2-1}\cos\varphi)^\ell (\sin\varphi)^{2\lambda-1} d\varphi \ , \tag{2}$$

with

$$M(\ell,\lambda) = \frac{\Gamma(\ell+2\lambda) \ \Gamma(\lambda+1/2)}{\sqrt{\pi} \ \Gamma(\ell+1) \ \Gamma(2\lambda) \ \Gamma(\lambda)} \ .$$

In the case of the Minkowski space D and λ equal 4 and 1/2 respectively and (1) becomes the Legendre expansion. In analogy with this case, the Gegenbauer polynomials are obtained, by integrating over irrelevant angles, from the generalized spherical functions which span the representation space of the $SO(D-1,1)$ group corresponding to unitary irreducible representations of its maximal compact subgroup. The normalization is such that the unitarity is given by

$$|f_\ell(s)|^2 \leq a_\ell(s) \leq 1 \ , \tag{3}$$

where $a_\ell(s) = Imf_\ell(s)$.

We can now use (2) to derive, in analogy with Martin's result [7] for $D = 4$, the following lower bound on $C_\ell^\lambda(x)$ (see [5] for details):

$$C_\ell^\lambda(x) \geq M(\ell,\lambda)u^\ell \int_0^{\varphi_0} (\sin\phi)^{2\lambda-1} d\varphi \ , \tag{4}$$

for any φ_0, $0 < \varphi_0 < \pi$, and $\lambda \geq 1/2$, where $u = x + \sqrt{x^2-1}\cos\varphi_0$.

We shall now assume that, for s fixed, the amplitude $F(s,t)$ is analytic in $\cos\theta$ in an ellipse with foci at $\cos\theta = \pm 1$ and the semimajor axis equal to $1 + R/s$. Further steps of the derivation are analogous to Martin 's derivation of the bounds in the $D = 4$ case. The resulting bounds on A have the form [5]

$$A(s,o) \ \leq \ K_1\left(\frac{1}{\sqrt{R}\cos\varphi_0}\right)^{D-2} s(lns)^{D-2} \ , \tag{5}$$

$$|A(s,t)| \ \leq \ K_2 \ s^{(7-D)/4}(lns)^{(D-1)/2} \ |\theta|^{(3-D)/2} \ , \tag{6}$$

$$|A(s,\cos\theta)| \ \leq \ K_3 \ s(lns)^{(D-1)/2} \ |t|^{(3-D)/4} \ , \tag{7}$$

where the constants K_i, $i = 1,2,3$, are independent of s and t; the φ_0- and R-dependence of the forward scattering bound has been written explicitly for further use. Since polynomi al boundedness has been assumed, the constants depend also on the parameters of the polynomi al bound.

Bounds of the same form are obtained also for the modulus $|F|$ of the scattering amplitude.

The case $D = 3$ ($\lambda = 0$) has to be considered separately because undefined expressions are met during the calculation. But all steps make sense with λ small positive, and the limit $\lambda \to 0_+$ can be carried out in the final expressions.

3. Discussion

Let us consider elastic scattering of two scalar particles in D-dimensional flat space-time in tree approximation. The partial-wave amplitude $f_\ell(s)$ can be expressed as

$$f_\ell^{(\lambda)}(s) = \frac{2^{D-3}}{\phi_\lambda(s)C_\ell^\lambda(1)} \int_{-s}^{0} \frac{dt}{s}\left(-\frac{t}{s} - \frac{t^2}{4s^2}\right)^{D/2-2} C_\ell^\lambda(1+\frac{2t}{s})F_\lambda(s,t) \ . \tag{8}$$

Due to the graviton term, the scattering amplitude will be dominated by s^2/t near $t = 0$, $s \to \infty$. On the Regge trajectory, we have

$$f_\ell^{(\lambda)}(s) \sim s \int_{-\varepsilon(s)}^{0} dt(-t)^{D/2-3} + \ ... \ , \quad \varepsilon(s) \sim \frac{1}{lns} \ . \tag{9}$$

It is interesting to discuss the D-dependence of this expression. Take first $D = 4$; then (9) exhibits infrared divergence and there is no partial-wave expansion. For $D \geq 5$, the expression is regular, but the integration leads to an unbounded s-dependence,

$$|f_\ell^{(\lambda)}(s)| \sim s(lns)^{-(D-4)/2} \quad ,$$

which at sufficiently high s violates unitarity (3).

A massive spin-2 term, $s^2/(t - m^2)$, is more appropriate for comparison, because we assumed a mass gap. But unitarity is violated also in this case: $|f_\ell^{(\lambda)}(s)| \sim s$.

Amati, Ciafaloni and Veneziano [3] have studied the high-energy behaviour of string loops to all orders; it turns out that the graviton pole still dominates at small $|t|$. The individual components of σ_{tot},

$$\sigma_{\text{tot}} = \sigma_{\text{el}} + \sigma_{\text{diff}} + \sigma_{\text{inel}} \quad ,$$

have, according to [3], the following high-energy dependence:

$$\begin{aligned} \sigma_{\text{tot}} &\sim s^{\frac{D-2}{D-4}} \quad , \\ \sigma_{\text{diff}} &\sim s \quad , \\ \sigma_{\text{inel}} &\sim (\ln s)^{D-2} \quad . \end{aligned} \tag{10}$$

One can expect that σ_{inel} will be the "softest" part of σ_{tot}, with the weakest influence of the $t = 0$ pole. Note in this connection that σ_{inel} obeys (and saturates) our bound (5), which we derived assuming there is no singularity at $t = 0$ for any finite s.

Formal replacement of R with $1/s$ in (5) leads to the following bound

$$\sigma \leq c \;\; s^{(D-2)/2}(lns)^{D-2} \, ,$$

which, for $D \geq 6$, is satisfied by Amati et al's σ_{tot} (see (10)) and, for $D = 4$, coincides with the Greenberg-Low bound [8]:

$$\sigma \leq c \;\; s \; (lns)^2 \quad .$$

Therefore to have consistency with the behaviour of σ_{tot}, we should assume that $R \sim \frac{1}{s}$ and $D \geq 6$. The requirement $D \geq 6$ comes independently also from the validity of partial wave expansions when massless particles are involved. The shrinkage of the analyticity domain, i.e. $R \sim \frac{1}{s}$, is particular to the nonlocal field theories (for details, see [5]). Thus the implication on the string theory is the indication that the latter is a nonlocal one.

References

[1] M. Soldate: Phys. Lett. **186B** (1987) 321.

[2] D.J. Gross, P. Mende: Phys. Lett. **197B** (1987) 129 and Nucl. Phys. **B303** (1988) 407.

[3] D. Amati, M. Ciafaloni, G. Veneziano: Phys. Lett. **197B** (1987) 81; Int. Journ. Mod. Phys. **3A** (1988) 1615 and SISSA preprint 121 EP (Sept. 88).

[4] I.J. Muzinich, M. Soldate: Phys. Rev. **D37** (1988) 359.

[5] M. Chaichian, J. Fischer: Nucl. Phys. **B303** (1988) 557 and in preparation.

[6] M. Froissart: Phys. Rev. **123** (1961) 1053;
A. Martin: Phys. Rev. **129** (1963) 1432; Nuovo Cinvoto **42A** (1966) 930.

[7] A. Martin, F. Cheung: Analyticity properties and bounds of the scattering amplitudes (Gordon and Breach, 1970).

[8] O.W. Greenberg, F.E. Low: Phys. Rev. **124** (1961) 2047.

III. FUNCTIONAL INTEGRATION AND CONFORMAL FIELD THEORY

A SHORT OVERVIEW OF MATHEMATICAL APPROACHES TO FUNCTIONAL INTEGRATION

by

Sergio Albeverio[*,♯,**], **Sylvie Paycha**[*], **Sergio Scarlatti**[***]

Fakultät für Mathematik, Ruhr–Universität Bochum,
D–4630 Bochum 1 (FRG)

[*] SFB 237 – Essen–Bochum–Düsseldorf
[♯] BiBoS – Research Centre
[**] CERFIM, Locarno
[***] Supported by a CNR–Research Grant, Italy

Abstract

In the first part of these lectures we give a short overview of mathematical techniques for defining and handling some functional integrals occurring in non relativistic quantum theory and in quantum field theory. We cover oscillatory type integrals (Feynman path integrals) as well as probabilistic integrals.
In the second more detailed part we discuss an approach to probabilistic integrals for bosonic strings. In this approach a heat kernel regularization procedure for determinants of elliptic operators is used. We present in details new developements concerning this procedure, in terms of approximations for transformed Gaussian measures on nuclear spaces.

0. Structure of the paper

This paper is based essentially on 3 lectures given at the Karpacz school. In the first part (corresponding to lectures given by the first author) a short overview of mathematical techniques to give a meaning and study heuristic path integrals for quantum theory is presented. This part is divided in 5 sections. The first one is an introduction, presenting the basic division in integrals of oscillatory type and integrals of probabilistic type. In the second section oscillatory type (Feynman's type) integrals are discussed. In section 3 probabilistic integrals are discussed, with applications in non relativistic quantum theory as well as in quantum field theory. In section 4 we give a brief introduction to methods of non standard analysis to handle functional integrals. In section 5 we discuss functional integrals in more geometrical settings.
The second part of the paper (corresponding to a lecture given by the second author) consists of two parts A and B. In part A we briefly present probabilistic type integrals for bosonic strings. In part B we take the opportunity to present a rather detailed discussion of new developments concerning the heat kernel regularization procedure for determinants of elliptic operators arising in the computation of probabilistic functional integrals for strings. This procedure is put in relation with operations performed on certain probabilistic infinite dimensional gaussian measures.

It is a pleasure to thank the organizers, and in particular Zbigniev Haba, for kind invitations to a very stimulating School. We are also grateful to Philippe Blanchard, Anne Boutet de Monvel, Hannes Brasche, Zdzislaw Brzezniák, Zbigniev Haba, Koichiro Iwata, Jürgen Potthof, Michael Röckner, Ludwig Streit, Boguslav Zegarlinski for the joy of collaboration and many interesting discussions. We thank E. Aich, M. Jarrath, R. Kirchhoff, H. Niehrenheim and C. Welge for their help in the setting of the manuscript.

DEDICATED TO THE MEMORY OF RAPHAEL HØEGH–KROHN (1938–1988)

I. QUANTUM (FIELD) THEORY

1. Introduction

Functional integration is in a way a strange domain. It is on the one side a heuristic tool, in the hands of physicists (and chemists) permitting to relate different quantities, and do heuristic computations (e.g. in perturbation theory). If one tries to justify the operations done one encounters most of the time inextricable mathematical problems: the integrals to be evaluated are not defined and sometimes even supposing they were, so the operations done on them defy mathematical justifications. On the other hand, there is a well defined mathematical theory of certain types of functional integrals, of which the Wiener integral introduced by Wiener in the early twenties is the prototype, but this theory is rarely referred to or used in connection with heuristic functional integrals. The gap between the mathematical theory and the physical praxis is enormous and in a sense a unique phenomenon, in the complex history of relations between mathematics and physics. It is a big challenge to try to reduce the gap; in our opinion, this is also, in the long run, absolutely necessary, if physics should not be reduced to a set of computational rules without basic justification and functional integration in the mathematical sense be deprived of its most interesting examples.

In these lectures we shall try to give a short overview of some mathematical approaches to functional integrals. We hasten to say that we can not be complete, due to lack of space and time. E.g. we concentrate ourselves to the discussion of functional integrals suited for theories concerning bosons, and leave completely outside our discussion the cases of functional integrals arising in the study of fermions (a fortiori we also do not discuss here supersymmetric functional integrals). For these omitted topics see e.g. [1] [0]). As well known there are two basic types of functional integrals, which we characterize as follows

A) Integrals of Feynman type: these are oscillatory integrals (associated with hyperbolic type of equations, pseudodifferential operators, Fourier integral operators, wave equations ...)

B) Integrals of Wiener type: these are probabilistic integrals (associated with elliptic and parabolic problems, potential theory, diffusion ...)

We can first characterize the difference between these 2 types of integrals by looking at their respective finite dimensional analogues:

A′) $I(f) = \int_{\mathbb{R}^n} f(x)\, e^{i\Phi(x)} dx$, with f, Φ real valued functions over $\mathbb{R}^n$. We remark that the integral is well defined for $f \in L^1(\mathbb{R}^n\,, dx)$ and Φ Lebesgue measurable, and that $e^{i\Phi(x)}dx$ is a well defined complex measure on any compact subset of $\mathbb{R}^n$, so that for f of compact support we can look upon $I(f)$ as the integral of f against the complex measure $e^{i\Phi(x)}dx$ on the support of f. For f not of compact support but locally integrable one can define $I(f)$ as $\lim_{K \nearrow \mathbb{R}^n} \int e^{i\Phi(x)} f(x)\,dx$, whenever the limit taken along increasing compacts exists and is independent of the increasing sequence of compacts. In this case $I(f)$ is a so called improper Lebesgue integral (improper refering to the fact that the integral might exist without absolute integrability of the integrand).

B′) $I(f) = \int_{\mathbb{R}^n} f(x)\,e^{-\Phi(x)}\,dx$, with f real valued, $\Phi(x) \in [0,\infty]$. The integral is well defined for Φ measurable and s.t. $f \in L^1(\mathbb{R}^n, e^{-\Phi}dx)$, which is certainly the case, due to $\Phi \geq 0$, whenever $f \in L^1(\mathbb{R}^n, dx)$. In this case $\mu(dx) = e^{-\Phi(x)}dx$ is a positive σ–finite measure on $\mathbb{R}^n$ (finite iff $e^{-\Phi} \in L^1(\mathbb{R}^n, dx)$). Obviously $I(f)$ can be looked upon as the integral of f against $\mu(dx)$.

We remark, anticipating our discussion of non–standard analytic methods, that both integrals $I(f)$ in A′), B′), when they exist, can be looked upon as standard parts of hyperfinite sums $\sum_{j \in {}^*\mathbb{Z}} \delta^d\, f(j\delta)\, e^{i\Phi(j\delta)}$ resp. $\sum_{j \in {}^*\mathbb{Z}} \delta^d\, f(j\delta)\, e^{-\Phi(j\delta)}$, with δ a positive infinitesimal (c.f. [2]).

Now let us comment separately on the two types of integrals in the infinite dimensional case. Let us make the convention to use the adjective heuristic for all proposed expressions which are not (or not yet) mathematically well defined (these are sometimes also called informal by mathematicians and formal by (mathematical) physicists). Sometimes (but not always) a heuristic expression has indeed also heuristic value (in the proper sense of the word heuristic), even though it can not be given rigorous meaning in the given context and /or by present mathematical tools. We take moreover here the basic attitude that "functional integration" should be or become a field of mathematical investigations as rigorous as the rest of mathematics. The challenge is in particular to give meaning as far as possible to the heuristic calculations made by physicists.

2. Oscillatory integrals

2.1 Configurational integrals, quantum mechanics and "time slicing"

This type of integrals were discussed heuristically (without mathematical definition) by Dirac 1932 and more explicitely by Feyman 1948 see references in [3]. They appear in connection with Schrödinger equation on $\mathbb{R}^d, t \geq 0$

$$i\,h \frac{\partial}{\partial t}\psi = H\psi \tag{2.1}$$

$$H \equiv -\frac{h^2}{2}\Delta + V \tag{2.2}$$

with initial condition $\psi(t=0) = \psi_0$ (we choose the mass to be 1; h is Planck's constant divided by 2π). Heuristically Schrödinger equation is solved by the heuristic "Feynman path integral"

$$\psi_t(x) = \text{``} \int_{\Gamma_x} e^{\frac{i}{h}S_t(\gamma)}\psi_0(\gamma(0))d\gamma\text{''} \tag{2.3}$$

where Γ_x is a "space of paths" γ s.t. $\gamma(t) = x$ and

$$S_t(\gamma) \equiv \frac{1}{2}\int_0^t |\dot{\gamma}(s)|^2 ds - \int_0^t V(\gamma(s))ds \tag{2.4}$$

the classical action along the path γ.

It is well known, see e.g. [3], that one way (in Dirac's spirit) to give a mathematical meaning to the right hand side of (2.3) is to use Lie–Trotter formula.

Let us assume V is bounded continuous, then H is a well defined self-adjoint operator in $L^2(\mathbb{R}^d, dx)$ on the definition domain of $-\Delta$. By Stone's theorem $(e^{-\frac{i}{h}tH}, t \in \mathbb{R})$ is a well defined unitary group and the solution of (2.1) is given by

$$\psi_t = e^{-\frac{i}{h}tH}\psi_0 \tag{2.5}$$

(both ψ_t and ψ_0 being interpreted as elements in $L^2(\mathbb{R}^d, dx)$). By Lie–Trotter formula, with $H_0 = -\frac{h^2}{2}\Delta$ one has

$$e^{-\frac{i}{h}tH} = s - \lim_{n\to\infty} \left(e^{-\frac{i}{h}\frac{t}{n}V}\, e^{-\frac{i}{h}\frac{t}{n}H_0}\right)^n \tag{2.6}$$

where the convergence is in the strong L^2-sense (hence denoted by $s - \lim$) (i.e. one has convergence on the Hilbert space for the operators applied to any vector in the Hilbert space).

It is well known that $e^{it\Delta}, t > 0$ has an integral kernel $p_{it}(x,y)$ s.t.

$$\left(e^{it\Delta} f\right)(x) = \int_{\mathbb{R}^d} p_{it}(x,y) f(y) dy \tag{2.7}$$

for all $f \in \mathcal{S}(\mathbb{R}^d)$ (Schwartz test function space) (the integral being in the usual Riemann or Lebesgue sense). In fact

$$p_{it}(x,y) = \frac{e^{i\frac{|x-y|^2}{2t}}}{(2\pi it)^{\frac{d}{2}}}, \quad t > 0\,. \tag{2.8}$$

If for $t = 0$ we interpret $p_{it}(x,y)dy$ in (2.7) as the Dirac measure $\delta(x-y)$ then (2.7) holds for all $t \geq 0$. Using (2.8) we have, for $\psi_0 \in \mathcal{S}(\mathbb{R}^n)$

$$\left(\left(e^{-\frac{i}{\hbar}\frac{t}{n}V} e^{-\frac{i}{\hbar}\frac{t}{n}H_0}\right)^n \psi_0\right)(x) = \int_{\mathbb{R}^{nd}} e^{\frac{i}{\hbar}\tilde{S}_t(\gamma_n)} \psi_0\left(\gamma_n(0)\right) d\gamma_n \tag{2.9}$$

with γ_n the polygonal (piecewise linear) path $[0,t] \to \mathbb{R}^d$ s.t. $\gamma_n(t) \equiv x$, $\gamma_n(j\frac{t}{n}) \equiv x_j, j = 0, \ldots, n$ and

$$\tilde{S}_t(\gamma_n) \equiv \sum_{j=1}^n \frac{|\gamma_n(j\frac{t}{n}) - \gamma_n((j-1)\frac{t}{n})|^2}{(\frac{t}{n})^2} - \sum_{j=1}^n V\left(\gamma_n(j\frac{t}{n})\right), \tag{2.10}$$

$$d\gamma_n \equiv \prod_{j=0}^{n-1} \left(2\pi i h \frac{t}{n}\right)^{-n\frac{d}{2}} dx_j\,. \tag{2.11}$$

The integral in (2.9) is in the improper Riemann or Lebesgue sense.
Introducing (2.9) into (2.6) and using (2.5), (2.4), we get

$$\psi_t(x) = s - \lim_{n\to\infty} \int_{\mathbb{R}^{nd}} e^{\frac{i}{\hbar}\tilde{S}_t(\gamma_n)} \psi_0\left(\gamma_n(0)\right) d\gamma_n \tag{2.12}$$

where the convergence is in the $L^2(\mathbb{R}^d, dx)$ sense.
The right hand side of this formula gives thus a mathematical sense to the heuristic Feynman path integral formula (2.3).

Remarks

1) This just means that we can define the r.h.s. $I(\psi_0)$ of (2.3) by the r.h.s. of (2.12). Obviously this does not yet imply we can now operate freely with the symbol on the r.h.s. of (2.3), handling as if they were "usual integrals": any operation we do has to be mathematically justified starting from the definition, i.e. using the fact that

$$I(\psi_0) = s - \lim_{n\to\infty} I_n(\psi_0) \tag{2.13}$$

with $I_n(\psi_0) = \int_{\mathbb{R}^{nd}} e^{\frac{i}{\hbar}\tilde{S}_t(\gamma_n)} \psi_0\left(\gamma_n(0)\right) d\gamma_n$.

2) This method of giving meaning to Feyman path integrals is often called "time slicing" or "polygonal path" method. It has been refined in various directions:
   a) V which are not necessarily bounded and continous: see e.g. [21], [9], [2].

   We note that the extension to arbitrary initial conditions $\psi_0 \in L^2(\mathbb{R}^d)$ is immediate, in the L^2–sense, $\psi_0 \to I(\psi_0)$ being continuous in the Hilbert space sense in ψ_0, since $I(\psi_0) = e^{-\frac{i}{\hbar}tH}\,\psi_0$ and $e^{-\frac{i}{\hbar}tH}$ is unitary.
   b) replacement of the approximation by polygonal paths γ_n "interpolating linearly in time $\frac{t}{n}$ between x_j and x_{j+1}" through an approximation by solutions γ_c of the classical equation of motion associated with Schrödinger equation "with final condition $\gamma_c(t) = x$" i.e.

$$\begin{cases} \ddot{\gamma}_c(t) = -\nabla V\left(\gamma_c(t)\right) \, , \, t \geq 0 \\ \gamma_c(t) = x \, . \end{cases} \tag{2.14}$$

   This has been done by Fujiwara [48].

Of course, as with Riemann or Lebesgue approximations of finite dimensional integrals, it is very useful to gather information on the operations which can be done with the limit quantities.

One would like to handle the latter "directly as integrals" (without always to have to take into account the fact that they have been defined as limits). Hence it is useful to have a set of proven rules to handle such "limit integrals". This can be provided by alternative definitions of the integrals. Let us at this point mention that historically (in work by Dirac and Feynman) one of the main purposes of introducing heuristic "path integrals" was to be able to "understand" the fact that classical mechanics is an approximation of quantum mechanics through a "stationary phase method" applied to "path integrals". In fact heuristically (and correctly so for the finite classical approximation $I_n(\psi_0)$) $I(\psi_0)$ is an oscillatory integral, and in the (of course mathematical, not to be understood in strict physical sense) limit $h \downarrow 0$ the leading contribution should be given (exploiting the analogy with finite dimensional oscillatory integrals) by the stationary (i.e. critical) points of the phase function S_t. By the classical Hamilton's principle the stationary points of S_t are precisely the solutions of Newton's equation of motion.

A mathematical justification of this procedure is possible, but cumbersome, if tried out directly on $\lim I_n(\psi_0)$ (because of the double limit procedure $n \longrightarrow \infty, h \downarrow 0$) (see [2], for comments on this). It is more appealing and profitable to look at alternative definitions of $I(\psi_0)$ which would permit the building of a method of stationary phase in infinite dimensions. One such definition is as follows:

2.2 The definition of oscillatory integrals and the method of stationary phase in infinite dimensions by the method of Fourier transforms.

This method has its origins in work by K. Ito (see references in [3]). It has been developed extensively by Albeverio and Høegh-Krohn in [3] (see also e.g. [4], [5], [6], [7], [8], [9], [10], [11], [12], [13], [14]).

Here is the idea: Split the action functional S_t in a kinetic part

$$S_t^0(\gamma) \equiv \frac{1}{2}\int_0^t |\dot{\gamma}(s)|^2 ds \tag{2.15}$$

and an interaction part

$$V_t(\gamma) \equiv \int_0^t v(\gamma(s))ds \,. \tag{2.16}$$

Remark that S_t^0 is quadratic in γ. It is suitable to introduce, instead of a space Γ_x of path ending at x as above, a linear space of paths Γ s.t. $\gamma(t) = 0$ (the passage can be achieved once a definition is available by a simple translation).

We would then like to define integrals of oscillatory type over a linear space of paths. Let us look again at the finite dimensional analogy A') with $\Phi(x) = \frac{1}{2}(x, Bx)$ with B a $n \times n$ symmetric matrix and $f(x)$ replaced by $(2\pi i)^{-\frac{n}{2}} f(x)$, for reasons which we will see in a while, i.e.

$$J(f) = \sqrt{\det B}(2\pi i)^{-\frac{n}{2}} \int_{I\!R^n} e^{\frac{i}{2}(x,Bx)} f(x)dx \tag{2.17}$$

(we have set here $h = 1$ for convenience). Let f be the Fourier transform of a some bounded measure on $I\!R^n$, i.e. $f(x) = \int e^{i\alpha x} d\mu_f(\alpha)$. We denote by $\mathcal{F}(I\!R^n)$ the space of Fourier transforms of bounded complex measures. By the fact that the Fourier transform of $e^{\frac{i}{2}(x,Bx)}$ in distributional sense is

$$e^{-\frac{i}{2}(\alpha,B^{-1}\alpha)} \tag{2.18}$$

we have as a special case of Parseval's formula for distributions, if $f \in L^1(I\!R^n, dx)$,

$$\int_{I\!R^n} e^{-\frac{i}{2}(\alpha,B^{-1}\alpha)} d\mu_f(\alpha)$$

$$= \sqrt{\det B}(2\pi i)^{-\frac{n}{2}} \int_{R^n} e^{\frac{i}{2}(x,Bx)} f(x)dx \,. \tag{2.19}$$

Let us set
$\tilde{I}_n(f) \equiv \tilde{\int}_{I\!R^n} e^{\frac{i}{2}(x,Bx)} f(x)dx \equiv \int_{I\!R^n} e^{\frac{i}{2}(x,Bx)} f(x)\sqrt{\det B}\frac{dx}{(2\pi i)^{\frac{n}{2}}}$, for $f \in L^1(I\!R^n dx) \cap \mathcal{F}(I\!R^n)$.

Since the l.h.s. of (2.19) also exists for $f \in \mathcal{F}(I\!R^n)$, it is natural to extend the definition of $\tilde{\int}_{I\!R^n}$ by setting

$$\tilde{I}_n(f) \equiv \tilde{\int}_{I\!R^n} e^{\frac{i}{2}(x,Bx)} f(x)dx = \int_{I\!R^n} e^{-\frac{i}{2}(\alpha,B^{-1}\alpha)} d\mu_f(\alpha) \tag{2.20}$$

for all $f \in \mathcal{F}(\mathbb{R}^n)$.

It is not difficult to show that the extension is continuous in $\mathcal{F}(R^n)$ (in the topology given by the total variation $\|f\|_0 \equiv \int d|\mu_f|) \equiv \|\mu_f\|$, so that if $f_k \in (\mathcal{F} \cap L^1)(\mathbb{R}^n)$ and $f_k \longrightarrow f$ in $\mathcal{F}(\mathbb{R}^n)$, then

$$\tilde{I}_n(f_k) \longrightarrow \tilde{I}_n(f) \text{ as } k \longrightarrow \infty.$$

We look upon $\mathcal{F}(\mathbb{R}^n)$ as the space of integrable functions for the integral $\tilde{I}_n$. Now let $\mathcal{H}$ be a separable Hilbert space with basis $\{e_n, n \in \mathbb{N}\}$ and scalar product $<,>$. Let B a bounded symmetric operator on $\mathcal{H}$ with bounded symmetric inverse B^{-1}. Let $\mathcal{F}(\mathcal{H})$ be the space of Fourier transforms of complex measures μ_f on $\mathcal{H}$ (it is a Banach algebra with norm $\|f\|_0 \equiv \sup\|\mu_f\|$). Then $\int_{\mathcal{H}} e^{-\frac{i}{2}(\alpha,B^{-1}\alpha)} d\mu_f(\alpha) \equiv \tilde{I}(f)$ is well defined. If f is a cylinder function (i.e. $f(\gamma) = \tilde{f}(< e_{i_1}, \gamma >, \cdots, < e_{i_k}, \gamma >)$ for some $\tilde{f} \in \mathcal{F}(\mathbb{R}^{kn}), i_1, \cdots, i_k \in \mathbb{N}$) and B commutes with the projection onto the closed subspace $V_{kn} \cong \mathbb{R}^{kn}$ generated by $e_{i_1}, \cdots, e_{i_k}$, then $\tilde{I}(f) = \int e^{-\frac{i}{2}(\alpha,(B^{-1})_{kn}\alpha)} d\mu_{\tilde{f}}(\alpha), = \int_{\mathbb{R}^n} e^{\frac{i}{2}(x,(B)_{kn})x} \tilde{f}(x)dx$, with $(B^{-1})_{kn}$ the restriction of B^{-1} to $V_{kn} \cong \mathbb{R}^{kn}$. Moreover if $f \in \mathcal{F}(\mathcal{H})$ is limit as $n \longrightarrow \infty$ of f_m of the above form, then $\tilde{I}(f) = \lim_{m\to\infty} \tilde{I}(f_m)$. It is then natural to introduce the notation

$$\tilde{\int_{\mathcal{H}}} e^{\frac{i}{2}(\gamma,B\gamma)} f(\gamma) d\gamma$$

for $\tilde{I}(f)$. Thus

$$\tilde{I}(f) \equiv \tilde{\int_{\mathcal{H}}} e^{\frac{i}{2}(\gamma,B\gamma)} f(\gamma) d\gamma = \int_{\mathcal{H}} e^{-\frac{i}{2}(\alpha,B^{-1}\alpha)} d\mu_f(\alpha), \tag{2.21}$$

where the integral on the r.h.s. is well defined. $\tilde{I}(f)$ is called the normalised integral of f (with respect to B). Basic properties of this integral have been established in [3], it is a linear continuous normalized ($\tilde{I}(1) = 1$) functional from $\mathcal{F}(\mathcal{H})$ into $\mathbb{C}$, which has natural transformation properties under translation and rotations. Moreover there is a natural "Fubini" iterated integrations theorem, see [3], as well as approximation formulae in terms of Gaussian integrals and limit convergence theorems [6]. [1])

Looking back to our formulae (2.9), (2.28) we see that with the substitutions ($h = 1$)

$$x \longleftrightarrow \gamma_{nd}, \frac{1}{2}(x, Bx) \longleftrightarrow S_t^0(\gamma_{nd})$$
$$f \longleftrightarrow \psi_0(\gamma_{nd}(0)) e^{-i\sum_j V(\gamma_{nd}(j\frac{t}{n}))}$$

we have

$$I_n(\psi_0) = \tilde{I}_{nd}(\psi_0).$$

Thus for $\psi_0 \in \mathcal{S}(\mathbb{R}^d)$, $\lim_{n\to\infty} \tilde{I}_n(\psi_0)$ exists and is equal $I(\psi_0)$. Hence

$$I(\psi_0) = \lim_{n\to\infty} \tilde{I}_n(\psi_0) = \lim_{n\to\infty} \int_{\mathbb{R}^{nd}} e^{-\frac{i}{2}(\alpha,B^{-1}\alpha)} d\mu_{f_n}(\alpha)$$

with $f_n(\gamma) = e^{-i\sum_{j=1}^{n} V(\gamma_n(j\frac{t}{n}))} \psi_0(\gamma_n(0))$.

Let us now introduce the Hilbert space $\mathcal{H}$ (Sobolev space) $H^{1,2}(\mathbb{R}^n)$ consisting of paths $\gamma : [0,t] \longrightarrow R^d$, $\gamma(t) = 0$ of finite kinetic energy (weakly differentiable with square integrable derivative).

The space $\mathcal{F}(\mathcal{H})$ of Fourier transforms of bounded complex measures on $\mathcal{H}$ is well defined (in fact a Banach algebra, if given the finite variation norm $\|f\|_0 \equiv \|\mu_f\|$). Let $B = \mathbb{1}$, $f \in \mathcal{F}(\mathcal{H})$.

Then

$$\int_{\mathcal{H}} e^{-\frac{i}{2}(\alpha, B^{-1}\alpha)} d\mu_f(\alpha)$$

is well defined. It is not difficult to see that f_x defined by

$$f_x(\gamma) = e^{-i\int_0^t V(\gamma(s)+x)ds} \psi_0(\gamma(0) + x)$$

for $v, \psi_0 \in \mathcal{F}(\mathbb{R}^d)$ is in $\mathcal{F}(\mathcal{H})$ (cfr. [3]).

From the above convergence it follows easily that

$$I(\psi_0) = \int_{\mathcal{H}} e^{-\frac{i}{2}(\alpha, B^{-1}\alpha)} d\mu_f(\alpha)$$

and thus the solution of Schrödinger equation is given by the well defined oscillatory integral

$$\int_{\mathcal{H}} e^{-\frac{i}{2}(\alpha, B^{-1}\alpha)} d\mu_f(\alpha).$$

The inclusion of an additional anharmonic term $\frac{1}{2}(x, A^2 x)$ in Schrödinger equation causes no problem. It suffices to replace $B = \mathbb{1}$ by B such that $(\gamma, B\gamma) = (\gamma, \gamma) - \int_0^t (\gamma, A^2\gamma)\, ds$.

Remark.

Various extensions of this theory have been given. In addition to the already given references, see also the references in [9], the work of H. Watanabe [14] and [47], [50].

Certainly a drawback of the above approach is the fact that integrable functions in the above sense cannot have singularities. Also one has problems when the configuration space is not $\mathbb{R}^d$ but say an open subset of $\mathbb{R}^d$, or a manifold (except for problems with special symmetries, like certain symmetric spaces see [15]). The main advantage of the above theory is that it has permitted to develop a rigorous method of stationary phase in infinite dimensions. There is indeed such a method for general oscillatory integrals in Hilbert spaces of the form

$$I(f) = \widetilde{\int_{\mathcal{H}}} e^{\frac{i}{2\epsilon}(\gamma, B\gamma)} e^{-\frac{i}{\epsilon}W(\gamma)} f(\gamma) d\gamma$$

with $W \in \mathcal{F}(\mathcal{H})$ and real valued, $f \in \mathcal{F}(\mathcal{H})$, in terms of asymptotic expansions in powers of ε for $\varepsilon \downarrow 0$ (Borel summable on the real line).

This has been developed in [4] 2), [16], [5],see also [7]. This "abstract" theory can be applied to the study of asymptotics from quantum mechanics to classical mechanics, see these references.

In particular the solution of Schrödinger's equation on $I\!R^d$ with initial conditions of the form $e^{-\frac{i}{h}\chi(x)}\varphi(x)$, χ smooth and real, φ smooth and of compact support, has been studied for $h \downarrow 0$ and asymptotic expansions in powers of h have been found.

Also a notable trace formula for Schrödinger operators has been proven [16], [7] 2). Let $H = -\frac{h^2}{2}\Delta + \frac{1}{2}(x, A^2 x) + V(x)$, with $V \in \mathcal{F}(I\!R^d)$, A positive definite and symmetric on $I\!R^d$ and let

$$\Theta_h(t) \equiv \int e^{-i\frac{t}{h}H}(x,x)dx$$

be the corresponding theta function (well defined in the sense of generalised functions in t). One has

$$\Theta_h(t) = \sum_n e^{-i\frac{t}{h}\lambda_n}$$

with λ_n the discrete eigenvalues of H. The trace formula proven from above rigorously defined infinite dimensional oscillatory integrals is

$$\Theta_h(t) = \sum_{\gamma_p} \frac{e^{\frac{i}{h}S_t(\gamma_p)}}{(2\pi i h)^{\frac{d}{2}}} \frac{e^{-i\frac{\pi}{2}m(\gamma_p)}}{\Delta(\gamma_p)}\{1 + 0(h)\},$$

where the sum is over the periodic classical orbits γ_p (solutions of Newton's equation with $\gamma_p(0) = \gamma_p(t)$), Δ being a known expression and $m(\cdot)$ the Maslov index.

This formula corresponds to Poisson formula resp. Selberg trace formula in the case of certain symmetric spaces [2]). It should also be seen as the analogue for Schrödinger operators of small time expansions for heat kernels on manifolds, see. e.g. [17], [60] and references therein. See also [18], [19], [20], [141], where exact formulae for the heat kernel for certain groups and symmetric spaces have been derived.

In the physical literature heuristic infinite dimensional oscillatory integrals, the famous "Feynman path integrals", have also been discussed for quantum fields, gauge fields and strings. These "integrals" are taken over paths γ which take values in infinite dimensional spaces. Little has been done in the mathematical sense to define these integrals in a direct way (i.e. not by analytic continuation-from probabilistic integrals, to be discussed below). In [3] there is a definition suitable for "scalar quantum fields", where $\mathcal{H}$ is replaced by a Banach space and $(\alpha, B^{-1}\alpha)$ replaced by a certain quadratic form $\Delta(\alpha,\alpha) \leq 0$ (an inverse of B).

For discussion of various analytic continuations of probabilistic integrals, relating them to oscillatory integrals see e.g. [3], [9]–[13], [21], [22] and references therein. It turns out that in all approaches the class of potentials covered in Schrödinger equation is essentially

$$\frac{1}{2}(x, A^2 x) + V(x)$$

with $V(x)$ a Fourier transform of some complex measure. To close our discussion of configurational oscillatory integrals let us mention an approach based on non standard analysis and an approach based on white noise calculus. As for the former, a word of caution is necessary: although many "starting points" of heuristic expressions can be given a meaning by non standard analysis, one should not take this as an excuse for operating freely on non rigorously defined expressions invoking non standard analysis as justification. In fact the preservation of operations, especially those exploiting heuristic symmetries, has to be justified separately. Moreover the coming back from non standard calculations to standard analysis has to be taken with care, see. e.g. the discussion of Feynman path integrals by non standard analysis in [2]. See also Sect. 4.
For the approach to Feynman path integrals based on "white noise calculus" see [55] and [128]. It would be interesting to see whether the class of potentials can be enlarged, maintaining the possibility of studying expansions in h.

2.3 Remarks on two other approaches to oscillatory integrals

2.3.1. Momentum space Feynman path integrals

The basic idea for this mathematical approach to Feynman path integrals is to use the interaction ("V")-part of the action S_t instead of the free (kinetic) part S_t^0 to form the basic measure. The paths are then paths of a complex jumping process. This is an approach introduced by Chebotarev-Maslov and worked out notably by Bertrand, Blanchard, Combe, Høegh-Krohn, B. Gaveau, Rideau, Rodriguez, Sirugue, Sirugue-Collin, see e.g. [23], [54] and references therein.

For ordinary quantum mechanics the advantage is to include potentials which might depend on momenta. However the same drawback as in 2.2 of not permitting singularities in the potential holds. A full fledged method of expansion in powers of h does not seem to have been given, see however [24], [161]. the mathematical momentum space Feynman path integrals when applied to the problem of quantum fields provide a powerful method for solving the associated infinite dimensional Schrödinger equation, however the construction of an associated Hilbert space is left open, see [25].

2.3.2 Phase space Feynman path integrals

We just want to mention here that Daubechies and Klauder have developed a mathematical method for having phase space integrals well defined for quantum mechanics (permitting inclusion of polynomial interactions, yet no singularities). It would be very interesting to have extensions of this framework to the case of quantum fields and of quantum theory on manifolds, see [26].
Moreover, as in 2.3.1, a method of stationary phase for this type of integrals has not yet been worked out.

Let us now draw some conclusions concerning the methods discussed in sections 2.2, 2.3. The main difficulty of applying these methods to interacting quantum fields is due to the presence of singularities in the interactions, see however [4] 1), [25] for a model where one can cope with them to a certain extent. As already mentioned, there is an enormous gap between Feynman path integrals theory as a heuristic not mathematically defined tool in field theory and the corresponding mathematical theory. For ordinary non relativistic mechanics [3]) however let us stress once more that a rigorous and powerful mathematical theory exists, which permits among other things to give a rigorous derivation of Borel summable asymptotic expansions in powers of h for quantum mechanical quantities in terms of classical ones [4]). Let us also mention that an alternative well known way to handle mathematically Feynman path integrals is to define them by analytic continuation from probabilistic integrals (the famous "Wick rotation" or Nakano-Schwinger-Symanzik-Nelson "Euclidean trick" in quantum field theory).

The idea itself occurred very early in the mathematical and physical literature, see references e.g. in [28], [29]. It consists simply in replacing $e^{\frac{i}{\hbar}S_t(\gamma)}$ in (2.3) by $e^{-\frac{1}{\hbar}S_t^E(\gamma)}$, with S_t^E the "Euclidean action" $S_t^E(\gamma) \equiv \frac{1}{2}\int_0^t |\dot\gamma(s)|^2 ds + \int_0^t V(\gamma(s))ds$.
This then replaces the non existing heuristic complex measure "$e^{\frac{i}{2}\int_0^t |\dot\gamma|^2 ds} d\gamma$ " by the heuristic measure "$e^{-\frac{1}{2}\int_0^t |\dot\gamma|^2 ds} d\gamma$", which can be given a mathematical meaning as Wiener measure (see [27] for a full justification of this fact from the point of view of nonstandard analysis).

Probabilistic integrals with respect to Wiener measure are of course a well established domain of the theory of stochastic processes, we shall return to them below. The above substitution amounts to replace Schrödinger equation by a corresponding heat equation[9]). Astonishingly enough analytic continuation back from the heat equation, as solved by probabilistic integrals, to the Schrödinger equation is relatively little discussed in the mathematical (and physical!) literature, see however [30], [31].

3. Probabilistic functional integrals

Generically these functional integrals are of the form mentioned at the end of the preceding section i.e. heuristically given by "$e^{-W(X)}dX$", with $W(X)$ some functional of the process or field X (typically an "action functional"). Let us split the discussion into subsections, according to the technique used and/or the domain of application.

3.1 Wiener measure and Feynman-Kač formula

We already mentioned that Wiener measure (heuristically "$e^{-\frac{1}{2}\int_0^t \dot{w}(s)^2 ds}dw$", w a path $[0,t] \to \mathbb{R}^d$ with $w(0) = 0$) permits to give a probabilistic solution of the heat equation

$$\frac{\partial}{\partial t}\psi_t = -H\psi_t, \quad x \in \mathbb{R}^d, t \geq 0 \quad H = -\frac{1}{2}\Delta + V.$$

In fact for V e.g. bounded and continuous (or, for $d = 3, V \in L^2(\mathbb{R}^3) + L^\infty(\mathbb{R}^3)$)

$$\psi_t(x) = E\left(e^{-\int_0^t V(w(s)+x)ds}\psi_0(w(t) + x)\right),$$

where E is expectation with respect to Wiener measure. This is the well known Feynman-Kač formula, see e.g.[32], [29], [68], [74], [125], [143], [145]. Wiener measure is the path space measure for a stochastic process (Brownian motion or Wiener process) $w(s), s \in [0,t]$ in $\mathbb{R}^d$ with $w(0) = 0$, which is Gaussian distributed with mean 0 and covariance $E(w_i(t)w_j(s)) = \delta_{ij} t \wedge s$ (where $t \wedge s = \min(t,s)$). The finite dimensional distributions of the process are given by

$$P(w(t_1) \in A_1, ..., w(t_n) \in A_n)) = \int_{A_1} ... \int_{A_n} \prod_{j=1}^{n} [2\pi(t_i - t_{i-1})]^{-\frac{d}{2}} e^{-\frac{|x_j - x_{j-1}|^2}{2|t_j - t_{j-1}|}} dx_j$$

with $t_0 \equiv 0 < t_1 < ... < t_n, w_0 \equiv 0$. We also remark that the derivative with respect to s of a fixed path $w(s)$ in the sense of distributions is Gaussian white noise (see e.g. [33]). The above probabilistic solution (Feynman-Kač formula) is good for handling singular potentials (roughly speaking any positive singularity is permitted, the corresponding contribution to the expectation being zero). Also the $h \downarrow 0$ limit of solutions of $h\frac{\partial}{\partial t}\psi_t = -H_n\psi_t, H_n = -\frac{h^2}{2}\Delta + V$, has been studied by a "saddle point method in infinite dimensions" (connected with the "method of large deviations"). For references see e.g. [34], [35], [57], [144].

Advantages of the method is that it permits estimates on eigenvalues and eigenfunctions. The main disadvantage concerning this approach to Schrödinger equation is its being indirect, relying on an (often difficult) analytic continuation (a minor drawback is that potentials which are unbounded below are in a sense less tractable than in the corresponding rigorous oscillatory integrals, cfr [3], [4]).

There has been numerous and powerful extensions of the method to the study of the heat equation on manifolds, see e.g. [20], [60]. Let us mention here shortly new developments connected with the Elworthy-Truman representation of the heat kernel on a Riemannian manifold

$$e^{t\Delta}(x,y) \equiv \sum_{\exp X_i = x} (2\pi t)^{-\frac{d}{2}} \theta_y(x_i)^{-\frac{1}{2}} e^{-\frac{l^{i2}}{2t}} E\left(e^{\frac{1}{2}\int_0^t \theta_y^{+\frac{1}{2}}(X_s^i)\Delta\theta_x^{-\frac{1}{2}}(X_s^i)ds}\right)(x,y)$$

where Δ is the Laplace-Beltrami operator, $\theta_y(\cdot)$, the Jacobian of the exponential map from the tangent space T_yM of M at y. The expectation is with respect to a process X_s^i started at time zero in 0 and ending at time t in X which is a diffusion with drift

$$-\frac{X_i}{t-s} - \frac{1}{2}\nabla \log \theta_y(X_i)$$

$l^i(x,y)$ is the length of the $i-th$ geodesics between x and y. The sum is over all X, s.t. $\exp X_i = x$. This has been exploited in discussions of when the semiclassical approximation is exact, see [84], [18], [19], [141], [145].

3.2 Quantum mechanical path integrals by processes associated with Dirichlet forms

For problems either with V more singular (e.g. no function but rather a distribution or a more singular mathematical object) or infinitely many degrees of freedom (infinite dimensional state space, instead of $I\!R^d$) it is useful to consider other realizations of both the generator H and of the process and associated functional integrals. This is the approach to quantum mechanics and associated functional integrals by Dirichlet forms. This has been described in details in numerous publications, see e.g. [36], [4] 3), [138] and references therein, we shall limit ourselves here to a description of the basic idea. Suppose first H is a Schrödinger operator with the infimum of its spectrum being an isolated eigenvalue (ground state energy), which we assume for simplicity to be at zero (if H is semibounded this can always be achieved by adding if necessary a suitable constant to H). Let φ be a corresponding eigenfunction ("ground state"); by classical results (at least when the potential in H is smooth) φ can be chosen to be strictly positive. We then have by a simple computation that H in $L^2(I\!R^d, dx)$ is unitarily equivalent with H_μ in $L^2(I\!R^d, \mu)$, with $d\mu(x) = \varphi(x)^2 dx$, the unitary transformation being simply $f \in L^2(dx) \to \frac{f}{\varphi} \in L^2(\mu)$. This is the so called ground state transformation (also called "Darboux transformation" in connection with inverse problems of partial differential equations, or "Hopf-Cole transformation" in connection with hydrodynamic equations, cfr. [37]).

A simple integration by parts gives $(H_\mu^{\frac{1}{2}} f, H_\mu^{\frac{1}{2}} f)_{L^2(\mu)} = \frac{1}{2}\int |\nabla f|^2 d\mu$ and H_μ is then precisely the self-adjoint operator associated with the Dirichlet form $\mathcal{E}(f,g) = \frac{1}{2}\int \nabla f \cdot \nabla g d\mu$, closure of its restrictions to $C_0^\infty(I\!R^d)$. Vice versa, given a Radon measure μ absolutely continuous with respect to Lebesgue measure, $d\mu = \varphi^2 dx$, and supposing φ is such that

the above quadratic form (pre Dirichlet form) $\mathcal{E}^\circ(f,g) \equiv \frac{1}{2}\int \nabla f \nabla g d\mu, f,g \in C_0^\infty(I\!R^d)$ is closable (see e.g. [40], [36], [88] for sufficient conditions) we obtain by closure a closed form $\mathcal{E}$ with associated self-adjoint positive operator H_μ. Finally H_μ is unitarily equivalent to an operator H in $L^2(I\!R^d, dx)$. It is heuristically $-\frac{1}{2}\Delta + V$, with "$V = \frac{1}{2}\frac{\Delta\varphi}{\varphi}$", but V might be too singular to be defined as multiplication operator or quadratic form on $L^2(I\!R^d,\mu)$. H which is well defined as well as H_μ is then a so called "generalized Schrödinger operator". E.g. for $\varphi(x) = \frac{e^{4\pi\alpha|x|}}{\sqrt{2}|x|}, \alpha \in I\!R$, we have H_μ and H well defined. H itself can only be written as $-\frac{1}{2}\Delta + V$ by nonstandard analysis, as standard part of $-\frac{1}{2}\Delta + \lambda_\alpha(\varepsilon)\delta_\varepsilon(x)$, with $\lambda_\alpha(\varepsilon) = -\varepsilon[\pi + \frac{8}{3}\alpha\varepsilon]\frac{\pi}{3}^2 + 8\pi^2\alpha^2, \varepsilon$ infinitesimal, δ_ε an "ε - approximation of the δ-function". These "Dirichlet quantum mechanical operators" H_μ are more general than Schrödinger operators and cover e.g. point interactions. Moreover to H_μ there is associated a well defined potential theory and a diffusion process X_t, see [38], [39].
One has then e.g.

$$(f, e^{-tH_\mu}g)_{L^2(\mu)} = E^\mu(f(X_0)g(X_t)), t \geq 0$$

for any $f,g \in L^2(\mu)$, with E^μ expectation with respect to the paths space measure of the time-homogeneous Markov process X_t with start measure μ. The right hand side is thus a functional integral expressing e^{-tH_μ}, hence also, by analytic continuation and unitary equivalence, the quantum mechanical group e^{-itH}. These new functional integrals are more adapted to covering singular potentials even worse than distributions (as in the above point interaction example) or infinite dimensional state space.
For recent progress on existence and structure of Dirichlet forms see e.g. [40], [41], [61], for recent progress on perturbations supported by lower dimensional sets (like above point interactions) see [2], [36], [42]-[44].

3.3 Probabilistic integrals in the case of quantum fields

Quantum fields can be thought of as quantum mechanical systems as the above ones but with infinitely many degrees of freedom, i.e. state space $I\!R^d$ replaced by an infinite dimensional one. Models for quantum fields can be described heuristically by analoga of the above path space measures (in 3.1) 3.2)):

$$d\mu_E =" e^{-S(X)}dX" .$$

As for non relativistic quantum mechanics their mathematical realization is either by Wiener measure via a Feynman Kač formula or by the path space measure of Dirichlet quantum mechanics. Let us start with the latter. The basic point of view is as follows: Non relativistic quantum mechanics on $I\!R$ is given in terms of an Hamiltonian $H = -\frac{1}{2}\frac{d^2}{dx^2} + V$ associated with a classical Newton equation of $\ddot{X} = -V'(x)$. Canonical relativistic quantum field theory should come from a "Newton" field equation. Heuristically, in analogy with non relativistic quantum mechanics over $I\!R^n$, where the

Hamiltonian can be written $H = -\sum_{i=1}^{n}(e_i \cdot \nabla_x)^2 + V(x), x \in \mathbb{R}^n$, e_i a basis of $\mathbb{R}^n$, we have for the Hamiltonian of canonical quantum field theory

$$H = -\frac{1}{2}\int_{\mathbb{R}^s} \frac{\delta^2}{\delta X(0,\vec{x})^2} d\vec{x} + \frac{1}{2}\int_{\mathbb{R}^s} X(0,\vec{x})(-\Delta_{\vec{x}})X(0,\vec{x})d\vec{x}$$
$$+ \int_{\mathbb{R}^s} V\left(X(0,\vec{x})\right) d\vec{x} \tag{3.1}$$

where s is the space dimension. But here the problem is the Hilbert space: whereas in quantum mechanics the Hilbert space is $L^2(\mathbb{R}^n, dx)$ here it should be L^2 with respect to a non existing infinite dimensional Lebesgue measure (see the discussion in footnote 1)). But here comes our "Dirichlet quantum mechanics" to help, in fact there the Hilbert space is $L^2(\mathbb{R}^n, d\mu)$ for some measure μ. L^2 spaces with respect to probability measures exists also over infinite dimensional spaces.
For $V(x) = x^2m^2$, H heuristically given by (3.1), is then realized as the self-adjoint operator associated with the closure of the pre-Dirichlet form $\frac{1}{2}\int |\nabla f|^2 \, d\mu$, defined on the space FC_b^∞ of smooth cylinder functions (of the form $f\left(X(0,\vec{x}), \vec{x} \in \mathbb{R}^s\right) = \tilde{f}\left(< e_{i_1}, X(0) >, ..., < e_{i_n}, X(0)\right)$, $X(0) \in \mathcal{S}'(\mathbb{R}^s)$, $(e_i, i \in \mathbb{N})$ basis in $L^2(\mathbb{R}^s, d\vec{x})$ consisting of elements in $\mathcal{S}(\mathbb{R}^s)$, with μ the Gaussian measure on $\mathcal{S}'(\mathbb{R}^s)$ with mean 0 and variance $(\sqrt{-\Delta_{\vec{x}} + m^2})^{-1}$. $X(0,\vec{x})$ is then an infinite dimensional process. That the above closure exists is well known, see [85], [86], [87]. There is indeed a general theory of Dirichlet forms over infinite dimensional spaces, see [40, 85, 88] and references therein (so general as to cover Souslin spaces, including thus $\mathcal{S}'(\mathbb{R}^s)$, Banach spaces etc.). The theory is a proper extension of the finite dimensional theory (over locally compact spaces) for which a standard reference is [89]. To the Dirichlet forms there are associated Markov semigroups and, in the case of forms of main interest for quantum fields, diffusion processes, analogous to the finite dimensional case. The Hamiltonians for canonical quantum fields are generators associated with "classical Dirichlet forms"

$$\frac{1}{2}\int |\nabla f|^2 \, d\mu \ , \tag{3.2}$$

with μ being a measure on $\mathcal{S}'(\mathbb{R}^s)$ associated with the quantum fields. For $s = 1$ one has then Hamiltonians for the scalar relativistic quantum fields with polynomical, exponential or trigonometric interactions (see [56], [85, 88]). Recently these have been realized also as Dirichlet forms associated with white noise calculus, see [90]. To make this into a constructive approach to relativistic quantum fields the problem remains of proving in an intrinsic way, without recourse to "Euclidean quantum field theory", the relativistic invariance, see [91]. Once this is proven, which is the case for the forms associated with Markov Euclidean fields [76], then the corresponding path space measure has the form

$$d\mu_E(X) = \text{"}e^{-S(X)}dX\text{"} \tag{3.3}$$

with $S(X)$ the (Euclidean) action along the space-time field $X(t, \vec{x}) \equiv X(x), x = (t, \vec{x}) \in \mathbb{R}^{s+1}$, given by

$$S(X) \equiv \frac{1}{2}\int_{\mathbb{R}^{s+1}} |\nabla X(x)|^2 \, dx + \int_{\mathbb{R}^{s+1}} V(X(x))dx \ . \tag{3.4}$$

According to wether V is an exponential trigonometric or polynomial function one speaks of models with exponential resp. trigonometric resp. polynomial interaction.
The construction of such measures raises well known difficulties, especially for $s > 1$. Once the path space measure is constructed one recovers the corresponding relativistic model by an analytic continuation in the time variable t. For $V(\lambda) = m^2\lambda^2$, m a constant ≥ 0, μ_E is mathematically realized as $N(0; (-\Delta + m^2)^{-1})$ (the normal distribution with mean zero and covariance the Yukawa resolvent $(-\Delta + m^2)^{-1}$). [5])

The associated (generalized) process (random field) is $X(x), x \in R^{s+1}$, with mean 0 and covariance $E(X(x)X(y)) = (-\Delta + m^2)^{-1}(x, y)$. X is $\mathcal{S}'(\mathbb{R}^{s+1})$-valued. The singularity on the diagonal makes it impossible to find continuous realizations of the process, which makes it difficult to define additive functionals of it like the heuristic objects $\int V(X(x))dx$ occurring in above formula. For $s = 1$ this can however be done in a variety of models, see e.g. [28, 56], [2] and one arrives at a class of interacting fields which have analytic continuation to relativistic fields.

The possibility of analytic continuation is best understood when realizing that all constructed fields have Euclidean invariance and the global Markov property (in particular the Markov property with respect to half spaces: intuitively splitting $\mathbb{R}^{s+1}$ as $(t \geq 0, \mathbb{R}^s) \cup (t < 0, \mathbb{R}^s)$ functions of the fields measurable in the upper half-space are independent of those measurable in the lower half-space, given all information on fields on the boundary $\{t = 0\}$).

This Markov property was postulated by Nelson in 1971 remained open for many years until it was definitely proven (1979 for trigonometric interactions [77]; 1984 for exponential interactions [78], [79]; 1988 for polynomial interactions [76]). Thus in one space dimension (s=1) a "full canonical picture" holds [91, 92], making the theory of quantum fields indeed just an infinite dimensional version of quantum mechanics. Yet some open problems remain, to mention a few: uniqueness (cfr. [92, 93], [80], [73], [75]); non relativistic limit (cfr. [94]), classical limit (cfr. [95])).

A case $s > 1$ (in fact $s = 3$) will be discussed below. We also remark that a method of stochastic quantization for quantum fields for $s = 1$ (and partially for $s = 2$) can be developed : see [83], [96], [147], [148]. See also [157], [159], [163], [164] for new aspects of the discussion concerning quantum fields and stochastic mechanics.

Finally let us remark that a new method of defining mathematically probabilistic functional integrals for quantum fields on manifolds[7)] has been very recently developed in [97]. It consists of suitably triangulating the Riemannian manifolds and introducing corresponding simplicial fields. Their convergence has been shown in the Gaussian as well as interacting case, e.g. with exponential interaction, for space-time dimension 2.

Let us give a very short sketch of these methods, which are expected to be also useful in connection which the problems mentioned in sect. 5 and in part II. Let K be a simplicial complex providing a smooth triangulation of a Riemannian manifold M of dimension d. Let $p_0, \ldots, p_n$ be the vertices of K. Let $C^q(K)$ resp. $C_q(K)$ be the space of corresponding q-cochains resp. chains. Let W be Whitney's mapping s.t. $W([p_i]) \equiv b_{p_i}$, where b_{p_i} is the barycentric coordinate of p_i, $W(\sigma^q) \equiv q! \sum_{i=0}^{q} (-i)^q b_{p_i} db_{p_0} \wedge \ldots \wedge d\hat{b}_{p_i} \wedge \ldots \wedge db_{p_q}$, $\sigma^q \in C^q(K)$, where $\hat{b}$ means b is omitted). One has then a natural scalar product in

$C^q(K)$ defined by $(\sigma^q, \sigma^{q'}) \equiv \int W(\sigma^q) * W(\sigma^{q'})$. Let Λ be the space of differential forms over M and let R be the DeRham map from Λ into $C(K) \equiv \bigcup_q C^q(K)$, defined by $\alpha \in \Lambda^q \longrightarrow R\alpha \equiv \sum \left(\int_\sigma \alpha\right) \sigma$.
Let δ^c be the boundary map from $C_{q+1}(K)$ into $C_q(K)$ and d^c its natural adjoint from $C^q(K)$ into $C^{q+1}(K)$. A natural combinatorial Laplacian Δ^c acting on elements of Λ is defined as the closure of $-\left(d^c + \delta^c\right)^2$. Consider now a sequence of triangulations $\{K_n,\ n \in \mathbb{N}\}$ with mesh $\downarrow 0$ as $n \to \infty$. By results of Dodziuk, Patodi and Müller, see the references in [97], one has convergence of eigenvalues and eigenfunctions of the Laplace operator Δ^c on M. Moreover for any $\alpha \in \Lambda^q$ one has $W_n R_n \to \alpha$ in the natural L^2 space on Λ.
Now let ϕ be defined by

$$\phi \equiv \sum \phi_i \sigma_i$$

where the components ϕ_i are in $\mathbb{R}$ and the sum is over all q-complexes of K. We look at ϕ as a simplicial field mapping $\sigma \in C^q(K)$ into $\mathbb{R}$ by

$$\phi(\sigma) = \sum \phi_i(\sigma_i, \sigma).$$

Let μ_q^c be the corresponding free q-simplicial measure i.e. the standard Gaussian measure associated with the quadratic form

$$\frac{1}{2}\left(d^c\phi, d^c\phi\right) + \lambda\left(\phi, H_q^c\phi\right) + m^2(\phi,\phi),$$

where H_q^c is the projection onto the q-harmonic cochains. It is proven in [97] that $\phi(R_n\alpha)$ (distributed with measure μ_q^c), converges as $n \to \infty$ to the continuum limit field $\phi(\alpha)$ which is the free field Gaussian measure with mean zero and covariance $\left(-\Delta + \lambda H_q + m^2\right)^{-1}$, in the sense that $\lim E_{\mu_q^c}\left(e^{i\phi(R_n\alpha)}\right) \longrightarrow E_{\mu_q}\left(e^{i\phi(\alpha)}\right)$ (where E_ν is expectation with respect to ν.)
In dimension $d = 2$ and for $q = 0$ one can also handle interactions. In fact the N-th Wick power $:\phi^N:(R_n f) \equiv \sum_i :\phi_i^N:(\sigma_i, R_n f)$ converges as $n \to \infty$, for all $f \in C_0^\infty(M)$, to the Wick power $:\phi^N:(f)$ in the continuum (measurable function with resp. to μ_q, the continuum analogue of μ_q^c).
Moreover a "simplicial exponential interaction" model, described by a simplicial approximation of the mathematical well defined Euclidean measure to the heuristic measure given by (3.3) (as constructed in [129]), can be introduced. For M compact it is defined by $d\mu^K(\phi) \equiv \dfrac{e^{-U_K(\phi)} d\mu_0^c(\phi)}{\int e^{-U_K} d\mu_0^c}$, with U_K a simplicial analogue of an exponential V in (3.4).
It is shown in [97] that μ^K converges weakly to μ_E. A discussion of several inequalities for simplicial fields and their limits is also given in [97].

4. Functional integrals by non standard analysis

Let us sketch briefly what non standard analysis is (see also e.g. [2], [81, 82]). It is analysis over an enlarged field $^*\mathbb{R}$ of numbers containing in particular infinitesimally small and large numbers. A model is as follows. Divide subsets of $\mathbb{N}$ into big and small, according to wheather they have measure 1 or 0, where "measure" is taken with respect to an additive (but not σ-additive) 0-1-valued set function m,s.t. $m(\phi) = 0$, $m(A) = 0$ for all finite $A, m(\mathbb{N}) = 1$ (such m exists by the axiom of choice, see also [2] for discussion). $^*\mathbb{R}$ is defined as the set of all sequences modulo m i.e. $^*\mathbb{R} = \mathbb{R}^{\mathbb{N}}$ / mod m.It is a field with respect to sum (+) and multiplication ($\cdot$) defined componentwise. $\mathbb{R}$ is embedded in $^*\mathbb{R}$ by the natural embedding associating to $r \in \mathbb{R}$ the "identical sequence" $r_i = r$ mod $m \forall i \in \mathbb{N}$, denoted by $^*r$. Linear order is introduced in $^*\mathbb{R}$ componentwise by setting $< f ><< g > \Longleftrightarrow \{i | f(i) < g(i)\}$ big, where $< f >, < g > \in {}^*\mathbb{R}$, f, g being representatives in $\mathbb{R}^{\mathbb{N}}$ of mod m equivalence classes. $^*\mathbb{R}$ is then an ordered field. The equivalence class mod m of the sequence $a_i = i$, $i \in \mathbb{N}$ is obviously larger than all elements $^*r$, with $r \in \mathbb{R}$ and is a so called positive infinite number ω in $^*\mathbb{R}$. $^*\mathbb{R}$ being a field we have also e.g. ω^3, $\exp \omega$ infinite in $^*\mathbb{R}$ and the inverse of ω is by definition (and meaning) an infinitesimal in $^*\mathbb{R}$. There is a natural injective homeomorphism st from the set F of non infinite elements of $^*\mathbb{R}$ to $\mathbb{R}$ s.t. $st < f > \equiv \sup\{r \in \mathbb{R} | {}^*r << f >\}$ and $F/I \cong \mathbb{R}$, where I is the ideal of infinitesimals. One can extend ("put a star") functions and sets, again componentwise (e.g. $(^*F)(< f >) \equiv < F(f(i)) >$). Elements of quantities having a star are then called internal: for them we can transfer "properties formulated on standard quantities" (see e.g. [2] , [82] for precise definitions). The following important saturation theorem holds and is the basis for an efficient measure theory in the non standard analysis setting:

Saturation theorem: If $A_n, n \in \mathbb{N}$ is a decreasing sequence of non void internal quantities then $\bigcap_n A_n \neq \phi$. (An example is provided by $\bigcap_{n \in \mathbb{N}} {}^*(0, \frac{1}{n})$ which consists of all positive infinitesimals).

By this theorem measure theory becomes very easy, especially for probability measures. One has e.g. the following Theorem by Loeb (1975):

Theorem: Let X be an internal set and $\mathcal{A}$ be the algebra of its internal subsets. Let ν be $^*[0,1]$–valued finitely additive measure on X with $\nu(X) = 1$. Then there exists a probability space $(X, \sigma(\mathcal{A})), L(\nu)$, with $\sigma(\mathcal{A})$ the σ–algebra generated by $\mathcal{A}$ and $L(\nu)$ a $[0; 1]$–valued (σ–additive) probability measure. $L(\nu)$ is the σ–additive extension on $\sigma(\mathcal{A})$ of the finitely additive measure $^0\nu$ defined by $^0\nu(A) = st\nu(A) \forall A \in \mathcal{A}$. ν is approximated by $^0\nu$ in the sense that to any $B \in \sigma(\mathcal{A})$ there exists $A_B \in \mathcal{A}$ with $st\nu(A_B) = L(\nu)(B)$.

The proof is an immediate application of the saturation theorem and Caratheodory's extension theorem. ∎

$L(\nu)$ is called the Loeb measure to ν.

As a consequence of this result one has important non standard constructions of standard measures. Perhaps most surprising: every σ–additive Radon measure can be realized as Loeb measure of some finitely additive internal measure, c.f. [2]. In fact measure theory is reduced essentially to counting (i.e. combinatorial) measures. E.g. one has

a) Lebesgue measure: we consider only the case of Lebesgue measure on $[0,1]$, for simplicity (for extensions see e.g. [2]). Here is its construction as Loeb measure. Let

$$X = \{{}^*0, \Delta t, {}^*2\Delta t, \ldots, \eta!\Delta t\} \equiv T$$

be a "hyperfinite model" of $[0,1]$, with η some positive infinite number and $\Delta t = (\eta!)^{-1}$. Define $\nu(A) \equiv \frac{|A|}{|T|}$, where $|\cdot|$ means number of elements. Then the corresponding Loeb measure $L(\nu)$ realizes Lebesgue measure λ in the sense that $L(\nu)\left(st^{-1}(C)\right) = \lambda(C)$ for every Borel measurable $C \subset [0,1]$. Moreover Lebesgue integral is realized by an integral with respect to $L(\nu)$, hence $\int f dx = \int \tilde{f} dL(\nu) = st \sum_{i \in T} \frac{\tilde{f}(i)}{|T|}$, with $\tilde{f}$ the function on T which corresponds to f. In this way Lebesgue integrals are reduced to "hyperfinite sums", modulo infinitesimals.

b) Wiener measure: By non standard analysis Brownian motion on $I\!R$ is a random walk on a hyperfinite model of $I\!R$ with infinitesimal steps. Let in fact $\Omega \equiv \{-1,+1\}^T$, with T as in a). We shall let Ω play the role of X in the Loeb theorem. Let $B(t,\omega) \equiv \sum_{0}^{t-\Delta t} \omega(s)\sqrt{\Delta t}$, with Δt a positive infinitesimal. Then $B(t,\omega)$ is a symmetric hyperfinite random walk, going in steps of time length Δt left or right by $\sqrt{\Delta t}$ with equal probability. Its values are in finite points of ${}^*I\!R$ and

$$st\, B(t,\omega) \equiv b(st\ t,\omega)$$

where b is an $I\!R$–valued Brownian motion indexed by the time line [0,1]. The path space measure of it is Wiener measure $(\Omega, \sigma(\mathcal{A}), L(P))$, with P the product of $|T|$ - Bernoulli $\left(\frac{1}{2}, \frac{1}{2}\right)$ - measures. This reduction of Wiener measure to discrete measures permits also to give combinatorial formulations of stochastic analysis (stochastic integrals, and differential equations ...), see e.g. [2], [82] and references therein. To name just a few recent important applications of this let us mention Lindstrøm's constructions of Laplacians on fractal sets [98], the theory of Dirichlet forms by non standard analysis [2], the proposal of discussing Feynman path integrals by non standard analysis [2], the construction of hyperfinite models of quantum fields and polymer measures [2]. As an important area in which non standard tools have not yet been sufficiently developed let us mention finite and infinite dimensional differential geometry. Its developments could have important applications in the construction of probability measures for gauge and topological fields, as well as in the construction of strings and superstrings, see also Part II.

5. Functional integrals for gauge fields

In this section we shall briefly mention some techniques and results for handling mathematically certain gauge fields and other related quantum fields.
Let us start by mentioning that the problem of construction of gauge fields consists actually in giving mathematical meaning to the following heuristic functional measure:

$$d\mu_E(A) = \text{“}e^{-S(A)}dA\text{”}$$

where

$$S(A) \equiv \frac{1}{2}\int_M F(A) \wedge^* F(A)dx \ ,$$

M being a Riemannian manifold, A a connection one form from TM to the Lie algebra of the Lie group G, F the corresponding curvature 2-form, dx the Riemannian volume element on M. In the case where G is abelian and $M \subset \mathbb{R}^d$, $d \leq 4$ the problem is solved by various means, e.g. by lattice approximations [108], [101–107], [109], [110], [149]. The case $d = 4$ can also be recovered as special case of quantum fields depending by quaternionic Cauchy-Riemann equations, see below. Also the case $M = \mathbb{R}^d$, $d \leq 4$, G non abelian has been discussed in [101–107],[111–115]. In [101–107] a general theory of Markov cosurfaces (over $\mathbb{R}^d$ for arbitrary d) has been developed, recovering expectations of Wilson loops associated to gauge fields as special cases for $d = 2$. In [111] an estimate for proving lower boundedness of a space cut-off Hamiltonian for $d \leq 4$ has been given. Let us finally mention work on mathematically defined probabilistic measures for Higgs-like fields over 2-dimensional space-time: Let φ be a “matter field” i.e. a measurable map for $\mathbb{R}^2$ into a complex vector space V. Let $A = (A_\mu, \mu = 0, 1)$ be a g - valued 1-form over $\mathbb{R}^2$ i.e. $A = \sum_{\mu=0}^1 A_\mu dx^\mu$. Let ρ be a unitary representation of G resp. g in V. Set $D_A\varphi \equiv d\varphi + \rho(A)\varphi$. The heuristic measure

$$\text{“}e^{-\frac{1}{2}\int[|D_A\varphi|^2+m^2|\varphi|^2]dx}e^{-\frac{1}{2}\int F\wedge^* F dx}d\varphi dA\text{”}$$

describes a Higgs - like field φ interacting with the gauge field A. Giving a meaning to this heuristic measure has been achieved for cases where F is a Poisson-like distributed field, see [112]. The Schwinger functions (moments) of the measure are then expressed essentially by a multiplicative curve integral m_C (holonomy operator) associated with A, Brownian motion and the curve C. E.g. for the "2 - point function" one has

$$S_2(x,y) \equiv E_{m_C}\left(\int_0^\infty e^{-m^2 t}E^t_{x,y}\left[\text{tr}(\rho(m_{bC^{-1}}))\right] dt\right),$$

the expectation E_{m_C} being with respect to the multiplicative curve integral $m_C \equiv$ “$T \exp \int_C A$”. The expectation $E^t_{x,y}$ is with respect to a Brownian bridge b started at 0 in x and ending at t in x. See [112] for details.

We conclude with a mentioning of mathematical functional integrals for a model in space-time dimension $d = 4$. The model has been introduced in [113] and developed in [114], [115] [8]). It is best formulated using the skew field $\mathbb{H}$ of quaternions as a suitable vector space isometric isomorphic with $\mathbb{R}^4$:

$I\!H = \{\hat{q} | \hat{q} = \sum_{\mu=0}^{3} q_\mu \tau_\mu\}$, τ_μ being the quaterionic units, with τ_0 = identity. The isometric isomorphism is given $\hat{q} \leftrightarrow q = (q_\mu,\ \mu = 0,1,2,3)$, q being the vector in $I\!R^4$ with components q_μ. Isometry is achieved by giving $I\!H$ the scalar product $< \hat{q}, \hat{q}' >= \sum q_\mu q'_\mu$. The following first order constant coefficients differential operators are the quaternionic analoga of the Cauchy - Riemann operators $\partial, \bar{\partial}$ in $\mathbb{C}$:

$$\partial = \partial_0 \tau_0 - \sum_{k=1}^{3} \tau_k \partial_k$$

$$\bar{\partial} = \partial_0 \tau_0 + \sum_{k=1}^{3} \tau_k \partial_k$$

We have $\partial\bar{\partial} = \bar{\partial}\partial = \tau_0 \cdot \Delta$.
Now let F be an infinitely divisible Euclidean generalized random field over $I\!R^4$ i.e. for $f \in \mathcal{S}(I\!R^4; I\!H)$:

$$E\left(e^{i<f,F>}\right) = e^{-\int_{I\!R^4} \psi(f(x))dx},$$

for a negative definite function ψ, Lévy characteristics, such that $\psi(b \cdot b^{-1}) = \psi(\cdot)\ \forall\ b \in$ Sp $(1) \equiv \{\hat{q} \in I\!H | |\hat{q}| = 1\}$, with $|\ |$ the norm in $I\!H$ (the latter property expresses rotation - invariance of the path space measure). ψ consists of a Gaussian component ψ_G and a non Gaussian one ψ_P. Let us also assume for simplicity, that $\psi_P(\lambda) = O\left(|\lambda|^{\frac{4}{3}+\varepsilon}\right)$ as $\lambda \downarrow 0$, for some $\varepsilon > 0$. Then it is shown in [114] and [115] that the $I\!R^4$-vector field $A(x)$ over $I\!R^4$ s.t.

$$E\left(e^{i<f,A>}\right) = e^{-\int_{I\!R^4} \psi(-\bar{S}*f(x))dx},$$

f is a suitable test function space, solves the stochastic partial differential equation $\partial A = F$, where $\bar{S} = \frac{\bar{x}}{|x|^4}$, where $\bar{x} = (x_0, -\vec{x})$, $(x_0, \vec{x}) = x$, is the fundamental solution $(\bar{\partial}^{-1})$. A is proper Euclidean invariant (in law). It is also "time reflection" invariant i.e. invariant in law under the operation $A(x) \to -\left(A_0(-\bar{x}), \vec{A}(-\bar{x})\right)$, iff the support of ψ_P is in $I\!R \backslash \{0\}$.

Remarks:

a) For $\psi_P = 0$ we have that A is a Euclidean free electromagnetic potential field. In certain gauges it is a Markov field and by analytic continuation one recovers relativistic fields, cf. [114], [115].

b) if $\psi_P \neq 0$ then work is in progress concerning Markov property, study of the generator of an associated Markov semigroup, study of additive functionals of the fields, see [114], [115]. In this case the underlying path space measure is of the form

$$\text{“}e^{-\int g(\partial A(x))dx} dA\text{”}$$

with g $SO(4)$ - invariant over $I\!H \cong I\!R^4$, with values in $I\!R$.

II. QUANTUM STRINGS

A. Probabilistic integrals for closed bosonic strings

The classical equations of motion for a closed bosonic string are given by a variational principle from the action, which in the context of string theory was introduced by Brink,Di Vecchia and Howe ([65]):

$$A(x,g) \equiv \frac{1}{2} \int_\Lambda \sqrt{g(\eta)} g^{ab}(\eta) \frac{\partial}{\partial \eta^a} x^\mu(\eta) \frac{\partial}{\partial \eta^b} x^\mu(\eta) d\eta \tag{A.1}$$

where Λ is a C^∞ compact riemannian surface without boundary of genus $p > 1$, $\eta \equiv (\eta^1, \eta^2)$ is the parameter on Λ, $(x^\mu, \mu = 1, \cdots, d)$ is a C^∞ embedding of the surface Λ into euclidean space time $\mathbb{R}^d$, $d > 2$ and $(g_{ab})_{a,b=1,2}$ is a C^∞ riemannian metric on Λ, g denoting the determinant of the matrix (g_{ab}) which is strictly positive and $(g^{ab})_{a,b=1,2}$ the inverse matrix of $(g_{ab})_{a,b=1,2}$. The heuristic probabilistic integral for the quantization of bosonic strings suggested by Polyakov is defined in the following way:

$$\int \prod_{i=1}^m V_{k_i,\eta_i}(x,g) d\mu_P[x,g] \tag{A.2}$$

where $V_{k_i,\eta_i}(x,g), i = 1, \cdots, m$ is a "vertex operator" for a string attached to the point η_i of Λ and with momentum k_i and where $\mu_P[x,g]$ is the underlying heuristic measure given by:

$$\begin{aligned} \mu_P[x,g] &\equiv \text{``}\frac{1}{Z}(\text{vol}\,_g(D))^{-1} e^{-A(x,g)} e^{-\mu^2 \int_\Lambda \sqrt{g(\eta)} d\eta} D_g[x] D[g]\text{''} \\ Z &\equiv \text{``} \int_{X \times M} (\text{vol}_g(D))^{-1} e^{-A(x,g)} e^{-\mu^2 \int_\Lambda \sqrt{g(\eta)} d\eta} D_g[x] D[g]\text{''}. \end{aligned} \tag{A.3}$$

X is here the space of C^∞ embeddings of Λ in $\mathbb{R}^d$ and M is the space of C^∞ riemannian metrics on Λ, $e^{-\mu^2 \int_\Lambda \sqrt{g(\eta)} d\eta}$ is the "cosmological term" and $\text{vol}_g(D)$ is the volume (taken w.r.t the metric g) of the group D of C^∞ diffeomorphisms on Λ.

In the Polyakov scheme, the formal measure $D[g]$ undergoes a transformation induced by a parametrization of the manifold of metrics in terms of three spaces namely, the subgroup D_0 of D of diffeomorphisms homotopic to identity, the group W of smooth Weyl transformations acting on Λ, i.e $W \equiv \{e^\varphi, \varphi \in C^\infty(\Lambda, \mathbb{R})\}$ and the Teichmüller space $\mathcal{T} \equiv M/W/D_0$; recall that every riemannian metric can be written as $g = f^* e^\varphi g_t$, $(f, \varphi, t) \in D_0 \times W \times \mathcal{T}$ where $\{g_t, t \in \mathcal{T}\}$ is a slice of the space of metrics parametrized by $\mathcal{T}$.

In the new variables (f, φ, t), the measure μ_P reads formally after integrating out the diffeomorphisms:

$$\mu_P[\varphi, t] = \frac{1}{Z} \det(P^*_{\bar{g}_t} P_{\bar{g}_t})^{\frac{1}{2}} \left(\frac{\det \Delta'_{\bar{g}_t}}{\int_\Lambda \sqrt{\bar{g}_t}(\eta) d\eta} \right)^{\frac{-d}{2}} e^{-\mu^2 \int_\Lambda \sqrt{\bar{g}_t}(\eta) d\eta} DW[t] D[\varphi] \tag{A.4}$$

where $\overline{g}_t \equiv e^{\varphi} g_t$ and where $DW[t]$ is the Weil Peterson measure on Teichmüller space.

In (A.4) there appear the formal determinants of the elliptic operators $P_{\overline{g}_t}{}^* P_{\overline{g}_t}$ and $\Delta'_{\overline{g}_t}$ on the compact surface Λ. In [64], we gave a detailed description of the factorization of the jacobian determinant for the transformation described above as a product of these two determinants using heat-kernel regularization methods.

An evaluation of these heat-kernel regularised determinants through a Seeley expansion yields the following heuristic expression of the measure $\mu_P[\varphi, t]$ in terms of the Liouville action $A^{g_t}_{\text{Liouv}}[\varphi] \equiv \frac{1}{2}\int_\Lambda \sqrt{g_t}(\eta)(\Delta_{g_t}\varphi)(\eta)\varphi(\eta)d\eta + \mu^2 \int_\Lambda \sqrt{g_t}(\eta)e^{\varphi(\eta)d\eta} - \frac{1}{2}\int_\Lambda \sqrt{g_t}(\eta)\varphi(\eta)d\eta$:

$$\mu_P[\varphi, t] =" \frac{1}{Z} e^{-\left(\frac{26-d}{48\pi}\right)A^{g_t}_{\text{Liouv}}[\varphi]} D[\varphi] DW[t]". \tag{A.5}$$

The Liouville action is essentially the analogue on the Riemannian surface Λ of the action associated with a scalar quantum field over $\mathbb{R}^2$ as described in sect. I.3 with $V(\lambda)$ in formula (3.4) taken to be an exponential function $e^{\alpha\lambda}$ for some real α (this is a particular case of Høegh-Krohn's or exponential interaction models, first studied in the Euclidean setting in [129]). The heuristic measure in (A.5) is thus a particular case of a Euclidean measure for a two-dimensional quantum field of the form (3.3) discussed in sect. 3. The latter measure has been defined mathematically in [129] for $\alpha^2 < 4\pi$ and in [130] for $\alpha^2 < 8\pi$ (and it has been shown to coincide with the free field measure for $\alpha^2 > 8\pi$ or $d \geq 3$ in [131], [132]).[6)] By adapting these methods to the case of Liouville model one arrives at a mathematical definition of the Liouville measure (A.5), under a restriction corresponding to $\alpha^2 < 4\pi$, namely $d \leq 13$. Moreover one has to require $2 < d$ for other reasons see [64] so that one arrives at the result that for dimensions $2 < d \leq 13$, the formal measure $\frac{1}{2} e^{-\left(\frac{26-d}{48\pi}\right)A^{g_t}_{\text{Liouv}}[\varphi]} D[\varphi]$ can be interpreted (see [64]) in terms of the well defined Liouville probability measure $\mu_{g_t}[\varphi]$ where for a general smooth metric g on Λ and for some positive λ:

$$\mu_g[\varphi] \equiv \left(\int e^{-\lambda V_\Lambda(\xi)} d\mu_{0,g}[\xi] \right)^{-1} e^{-\lambda V_\Lambda(\xi)} d\mu_{0,g}[\xi]. \tag{A.6}$$

Here $\mu_{0,g}$ is the Gaussian measure with mean zero and covariance operator $(\Delta'_g)^{-1}$ and ξ is the corresponding generalised stochastic process so that for C^∞ functions h, k on Λ we have:

$$\int \xi(h)\xi(k) d\mu_{0,g}[\xi] = \int \sqrt{g}(\eta) h(\eta)((\Delta'_g)^{-1} k)(\eta) d\eta \tag{A.7}$$

and $V_\Lambda(\xi) \equiv \, : e^{\sqrt{\frac{48\pi}{26-d}}\xi} :_g$ - the double points indexed by g denoting the Wick-ordering w.r.t the covariance $(\Delta'_g)^{-1}$ -(see [64])[10)].

Two essential ingredients for a mathematical description of the Polyakov measure for bosonic strings (in above "non critical dimensions d") are therefore

a)- a precise description of heat-kernel regularized determinants of elliptic operators on compact surfaces

b)- a precise description of the Liouville measure.

Both these aspects have been studied in [64], a) using tools of functional analysis, and b) using probabilistic methods.

In the following, we show that one can also approach a) using probabilistic tools; we give a probabilistic intepretation of the heat-kernel regularization procedure of determinants of elliptic operators on compact surfaces. More precisely, we interpret this procedure in terms of an approximation for transformed Gaussian measures on nuclear spaces under the action of an elliptic operator. We report about this point of view already suggested in [72] quite in details since to our knowledge, it had not been remarked elsewhere.

In B.1, we review classical results on elliptic operators on compact boundaryless surfaces. In B.2 we describe an approximation procedure for gaussian measures with covariance defined by elliptic operators using heat-kernel regularisations.Finally, in B.3, we prove a theorem concerning the equivalence of these approximating measures. We show that they are equivalent whenever the underlying riemannian metrics on the manifold coincide and that in this case, the Radon Nikodym derivative is expressed in terms of the heat-kernel regularized determinants of the elliptic operators.

B. A probabilistic interpretation of the heat-kernel regularization procedure for determinants of elliptic operators

1. Elliptic operators on compact manifolds.

Let E be a smooth vector bundle based on the boundaryless C^∞ 2–dimensional real compact manifold Λ with fibres of finite dimension and let $C^\infty(E)$ be the vector space of smooth sections of E. Let $H \equiv L^2(E)$ be the closure of $C^\infty(E)$ w.r.t. the L^2 scalar product defined on it (see e.g. [66]). The bundle E itself will be realized as a tensor product of tensor bundles on Λ so that the scalar product on E is naturally induced by a C^∞ Riemannian metric g on Λ. We shall denote it by $< \cdot, \cdot >_g$. Note that for two different C^∞ Riemannian metrics g_1 and g_2 the two scalar products $< \cdot, \cdot >_{g_1}$ and $< \cdot, \cdot >_{g_2}$ induce equivalent norms on $C^\infty(E)$ so that the closure H is independent of the metric. Let $\mathrm{Ell}_g^+(E)$ denote the space of positive elliptic self-adjoint (w.r.t. $< \cdot, \cdot >_g$) operators on H. Since Λ is compact, they have purely discrete spectrum $0 \leq \alpha_1(g) \leq \ldots \leq \alpha_n(g) \to +\infty$ each of which has finite multiplicity and the asymptotic behavior is given by $\alpha_n(g) \sim Cn^{\frac{d}{2}}$ where d is the order of the elliptic operator, C is a positive constant (see e.g. [66]). Moreover for $A \in \mathrm{Ell}_g^+(E)$, the kernel Ker A of A is of finite dimension and we can define the orthogonal projection P_A onto Ker A w.r.t the scalar product $< \cdot, \cdot >_g$.Let us set

$$A' \equiv A + P_A$$

This defines a strictly positive self adjoint operator on H.

We introduce a heat-kernel cut-off function which is related to the usual heat-kernel cut-off procedure used to define infinite dimensional determinants (see [64]).

For $\varepsilon > 0,\ \lambda \in I\!R^+/\{0\}$, we set:

$$h_\varepsilon(\lambda) \equiv \exp\left(-\int_\varepsilon^{+\infty} e^{-t\lambda}\frac{dt}{t}\right). \tag{B.1.1}$$

For A in $\mathrm{Ell}_g^+(E)$, we can define through the spectral theorem the operator $h_\varepsilon(A')$. Let also $J_{1,g}(H)$ be the set of compact operators A on H such that $\mathrm{Tr}_g\,(A^*A)^{\frac{1}{2}} < \infty$, whereby the trace Tr_g is taken with respect to the scalar product $< \cdot, \cdot >_g$ and A^* denotes the adjoint of A w.r.t. this scalar product. For A in $\mathbb{1} + J_{1,g}(H)$ (i.e. $A \equiv \mathbb{1} + \mathcal{C},\ \mathcal{C} \in J_{1,g}(H)$), we can define the determinant as in [70], [68], see also [64] which for a strictly positive self adjoint operator $A \in \mathbb{1} + J_{1,g}(H)$, coincides with the infinite product of eigenvalues of the operator A. Let us recall here a result of [64] which enables us to define heat-kernel regularized operators.

Lemma 1: For a C^∞ Riemannian metric g on Λ, let $A \in \mathrm{Ell}_g^+(E)$. Then for $\varepsilon > 0,\ h_\varepsilon(A') \in \mathbb{1} + J_{1,g}(H)$. Moreover $h_\varepsilon(A')$ is invertible and $h_\varepsilon(A')^{-1} \in \mathbb{1} + J_{1,g}(H)$.

Proof: Using the fact that the asymptotic behavior of the eigenvalues is given by $\alpha_n(g) \sim Cn^\alpha, \alpha, C > 0$ and that the $\alpha_n(g)$ are ordered in increasing order, we first show that the series $\sum_{n=1}^{\infty} |\log h_\varepsilon((\alpha_n))(g))|$ converges using the dominated convergence theorem. From the behavior at infinity of $\alpha_n(g)$, we obtain a lower bound estimate $\alpha_n(g) \geq C_1 n^\alpha$, for some $C_1 > 0$ and n large enough.The monotonicity of the function h_ε yields then $h_\varepsilon(\alpha_n(g)) \geq h_\varepsilon(C_1 n^\alpha)$,and hence $\log h_\varepsilon(\alpha_n(g)) \geq \log(C_1 n^\alpha)$ for n large enough. Since $\log h_\varepsilon(\lambda) \leq 0, \forall \lambda > 0$, from this inequality follows that $|\log h_\varepsilon(\alpha_n(g))| \leq |\log h_\varepsilon(C_1 n^\alpha)|$ for large enough n.On the other hand, from the definition of h_ε, follows the estimate $|\log h_\varepsilon(\lambda)| \leq \frac{1}{\varepsilon} e^{-\varepsilon\lambda}$ for $\lambda \geq 1$.Hence we have, for large enough n, $|\log h_\varepsilon((\alpha_n)(g))| \leq \frac{1}{\varepsilon} e^{-\varepsilon C_1 n^\alpha}$. This yields the absolute convergence of the series with general term $|\log h_\varepsilon(\alpha_n(g))|$ and in particular $\lim_{n\to+\infty} h_\varepsilon((\alpha_n)(g)) = 1$. Hence, setting $\gamma_n(g) \equiv 1 - h_\varepsilon((\alpha_n)(g))$, we have $\lim_{n\to+\infty} \gamma_n(g) = 0$ and $|\log h_\varepsilon((\alpha_n)(g))| \sim |\gamma_n(g)|$ in a neighborhood of $+\infty$. The above result implies that $\sum_{n=1}^{\infty} |\gamma_n(g)| < \infty$ and hence that $h_\varepsilon(A')$ lies in $\mathbb{1} + J_{1,g}(H)$. This proves the first part of the lemma.

Let us check the invertibility of $h_\varepsilon(A')$. We have that $h_\varepsilon((\alpha_n)(g))$ is invertible for every $n \in I\!N$ (which yields the invertibility of $h_\varepsilon(A)$) and $h_\varepsilon((\alpha_n)(g))^{-1} = 1 - \gamma_n(g) + o(\gamma_n(g))$ (where $o(\gamma_n(g)) \equiv \varepsilon(\gamma_n(g))\gamma_n(g)$with $\lim_{n\to+\infty} \varepsilon(\gamma_n(g)) = 0$) as n goes to $+\infty$. Setting $\delta_n(g) \equiv (h_\varepsilon(\alpha_n)(g))^{-1} - 1$, we have $\sum_{n=1}^{\infty} |\delta_n(g)| < \infty$,which shows that $(h_\varepsilon(A'))^{-1}$ lies in $\mathbb{1} + J_1(H)$.

■

2. Approximation of gaussian measures with covariance defined by elliptic operators

Let g be a C^∞ Riemannian metric on Λ.Since the C^∞ Whitney topology on $C^\infty(E)$ can be described by a countable family of semi norms on $C^\infty(E)$, namely a family of Sobolev seminorms,the space $C^\infty(E)$ is a metrizable nuclear space. By the Minlos theorem (see e.g [71] Th 20.1 PartA, see [67] in the Hilbert case), there is a unique gaussian measure μ_g defined on the topological dual $C^\infty(E)^*$ of $C^\infty(E)$ by the characteristic function:

$$\chi_g(h) \equiv e^{-\frac{1}{2}<h,h>_g} \qquad (B.2.1)$$

which is a continuous function on $C^\infty(E)$ w.r.t the C^∞ Whitney toplogy.

Let $A \in \mathrm{Ell}_g^+(E)$ and let f be a strictly positive and bounded continuous function on $I\!R^+/\{0\}$ such that $f(A')$ (defined by the spectral theorem) maps $C^\infty(E)$ into itself.Let us call $\mathcal{F}_A$ the class of such functions.Then the function $h \to \chi_g(f(A')h)$ is a continuous positive definite function on $C^\infty(E)$ (positive definite because $f(A')$ is strictly positive self adjoint operator), so that the map $C^\infty(E) \to I\!R$, $h \to \chi_g(f(A')h)$ defines the characteristic function of a unique measure $\mu_g^{f(A')}$ on $C^\infty(E)^*$ (see e.g Th. 16.2 [71]).

In the following, we define a family of measures which approximate the measure $\mu_g^{A'}$. For this, we first introduce the family of functions from $I\!R^+/\{0\}$ into $I\!R^+/\{0\}$:

$$f_\varepsilon(\lambda) \equiv \exp\left(-\int_\varepsilon^{\frac{\varepsilon}{\lambda}} e^{-t\lambda} \frac{dt}{t}\right). \qquad (B.2.2)$$

We have:

$$\lim_{\varepsilon\to 0} f_\varepsilon(\lambda) = \lambda \quad \forall \lambda \in I\!R^+/\{0\} \tag{B.2.3}$$

and

$$\frac{f_\varepsilon(\lambda)}{h_\varepsilon(\lambda)} = e^{C_\varepsilon} \tag{B.2.4}$$

where

$$C_\varepsilon \equiv \int_\varepsilon^{+\infty} e^{-t} \frac{dt}{t}$$

is a strictly positive constant.
In particular, for an operator A in $\mathrm{Ell}_g^+(E)$, we can define $f_\varepsilon(A')$ for $\varepsilon > 0$ using the spectral theorem and we have from (B.2.4), $f_\varepsilon(A') = e^{C_\varepsilon} h_\varepsilon(A')$. Moreover we have:

Lemma 2: For a smooth riemannian metric g on Λ and an operator A in $\mathrm{Ell}_g^+(E)$, for $\varepsilon > 0$, the functions f_ε and h_ε belong to $\mathcal{F}_A$ and the corresponding operators $f_\varepsilon(A')$ and $h_\varepsilon(A')$ define homeomorphisms of $C^\infty(E)$ equipped with the $C^\infty(E)$ Whitney topology.

Proof:To check that f_ε and h_ε are in $\mathcal{F}_A$, it is sufficient to verify that the corresponding operators leave the space $C^\infty(E)$ stable since they are both continuous and bounded.

We first study the operator $\int_\varepsilon^{+\infty} e^{-tA'} \frac{dt}{t}$; we prove it leaves $C^\infty(E)$ stable and that it is continuous w.r.t the C^∞ Whitney topology. The integrand has both these properties since $e^{-tA'}$ is a smoothing operator (see e.g [66]). To prove that the integral also satisfies these properties, it is sufficient - using the theory of Hilbert space valued integrals - to give an upper bound estimate of the H^k norm of the operator $e^{-tA'}$ by a function $\alpha(t)$ which has the property that $\int_\varepsilon^\infty \alpha(t)\frac{dt}{t} < \infty$. This estimate is given by the diffusive property of the operator $e^{-tA'}$ since we have $\| e^{-tA'} \|_{H^k} \leq Ce^{-t\delta}$ (where $\| \cdot \|_{H^k}$ denotes the Sobolev operator norm of order k) with $C > 0$ and where $\delta \equiv \inf_{n\in I\!N}\{\lambda_n\} > 0$, $(\lambda_n)_{n\in I\!N}$ denoting the eigenvalues of the operator A'. From this estimate it then it easily follows that the operator $\int_\varepsilon^\infty e^{-tA'}\frac{dt}{t}$ leaves every space $H^k(E)$ stable and is continuous as an operator on that space. Since the C^∞ Whitney topology is given by a family of Sobolev norms, this yields the stability of the space $C^\infty(E)$ under this operator and the continuity of the operator w.r.t the C^∞ Whitney toplogy.

This combined with the definition of $f_\varepsilon(A')$ and of $h_\varepsilon(A')$ yields in turn the stability of the space $C^\infty(E)$ under the action of these operators as well as the continuity w.r.t the C^∞ Whitney topology. The inverse operator of $h_\varepsilon(A')$ reads $e^{\int_\varepsilon^{+\infty} e^{-tA'}\frac{dt}{t}}$ so that the same stability and continuity results hold for this inverse and for the inverse of $f_\varepsilon(A')$ We can therefore conclude that both $h_\varepsilon(A')$ and $f_\varepsilon(A')$ are homeomorphisms of $C^\infty(E)$ for fixed $\varepsilon > 0$. ∎

Lemma 3:For a C^∞ metric g on Λ, and an operator A in $\mathrm{Ell}_g^+(E)$, A' defines a homeomorphic operator on $C^\infty(E)$.

Proof: The operator A' is an elliptic operator and hence takes a smooth vector into a smooth vector and is continuous w.r.t the C^∞ Whitney topology. Its inverse is also continuous since as in [66] (see Lemma 1.6.6 of the second reference listed under [66]), it can be shown that for any strictly positive k, we have:

$$\| (A')^{-1}u \|_{H^k} \leq C \| u \|_{H^k}$$

for some $C > 0$, for all $u \in H^k(E)$ and in particular for all $u \in C^\infty(E)$. (We have used here the notations of the proof of the previous lemma). The C^∞ Whitney topology can also be defined by the family of Sobolev norms $\| \cdot \|_{H^k}$, $k \in \mathbb{N}$ so that this inequality shows the continuity w.r.t the C^∞ Whitney topology. ∎

Since the operators $f_\varepsilon(A')$, $\varepsilon > 0$ (resp. A') are homeomorphisms on $C^\infty(E)$,so are the corresponding dual operators $f_\varepsilon(A')^*$ (resp.$(A')^*$) on the topological dual (w.r.t the C^∞ Whitney topology) space $C^\infty(E)^*$. We can therefore define the image measures on $C^\infty(E)^*$:

$$f_\varepsilon(A')^* \mu_g \equiv \mu_g \circ (f_\varepsilon(A')^*)^{-1}$$

(resp.$(A')^* \mu_g \equiv \mu_g \circ [(A')^*]^{-1}$).
These measures have characteristic functions $\chi_g(f_\varepsilon(A')\cdot)$ (resp. $\chi_g(A'\cdot)$) and therefore coincide with the measure $\mu_g^{f_\varepsilon(A')}$ (resp. $\mu_g^{A'}$) defined above.

Proposition 4:For a C^∞ Riemannian metric g on Λ, let A be in $\mathrm{Ell}_g^+(E)$. Then, the family of image measures $f_\varepsilon(A')^*\mu_g$ converges weakly to the measure $(A')^*\mu_g$.

Proof: This follows from (B.2.3). Indeed by (B.2.3), we have $\lim_{\varepsilon \to 0} < f_\varepsilon(A')h, f_\varepsilon(A')h >_g$ $=< A'h, A'h >_g$ for $h \in C^\infty(E)$ and hence the characteristic function $\chi_g(f(A')\cdot, f_\varepsilon(A')\cdot)$ of $f_\varepsilon(A')^*\mu_g$ converges to the characteristic function $\chi(A'\cdot, A'\cdot)$ of $(A')^*\mu_g$. From this in turn follows that $\int_{C^\infty(E)} \mathcal{F}(\xi)(f_\varepsilon(A')^*\mu_g)(d\xi) \to \int_{C^\infty(E)} \mathcal{F}(\xi)((A')^*\mu_g)(d\xi)$, where $\mathcal{F}$ varies in the separating family of continuous functionals on $C^\infty(E)^*$ of the form $\xi \to e^{i\xi(h)}$ for $h \in C^\infty(E)$. Since $C^\infty(E)$ is a separable Fréchet space, it is a Polish space (see [69] p.110) and this convergence property implies the weak convergence of measures $(f_\varepsilon(A'))^*\mu_g$ to $(A')^*\mu_g$ (see [69] Appendix).
∎

In the next paragraph, we investigate the equivalence properties of the various image measures constructed above. Although the transformed measures $A'^*\mu_g$ and $B'^*\mu_g$ are not equivalent in general for two different elliptic operators A and B in $\mathrm{Ell}_g^+(E)$, we shall show that the approximating transformed measures $f_\varepsilon(A')^*\mu_g$ and $f_\varepsilon(B')^*\mu_g$ are equivalent for any $\varepsilon > 0$ and their Radon Nikodym derivative is defined in terms of their heat-kernel regularized determinants.

3. Equivalence of image Gaussian measures

Theorem 5: Let g_1 and g_2 be C^∞ Riemannian metrics on Λ and let $A_1 \in \mathrm{Ell}_{g_1}^+(E)$, $A_2 \in \mathrm{Ell}_{g_2}^+(E)$. The image measures $f_\varepsilon(A_1)^*\mu_{g_1}$ and $f_\varepsilon(A_2)^*\mu_{g_2}$ are equivalent if and only

if $g_1 = g_2$.
If $g \equiv g_1 = g_2$, the Radon Nikodym derivative is given by:

$$\frac{f_\varepsilon(A_2')^*\mu_g}{f_\varepsilon(A_1')^*\mu_g}(h) = \frac{\det\, h_\varepsilon(A_2')}{\det\, h_\varepsilon(A_1')} \cdot e^{-\frac{1}{2}<(f_\varepsilon(A_2')^{-2} - f_\varepsilon(A_1')^{-2})h,h>_g}$$

Remark: The Jacobian operator is defined in terms of heat-kernel cut-off determinants $\det\, h_\varepsilon(A')$ and coincides with the Jacobian determinant that occurs in the Radon Nikodym derivative of the same measure μ_g transformed under the heat kernel operators $h_\varepsilon(A_i)$, $i = 1, 2$.

The proof of Theorem 5 uses classical results concerning the equivalence of Gaussian measures on nuclear spaces (see e.g. Th. 10.1 [71] Part A). Before we actually prove the theorem, we shall show a lemma which will be useful in the course of the actual proof.

Lemma 6: For a $C^\infty(E)$ metric g on Λ, the scalar product $< \cdot,\cdot >_g$ extends in a unique way to a scalar product on the topological dual $C^\infty(E)^+$ of $C^\infty(E)$ w.r.t the L^2 topology induced by $< \cdot,\cdot >_g$. The space $(C^\infty(E)^+, < \cdot,\cdot >_g)$ defines a Hilbert space.
Let B be a bounded operator on $H = L^2(E)$, and let B^+ denote its unique extension to $C^\infty(E)^+$. Then if B is a trace class operator in H with smooth eigenvectors, B^+ is trace-class in $C^\infty(E)^+$.
Proof of Lemma 6: Since the topology induced by $< \cdot,\cdot >_g$ is independent of the metric g, so is the space $C^\infty(E)^+$. We equip $C^\infty(E)^+$ with the strong dual topology, i.e with the topology induced by the dual norm

$$\|\xi\|_g \equiv \sup_{h\in C^\infty(E), <h,h>_g = 1} |\xi(h)| \qquad (B.3.1)$$

As before,this topology does not depend on the metric g underlying the scalar product $< \cdot,\cdot >_g$.
The space $C^\infty(E)^+$ can be viewed as a pre-Hilbert space in the following sense.The space $C^\infty(E)/M$ with $M \equiv \{h \in C^\infty(E), < h,h >_g = 0\}$ can be embedded in $C^\infty(E)^+$ by the map:

$$i : C^\infty(E) \to C^\infty(E)^+$$
$$h \to i_h : C^\infty(E) \to I\!R, \quad k \to < h,k >_g$$

since

$$i_h = i_{\tilde h} \Longleftrightarrow < h,k >_g = < \tilde h, k >_g \;\forall k \in C^\infty(E) \Longleftrightarrow h - \tilde h \in M.$$

This embedding i is isometric w.r.t the dual norm in $C^\infty(E)^+$ since $\sup_{\substack{k\in C^\infty(E)\\ <k,k>_g=1}} i_h(k)$ $= < h,h >_g$.
Moreover $i(C^\infty(E)/M)$ is dense in $C^\infty(E)^+$.Indeed, let $(e_n)_{n\in I\!N}$ be an orthonormal basis of the closure H of $C^\infty(E)$ w.r.t the scalar product $< \cdot,\cdot >_g$ and let H_n be the finite dimensional subspace of $C^\infty(E)$ spanned by $\{e_1, \cdots, e_n\}$.A map $\xi \in C^\infty(E)^+$ induces a

continuous linear map from H_n to $\mathbb{R}$ so that for each $k \in H_n$ there exists $h_n \in H_n$ such that $\xi(k) = < h_n, k >_g$.From this definition of h_n it follows in particular that $h_m = P_m h_n$ for $m < n$ where P_l is the orthogonal projection onto H_l , $l \in \mathbb{N}$ and that the sequence $(h_n)_{n\in\mathbb{N}}$ converges to an element $h \in H$. The sequence of maps $\xi_n : C^\infty(E) \to \mathbb{R}$, $k \to < h_n, k >_g$ converges to ξ for the strong topology on $C^\infty(E)^+$ for we have:

$$\begin{aligned}\|\xi - \xi_n\|_g &= \sup_{\substack{<k,k>_g=1 \\ k\in C^\infty(E)}} |(\xi - \xi_n)(k)| \\ &= \sup_{\substack{<k,k>_g=1 \\ k\in C^\infty(E)}} \lim_{m\to\infty} |(\xi - \xi_n)(P_m(k))| = < h - h_n, h - h_n >_g\end{aligned}$$

which proves the density property.

As a consequence of this dense embedding, the scalar product $< \cdot, \cdot >_g$ defined on $C^\infty(E)$ extends to a scalar product on $C^\infty(E)^+$ and we have $\xi(h) = < \xi, h >_g$, $\xi \in C^\infty(E)^+, h \in C^\infty(E)$ and $C^\infty(E)^+$ can be seen as a pre-Hilbert space equipped with this extended scalar product. The strong norm $\| \cdot \|_g$ defined in (B.3.1) then coincides with the norm induced by $< \cdot, \cdot >_g$.Since $C^\infty(E)^+$ is closed w.r.t the strong norm, it is also closed w.r.t the norm induced by the scalar product $< \cdot, \cdot >_g$. The space $(C^\infty(E)^+, < \cdot, \cdot >_g)$ is in fact a Hilbert space; indeed any Cauchy sequence $(f_n)_{n\in\mathbb{N}}$ for the strong convergence converges weakly to a functional on $C^\infty(E)$ since for any $h \in C^\infty(E)$, $f_n(h)$ is a Cauchy sequence in $\mathbb{R}$ and hence converges to an element denoted by $f(h)$. On the other hand, the sequence $(f_n)_{n\in\mathbb{N}}$ being strongly Cauchy, there exists a subsequence f_{n_k} which converges strongly and hence weakly to an element $\tilde{f}$ of $C^\infty(E)^+$. From this it follows that the two limits f and $\tilde{f}$ coincide and that the entire sequence (f_n) converges strongly to f.

Any linear operator from $C^\infty(E)$ into itself and continuous w.r.t the topology induced by $< \cdot, \cdot >_g$ can then be extended uniquely to an operator on $C^\infty(E)^+$.In particular, a bounded linear operator $\mathcal{B}$ on H that leaves $C^\infty(E)$ stable extends to an operator $\mathcal{B}^*$ on $C^\infty(E)^+$.The adjoint of the operator B on H then coincides with the restriction of B^* to its domain in H.Moreover, if B defined on H has purely discrete spectrum with corresponding eigenvectors in $C^\infty(E)$, then B^* is trace class whenever $\mathcal{B}$ is trace class since the orthonormal basis of eigenvectors for B is also an orthonormal basis of eigenvectors for B^* as follows from the isometric property of i. ∎

Proof of Theorem 5:

Returning now to the proof of Theorem 5, we first write the characteristic functions of $f_\varepsilon(A_i')^*\mu_{g_i}$, $i \in \{1,2\}$ in terms of the scalar product $< \cdot,\cdot >_{g_i}$ and we use a classical criterium (see e.g Th10.1 [71] Part B) to compare the corresponding measures.

Since the scalar products $< \cdot,\cdot >_{g_i}$ induce the same topology on $C^\infty(E)$, we can define a continuous linear operator U from $C^\infty(E)$ into $C^\infty(E)^+$ such that $< h,k >_{g_2} = < Uh,k >_{g_1}$ $\forall h,k \in C^\infty(E)$.Exchanging the role of h and k, we see that U is a symmetric operator, moreover it is positive definite since $< h,h >_{g_2} = < Uh,h >_{g_1} > 0 \quad \forall h \in C^\infty(E)/\{0\}$.Note that U can be looked upon as a bounded strictly positive symmetric operator on H so that $\sqrt{U}$ is well defined on H and hence from $C^\infty(E)$ to $C^\infty(E)^+$.It extends to a linear positive definite homeomorphic operator on $C^\infty(E)^+$. The operators $f_\varepsilon(A_i{}')$, $i = 1,2$ map $C^\infty(E)$ into $C^\infty(E)$ (by Lemma 2) .Since $h_\varepsilon(A_i{}') \in \mathbb{1} + J_1(H)$, it is invertible and $(h_\varepsilon(A_i{}'))^{-1}$ is also a bounded operator on H (by Lemma 1) so that $(f_\varepsilon(A_i{}'))^{-1}$ (see formula (B.2.4)) is also a continuous linear operator from $C^\infty(E)$ into $C^\infty(E)$ w.r.t the L^2 topology and thus $f_\varepsilon(A_i{}')$ extends to a positive definite homeomorphic operator on $C^\infty(E)^+$.

The characteristic functions $\chi_i(h)$ of $(f_\varepsilon(A_i{}'))^*\mu_{g_i}$ read $\chi_1(h) = e^{-\frac{1}{2}<f_\varepsilon(A_1{}')h, f_\varepsilon(A_1{}')h>_{g_1}}$ and $\chi_2(h) = e^{-\frac{1}{2}<\sqrt{U}(f_\varepsilon(A_2'))h, \sqrt{U}(f_\varepsilon(A_2'))h>_{g_1}}$.Applying Th 10.2 [71] Part B and using the fact that $f_\varepsilon(A_1{}')$ and $\sqrt{U}f_\varepsilon(A_2')$ are positive definite homeomorphic linear operators on $C^\infty(E)^+$, we see that the Gaussian measures $f_\varepsilon(A_1')^*\mu_{g_1}$ and $f_\varepsilon(A_2')^*\mu_{g_2}$ are equivalent if and only if the operator $T \equiv \mathbb{1} - \left(\sqrt{U}f_\varepsilon(A_2')(f_\varepsilon(A_1'))^{-1}\right)^* \sqrt{U}f_\varepsilon(A_2')(f_\varepsilon(A_1'))^{-1}$ is a Hilbert Schmidt operator on $C^\infty(E)^+$ (w.r.t. the scalar product $< \cdot,\cdot >_{g_1}$).Since the operators $\sqrt{U}$ and $f_\varepsilon(A_i')$ are self adjoint on H, we have $T = \mathbb{1} - f_\varepsilon(A_2')f_\varepsilon(A_1')^{-1}Uf_\varepsilon(A_2')f_\varepsilon(A_1')^{-1}$.

From the relation $f_\varepsilon(A') = e^{C_\varepsilon}h_\varepsilon(A')$ it then follows that

$$T = \mathbb{1} - h_\varepsilon(A_2')(h_\varepsilon(A_1'))^{-1}Uh_\varepsilon(A_2')h_\varepsilon(A_1')^{-1} \qquad (B.3.2)$$

We now prove that the equivalence:

$$\mathbb{1} - T \quad \text{is Hilbert-Schmidt on} \quad C^\infty(E)^+ \Longleftrightarrow \mathbb{1} - U \text{ is traceclass on } C^\infty(E)^+$$

The proof essentially relies on the ideal property of the space of trace-class operators in the space of bounded operators and lemma 1.

Let us first prove the implication from left to right. Since $h_\varepsilon(A_1') \in \mathbb{1} + J_{1,g_1}(H)$, its inverse also belongs to $\mathbb{1} + J_{1,g_1}(H)$ (see Lemma 1), so that by Lemma 6, the operator $\mathbb{1} - h_\varepsilon(A_1')$ is trace-class as an operator on $C^\infty(E)^+$.

On the other hand, since $h_\varepsilon(A_2') \in \mathbb{1} + J_{1,g_2}(H)$,we have $\sqrt{U}h_\varepsilon(A_2')(\sqrt{U})^{-1} \in \mathbb{1} + J_{1,g_1}(H)$, where we used the relation between the scalar products defined in terms of g_1 and g_2 so that $h_\varepsilon(A_2')$ is of the form $\mathbb{1} + \sqrt{U}^{-1}C\sqrt{U} \quad (*)$ with $C \in J_{1,g_1}(H)$.

Both the operators U and U^{-1} are bounded invertible operators on H since $< \cdot,\cdot >_{g_1}$ and $< \cdot,\cdot >_{g_2}$ are equivalent. Using the fact that $\mathbb{1} + J_{1,g_1}(H)$ is an ideal in the space of bounded linear operators on H , we see that from the fact that $h_\varepsilon(A_2')$ has the above form (*),it follows that the operator $h_\varepsilon(A_2')$ itself lies in $\mathbb{1}+ J_{1,g_1}(H)$ and hence that $\mathbb{1}- h_\varepsilon(A_2')$ is a traceclass operator on $C^\infty(E)^+$ using again the results of Lemma 6.

The fact that the operator $\mathbb{1} - T$ is Hilbert-Schmidt then implies -using again the ideal property- that $\mathbb{1} - U$ is a traceclass operator on $C^\infty(E)^+$, since we have by (3.2), $U = h_\varepsilon(A_1')h_\varepsilon(A_2')^{-1}(\mathbb{1} - T)h_\varepsilon(A_1')h_\varepsilon(A_2')^{-1}$.

Conversely, if $\mathbb{1}- U$ is trace-class, then from (B.3.2) and applying the same arguments as above, we find that $\mathbb{1} - T$ is trace-class and hence Hilbert-Schmidt.

In the case of the equivalence of the measures, namely when $\mathbb{1} - U$ is Hilbert-Schmidt, we have $U = \mathbb{1}$ and hence $g_1 = g_2$.Indeed, otherwise we would have $< h,k >_{g_2}=< h,k >_{g_1} + < Vh,k >_{g_1}$, where V is a non zero Hilbert-Schmidt operator.Let us consider the case when E is the bundle of zero forms on Λ.This would imply that the multiplication operator $h \to (1- \frac{\sqrt{g_2}}{\sqrt{g_1}})h$ is a Hilbert-Schmidt operator, which clearly only holds when $g_1 = g_2$. The same argument easily generalizes to the case of a more general bundle

The Radon Nikodym derivative is then given by the classical formula on densities of Gaussian measures, see e.g. [71], Th. 10.1 Part B.
■

4. Conclusion

This last theorem shows how the regularized heat-kernel determinants of elliptic operators can be interpreted as Jacobian determinants arising in the Radon-Nikodym derivatives of transformed Gaussian measures under the action of an elliptic operator. The proposition 2 showed that these Gaussian measures approximate (in a weak sense) Gaussian measures obtained as the image under an elliptic operator of the canonical generalized Gaussian measure on a nuclear space. This gives an interpretation of the heat-kernel regularization procedure for Jacobian determinants in terms of an approximation procedure for Gaussian measures.

The advantage of this point of view is that instead of extracting a finite part of the jacobian determinants after the transformation of the measure, the regularized determinants appear here naturally as jacobian determinants for the transformation measures.

Footnotes

[0]) Also within the "bosonic case" we can only discuss a few topics. Among the omitted (or only very partially treated) topics, let us mention mathematical path integral methods (of oscillatory and probabilistic type) for relativistic single particles (Dirac and Klein-Gordon equations), see e.g. [1] 5), [23], [49], [51], [53], [125], [126], for stochastic Schrödinger operators, see e.g. [63], [144], [160], for polymers, see e.g. [2]. Other areas where path integral methods are used are statistical mechanics, see e.g. [116], [117], [155], [156], [160] and hydrodynamics, see e.g. [118]. Obviously omission of a topics should not be understood in any sense to mean that there are no mathematically valuable results in that area. On the contrary: there are entire topics which are left out only because going into them would have required much more space and time. E.g. a forced major omission concerns mathematical derivations of topological results using probablistic functional integrals, see e.g. [63], [119–121], [136], [137], [155] we hope to return to this and other topics on a future occasion.

[1]) Of course the notation $\tilde{\int} \cdot d\gamma$ is just "mnemonic", the real definition is by the latter term in (2.21) (integration with respect to μ_f). The notation $\tilde{\int} \cdot d\gamma$ is suggestive of transformation properties of the integral, as it is all notation on paths integrals as integrals against a heuristic "infinite dimensional Lebesgue measure" used in the physical literature. That an infinite dimensional Lebegue can not exist as a σ-additive measure in the usual sense of (abstract) measure theory is well known. In fact the measure of a ball of radius 2 centerered at the origin should be larger than the sum of measures of the sets $U_i = \{x \in I\!R^n \mid |x - e_i| \leq \frac{1}{2}\}$, which all have equal positive measure, their union having then measure $+\infty$. Then a ball of radius 2 has already infinite volume, which shows that a "flat Lebesgue" (rotation and translation invariant) measure cannot exist in the usual sense, see also [67], [71]. Further references for probabilistic integration in infinite dimensions are e.g. [58], [59], [60], [62], [139], [150-153].

[2]) Heuristic related formulae have appeared in the physical literature, see e.g. references in [140]. It is certainly a challenge to prove mathematically such formulae.

[3]) There is however also mathematical work on infinite dimensional oscillatory integrals for relativistic one particle theories (Dirac and Klein-Gordon equations), see footnote [0]).

[4]) Unfortunately most physics books have a tendency to ignore this fact and specific mathematical problems with Feynman path integrals (besides perhaps mentioning a generic being "ill defined"). A notable exception is however [45]. Also mathematical books on functional integration, e.g. [28], [29], have just short mentioning of rigorously defined Feynman path integrals (exceptions are [9] and obviously [3], [22], which have the mathematical theory of Feynman path integrals as own topics).

[5]) $N(0, B^{-1})$ is an infinite dimensional analogue of $\sqrt{\text{Det } B} e^{-\frac{1}{2}xBx} \frac{dx}{(2\pi)^{\frac{n}{2}}}$.

[6]) The Høegh-Krohn or exponential interaction model is one of the best studied models of constructive quantum field theory [129–133], [142]. Uniqueness of Gibbs states and global Markov property have been established [78], [79]. The Dirichlet forms associated with the time zero measure ("ground state Dirichlet forms") as well as the Euclidean measure μ_E have been studied [123], [124], [75], [90]. In particular a strong characterization of the Dirichlet form (in terms of a weighted Sobolev space in infinite dimensions) is known [123], [75].

[7]) For the relevance of discussions of probabilistic measures for quantum fields on manifolds see e.g. [127], [150], [162] and references therein.

[8]) For a partial extension to $d = 8$ see [134], [135].

[9]) For a new point of view concerning phenomena arising in the study of the heat equation corresponding to those encountered in the study of the Schrödinger equation see [158] and references therein.

[10]) Actually one should add to V_Λ in (A.6) a curvature term, which however being linear in the field ξ can be easily taken into account, see [64] for details.

References

[1] 1) F.A. Berezin: *The method of second quantization*, Academic Press, New York (1966)

2) J. Lott: *Supersymmetric path integrals*, Commun.Math.Phys. 108, 605-629 (1987)

3) A. Jaffe: A. Lesniewski: *Supersymmetric quantum fields and infinite dimensional analysis*, 1987 Cargèse Lecture Notes

4) P.A. Meyer: *Eléments de probabilités quantiques*, Sém. Prob. XX, XXI, XXII, Lect. Notes Math. 1204, 1247, 1321, Springer, Berlin (1986)

5) P. Krée: *La théorie des distributions en dimension quelconque et l'intégration stochastique*, in Stochastic Analysis and related topics, Lect. Notes Math. 1316, Springer, Berlin (1989)

6) Z. Haba: *Stochastic equations for super symmetric gauge fields*, Proc. Ahrenshoop Symp., DDR (Nov. 1985)

7) Z. Haba: *Supersymmetry and stochastic processes*, pp. 101–112 in "Stochastic Mechanics and Stochastic Processes", Proc. Swansea 1986, Edts. A. Truman, I.M. Davies, Lect. Notes Maths. 1325, Springer, Berlin (1988)

8) J. Kupsch: *Functional integration for Euclidean Dirac fields*, Ann. Inst. H. Poincaré 50, 143-160 (1989)

9) A. Rogers: *Fermionic path integration and Grassmann Brownian motion*, Commun. Math. Phys. 113, 353–368 (1987)

10) L. Accardi, K. R. Pathasarathy:*Stochastic calculus on local algebras*, pp. 9-23 in Edts. L. Accardi, W. von Waldenfels, Quantum Probability and Applications II, Lect. Notes Maths. 1136, Springer Berlin (1985)

11) L. Accardi: *A mathematical theory of quantum noise*, pp. 427-444 in Proc. 1st-World Congress of the Bernoulli Society, Edts. Yu. A. Prokorov, V. V. Sazonov, VNU Science, Utrecht (1987)

[2] S. Albeverio, J.E. Fenstad, R. Høegh-Krohn, T. Lindstrøm: *Non standard methods in stochastic analysis and mathematical physics*, Academic Press, Orlando (1986)

[3] S. Albeverio, R. Høegh-Krohn: *Mathematical Theory of Feynman Path Integrals*, Lect. Notes Maths. 523, Springer, Berlin (1976)

[4] 1) S. Albeverio, R. Høegh-Krohn: *Feynman path integrals and the corresponding method of stationary phase*, pp.3–57 in Proceedings of the Marseille International Colloquium on "Mathematical Problems in Feynman Path Integrals", May 1978 Edts. S. Albeverio, Ph. Combe, R. Høegh-Krohn, G. Rideau, M. Sirugue-Collin, R. Stora, Lecture Notes in Physics 106, Springer, (1979)

2) S. Albeverio, R. Høegh-Krohn: *Oscillatory integrals and the method of stationary phase in infinitely many dimensions, with applications to the classical limit of quantum mechanics I*, Invent. Math. 40, pp.59–106 (1977)

3) S. Albeverio: *Some recent developments and applications of path integrals*, pp. 3–32 in "Path Integrals from meV to MeV", Edts. M.C. Gutzwiller, A. Inomata, J.R. Klauder, L. Streit, World Scient., Singapore (1986)

4) S. Albeverio, Ph. Blanchard, R. Høegh-Krohn: *Some applications of functional integration*, in "Mathematical Problems in Theoretical Physics", Edts. R. Schrader, R. Seiler, D.A. Uhlenbrode, Lect. Notes Phys. 153, Springer, Berlin (1982)

[5] 1) J. Rezende: *The method of stationary phase for oscillatory integrals*, Commun. Math. Phys. 101, 187–206 (1985)

2) J. Rezende, Ph. D. Thesis, Bielefeld (1985)

[6] D. Elworthy, A. Truman: *Feynman maps, Cameron - Martin formulae and anharmonic oscillators*, Ann. Inst. H. Poincaré 41, 115–142 (1984)

[7] 1) S. Albeverio, Z.D. Brzezniák: in preparation

2) S. Albeverio, A.M. Boutet de Monvel, G.Lebeau: in preparation

[8] 1) T. Zastawniak: *Approximation of Feynman path integrals by integrals over finite dimensional spaces*, Bull. Pol. Acade. Sci. Math. 34, 355–372 (1986)

2) T. Zastawniak: *The equivalence of two approaches to the Feynman integral for the anharmonic oscillator*, Cracow Preprint (1986) J. Jagell. Univ.

[9] P. Exner: *Open quantum systems and Feynman integrals*, D. Reidel, Dordrecht (1985)

[10] R.H. Cameron, D.A. Storvick: *A simple definition of the Feynman integral, with applications*, Mem. AMS, No. 288, 46, 1–46 (1983)

[11] G.W. Johnson, M.L. Lapidus: *Generalized Dyson Series, generalized Feynman integrals and Feynman's operational calculus*, Mem. Amer. Matrh. Soc. 62, 1–78 (1986)

[12] G. Kallianpur, C. Bromley: *Generalized Feynman integrals using analytic continuation in several complex variables*, pp.217–267 in "Stochastic Analysis and Applications", Ed. M.A. Pinsky, M. Dekker, New York (1984)

[13] Y.Z. Hu, P.A. Meyer: *Chaos de Wiener et intégrale de Feynman*, pp. 51–71 in Sém. Prob. XXII, Lect. Notes Math. 1321, Springer, Berlin (1988)

[14] H. Watanabe: *Feynman-Kac formula associated with a system of partial differential operators*, J. Funct. Anal. 65, 204–235 (1986)

[15] S. Albeverio: in preparation

[16] S. Albeverio, Ph. Blanchard, R. Høegh-Krohn: *Feynman path integrals and the trace formula for Schrödinger operators*, Comm. Math. Phys. 83, 49–76 (1982)

[17] S. Albeverio, T. Arede: *The relation between quantum mechanics and slassical mechanics: a survey of some mathematical aspects*, pp.37–76 in "Stochastic behaviour in quantum systems", Edts. G Casati, J. Ford, Plenum, New York (1985)

[18] T. Arede: *Equação do calor em grupos de Lie e alguns espaços simétricos*, Ph. D. Thesis, Lisboa (1989); and paper in preparation

[19] S. Albeverio, T. Arede, Z. Haba: *On left invariant Brownian motion an Heat kernels of nilpotent Lie groups*, SFB 237-Preprint, to appear in J. Math. Phys. (1989)

[20] E.B. Davies: *Heat kernels and spectral theory*, Cambridge University Press (1989)

[21] E. Nelson: *Feynman integrals and the Schrödinger equation*, J. Math. Phys. I, pp.332–343 (1984)

[22] P. Maldowney: *A general theory of integration in function spaces, including Wiener and Feynman integration*, Longman, Essex (1987)

[23] Ph. Blanchard, Ph. Combe, M. Sirugue, M. Sirugue-Collin: *Jump processes and introduction and some applications to quantum physics*, supl. Rend. Civ. Mat. Palermo, Ser. II, No. 17, 47–104 (1987)

[24] Ph. Blanchard, M. Sirugue: *Large deviations for classical paths: Hamiltonian flows as classical limits of quantum flows*, Commun. Math. Phys. 101, 173–185 (1985)

[25] S. Albeverio, Ph. Blanchard, Ph. Combe, R. Høegh-Krohn, M. Sirugue: *Local relativistic invariant flow for quantum fields*, Commun. Math. Phys. 90, 329–351 (1983)

[26] I. Daubechies, J.R. Klauder, *True measures for real time path integrals*, pp. 425–432 in Proc. Ref. [4], 3).

[27] N. Cutland: *Infinitesimals in action*, J. London Math. Soc. (2) 35, 202–216 (1987)

[28] J. Glimm, A. Jaffe: *Quantum Physics: A Functional Point of View*, Springer, Berlin (1982)

[29] B. Simon: *Functional Integration and Quantum Physics*, Acad. Press, New York (1979)

[30] T.A. Osborn, R. Wong: *Schrödinger spectral kernels: high order asymptotic expansions*, J. Math. Phys. 24, 1487–1501 (1983)

[31] B. Simon: *Schrödinger semigroups*, Bull. An. Math. Soc. 7, 447–526 (1982)

[32] S. Albeverio, R. Høegh-Krohn: *Uniqueness of the physical vacuum and the Wightman functions in the infinite volume limit for some non-polynomal interactions*, Commun. Math. Phys. 30, 171–200 (1973)

[33] T. Hida: *Brownian motion*, Springer, New York (1980)

[34] G. Jona-Lasinio: *Stochastic processes and quantum mechanics*, Astérisque 132, 203–216 (1985)

[35] F. Martinelli, E. Scoppola: *"Stochastic Processes Physics and Geometry"* to appear in Proc. 2nd Int. Ascona-Locarno-Como, Ed. S. Albeverio, G. Casati, U. Cattaneo, D. Merlini, R. Moresi, World Scient. (1989)

[36] S. Albeverio, J. Brasche, M. Röckner: *"Dirichlet forms and generalized Schrödinger operators"*, to appear in "Schrödinger operators", Proc. Søndeborg, Edts. A. Jensen, H. Holden, Lect. Notes Maths., Springer, Berlin (1989)

[37] S. Albeverio, Ph. Blanchard, R. Høegh-Krohn: *Reduction of nonlinear problems to Schrödinger heat equations: formation of Kepler orbits, singular solutions for hydrodynamical equations*, pp.189–206 in " Stochastic Aspects of Classical and Quantum Systems", Proc. Marseille 1983 (Ed. S. Albeverio, Ph. Combe, M. Sirugue-Collin), Lect. Notes Maths. 1109, Springer, Berlin (1985)

[38] S. Albeverio, R. Høegh-Krohn, L. Streit: *Energy forms, Hamiltonians and distorted Brownian paths*, Journal of Mathematical Physics 18, 907–917 (1977)

[39] M. Fukushima: *Energy forms and diffusion processes*, pp.65–98 in "Mathematics and Physics", Lect. on Recent Results, Vol.1, Ed. L. Streit, World Scient., Singapore (1985)

[40] S. Albeverio, M. Röckner: *1) New developments in theory and applications of Dirichlet forms*, to appear in Proc. 2nd Int. Ascona-Locarno-Como, Ed. S. Albeverio,

G. Casati, U. Cattaneo, D. Merlini, R. Moresi, World Scient. (1989); 2) Papers in preparation

[41] S. Albeverio, Zhiming Ma: *Nowhere Radon smooth measures, perturbations of Dirichlet forms and singular quadratic forms*, to appear in Proc. Bad Honnef Conf. 1988, Ed. Christopeit, Lect. Notes Control and Inform. Sciences, Springer, Berlin (1989)

[42] S. Albeverio, F. Gesztesy, R. Høegh-Krohn, H. Holden: *Solvable models in quantum mechanics*, Springer, Berlin (1988)

[43] S. Albeverio, J.E. Fenstad, R. Høegh-Krohn, W. Karwowski, T. Lindstrøm: to appear in Mem. Volume R. Høegh-Krohn

[44] J. Brasche, Ph. D. Thesis, Bielefeld (1989), and paper in prep.

[45] 1) L.S. Schulman: *Introduction to the path integral*, in "Path Summation: Achievements and Goals", Edts. S.O. Lundquist, A. Ranfagni, V. Sa-yakanit, L.S. Schulman, World Scientific, Singapore (1988)

2) L.S. Schulman: *Techniques and Applications of Path Integration*, J. Wiley, New York (1981)

[46] 1) V.P. Maslov: *The Kolmogorov-Feller equation and the probabilistic model of quantum mechanics*, J. Sov. Math. 23, 2534–2553 (1983)

2) V.P. Maslov: *Asymptotic and geometric quantization*, Russ. Math. Surv. 39, 133–205 (1984)

3) V.P. Maslov, A.M. Chebotarev: *Second term of the logarithmic asymptotics of path integrals*, Itogi Nauk i Tekn., Ser. Theor. Ver. 19, 2580–2598 (1982)

[47] V. Mandrekar: *Some remarks on various definitions of Feynman integral*, pp. 170–177 in Levt. Notes Math., Ed. Beck, K. Jacob, (1983)

[48] 1) D. Fujiwara: *A construction of the fundamental solution for the Schrödinger equation*, J. Analyse Math. 35, 41–96 (1979)
2) D. Fujiwara: *Remarks on convergence of the Feynman path integrals*, Duke Math. J.47, 559-600 (1980)

[49] T. Ichinose, H. Tamura: *The Zitterbewegung of a Dirac particle in two dimensional space-time*, J. Math. Phys. 29, 103–109 (1988)

[50] G. Kallianpur, D. Kannan, R.L. Karandikar: *Analytic and sequential Feynman integrals on abstract Wiener and Hilbert spaces, and a Cameron-Martin formula*, Ann. Inst. H. Poincaré, Prob. Th. 21, 323–361 (1985)

[51] H. Watanabe, Y. Ito: *A construction of the fundamental solution for the relativistic wave equation I*, Tokyo J. Math. 7, 99–117 (1984)

[52] G. F. De Angelis, G. Jona.Lasinio, M. Sirugue: *Probalistic solution of Pauli type equations*, J. Phys. A 16, 2433-2444, (1983)

[53] B. Gaveau, T. Jacobson, M. Kac, L.S. Schulman: Phys. Rev. Lett. 53, 419 (1984)

[54] J. Bertrand, B. Gaveau, G. Rideau: *Quantum fields and Poisson processes; interaction of a cut-off Boson field with a quantum particle*, Ed. L. Accardi, W.V. Waldenfels, Lect. Notes Math. 1136 (1985)

[55] T. Hida, L. Streit: *Generalized Brownian functionals and the Feynman integral*, Stoc. Proc. Appl. 16, 55–69 (1983)

[56] S. Albeverio, R. Høegh-Krohn: *Diffusion fields, quantum fields, and fields with values in Lie groups*, pp. 1–98 in "Stochastic Analysis and Applications", Edt. M.A. Pinsky, M. Dekker, New York (1984)

[57] H. Doss: *Démonstration probabiliste de certains développements asymptotiques quasi-classiques*, Bull. Sci. Math. 109, 179–208 (1982)

[58] S. Kusuoka: *The non linear transformation of Gaussian measure on Banach space and absolute continuity*, J. Fac. Sci. Univ. Tokyo Sect. 1A29, 567–596 (1982)

[59] A. Arai: *On dimensional reduction of functional measures*, J. Funct. Anal. 77, 372–402 (1988)

[60] K.D. Elworthy: *Path integrals on manifolds*, pp. 47–90 in "Mathematical Aspects of Superspace", Edt. H.J. Seifert, C.J.S. Clarke, A. Rosenblum, D. Reidel, Dordrecht (1984)

[61] D. Pantic: *Stochastic calculus on distorted Brownian motion*, J. Math. Phys. 29, 207–210 (1988)

[62] Yu.I. Daletskii, *Infinite dimensional elliptic operatores and parabolic equations connected with them*, Russ. Math. Surv. 22, 4–54, 1(1967)

[63] H.I. Cycon, R.G. Froese, W. Kirsch, B. Simon: *Schrödinger operators*, Springer, Berlin (1987)

[64] S. Albeverio, R. Høegh–Krohn, S. Paycha, S. Scarlatti: *A global and stochastic analysis approach to bosonic strings and associated quantum fields*, Preprint SFB N° 47, Bochum

[65] L. Brink, P. Di Vecchia, P. Howe: *A locally supersymmetric and reparametrization invariant action for the spinning string*, Phys. Letters B 65, 471 (1971)

[66] P. Gilkey: *The index theorem and the heat equation*, Publish or Perish (1974)
P. Gilkey: *Invariance theory. The heat equation and the Atyah–Singer Index theorem*, Publish or Perish (1984)

[67] H.H. Kuo: *Gaussian measures in Banach spaces*, Springer (1975)

[68] M. Reed, B. Simon: *Methods of Modern Mathematical Physics*, Vol. II, Academic Press (1975)

[69] L. Schwartz: *Radon measures*, Oxford University Press (1973)

[70] B. Simon: *Trace ideals and their applications*, Cambridge University Press (1979)

[71] Y. Yamasaki: *Measures on infinite dimensional spaces*, Series in pure mathematics, Vol. 5, World Scientific (1985)

[72] S. Paycha: *Une interprétation probabiliste du procédé de régularisation par la méthode de la chaleur*, Bochum Preprint (1989)

[73] S. Albeverio: *Some new developments concerning Dirichlet forms, Markov fields and quantum fields*, pp. 250–259 in IXth International Congress on Mathematical Physics, Edts. B. Simon, A. Truman, I.M. Davies, Adam Hidger, Bristol (1989)

[74] J.A. Van Casteren, M. Demuth: *On differences of heat semigroups*, Antwerpen, Berlin Preprint (1988)

[75] S. Albeverio, S. Kusuoka, M. Röckner: *On partial integration in infinite dimensional Spaces and applications to Dirichlet forms*, to appear in J. London Math. Soc.

[76] S. Albeverio, R. Høegh–Krohn, B. Zegarlinski: *Uniqueness and global Markov property for Euclidean fields; the case of general polynomial interactions*, Commun. Math. Phys. 123, 377–424 (1989)

[77] S. Albeverio, R. Høegh–Krohn: *Uniqueness and global Markov property for Euclidean fields. The case of trigonometric interactions*, Commun. Math. Phys. 68, 95–128 (1979)

[78] R. Gielerak: *Verification of the global Markov property in some class of strongly completed exponential interactions*, J. Math. Phys. 24, 3447–355 (1983)

[79] B. Zegarlinski: *Uniqueness and global Markov property for Euclidean fields: the case of general exponential interactions*, Commun. Math. Phys. 96, 195–221 (1984)

[80] M. Takeda: *On the uniqueness of the Markovian self–adjoint extension*, pp. 319–325 in "Stochastic Processes – Mathematics and Physics II", Edts. S. Albeverio, Ph. Blanchard, L. Streit, Lect. Notes Math. 1250, Springer, Berlin (1987)

[81] S. Albeverio: *Applications of nonstandard analysis in mathematical physics*, pp. 182–220 in "Nonstandard Analysis and its Applications", Ed. N. Cutland, Cambridge University Press, Cambridge (1988)

[82] N. Cutland, Ed.: *Nonstandard Analysis and its Applications*, Cambridge University Press, Cambridge (1988)

[83] 1) G. Jona-Lasino, P.K. Mitter: *On the stochastic quantization of field theory*, Commun. Math. Phys. 101, 409-436 (1985)

2) G. Jona-Lasinio, C. Parinello: *Large deriations theory and stochastic gauge fixing*, Roma Preprint (1989), to appear in Proc. 2nd Int. Ascona-Locarno- Como Conf. "Stochastic Processes-Physics and Geometry", Edts. S. Albeverio, G. Caseti, U. Cattaneo, D. Merlini, R. Moresi, World Scient. (1989)

3) G. Jona-Lasinio, P. K. Mitter: *Large deriation estimates in the stochastic quantization of* φ_2^4, *Roma Preprint (1989)*

[84] D. Elworthy: Stochastic Differential Equations On Manifolds, Cambridge University Press (1982)

[85] S. Albeverio, M. Röckner: Classical Dirichlet forms on topological vector spaces - the construction of the associated diffusion processes, BiBoS - SFB 237 Preprint (1988), to appear in Prob. theory and Rel. Fields

[86] M. Röckner: Traces of harmonic functions and a new path space for the free quantum field, J. Ffunct. Anal. 79, 211–249 (1988)

[87] T. Kolsrud: Gaussian random fields, infinite dimensional Ornstein - Uhlenbeck processes, and symmetric Markov processes, Acta Appl. Math. 12, 237-263 (1988)

[88] S. Albeverio, M. Röckner: Classical Dirichlet forms on topological vector spaces – closability and a Cameron - Martin formula, BiBoS - SFB 237 Preprint (1988), to appear in J. Funct. Anal.

[89] M. Fukushima: Dirichlet forms and Markov processes, Kodansha, Tohyo (1980)

[90] 1) *S. Albeverio, T. Hida, J. Potthoff, M. Röckner, L. Streit: Dirichlet forms in terms of white noise analysis I – Construction and QFT examples, BiBoS Preprint (1989)*

2) *S. Albeverio, T. Hida, J. Potthoff, M. Röckner, L. Streit: Dirichlet forms in terms of white noise analysis II – Closability and Diffusion Processes, BiBoS Preprint (1989)*

[91] *S. Albeverio, R. Høegh-Krohn, M. Röckner: in preparation*

[92] *S. Albeverio, R. Høegh-Krohn: Quasi invariant measures, symmetric diffusion processes and quantum fields, pp. 11–59 in "Proceedings of the International Colloquium of Mathematical Metods of Quantum Field Theory", Editions du CNRS. 1976, (Colloques Internationaux du Centre National de la Recherche Scientifique, No. 248).*

[93] *S. Albeverio, R. Høegh-Krohn: Diffusion, quantum fields and groups of mappings, pp. 133-145 in Proc. Katata Workshop and Kyoto Conf., Functional Analysis in Markov Processes, Ed. M. Fukushima, Lect. Notes in Maths. 923, Springer (1982)*

[94] *J. Dimock: Commun. Math. Phys. 57, 51–66 (1977)*

[95] *M.J. Donald: The classical limit of $P(\varphi)_2$ quantum field theory, Commun. Math. Phys. 71, 153–165 (1981)*

[96] *S. Albeverio, M. Röckner: Dirichlet forms, quantum fields and stochastic quantisation, pp. 1–21 in "Stochastic Analysis, path integration and dynamics", Emanations from "Summer Stochastics", Warwick 1987, Edts. D. Elworthy, J.C. Zambrini, Pitman Rs. Notes, Longman, Harlow (1989).*

[97] *S. Albeverio, B. Zegarlinski: Construction of convergent simplicial approximations of quantum fields on Riemannian manifolds, SFB 237 - Preprint (1989)*

[98] *T. Lindstrøm: Brownian motion on nested fractals, Oslo Preprint (1988)*

[99] *L. Gross, Ch. King, A. Sengupta: Two dimensional Yang-Mills theory via stochastic differential equations, Cornell University Preprint (1989)*

[100] *B. Driver: YM_2: Continuum expectations, lattice convergence and lassos, Commun. Math. Phys. 123, 575–616 (1989)*

[101] *S. Albeverio, R. Høegh-Krohn, H. Holden: Markov surfaces and gauge fields, pp. 211–231 in Proc. Schladming 84, 23. Intern. Universitätswochen für Kernphysik Acta Phys. Suppl. XXVI (1984): "Stochastic Methods and Computer Techniques in Quantum Dynamics", Edts. H. Mitter, L. Pittner.*

[102] *S. Albeverio, R. Høegh-Krohn, H. Holden: Markov processes on infinite dimensional spaces, Markov fiels and Markov cosurfaces, pp. 11–40 in "Stochastic Space-Time Models and Limit Theorems", Ed. L. Arnold, P. Kotelenez, D. Reidel, Dordrecht (1985).*

[103] *S. Albeverio, R. Høegh-Krohn: Brownian motion, Markov cosurfaces, Higgs fields, pp. 95–104 in "Fundamental Aspects of Quantum Theory", Proc. Como. Conf. 1985, Ed. V. Gorini, A. Frigerio, Plenum Publ., New York (1986)*

[104] *S. Albeverio, R. Høegh-Krohn, H. Holden: Random fields with values in Lie groups and Higgs fields, pp. 1–13 in "Stochastic Processes in Classical and Quantum Systems", Proceedings, Ascona 1985, Edts. S. Albeverio, G. Casati, D. Merlini, Lect. Notes Physics 262, Springer, Berlin (1986)*

[105] S. Albeverio, R. Høegh-Krohn, H. Holden: Stochastic Lie group-valued measures and their relation to stochastic curve integrals, guage fields and Markov cosurfaces, pp. 1–24 in "Stochastic Processes in Mathematics and Physics", Proc. BiBoS I, Edts. S. Aleverio, Ph. Blanchard, L. Streit, Lect. Notes Maths. 1158, Springer (1986)

[106] S. Albeverio, R. Høegh-Krohn, H. Holden: Representation and construction of multiplicative noise, Stockholm–BiBoS–Preprint (1987), to appear in J. Funct. Anal..

[107] 1) S. Albeverio, R. Høegh-Krohn, H. Holden: Stochastic multiplicative measures, generalized Markov semigroups and group valued stochastic processes and fields, J. Funct. Anal. 78, 154–184 (1988).

2) S. Albeverio, R. Høegh-Krohn, H. Holden, T. Kolsrud Representation and construction of multiplicative noise, Stockholm Preprint (1987), to appear in J. Funct. Anal. 1989

[108] H.G. Dosch, V.F. Müller: Lattice gauge theory in two space–time dimensions, Fortschr. d. Phys. 27, 547–559 (1979)

[109] L. Gross: Convergence of the $U(1)_3$ lattice gauge theory to its continuum limit, Commun. Math. Phys. 92, 137–162 (1983)

[110] B.K. Driver: Convergence of the $U(1)_4$ lattice gauge theory to its continuum limit, Commun. Math. Phys. 110, 479–501 (1987)

[111] T. Balaban: Convergent renormalization expansions for lattice gauge theories, Commun. Math. Phys. 119, 243–285 (1988)

[112] 1) S. Albeverio, R. Høegh-Krohn, H. Holden, T. Kolsrud: Construction of quantised Higgs-like fields in two dimensions, Phys. Letts. B 222, 263–268 (1989). 2) S. Albeverio, R. Høegh-Krohn, H. Holden, T. Kolsrud: A covariant Feynman-Kac formula for unitary bundles over Euclidean space, to appear in Proc. Tento Conf. "Stochastic partial differential equations", Edts. G. Da Prato, L. Tubaro, Lect. Notes Maths., Springer (1989)

[113]1) S. Albeverio, R. Høegh-Krohn: Euclidean Markov fields and relativistic quantum fields from stochastic partial differential equations in four dimenions, Phys. Letts. B 117, 175–179 (1986).

2) S. Albeverio, R. Høegh-Krohn: Quaternionic non abelian relativistic quantum fields in four space–time dimensions, Phys. Letts. B 189, 329–336 (1987).

[114] S. Albeverio, R. Høegh-Krohn, K. Iwata: Covariant Markovian random fields in four space–time dimensions with nonlinear electromagnetic interaction, pp. 69–83 in "Applications of Self-Adjoint Extensions in Quantum Physics", in Proc. Dubna Conf. 1987, Edts. P. Exner, P. Seba, Lec. Notes in Physics 324, Springer, Berlin (1989).

[115] S. Albeverio, K. Iwata, T. Kolsrud: Random fields as solutions of the inhomogeneous quaternionic Cauchy-Riemann equation I. Invariance and Analytic Continuation, SFB 237-Preprint (1989)

[116] S. Albeverio, R. Høegh-Krohn: Homogeneous random fields and statistical mechanics, Journal of Functional Analysis, 19, 242–272 (1975)

[117] B.Zegarlinski: On equivalence of spin and field pictures for lattice systems, Bochum Preprint (1989)

[118] S. Albeverio, A.B. Cruzeiro: Global flows with invariant (Gibbs) measures for Euler and Navier-Stokes two dimensional fluids, BiBoS-SFB 237-Preprint (1988) (to appear in Commun. Math. Phys.).

[119] J.M. Bismut: Index theorem and the heat equation, pp. 491–504 in Proceedings ICM 1986, Berkeley (1986)

[120] E. Getzler: The degree theory of the Nicolai maps, pp. 431–438 in Proc VIIIth Int. Congress on Mathematical Physics, Edts. M. Mebkhout, R. Sénéor, World Scient., Singapore (1987)

[121] S. Kusuoka: Degree theorem in certain Wiener Riemannian manifolds, pp. 93–108 in M. Métivier, S. Watanabe, Edts., Stochastic Analysis, Paris 87, Lect. Notes Maths. 1322, Springer, Berlin (1988)

[122] G.E. De Angelis, D. De Falco, F. Guerra: Scalar quantum electrodynamics on the lattice as classical statistical mechanics, Commun. Math. Phys. 57, 201–212 (1977)

[123] S. Albeverio, R. Høegh-Krohn: Dirichlet forms and diffusion processes on rigged Hilbert spaces, Zeitschrift für Wahrscheinlichkeitstheorie und verwandte Gebiete 40, 1–57 (1977)

[124] S. Albeverio, S. Kusuoka: Maximality of infinite dimensional Dirichlet forms and R. Høegh-Krohn's model of quantum fields, to appear in Memorial Volume for Raphael Høegh-Krohn.

[125] Z. Brzezniak: Some remarks on Feynman-Kac formulae, SFB 237-Preprint (1989)

[126] T. Zastawniak: The non existence of path-space measure for the Dirac equation in four space-time dimensions, to appear in J. Math. Phys. (1989)

[127] G.F. De Angelis, D. De Falco, G. Di Genova: The stochastic versus the Euclidean approach to quantum fields on a static space-time, pp. 397–402 in "Fundamental aspects of quantum theory", Edts. A. Frigerio, V. Gorini, Plenum Press (1986)

[128] D. De Falco, D. Khandekar: Applications of white noise calculus to the computation of Feynman path integrals, Stoch. Process Appl., 29, 257 (1988)

[129] S. Albeverio, R. Høegh-Krohn: The Wightman axioms and the mass gap for strong interactions of exponential type in two-dimensional space-time, Journal of Functional Analysis, 16, pp. 39–82, (1974)

[130] S. Kusuoka, to appear in Memorial Volume for R. Høegh-Krohn

[131] S. Albeverio, G. Gallavotti, R. Høegh-Krohn: Some results for the exponential interaction in 2 or more dimensions, Comm. Math. Phys. 70, 187-192 (1979)

[132] E.P. Osipov: On triviality of the $:\exp \lambda\varphi:_4$ quantum field theory in a finite volume, Repts. Math. Phys. 20, 111–116 (1984)

[133] J. Fröhlich, Y.M. Park: Remarks on exponential interactions and the quantum Sine - Gordon equation in two space-time dimensions, Helv. Phys. Acta 50, 315–329 (1977)

[134] E.P. Osipov: Octavic Markov cosurface and relativistic quantum fields in eight-dimensional space-time, Novosibirsk Preprint (1988)

[135] E.P. Osipov: Euclidean Markov fields from stochastic partial differential equations in eight-dimensional space, Novosibirsk Preprint (1987), 1988

[136] T.E. Duncan: The heat equation, the Kac formula and some index formula, pp.19–23 in "Partial Differential Equations and Geometry", Proc. Park City Conf., Ed. C.I. Byrnes, M. Dekker, New York (1977)

[137] R. Léandre: Applications quantitatives et géométriques du calcul de Malliavin, pp. 109-133 in same Proceedings as in Ref. [121]

[138] L. Streit: Quantum theory and stochastic processes - some contact points, pp. 197–213 in "Stochastic processes and their applications", Edts. K. Ito, T. Hida, Lect. Notes Maths. 1203, Springer, Berlin (1986)

[139] H. Föllmer: Von der Brownschen Bewegung zum Brownschen Blatt: Einige neue Richtungen in der Theorie der stochastischen Prozesse, pp. 159–190 in "Perspectives in Mathematics", Anniversary of Oberwolfach 1984, Birkhäuser, Basel (1985)

[140] 1) M.C. Gutzwiller: The path integral in chaotic Hamiltonian systems, pp. 119–136 in "Path Integrals from meV to MeV", Bielefeld Encounters 1985, Edts. M.C. Gutzwiller, A. Inomata, J.R. Klauder, L. Streit, World Scientific, Singapore (1986)

2) A. Voros: Schrödinger equation from $O(h)$ to $O(h^{\infty})$, pp. 173–195 in same Proceedings as 1)

[141] K.D. Elworthy: The method of images for the heat-kernel of S^3, to appear in Proc. 2. Ascona-Locarno-Como Conference, "Stochastic Processes– Physics and Geometry", Edts. S. Albeverio, G. Casati, V. Cattaneo, D. Merlini, R. Moresi, World Scient., Singapore (1989)

[142] B. Simon: The $P(\varphi)_2$ Euclidean (Quantum) Field Theory, Princeton University Press, Princeton (1974)

[143] J.A. van Casteren: Generators of strongly continuous semigroups, Res. Notes Maths., Pitman, Boston (1985)

[144] R. Carmona: Random Schrödinger Operators, pp. 1–125 in R. Carmona, H. Kesten, J.B. Walsh, Ecole d'Eté de Probabilité de Saint-Flour XIV – 1984, Lect. Notes Maths. 1180, Springer Berlin (1986)

[145] K.D. Elworthy, M.N. Ndumu, A. Truman: An elementary inequality for the heat kernel on a Riemannian manifold and the classical limit of the quantum partition function, pp. 84–99 in K.D. Elworthy, Ed., From local times to global geometry, control and physics, Pitman Res. Notes, Longman, Harlow (1986)

[146] S. Kusuoka: Dirichlet forms and diffusion processes in Banach space, J. Fac. Science Univ. Tokyo, Sec. 1A29, 79–95 (1982)

[147] V.S. Borkar, R.T. Chari, S.K. Mitter: Stochastic quantization of field theory in finite and infinite volume, J. Funct. Anal. 81, 184-206 (1988)

[148] C.R. Doering: Nonlinear parabolic stochastic differential equations with additive colored noise on $\mathbb{R}^d \times \mathbb{R}_+$: a regulated stochastic quantization, Commun. Math. Phys. 109, 537 (1987)

[149] S. Klimek, W. Kondracki: A construction of two-dimensional quantum chromodynamics, Commun. Math. Phys. 113, 389– 402 (1987)

[150] F. Guerra: Quantum field theory and probability theory. Outlook on new possible developments, pp. 214–243 in " Trends and Developments in the Eighties", Edts. S. Albeverio, Ph. Blanchard, World Scient., Singapore (1985)

[151] P.L. Chow: Stochastic differential equations in infinite dimensions, Wayne Preprint (1987)

[152] Y. Rozanov: Markov Random Fields, Springer, Berlin (1982)

[153] 1) Yu.M. Berezanskii: Self-adjoint operators in spaces of functions of infinitely many variables, AMS, Providence (1986)

2) Yu.M. Berezanskii, Yu.G. Kondratsev: Spectral methods in infinite dimensional analysis, 1988 (in russ.)

[154] H. Hogreve, W. Müller, J.Potthoff, R. Schrader: A Feynman-Kac formula for the quantum Heisenberg ferromagnet, Berlin Preprint (1989)

[155] J. Ginibre: Some applications of functional integration in statistical mechanics, pp. 327-427 in "Statistical Mechanics and Quantum Field Theory" Edts. C. De Witt, R. Stora, Gordon-Breach, New York (1971)

[156] S. Albeverio, R. Høegh-Krohn, D. Merlini: Euler flows, associated generalized random fields and Coulomb systems, pp. 197-215 in Edts. S. Albeverio "Infinite dimensional analysis and stochastic processs", Res. Notes Maths. Pitman, Boston (1985)

[157] E. A. Carlen: Stochastic mechanics of free scalar fields, pp. 40-60 in "Stochastic Mechanics and Stochastic Processes", Edts. A. Truman, J. M. Davies, Lect. Notes Mathematics 1325, Springer, Berlin (1988)

[158] S. Albeverio, J. C. Zambrini, K. Yasue: Euclidean quantum mechanics: analytic approach, to appear in Ann. Inst. H. Poincaré (Physique Théorique)

[159] E. Nelson, Lectures at Ecole d' Eté de Probabilités, St. Flour (1987)

[160] I. M. Lifshitz, S. A. Gredeskul, L. A. Pastur, Introduction to the theory of disordered systems, Nauka, Moscow (1982)

[161] V. P. Maslov, S. I. Chernyckh: Global exact asymptotics of large deviations probabilities for diffusion and jump processes, pp. 641-649 in same Proceedings as in Ref [1], 11).

[162] Z. Haba: Quantum fields on a Riemann surface, SFB 237-Preprint, Bochum (1989)

[163] G. F. Dell'Antonio: Methods of stochastic stability and properties of the Gribov horizon in the stochastic quantization of gauge theories, ISAS (Trieste)-Preprint (1989), to appear in "Stochastic Processes-Physics and Geometry", Proc. 2nd Int. Ascona-Locarno-Como Conf., Edts. S. Albeverio, G. Casati, U. Cattaneo, D. Merlini, R. Moresi, World Scient., Singapore (1989)

[164] Ph. Blanchard, Ph. Combe, W. Zheng: Physical and mathematical aspects of stochastic mechanics, Lect. Notes in Phys. 281, Springer, Berlin (1987)

CONSTRUCTIVE CONFORMAL FIELD THEORY

K. Gawędzki

C.N.R.S., I.H.E.S., Bures-sur-Yvette, France

1. Conformal fields in geometric background

The main domains of application of quantum field theory have been traditionally particle physics and statistical mechanics. In the latter, the special role is played by scale invariant theories describing the asymptotic behavior at large distances of critical models of statistical mechanics. In quantum field theory, scale invariance implies a richer conformal invariance [1,2]. In two (space-time) dimensions the conformal symmetry is infinite-dimensional and imposes strong restrictions on possible realizations of quantum fields allowing in many cases exact solution of the models [3]. In the present lectures, we shall discuss a class of such soluble two-dimensional models called the coset theories. They all derive from the theory, known under the name of Wess-Zumino-Witten (WZW) model [4-7], in which fields take values in a group manifold.

In order to expose the geometric content of the conformal symmetry, it will be covenient to consider (euclidean) quantum field theory in background geometry of a Riemann surface Σ equipped with a metric γ compatible with the complex structure. Fields Φ will be maps from Σ to a target space M equipped with a measure $d\mu$. Given a local action functional $S(\Phi)$, a finite set of points ξ_l in Σ and functions ϕ_l on M, we may write down a formal functional integral expression for the Green functions of the quantum field theory:

$$(1) \qquad \Gamma((\xi_l);(\phi_l)) = \left(\int \prod_l \phi_l(\Phi(\xi_l))\ e^{-S(\Phi)} \prod_\xi d\mu(\Phi(\xi)) \right)_{ren} .$$

Making (some, e.g. perturbative) sense out of the right hand side requires, as a rule, a renormalization - hence the notation.

The important property of the Green functions should be their

General covariance:

$$(2) \qquad \Gamma((D(\xi_l));(\phi_l))|_{D_*\gamma} = \Gamma((\xi_l);(\phi_l))|_\gamma$$

where D is a diffeomorphism of Σ.

Among all generally covariant two-dimensional quantum field theories, the conformally invariant ones (CFT's) are characterized by the (projective) invariance under local rescalings of the metric often called

Weyl invariance:

$$(3) \qquad \Gamma((\xi_l);(\phi_l))|_{e^\sigma\gamma} = exp\left[\frac{ic}{24\pi} \int_\Sigma (\frac{1}{2}(\partial\sigma)(\bar\partial\sigma) + \mathcal{R}\sigma) - \sum_l \Delta_l\ \sigma(\xi_l) \right] \Gamma((\xi_l);(\phi_l))|_\gamma .$$

$\mathcal{R}$ is the metric curvature form on Σ normalized so that $\int \mathcal{R} = 4\pi i(genus - 1)$, numbers c and Δ_l are, respectively, the (Virasoro) central charge and the conformal weights of fields $(\phi_l(\Phi))_{ren}$. (In fact the precise form of the exponential factor on the right hand side of eq. (3) is partly a matter of the renormalization scheme. We have also chosen to describe the Green functions as metric-dependent scalar functions rather than as $(\Delta_l, \overline{\Delta}_l)$-forms on Σ. This is possible only for spinless fields for which $\Delta_l = \overline{\Delta}_l$.)

Except for free fields and/or their modifications where the fields take values in tori or their orbifolds, there are, strictly speaking, no known cases in which the functional integrals (1) can be rigorously controled and the resulting Green functions can be shown to have the desired properties (2) and (3). One can however describe highly plausible scenarios of such situations. The simplest case is that of the scaling limits of $P(\Phi)_2$ theories for $P(.)$ a bounded below even polynomial of the real variable [8]. The $P(\Phi)_2$ theories have the Green functions given as

$$(4)\qquad \Gamma((\xi_l))_P = \int \prod_l \Phi(\xi_l) \ exp[-\int_\Sigma : P(\Phi) : \sqrt{\gamma} - (2\pi)^{-1} i \int_\Sigma (\partial\Phi)(\bar{\partial}\Phi)] \ \prod_l d\Phi(\xi) \, .$$

The functional integrals on the right hand side can be rigorously defined. If certain care is taken in defining the Wick-ordered polynomials $: P(\Phi) :$ of the field, the resulting Green functions will obey (2) but not (3). One expects however [8] that the scaling limits

$$(5)\qquad \lim_{\Theta\to\infty} \Theta^{\eta n/2} \, \Gamma((\xi_l))_{P_\Theta} \longrightarrow \Gamma_*((\xi_l))$$

exist for special choices of P_Θ and give one new CFT obeying (2) and (3) for each new even degree of P. In the spherical (=planar) or toroidal geometry, the scaling limits of the $P(\Phi)_2$ theories should coincide with the continuum limits of lattice exactly soluble models of statistical mechanics [9]. In the simplest case, the Φ^4_2 theory scaling limit should coincide with the continuum limit of the Ising model at its second order phase transition point.

It is convenient to describe the special features of CFT's using the formalism of Green functions with the energy-momentum-tensor insertions. Let z be a local complex coordinate on the surface Σ. Choose a metric $\gamma =' dz \otimes d\bar{z}$ on the domain of z. Then, due to (3),

$$(6)\qquad \frac{\delta}{\delta\gamma^{z\bar{z}}(z)} \, \Gamma((\xi_l);(\phi_l)) = 0$$

and, due to (2), functions

$$(7)\qquad \prod_j (2\pi \frac{\delta}{\delta\gamma^{zz}(z_j)}) \, \Gamma((\xi_l);(\phi_l)) \equiv \Gamma((z_j),(\xi_l);(T),(\phi_l))$$

are analytic in z_j's as long as those do not coincide among themselves nor with ξ_l's. Moreover the following short distance expansions [3] follow (we apologize for the unconventional and clumsy notation)

$$(8)\qquad \Gamma(z_1,z_2,(\xi_l);T,T,(\phi_l)) = \frac{c/2}{(z_1-z_2)^4}\,\Gamma((\xi_l);(\phi_l)) + \frac{2}{(z_1-z_2)^2}\,\Gamma(z_2,(\xi_l);T,(\phi_l)) + \frac{1}{z_1-z_2}\,\partial_{z_2}\Gamma(z_2,(\xi_l);T,(\phi_l)) + \text{terms regular when } z_1\to z_2\neq\xi_l$$

and

$$(9)\qquad \Gamma(z,(\xi_l);T,(\phi_l)) = \frac{\Delta_l}{(z-\xi_l)^2}\,\Gamma((\xi_l);(\phi_l)) + \frac{1}{z-\xi_l}\,\partial_{\xi_l}\Gamma((\xi_l);(\phi_l)) + \text{terms regular when } z\to\xi_l\,.$$

Eqs (6) - (9) constitute an equivalent way to describe the conformal invariance of a quantum field theory. In particular, eq. (8) is the euclidean transcription of the action of the Virasora algebra (with central charge c) in the space of states of the theory.

Our strategy for constructing and analysing CFT's will not consist of giving a rigorous meaning to the formal functional integration. Instead, we shall try to show that symmetry properties, formally following from the functional-integral expressions, in some cases determine the Green functions completely. To this end we shall start not from $P(\Phi)_2$ models but from a theory which has a rich symmetry built in already on the classical level. This is the case of the Wess-Zumino-Witten (WZW) model of quantum field theory with the Kac-Moody algebra symmetry extending the Virasoro algebra one. The WZW model appears to play the role of the generating theory for a whole class of new (coset) CFT's. Among those, one can also find the theories which seem to describe the scaling limits of the $P(\Phi)_2$ models.

2. Wess-Zumino-Witten model

The WZW theory [4-7] is a version of a two-dimensional sigma model in which fields $\Phi \equiv g$ take values in a compact Lie group G (or more generally in its complexification $G^{\mathbf{C}}$) . The (euclidean) action functional is given by

$$(10) \qquad S(g) = -(4\pi)^{-1} i \int_{\Sigma} tr(g^{-1}\partial g)(g^{-1}\bar{\partial} g) - (12\pi)^{-1} i \int_{B} tr(\tilde{g}^{-1} d\tilde{g})^3$$

where tr stands for the Killing form on the Lie algebra $\mathcal{G}$ of G and $\tilde{g}$ extends g to a three-dimensional manifold B with boundary Σ (for simplicity, we shall assume G to be simple, connected and simply connected). $S(g)$ is defined by (10) up to multiples of $2\pi i$ if the Killing form is normalized so that for the highest root ϕ of $\mathcal{G}$ $tr\phi^2 = 2$.

We shall also need the version of the WZW model with a coupling to external gauge field A with values in $i\mathcal{G}$. Let us write $A = A^{10} + A^{01}$, $A^{10} = -(A^{01})^*$, for the decomposition of A into the dz and $d\bar{z}$ parts. We shall introduce the coupled action

$$(11) \quad S(g, A)$$
$$= S(g) + (2\pi)^{-1} i \int_{\Sigma} tr(A^{10} g^{-1}\bar{\partial} g + (g\partial g^{-1})A^{01} + gA^{10}g^{-1}A^{01} - A^{10}A^{01}) .$$

Given a family of functions (ϕ_l) on G , one can define formally Green functions for the WZW theory coupled to external gauge field by adapting (1). We may limit ourselves to functions ϕ_l given by matrix elements $(w, D_{R_l}(.)v)$ of irreducible representations R_l of G acting in (finite dimensional) Hilbert spaces V_l . Let $\bar{V}_l$ denote the complex conjugate space to V_l carrying the representation $\bar{R}_l$. We may consider the Green functions with values in $(\bigotimes_l V_l) \otimes (\bigotimes_l \bar{V}_l)$ given formally by

$$(12) \qquad \Gamma((\xi_l); (R_l); A) = \left(\int \bigotimes_l D_{R_l}(g(\xi_l)) \, e^{-kS(g,A)} \prod_{\xi} dg(\xi) \right)_{ren}$$

where dg denotes the Haar measure on G and we have introduced a (positive) coupling constant k quantized (due to the ambiguity of S) to integral values.

An important symmetry property of the WZW model is a consequence of the transformation properties of the action (11) under chiral gauge transformations

$$A^{10} \longmapsto {}^{h}A^{10} \equiv (h^*)^{-1} A^{10} h^* + (h^*)^{-1} \partial h^* ,$$
$$A^{01} \longmapsto {}^{h}A^{01} \equiv h A^{01} h^{-1} + h \bar{\partial} h^{-1}$$

where $h:\ \Sigma \longrightarrow G^{\mathbf{C}}$. Namely

$$(13)\qquad S(hgh^*,\ {}^hA) \ = \ S(g,A) - S(h^{-1}(h^*)^{-1},A)\ .$$

Inserting (13) to (12) and using the (analitically continued) invariance of the formal product of Haar measures, we obtain the chiral Ward identity expressing the chiral gauge symmetry property of the quantum WZW model:

$$(14)\quad \Gamma((\xi_l);(R_l);\ {}^hA)$$
$$= \ e^{kS(h^{-1}(h^*)^{-1},A)}\ (\bigotimes_l D_{R_l}(h(\xi_l)))\otimes(\bigotimes_l D_{\bar{R}_l}(h(\xi_l)))\ \ \Gamma((\xi_l);R(_l);A)\ .$$

A priori the chiral gauge symmetry (14) could be destroyed by the renormalization. The point is however that the model needs only multiplicative wave function renormalization which leaves (14) intact. Following our strategy, let us then assume (14) and explore its consequences. For simplicity, we shall consider the case $\Sigma = \mathbf{C}P^1$ with the standard local coordinate z although much of the discussion may be generalized to the case of any Riemann surface, see [10,11].

Let us perform the first order calculation with $h = e^{\delta\Lambda}$. Introducing the Green functions with current insertions

$$(15)\qquad \prod_j(\pi\frac{\delta}{\delta A^{a_j}_{\bar{z}}(z_j)})\ \Gamma((\xi_l);(R_l);A)\ \equiv\ \Gamma((z_j),(\xi_l);(J^{a_j}),(R_l);A)\ ,$$

we get from (14)

$$(16)\qquad \int\Gamma(z,(\xi_l);J^a,(R_l);A)\ (\partial_{\bar{z}}\delta\Lambda^a(z) - if^{abc}\delta\Lambda^b(z)A^c_{\bar{z}}(z))\ d^2z$$
$$= -\left(k/2\int\partial_z(\delta\Lambda^a)A^a_{\bar{z}}\ d^2z\ +\ \pi\sum_l\delta\Lambda^a(\xi_l)t^a_l\right)\Gamma((\xi_l);(R_l);A)\ +\ \dots$$

where t^a are generators of the Lie algebra $\mathcal{G}$, $tr\ t^at^b = \delta^{ab}/2$, $t^a_l \equiv dR_l(t^a)$ acting on the V_l-component of the Green functions and " . . . " stands for terms proportional to $\delta\Lambda^*$ or to A^{10} .

To strip eq. (16) of arbitrary $\delta\Lambda$, we may put

$$\delta\Lambda^a(z)\ =\ 1/\pi\ \delta^{aa_0}\ (z-w)^{-1}$$

so that

$$\partial_{\bar{z}} \delta \Lambda^a(z) \;=\; \delta^{aa_0}\, \delta^2(z-w) \,.$$

This gives

$$\text{(17)} \qquad \Gamma(w,(\xi_l);J^{a_0},(R_l);A) \;+\; i/\pi\, f^{aa_0c} \int \Gamma(z,(\xi_l);J^a,(R_l);A) \cdot (w-z)^{-1} A^c_{\bar{z}}(z)\, d^2z$$
$$= \Big((2\pi)^{-1} k \int (z-w)^{-2} A^{a_0}_{\bar{z}}(z)\, d^2z \;+\; \sum_l (w-\xi_l)^{-1}\, t_l^{a_0} \Big) \cdot \Gamma((\xi_l);(R_l);A) \;+\; \ldots$$

For $A=0$, this reduces to

$$\text{(18)} \qquad \Gamma(w,(\xi_l);J^a,(R_l);0) \;=\; \sum_l (w-\xi_l)^{-1}\, t_l^a\, \Gamma((\xi_l);(R_l);0) \,.$$

Differentiating eq. (17) over $A^b_{\bar{z}}(z)$ at $A=0$ and using (18), we obtain

$$\text{(19)} \qquad \Gamma(w,z,(\xi_l);J^a,J^b,(R_l);0) \;+\; \Big(i f^{abc} \sum_l (w-z)^{-1}(z-\xi_l)^{-1}\, t_l^c$$
$$-\, k/2\, \delta^{ab}(z-w)^{-2} \;-\; \sum_{l,l'} ((w-\xi_l)(z-\xi_{l'}))^{-1}\, t_l^a\, t_{l'}^b \Big)\, \Gamma((\xi_l);(R_l);0) \;=\; 0 \,.$$

Up to now, we have used only the Ward identity of the chiral gauge symmetry. The other crucial input in the quantization of the WZW model is the relation of the gauge symmetry to the conformal one given by the Sugawara construction [12,13] of the energy-momentum tensor:

$$\text{(20)} \qquad T(z) \;=\; lim_{w\to z}\, (k+g^\vee)^{-1} \Big(J^a(w)J^a(z) \;-\; k/2\; dim(G)(w-z)^{-2} \Big)$$

understood as the equality of the Green-functions with T and J insertions at zero external gauge field. Eq. (20) is a quantum version [13] of the classical identity [12] $T_{cl} = k^{-1}(J^a_{cl})^2$. The shift $k \mapsto k+g^\vee$, where $g^\vee$ is the dual Coxeter number of the group (equal N for $SU(N)$), as well as the need to subtract the singularity in (20), is due to the quantum effects which introduce rather minor corrections to the semiclassical result for the highly symmetric WZW theory. It may be checked with the help of the chiral-gauge-symmetry Ward identities that definition (20) leads to the Green functions with the energy-momentum-tensor insertions which obey (8) and (9) with

$$(21) \qquad c = k\, dim(G)/(k+g^\vee) \qquad \text{and} \qquad \Delta_l = c_2(R_l)/(k+g^\vee)$$

where $c_2(R_l)$ stands for the quadratic Casimir of the representation R_l [5] .

From (19) and (20), we obtain

$$(22) \qquad \Gamma(z,\xi_l);T,(R_l);0) \\ = (k+g^\vee)^{-1} \sum_{l,l'}((z-\xi_l)(z-\xi_{l'}))^{-1}\, t_l^a\, t_{l'}^a\, \Gamma((\xi_l);(R_l);0)\ .$$

According to (9), the left hand side has, as a function of z , poles at points ξ_l . Comparison of the residues on the both sides gives:

$$(23)\ \partial_{\xi_l}\Gamma((\xi_l);(R_l);0) = 2(k+g^\vee)^{-1} \sum_{l\neq l'}(\xi_l-\xi_{l'})^{-1}\, t_l^a\, t_{l'}^a\, \Gamma((\xi_l);(R_l);0)\ .$$

This is the celebrated Knizhnik-Zamolodchikov (KZ) equation [5].

The KZ equation has been the basic tool in attempts to find explicitly the Green functions of the WZW model [5,14]. As noticed in [15], it has an interpretation of the covariant horizontality condition with respect to the non-abelian connection

$$(24) \qquad \Omega((\xi_l)) = -2(k+g^\vee)^{-1}\sum_{l,l'}(\xi_l-\xi_{l'})^{-1}\, t_l^a\, t_{l'}^a\, d\xi_l$$

on the space of sequences (ξ_l) of non-coinciding points in the complex plane. This connection is flat and its holonomy gives rise to a representation of Artin's braid group [16], which is the fundamental group of the space of (ξ_l)'s. As shown in [15], this representation can be also described in terms of the R-matrices of the quantum deformation G_q of group G [17-19] for the (exceptional) values of the deformation parameter $q = e^{2\pi i/(k+g^\vee)}$. R-matrices relate the tensor products of two quntum group representations in different order and are q-deformations of the permutation of two representation spaces. The relation of quantum groups to

CFT in general and specifically to the WZW model has been recently the topic of a very active research [20-25] which we shall not discuss here. We shall take however a fresh look at the connection (24) below, analysing its metric properties related to the unitarizability of representations of the algebra generated by the R matrices at special values of q .

3. Holomorphic Ward identities

One of the characteristic features of the two-dimensional CFT's is the separation of left- and right-moving modes. In the euclidean version of the theory, this translates into a factorization of the Green functions into sesquilinear combinations of expressions holomorphic in the insertion points and the moduli parameters of the surface. If the combinations are finite, the CFT is called rational. In more geometric language, the Green functions should be expressible as scalar products of holomorphic sections of vector bundles over the space of punctured Riemann surfaces, the dimension of the bundle being finite for rational CFT's [26]. We shall explain now how to obtain such a picture in the WZW theory.

The first step is to write down the holomorphic version of the Ward identity expressed by (14):

$$\gamma({}^{h}A^{01}) = e^{kS(h^{-1},A^{01})} \bigotimes_{l} D_{R_l}(h(\xi_l)) \ \gamma(A^{01}) \ . \tag{25}$$

It is an equation for maps taking values in $\bigotimes_l V_l$, depending analytically on $\mathcal{G}^{\mathbf{C}}$-valued (0,1)-forms A^{01} . $S(h,A^{01})$ in (25) is as in (11) but with A^{10} set to zero. We shall call solutions of (25) current blocks and shall denote their space by $W(\Sigma;(\xi_l);(R_l))$. This is a complex vector space of finite dimension (below, we shall see this explicitly for $\Sigma = \mathbf{C}P^1$) .

Following Witten [27], see also [28], one can identify the space of current blocks with the space of Schrödinger-picture states of the three dimensional topological gauge theory with the Chern-Simons action quantized in the geometry $\Sigma \times \mathbf{R}^1$ in the presence of Wilson lines $(\{\xi_l\} \times \mathbf{R}^1)$ in representations (R_l) . Indeed. The holomorphic Ward identity (25), when written in the infinitesimal form (16), becomes

$$F^a(A)(z)\,\gamma(A^{01}) \;=\; 2\pi/k\,\sum_l \delta(z-\xi_l)\,t_l^a\,\gamma(A^{01}) \tag{26}$$

where

$$F(A) \;=\; \partial_z A_{\bar z} - \partial_{\bar z} A_z + [A_z, A_{\bar z}]$$

with $A_{\bar z}$ acting on γ by multiplication and $A_z^a = -(2\pi/k)\delta/\delta A_{\bar z}^a$. Eq. (26) is the constraint relation imposed on analytic functionals $\gamma(.)$ in the Bargmann quantization of the Chern-Simons theory, see [28].

Relations (25) and (26) are equivalent, but the global form (25) is easier to analyse. It fixes the behavior of γ along the orbits of the group $\mathbf{G}^{\mathbf{C}}$ of chiral gauge transformations. The simplest situation occurs for $\Sigma = \mathbf{C}P^1$ where the orbit of $A^{01} = 0$ is open and dense in the space $\mathbf{A}^{01}$ of $\mathcal{G}^{\mathbf{C}}$-valued (0,1)-forms (with C^∞ topology). Thus on $\mathbf{C}P^1$, $\gamma(0)$ fixes the current block γ completely. As follows from (25) with $A^{01} = 0$ and $h = const.$, $\gamma(0) \in Inv_G(\otimes V_l)$ i.e. $\gamma(0)$ is a G-invariant tensor in $\otimes V_l$. We have then an embedding

$$W(\mathbf{C}P^1;(\xi_l);(R_l)) \;\hookrightarrow\; Inv_G(\otimes V_l) \tag{27}$$

which in general is not onto. The reason is that the initial value $\gamma(0)$ defines via (25) $\gamma(A^{01})$ only on the open dense subset of $A^{01} = h\partial h^{-1}$ for $h: \mathbf{C}P^1 \rightarrow G^{\mathbf{C}}$ and $\gamma(A^{01})$ might show a singular behavior when A^{01} approaches the boundary of this subset. In fact the boundary is stratified into complex manifolds of finite codimension. By the Hartogs theorem, the singularities may arise only when we approach a codimension one stratum. There is one such dangerous stratum composed of the orbit of the gauge field

$$A_0^{01} \;=\; \begin{cases} g_0\bar\partial g_0^{-1} & \text{for} \quad |z| \le 1\,, \\ 0 & \text{for} \quad |z^{-1}| \le 1 \end{cases} \tag{28}$$

where

$$g_0 \;=\; R_\phi \begin{pmatrix} (1-f(|z|^2)^2)^{1/2} & -iz^{-1}f(|z|^2) \\ -izf(|z|^2) & (1-f(|z|^2)^2)^{-1/2} \end{pmatrix},$$

$f(\sigma)$ interpolates smoothly between 0 for σ close to 0 and 1 for σ close to 1 and R_ϕ is the embedding of $SL(2,\mathbf{C})$ into G generated by the highest root ϕ of $\mathcal{G}$ and its step generators $e_{\pm\phi}$. Consider a one- (complex) parameter family of chiral gauge transformations h_t , $t \neq 0$,

$$(29) \qquad h_t = \begin{cases} R_\phi\begin{pmatrix} 1 & 0 \\ itz & 1 \end{pmatrix} g_0 \, R_\phi\begin{pmatrix} t & 0 \\ iz & t^{-1} \end{pmatrix} & \text{for } |z| \leq 1 , \\ R_\phi\begin{pmatrix} 1 & -it^{-1}z^{-1} \\ 0 & 1 \end{pmatrix} & \text{for } |z^{-1}| \leq 1 . \end{cases}$$

The gauge field

$$(30) \qquad A_t^{01} = h_t \bar{\partial} h_t^{-1}$$

is clearly analytic in t also at $t = 0$ and at this value of the parameters coincides with (28).

By virtue of (25), for $t \neq 0$,

$$(31) \qquad \gamma(A_t^{01}) = e^{kS(h_t^{-1})} \bigotimes_l D_{R_l}(h_t(\xi_l)) \;\; \gamma(0) .$$

In order to describe the image of embedding (27) (that is the so called <u>fusion rules</u>), we have then to check for what $\gamma(0)$ the right hand side of (31) has a limit when $t \to 0$. Let us consider the three-point function case. By the Möbius invariance, for three points ξ_l the result will not depend on the position of the points and we may take $\xi_0 = 0$, $\xi_1 = 1$, $\xi_2 = \infty$. Decomposing the representations R_l with respect to $R_\phi(SL(2,\mathbf{C}))$:

$$V_{R_{l_0}} \otimes V_{R_{l_1}} \otimes V_{R_{l_2}} = \sum_{j_0,j_1,j_2} W_{j_0,j_1,j_2} \otimes V_{j_0}^{SL(2)} \otimes V_{j_1}^{SL(2)} \otimes V_{j_2}^{SL(2)} ,$$

where W_{j_0,j_1,j_2} are the multiplicity spaces, we may write for $\gamma(0) \in Inv_G(\otimes V_l)$

$$\gamma(0) = \sum_{j_0,j_1,j_2} w_{j_0,j_1,j_2} \otimes v_{j_0,j_1,j_2}$$

with v_{j_0,j_1,j_2} in the zero- or one-dimensional space $Inv_{SU(2)}(\otimes V_{j_l}^{SL(2)})$. Now

$$\bigotimes_l D_{R_l}(h_t(\xi_l)) \;\; \gamma(0)$$
$$= \sum_{j_0,j_1,j_2} w_{j_0,j_1,j_2} \otimes \left[D_{j_0}\begin{pmatrix} t & 0 \\ 0 & t^{-1} \end{pmatrix} \otimes D_{j_1}\begin{pmatrix} 1 & -it^{-1} \\ 0 & 1 \end{pmatrix} \otimes D_{j_2}(1) \right] v_{j_0,j_1,j_2} .$$

We may put

$$v_{j_0,j_1,j_2} = \sum_{m_0+m_1+m_2=0} (-1)^{j_2-m_2} < j_0, m_0; j_1, m_1|j_2, -m_2 > |m_0, m_1, m_2 >$$

in the magnetic number basis where $< j_0, m_0; j_1, m_1|j_2, m_2 >$ are the $SU(2)$ Clebsch-Gordan coefficients. It is straightforward to see that the most singular component of

$$\left[D_{j_0}\begin{pmatrix} t & 0 \\ 0 & t^{-1} \end{pmatrix} \otimes D_{j_1}\begin{pmatrix} 1 & -it^{-1} \\ 0 & 1 \end{pmatrix} \otimes D_{j_2}(1)\right] v_{j_0,j_1,j_2}$$

is proportional to

$$t^{j_0+j_1+j_2} < j_0, -j_0; j_1, j_0 - j_2|j_2, -j_2 > |-j_0, j_1, j_2 > .$$

Since a direct computation shows that $e^{kS(h^{-1})} = \mathcal{O}(t^k)$, we infer that $\gamma(0)$ is in the image of (27) if and only if

(32) $$\gamma(0) = \sum_{j_0+j_1+j_2 \leq k} w_{j_0,j_1,j_2} \otimes v_{j_0,j_1,j_2} .$$

This gives the three-point fusion rules [6,29,30]. Notice that, since also $j_0 \leq j_1 + j_2$ etc. for non-vanishing terms, $j_l \leq k/2$. Let us still mention another consequence of (32). Call a representation V_l integrable if its highest weight λ_l satisfies $tr\lambda_l\phi \leq k$. Let now $v_{l_0}^+$ be the highest weight vector of a non-integrable representation V_{l_0}. It follows from (32) (with the restriction $j_l \leq k/2$) that $v_{l_0}^+ \otimes (\bigotimes_{l\neq l_0} V_l)$ is orthogonal to $\gamma(0)$. Acting on $v_{l_0}^+$ with lowering generators of the Lie algebra and using the G-invariance of $\gamma(0)$, we see by induction that $\gamma(0)$ is orthogonal to all vectors in $\otimes V_l$ and thus zero. We infer that there may exist non-trivial three-point current blocks only if all representations R_l are integrable. This is still true for an arbitrary number of insertion points [6].

4. From holomorphic to real theory

As $W(\Sigma;(\xi_l);(R_l))$ may be viewed as the space of quantum states in the Bargmann quantization of the three-dimensional Chern-Simons gauge theory, it is natural to equip it with the Bargmann scalar product $\ll .,. \gg$ given formally by

$$(33)\qquad \|\gamma\|^2 = \left(\int |\gamma(A^{01})|^2 \, e^{(2\pi)^{-1}ik\int trA^{10}A^{01}} \, DA\right)_{ren}$$

where $|\,.\,|$ comes from the scalar product in the representation space $\otimes V_l$. We have anticipated the necessity to renormalize the functional integral on the right hand side in order to make it well defined. For the two-dimensional WZW theory, scalar product (33) provides the information which allows to express the Green functions (12) by the current blocks. Namely, if (γ_n) is a basis of $W(\Sigma;(\xi_l);(R_l))$ then

$$(34)\quad \Gamma((\xi_l);(R_l);A) = e^{(2\pi)^{-1}ik\int trA^{10}A^{01}} \sum_{m,n} (\mathcal{H}^{-1})^{mn} \, \gamma_m(A^{01}) \otimes \overline{\gamma_n(A^{01})}$$

where

$$(35)\qquad \mathcal{H}_{mn} = \ll \gamma_m, \gamma_n \gg \; .$$

Notice that the right hand side is independent of the choice of basis (γ_n) . Expression (34) leads to the correct dependence of the Green functions on the insertion points and on the metric on the surface (the right conformal Ward identities). Let us explain briefly this important point which has a nice geometric interpretation.

The spaces $W(\Sigma;(\xi_l);(R_l))$ form a (finite-dimensional) vector bundle $\mathcal{W}$ over the space $\mathcal{M} = \{(J;(\xi_l))\}$ where J runs through the complex structures on surface Σ . $\mathcal{M}$ is a complex manifold with the anti-holomorphic tangent vectors given by variations

$$(36)\qquad \delta J = \overline{\delta\mu}\,, \qquad \delta\xi_l = \delta\bar{z}_l$$

where z is a complex coordinate corresponding to J and $\delta\mu = \delta\mu^z_{\bar{z}} \frac{\partial}{\partial z} \otimes d\bar{z}$. Bundle $\mathcal{W}$ has a structure of a holomorphic vector bundle with (local) holomorphic sections given by maps

$$(J;(\xi_l);A) \longmapsto \gamma(J;(\xi_l);A)$$

analytic in $A^{01} = (A - iAJ)/2$, satisfying Ward identity (25) for fixed $(J;(\xi_l))$ and such that

$$(37) \qquad \delta\gamma = -\left((8\pi)^{-1}k\int_\Sigma trA^{01}(A^{01}\overline{\delta\mu}) + \sum_l D_{R_l}(A^{01}(\xi_l)\delta\bar{z}_l)\right)\gamma$$

under variations (36). Scalar product (33) defines (formally) a hermitian structure on $\mathcal{W}$ and this, together with the structure of a holomorphic bundle, induces in turn a connection $(\nabla, \bar{\nabla} = \bar{\partial})$ such that

$$(38) \qquad \partial \ll \gamma', \gamma \gg = \ll \gamma', \nabla\gamma \gg + \ll \bar{\partial}\gamma', \gamma \gg .$$

Before discussing how to give precise sense to (33), we shall motivate relation (34) by formal considerations ignoring renormalization and all that. Under the holomorphic variation

$$(39) \qquad \delta J = \delta\mu , \qquad \delta\xi_l = 0 ,$$

the change of the Green function given by (34) is

$$(40) \qquad \delta\Gamma((\xi_l);(R_l);A) = e^{(2\pi)^{-1}ik\int trA^{10}A^{01}} \sum_{m,n} (\mathcal{H}^{-1})^{mn} \Big((4\pi)^{-1}k\int trA^{10}$$
$$\cdot (A^{10}\delta\mu)\ \gamma_m(A^{01})\otimes\overline{\gamma_n(A^{01})} + \delta\gamma_m(A^{01})\otimes\overline{\gamma_n(A^{01})} + \gamma_m(A^{01})\otimes\delta\overline{\gamma_n(A^{01})}\Big)$$

provided that

$$(41) \qquad \nabla_{\delta\mu}\gamma_n = \bar{\nabla}_{\overline{\delta\mu}}\gamma_n = 0$$

so that the variation of $\mathcal{H}$ vanishes. Now, by (38),

$$(42) \qquad \ll \gamma', \nabla_{\delta\mu}\gamma \gg = \delta\int <\gamma'(A),\gamma(A)>\ e^{(2\pi)^{-1}ik\int trA^{10}A^{01}}\ DA$$
$$- \int <\bar{\nabla}_{\overline{\delta\mu}}\gamma'(A),\gamma(A)>\ e^{(2\pi)^{-1}ik\int trA^{10}A^{01}}\ DA$$

and by (37)

$$(43) \qquad \bar{\nabla}_{\overline{\delta\mu}}\gamma \equiv \bar{\partial}_{\overline{\delta\mu}}\gamma = \bar{\delta}\gamma + (8\pi)^{-1}k\int trA^{01}(A^{01}\overline{\delta\mu})\,\gamma .$$

Differentiating in the first term on the right hand side of (42) under the integral sign and using (43), we obtain

$$(44) \ll \gamma', \nabla_{\delta\mu}\gamma \gg = \int < \gamma'(A)\,,\ \delta\gamma(A)\ +\ (8\pi)^{-1}k\int trA^{10}(A^{10}\delta\mu)\ \gamma(A) > \cdot\, e^{(2\pi)^{-1}ik\int trA^{10}A^{01}}\ DA$$

$$= \int < \gamma'(A)\,,\ \delta\gamma(A)\ +\ \pi i(k)^{-1}\int tr(\delta^2/\delta A_{\bar z}^2)\delta\mu_{\bar z}^{z}d^2z\ \gamma(A) + i\int trA_z(\delta/\delta A_{\bar z})\delta\mu_{\bar z}^{z}d^2z\ \gamma(A) > \ e^{(2\pi)^{-1}ik\int trA^{10}A^{01}}\ DA$$

where $\delta/\delta A_{\bar z} \equiv t^a\delta/\delta A_{\bar z}^a$. To get the last equality we have integrated by parts and used $(\delta/\delta A_z)\gamma' = 0$. As a result,

$$(45) \qquad \nabla_{\delta\mu}\gamma\ =\ \Big(\delta\gamma\ +\ \pi i k^{-1}\int tr(\delta^2/\delta A_{\bar z}^2)\delta\mu_{\bar z}^{z}d^2z\ \gamma + i\int trA_z(\delta/\delta A_{\bar z})\delta\mu_{\bar z}^{z}d^2z\ \gamma\ \Big)_{ren}$$

since the expression inside $(\ .\ )_{ren}$ may be checked to be a current block provided that we ignore a short distance singularity in $\delta^2/(\delta A_{\bar z}(z)\delta A_{\bar z}(z'))$ which makes it ill-defined. Using (41) together with (43) and (45), we obtain from (40) after a short algebra:

$$(46) \quad \delta\Gamma((\xi_l);(R_l);A)\ =\ \Big(-\pi i k^{-1}\int tr(\delta^2/\delta A_{\bar z}^2)\delta\mu_{\bar z}^{z}d^2z\ \ \Gamma((\xi_l);(R_l);A)\Big)_{ren}$$

which is nothing else but the classical form [12] of the Sugawara construction for the energy-momentum tensor of the WZW theory.

The renormalized version of (45) is more subtle:

$$(47) \quad \nabla_{\delta\mu}\gamma\ =\ \delta\gamma\ +\ \pi i(k+g^\vee)^{-1}\int\lim_{\epsilon\to 0}\Big(tr\ ad_{exp[\int_{z-\epsilon}^{z+\epsilon}A^{01}]}\big(\frac{\delta}{\delta A_{\bar z}(z+\epsilon)}\big) \cdot\frac{\delta}{\delta A_{\bar z}(z-\epsilon)}\ -\ \frac{k\ dim(G)}{16\pi^2\epsilon^2}\Big)\delta\mu_{\bar z}^{z}(z)d^2z\ \gamma + i\int trA_z\frac{\delta}{\delta A_{\bar z}(z)}\delta\mu_{\bar z}^{z}(z)d^2z\ \gamma\ -\ \frac{ic}{24\pi}\int t_{zz}(z)\delta\mu_{\bar z}^{z}d^2z\ \gamma$$

where c is given by (21) and t_{zz} depends on the metric $\gamma_{z\bar z}dz\otimes d\bar z$ on the surface:

$$(48) \qquad t_{zz} = \partial_z^2 ln\gamma_{z\bar{z}} - (1/2)(\partial_z ln\gamma_{z\bar{z}})^2 \ .$$

When used in (40), eq. (47) results in the renormalized version of (46):

$$(49) \quad \delta\Gamma((\xi_l);(R_l);A) = -\pi i(k+g^\vee)^{-1} \int \lim_{\epsilon\to 0} \Big(tr\ ad_{exp[\int_{z-\epsilon}^{z+\epsilon} A]} \big(\frac{\delta}{\delta A_{\bar{z}}(z+\epsilon)}\big) \cdot \frac{\delta}{\delta A_{\bar{z}}(z-\epsilon)} - \frac{k\ dim(G)}{16\pi^2\epsilon^2} - \frac{k\ dim(G)}{24\pi^2} t_{zz}\Big) \delta\mu_{\bar{z}}^{z} d^2z \ \Gamma((\xi_l);(R_l);A)$$

which is the correct version of the Sugawara construction, compare (20). As we see, the latter is nothing else but a formula for a connection (which in fact appears to be projectively flat) on the bundle $\mathcal{W}$. We could use this (well defined) connection in order to make sense out of the right hand side of (33). Instead, we shall attempt a direct computation of the functional integral occurring there in the spherical and toroidal geometry.

5. Coulomb gas representation

Since the Ward identity (25) determines the behavior of current blocks along the orbits of the group $\mathbf{G^C}$ of chiral gauge transformations, we may try to use this information inside the DA integral in (33) by decomposing it into the integral over $\mathbf{G^C}$ and over the orbit space $\mathbf{A}^{01}/\mathbf{G^C}$, i.e. over the moduli space [31] of holomorphic $G^{\mathbf{C}}$-bundles. This is achieved by parametrizing

$$(50) \qquad A^{01} = {}^{h^{-1}}A_u^{01}$$

where $h \in \mathbf{G^C}$ and $u \mapsto A_u^{01}$ picks up (locally) one gauge field from each orbit. If the genus of the surface is >1 then (50) is a good parametrization on an open dense set in $\mathbf{A}^{01}$ and we may take A_u flat [32,33]. Let

$$(\omega_j) \ , \ j = 1,...,J \equiv dim(G)(genus-1) \ ,$$

be a basis of $\mathcal{G}^{\mathbf{C}}$-valued (1,0)-forms on Σ satisfying

$$\bar{\partial}_{A_u^{01}} \omega_j \equiv \bar{\partial} \omega_j + ad_{A_u^{01}}(\omega_j) = 0 .$$

ω_j may be treated as cotangent vectors to $\mathbf{A}^{01}/\mathbf{G}^{\mathbf{C}}$ as, under the natural pairing, they anihilate vectors tangent to $\mathbf{G}^{\mathbf{C}}$-orbits at gauge fields A^{01} . The volume on $\mathbf{A}^{01}$ decomposes as

$$(51) \qquad DA = e^{2g^\vee S(hh^*, A_u)} \, det'\Big((\bar{\partial}_{A_u^{01}})^* \bar{\partial}_{A_u^{01}} \Big) \cdot det_{j,j'} \Big((\omega_j, \omega_{j'}) \Big)^{-1} \; Dh \;\; \omega_1 ... \omega_J \bar{\omega}_1 ... \bar{\omega}_J$$

where det' denotes the determinant on the subspace orthogonal to the zero modes (we shall define it via the ζ-function regularization). Relation (51) may be shown using the chiral anomaly describing the behavior of $det'((\bar{\partial}_{A^{01}})^* \bar{\partial}_{A^{01}})$ under the chiral gauge transformations [34,35].

On the other hand, by virtue of (25),

$$(52) \quad |\gamma(A^{01})|^2 \; e^{(2\pi)^{-1} ik \int tr A^{10} A^{01}} = < \gamma(A_u^{01}) \, , \, \bigotimes_l D_{R_l}(hh^*(\xi_l))^{-1} \; \gamma(A_u^{01}) > \; e^{(2\pi)^{-1} ik \int tr A_u^{10} A_u^{01}} \; e^{kS(hh^*, A_u)} \; .$$

With the use of (51) and (52), eq. (33) becomes

$$(53) \quad \|\gamma\|^2 = \Big(\int < \gamma(A_u^{01}) \, , \, \bigotimes_l D_{R_l}(hh^*(\xi_l))^{-1} \; \gamma(A_u^{01}) > \; e^{(2\pi)^{-1} ik \int tr A_u^{01} A_u^{01}} \cdot e^{(k+2g^\vee) S(hh^*, A_u^{01})} \; det'((\bar{\partial}_{A_u^{01}})^* \bar{\partial}_{A_u^{01}}) \; det_{j,j'}((\omega_j, \omega_{j'}))^{-1} \; D(hG) \; \omega_1 ... \bar{\omega}_J \Big)_{ren} \; .$$

As we see, the scalar product (33) of current blocks is expressible in terms of correlation functions of a $G^{\mathbf{C}}/G$-valued (non-unitary) sigma model (the G-valued contributions to h , corresponding to standard gauge group degrees of freedom, factor out in (53)). The action of the $G^{\mathbf{C}}/G$-valued model is of the WZW type in external (flat) gauge field but since $G^{\mathbf{C}}/G$ is contractible, it looses its topological character.

On $\mathbf{C}P^1$, parametrizing the orbit of the zero gauge field by $h \in \mathbf{G}^{\mathbf{C}}$, $h(\xi_0) = 1$, we get

$$\|\gamma\|^2 = \Big(\int < \gamma(0) \, , \, \bigotimes_l D_{R_l}(hh^*(\xi_l))^{-1} \, \gamma(0) > \tag{54}$$

$$\cdot \, e^{(k+2g^\vee)S(hh^*)} \;\; \delta(hG(\xi_0)) \;\; D(hG) \;\; (det'(-\Delta)/Area)^{dim(G)}\Big)_{ren}$$

where Δ is the scalar Laplacian on $\mathbf{C}P^1$. Similarly on the torus one obtains a version of (53). Now the point is that in these two simplest topologies, if we parametrize $G^{\mathbf{C}}/G$ by the Borel subgroup of $G^{\mathbf{C}}$, the hG-field functional integral becomes an iterative Gaussian one and may be explicitly evaluated. For simplicity, let us concider the case $G = SU(2)$ in the spherical topology. Parametrizing

$$hSU(2) = \begin{pmatrix} e^{\phi/2} & e^{-\phi/2}v \\ 0 & e^{-\phi/2} \end{pmatrix} SU(2)$$

for ϕ real and v complex, we obtain:

$$S(hh^*) = -(2\pi)^{-1}i\int(\partial\phi)(\bar\partial\phi) \; - \; (2\pi)^{-1}i\int e^{-2\phi}(\partial\bar v)(\bar\partial v) \; , \tag{55}$$

$$d(hG) = e^{-2\phi} \; d\phi \; d^2v \; . \tag{56}$$

The insertion $< ... >$ in (54) becomes a polynomial in $e^{\pm\phi(\xi_l)}$, $v(\xi_l)$ and $\bar v(\xi_l)$.

First observe that the $v-$integral is Gaussian. Placing point ξ_0 at infinity, we obtain

$$\int exp[-(2\pi)^{-1}i(k+4)\int e^{-2\phi}(\partial\bar v)(\bar\partial v)] \;\; \delta(v(\xi_0))\prod_\xi e^{-2\phi(\xi)}d^2v(\xi) \tag{57}$$

$$= \; exp[i\pi^{-1}\int(\partial\phi)(\bar\partial\phi) + (2\pi i)^{-1}\int\phi\mathcal{R}] \;\; (det'(-\Delta)/Area)^{-1}$$

by the chiral anomaly ($\mathcal{R}$ is the metric curvature form of $\mathbf{C}P^1$ with the integral equal to $-4\pi i$) ,

$$\langle v(z_1)\bar v(z_2)\rangle_v \; = \; (\pi(k+4))^{-1}\int e^{2\phi(y)}((z_1-y)(\bar z_2-\bar y))^{-1} \, d^2y \; , \tag{58}$$

for the two-point function of v . The subsequent ϕ integral becomes also Gaussian:

$$(59) \qquad \|\gamma\|^2 = \Big(\int \prod_a (\int d^2 y_a) \;\; polyn.(e^{\pm\phi(\xi_l)}, e^{2\phi(y_a)}) \cdot exp[-(2\pi)^{-1} i(k+2) \int (\partial\phi)(\bar\partial\phi) + (2\pi i)^{-1} \int \phi\mathcal{R}] \;\; \delta(\phi(\infty)) \; D\phi \cdot (det'(-\Delta)/Area)^2 \Big)_{ren}$$

where the polynomial under the integral has coefficients depending on ξ_l 's and y_a 's . Eq. (59) has the interpretation of a Coulomb gas correlation function. The (imaginary) charges are located in the insertion points ξ_l and points y_a . The correlation function is integrated over the positions of the latter ones playing the role of screening charges. Besides, there is a vacuum charge smeared over the space represented by the $\mathcal{R}$-term in (59). Expression (59) is in some sense the dual form (for the $G^{\mathbf{C}}/G$-valued rather than for the G-valued WZW model) of the recently discussed Coulomb gas representations of the current algebras [36-40].

The renormalization needed to give the precise sense to (59) is now the straightforward multiplicative one leading to the Wick ordering of "vertex operators" $e^{q\phi(\xi)}$ (in the most divergent terms on the right hand side of (59) and killing the other terms). The resulting Gaussian integral is easy to evaluate and gives a finite-dimensional integral expression for the scalar product of current blocks on CP^1 :

$$(60) \qquad \|\gamma\|^2 = \prod_a (\int d^2 y_a) \;\; \mathcal{P}_{\gamma(0)}((\xi_l);(y_a))$$

where function $\mathcal{P}_{\gamma(0)}$ is explicitly calculable and is a combination of products of powers of absolute values of differences between points ξ_l, y_a . The natural question to ask is whether the integral in (60) converges. In fact it is easy to see that the only problem might come from the behavior at infinity. Trial calculations [41] support the natural conjecture that the integral (60) is finite if and only if $\gamma(0)$ satisfies the fusion rules.

The whole discussion generalizes to the case of any group and to the toroidal geometry [41] allowing to express the scalar product (33) of current blocks and, due to (34), also the Green functions of the WZW models on sphere or torus, by finite dimensional integrals. This way, we obtain an explicit solution for the WZW models in those two geometries. Extension of the Coulomb gas representation to the case of higher genus surfaces remains however an open problem.

6. Coset models

There is a simple way, called the coset construction [42-50], to generate new CFT's from the group G-valued WZW model. If H is a (connected) subgroup of G then we may consider the functional integral

$$(61)\qquad \left(\int \ldots e^{-kS(g,A_H)}\ Dg\ DA_H\right)_{ren}$$

where the action is as in (11) and A_H runs through the connections with values in the Lie algebra of H. The insertions in (61) should be H-group gauge invariant. The simplest example is to take linear maps

$$t_l : V^G_{R_l} \to V^H_{r_l}$$

between the G and H representation spaces, intertwining the action of H, and to consider the insertion

$$(62)\qquad \prod_l tr_{V_{r_l}}(t_l D_{R_l}(g(\xi_l))t_l^*)$$

where t_l^* is the adjoint map. The Green functions $\Gamma_{G/H}((\xi_l);(t_l))$ given formally by evaluation of (61) with insertion (62) can be expressed using bases (γ^G_n) and (γ^H_i) of current blocks of, respectively, group G and group H WZW models (at the corresponding values of the coupling constant k). Notice that if γ^G is a G-current block then $(\bigotimes_l t_l)\gamma^G$, when restricted to the gauge fields A^{01}_H, becomes an H-current block. Let us introduce the branching matrices [51] (b_{ni}),

$$(\bigotimes_l t_l)\ \gamma^G_n(A^{01}_H) = \sum_i b_{ni}\ \gamma^H_i(A^{01}_H)\ .$$

Then

$$(63)\qquad \Gamma_{G/H}((\xi_l);(t_l)) = \sum_{\substack{m,n\\ i,j}} (\mathcal{H}_G^{-1})^{mn}\ (\mathcal{H}_H)_{ji}\ b_{mi}\ \overline{b_{nj}}\ ,$$

where

$$(\mathcal{H}_G)_{mn} = \ll \gamma^G_m, \gamma^G_n \gg$$

and

$$(\mathcal{H}_H)_{ij} = \ll \gamma_i^H, \gamma_j^H \gg ,$$

compare eqs. (34), (35). The Coulomb gas representation for the scalar products of WZW current blocks in the spherical and toroidal topologies allows then to obtain finite dimensional integral expressions for the Green functions (63).

The classical example of a coset theory [42,43] is to take $G = SU(2) \times SU(2)$ with H equal to the diagonal $SU(2)$. Since G is not simple, we may choose the coupling constant for the WZW theory based on it independently for both $SU(2)$ factors and we set it to be k and 1 respectively. The coset models obtained this way give the unitary theories in the series of the minimal models introduced by Belavin-Polyakov-Zamolodchikov [3] at the start of the last years revival of CFT. The Hilbert space of these theories is a finite combination of unitary representations of two copies of Virasoro algebras with central charge $c = 1 - 6/((k+2)(k+3))$ [52,53].

As the simplest case of the Green functions, we may consider (63) with R_l's corresponding to spins $(1/2, 0)$ and t_l's being the identity map on spin $1/2$ representation of the diagonal $SU(2)$. Eq. (63) combined with the Colomb gas representation of the previous section allows then to express the spherical Green functions

$$(64) \qquad \Gamma((\xi_l)) = \Big(\int \prod_l tr_{V_{1/2}} g(\xi_l) \; e^{-kS(g,A)-S(g',A)} \; Dg \; Dg' \; DA\Big)_{ren}$$

by finite-dimentional integrals. For four-point functions, the result may be written in terms of four-linear combinations of hypergeometric functions [54,49]. Arguments have been advanced [54,55] in suport of the conjecture that the Green functions (64) coincide with a scaling limit (5) of the Green functions (4) of the $P(\Phi)_2$ models with P of degree $2k+2$. It seems remarkable that, although a detailed rigorous control of these scaling limits have eluded the attempts of constructive field theoriests, we seem to have now at our disposal the exact expressions for the limiting Green functions. For a constructivist, it may be tempting to produce a clean proof that one obtains this way an example of a theory satisfying all (massless) quantum field theory axioms. More importantly, one may envisage a rigorous exploration of massive field theories close to scale invariant ones.

The coset models give general examples of rational conformal field theories, i.e. such that the Green functions are finite combinations of products of holomorphic and antiholomorphic expressions. It has been conjectured [50] that all rational CFT's may be obtained by (versions of) the coset construction. In any case, the coset models are a useful laboratory for the algebraization of CFT since they provide examples of symmetry algebras extending the Virasoro one [56].

References

[1] A. M. Polyakov, *Conformal symmetry of critical fluctuations*, JETP Lett. 12, 381-383 (1970)

[2] M. Lüscher, G. Mack, unpublished manuscript 1976

[3] A. A. Belavin, A. M. Polyakov, A. B. Zamolodchikov, *Infinite conformal symmetry in two-dimensional quantum field theory*, Nucl. Phys. B 241, 333-380 (1984)

[4] E. Witten, *Non-abelian bosonization in two dimensions*, Commun. Math. Phys. 92, 455-472 (1984)

[5] V. Knizhnik, A. B. Zamolodchikov, *Current algebra and Wess-Zumino model in two dimensions*, Nucl. Phys. B 247, 83-103 (1984)

[6] D. Gepner, E. Witten, *String theory on group manifolds*, Nucl. Phys. B 278, 493-549 (1986)

[7] G. Felder, K. Gawędzki, A. Kupiainen, *Spectra of Wess-Zumino-Witten models with arbitrary simple groups*, Commun. Math. Phys. 117, 127-158 (1988)

[8] J. Glimm, A. Jaffe, *Quantum physics. A functional integral point of view*, Springer, Berlin, Heidelberg, New-York 1987

[9] V. Pasquier,*Continiuum limit of lattice models built on quantum groups*, Nucl. Phys. B 295, 491-510 (1988)

[10] D. Bernard, *On the Wess-Zumino-Witten models on the torus*, Nucl. Phys. B 303, 77-93 (1988)

[11] D. Bernard, *On the Wess-Zumino-Witten models on Riemann surfaces*, Nucl. Phys. B 309, 145-174 (1988)

[12] H. Sugawara, *A field theory of currents*, Phys. Rev. 170, 1659-1662 (1968)

[13] I. B. Frenkel, *Two constructions of affine Lie algebra representations and Boson-Fermion correspondence in quantum field theory*, J. Func. Anal. 44, 259-327 (1981)

[14] P. Christe, R. Flume, *The four-point correlations of all primary operators of the d=2 conformal invariant SU(2) σ-model with Wess-Zumino term*, Nucl. Phys. B 282, 466-494 (1987)

[15] T. Kohno, *Linear representations of braid groups and classical Yang-Baxter equations*, Contemp. Math. 78, 339-363 (1988)

[16] E. Artin, *The theory of braids*, in *Collected Papers*, pp. 491-498, Addison-Wesley, Reading MA, Palo Alto, London,Dallas, Atlanta 1965

[17] V. G. Drinfel'd, *Quantum groups*, in *Proceedings of ICM, Berkeley 1986*, pp. 798-820

[18] M. Jimbo, *A q-difference analogue of $\mathcal{U}(g)$ and Yang-Baxter equation*, Lett. in Math. Phys. 10, 63-69 (1985)

[19] S. L. Woronowicz, *Twisted SU(2) group, an example of non-commutative differential calculus*, Publ. RIMS, Kyoto Univ. 23, 117-181 (1987)

[20] V. Pasquier, H. Saleur, *Common structures between finite systems and conformal field theories through quantum groups*, Saclay SPhT/89-031 preprint

[21] G. Moore, N. Yu. Reshetikhin, *A comment on quantum group symmetry in conformal field theory*, IASSNS-HEP-89/18 preprint

[22] D. Buchholz, I. B. Frenkel, G. Mack, I. Todorov, private communication

[23] E. Witten, *Gauge theories, vertex models and quantum groups*, IASSNS-HEP-89/32 preprint

[24] L. Alvarez-Gaumé, C. Gomez, G. Sierra, *Duality and quantum groups*, CERN-TH.5369/89 preprint

[25] G. Felder, J. Fröhlich, G. Keller, *On the structure of unitary conformal field theory II: Representation theoretic approach*, ETH-TH/89-12 preprint

[26] D. Friedan, S. Shenker, *The analytic geometry of two-dimensional conformal field theory*, Nucl. Phys. B 281, 509-545 (1987)

[27] E. Witten, *Quantum field theory and the Jones polynomial*, IASSNS-HEP-88/33 preprint

[28] S. Elitzur, G. Moore, A. Schwimmer, N. Seiberg, *Canonical quantization of the Chern-Simons-Witten theory*, IASSNS-HEP-89/20 preprint

[29] G. Felder, K. Gawędzki, A. Kupiainen, *The spectrum of Wess-Zumino-Witten models*, Nucl. Phys. B 299, 355-366 (1988)

[30] K. Gawędzki, *Wess-Zumino-Witten conformal field theory*, IHES/P/89/06 preprint

[31] M. F. Atiyah, R. Bott, *The Yang-Mills equations over Riemann surfaces*, Phil. Trans. R. Soc. Lond. A 308, 523-615 (1982)

[32] M. S. Narasimhan, C. S. Seshadri, *Stable and unitary vector bundles on a compact Riemann surface*, Ann. Math. 82, 540-567 (1965)

[33] S. K. Donaldson, *A new proof of a theorem of Narasimhan and Seshadri*, J. Diff. Geom. 18, 269-277 (1987)

[34] A. M. Polyakov, P. B. Wiegmann, *Theory of non-abelian Goldstone bosons in two dimensions*, Phys. Lett. B 131, 121-126 (1983)

[35] J. B. Bost, *Fibrés déterminants, déterminants régularisés et mesures sur les espaces de modules des courbes complexes*, Astérisque 152-153, 113-149 (1987)

[36] M. Wakimoto, *Fock representations of the affine Lie algebra* $A_1^{(1)}$, Commun. Math. Phys. 104, 605-609 (1986)

[37] A. B. Zamolodchikov, talk given in Montreal 1988

[38] B. L. Feigin, E. V. Frenkel, *Representations of affine Kac-Moody algebra and bosonization*, preprint 1989

[39] A. Gerasimov, A. Marshakov, M. Olshanetsky, S. Shatashvili, *Wess-Zumino-Witten model as theory of free fields*, preprint 1989

[40] D. Bernard, G. Felder, *Fock representations and BRST cohomology in SL(2) current algebra*, ETH-TH/89-26 preprint

[41] K. Gawędzki, *Quadrature of conformal field theories*, IHES/P/89/32 preprint

[42] P. Goddard, A. Kent, D. Olive, *Virasoro algebras and coset space models*, Phys. Lett. B 152, 88-92 (1985)

[43] P. Goddard, A. Kent, D. Olive, *Unitary representations of the Virasoro and super-Virasoro algebras*, Commun. Math. Phys. 103, 105-119 (1986)

[44] M. R. Douglas, *G/H conformal field theory*, CALT-68-1453 preprint

[45] K. Bardakci, E. Rabinovici, B. Säring, *String models with* $c < 1$ *components*, Nucl. Phys. B 299, 151-182 (1988)

[46] K. Gawędzki, A. Kupiainen, *G/H conformal field theory from gauged WZW model*, Phys. Lett. B 215, 119-123 (1988)

[47] D. Karabali, Q. Park, H. J. Schnitzer, Z. Yang, *A GKO construction based on a path integral formulation of gauged Wess-Zumino-Witten actions*, Phys. Lett. B 216, 307-312 (1989)

[48] P. Bowcock, *Canonical quantisation of the gauged Wess-Zumino model*, Nucl. Phys. B 316, 80-100 (1989)

[49] K. Gawędzki, A. Kupiainen, *Coset construction from functional integrals*, Nucl. Phys. B 320, 625-668 (1989)

[50] G. Moore, N. Seiberg, *Taming the conformal Zoo*, Phys. Lett. B 220, 422-430 (1989)

[51] V. G. Kac, M. Wakimoto, *Modular and conformal invariance constraints in representation theory of affine algebras*, Adv. Math. 70, 156-236 (1988)

[52] D. Friedan, Z. Qiu, S. Shenker, *Conformal invariance, unitarity and critical exponents in two dimensions*, Phys. Rev. Lett. 52, 1575-1578 (1984)

[53] D. Friedan, Z. Qiu, S. Shenker, *Details of the non-unitarity proof for highest weight representations of the Virasoro algebra*, Commun. Math. Phys. 107, 535-542 (1986)

[54] A. B. Zamolodchikov, *Conformal symmetry and multicritical points in two-dimensional quantum field theory*, Yad. Fiz. 44, 821-827 (1986)

[55] A. B. Zamolodchikov, *Renormalization group and perturbation theory near fixed points in two-dimensional field theory*, Yad. Fiz. 46, 1819-1831 (1987)

[56] A. B. Zamolodchikov, *Infinite additional symmetries in two-dimensional conformal quantum field theory*, Teor. Mat. Fiz. 45, 347-359 (1985)

Krzysztof Gawędzki, Institut des Hautes Etudes Scientifiques, 35 route de Chartres, 91440 Bures-sur-Yvette, France

SELECTED TOPICS IN CONFORMAL FIELD THEORY

Al.B.Zamolodchikov

Institute of Theoretical and Experimental Physics,
Moscow, USSR

1. Introduction to Conformal Field Theory

One of the ways the 2D conformal field theory (CFT) appears in physics is the two-dimensional critical phenomena. Considering a 2D statistical system which exhibits the second order phase transition one observes that on approaching the phase transition point T_c the characteristic size of the order parameter fluctuations - correlation length R_c - infinitely increases. In this case the long range fluctuations are not sensitive to fine details of microscopic structure and their dynamics is effectively described by the field theory. As a rule this theory enjoys certain additional symmetries which are not present at the microscopic level - namely, rotational and translational symmetry. Therefore in this scaling limit we are dealing with the Euclidean relativistic 2D field theory.

Strictly at $T = T_c$ the correlation length is infinite and the corresponding field theory becomes massless. This limit corresponds to the ultraviolet limit of the scaling field theory. The group of space-time symmetries enlarges again and due to the absence of mass parameter this massless theory is invariant under scale transformations $x^\mu \to \lambda x^\mu$. These general ideas of scaling [1,2] provide link between critical phenomena and relativistic field theory [3,4].

Polyakov suggested [5] that the critical fluctuations possess not only scale but conformal invariance. Conformal transformations are those coordinate transformations

$$y^\mu = y^\mu(x) \tag{1.1}$$

which conserve angles between vectors but locally change the scale

$$ds^2 = g_{\mu\nu}dy^\mu dy^\nu = \rho(x)g_{\mu\nu}dx^\mu dx^\nu \tag{1.2}$$

For the infinitesimal transformations $x^\mu \to x^\mu + \varepsilon^\mu(x)$ this condition acquires the form

$$\partial_\mu \varepsilon_\nu(x) + \partial_\nu \varepsilon_\mu(x) = g_{\mu\nu}\partial^\lambda \varepsilon_\lambda(x) \tag{1.3}$$

In the framework of the field theory it can be shown that in translationaly and rotationaly invariant systems conformal symmetry is a consequence of the scale symmetry. This means that an effective field theory of critical fluctuations is a CFT.

When the field theory becomes massless the standard approach based of the space of asymptotic states becomes unappropriate (at least in 2D) due to the lack of a scale parameter. Suitable tool in this situation is the algebra of operator product expansions (OPE), invented by Kadanoff [6], Polyakov [7] and Wilson [8]. The space of asymptotic states in this approach is substituted by the space of local states, or local fields. The object to study is a set of correlation functions of local fields $A_i(x)$

$$\langle A_1(x_1)A_2(x_2)\ldots A_N(x_N)\rangle \qquad x_i \in \mathbb{R}^2 \tag{1.4}$$

In Lagrangean theory these are something like functional integrals

$$\int A_1(x_1)\ldots A_N(x_N)\exp\{-H[\varphi]\}\mathcal{D}\varphi \tag{1.5}$$

over some set of "fundamental fields" φ; $H[\varphi] = \int\mathcal{H}(\varphi(x),\partial_\mu\varphi(x))d^2x$ is the local action, and $A_i(x_i)$ - some local functions of fields φ and their space derivatives. Nevertheless in the following we shall not use this definition and follow the nonlagrangean approach by Polyakov, based on OPE.

In this approach term "local fields" means a set of restrictions imposed on the analytical properties of the correlation functions. Namely, they are single valued functions of each argument x_i (space position of each field) real analytic everywhere in $\mathbb{R}^2$ except points x_j, $j\neq i$, where other fields are situated.

Fields are elements of an infinite dimensional vector space $\mathcal{A}$ with countable basis $\{A_j,\ j=0,\ 1,\ 2,\ \ldots\}$ which is complete in the sense of the OPE described below. In the Lagrangean field theory the space $\mathcal{A}$ consists of the "fundamental field" φ and all local composite fields like $:\varphi^n:$, $:\partial_\mu\varphi\ \varphi^n:$, etc., including their space derivatives. The simplest representative is the identity operator $I = A_0$.

Operator Product Expansions. Space $\mathcal{A}$ is an algebra w.r.t. OPE [6-10]. Product of two basis vectors A_i, A_j, situated at different space points x_1 and x_2 may be expanded in this basis

$$A_i(x_1)A_j(x_2) = \sum_k C^k_{ij}(x_1,x_2)A_k(x_2), \qquad (1.6)$$

where $C^k_{ij}(x_1,x_2)$ are complex coefficients, which are called the structure functions of the operator algebra. This relation means equation between correlation functions

$$\langle A_i(x_1)A_j(x_2)B_1(y_1)\ldots B_N(y_N)\rangle = \sum_k C^k_{ij}(x_1;x_2)\langle A_k(x_2)B_1(y_1)B_N(y_N)\rangle, \qquad (1.7)$$

where sum is convergent provided $|x-x|<\min|x_2-y_j|$ and $B_i(y_i)$ are any local fields in the theory. Note that the OPE relation (1.6) means that A_i themselves act in $\mathcal{A}$ as linear operators and so they play on the same footing the role of operators and states.

Turn to the space-time symmetries of the theory. $\mathcal{A}$ contains a symmetric tensor $T_{\mu\nu}(x)$ which satisfy the conservation equation of the following form (in the flat space)

$$\partial_\mu T^{\mu\nu}(x) = 0 \tag{1.8}$$

In 2D it is convenient to use complex coordinates

$$z = x^1+ix^2; \qquad \bar{z} = x^1-ix^2, \tag{1.9}$$

In components $T = T_{zz}$; $\bar{T} = T_{\bar{z}\bar{z}}$; $\Theta = T_{z\bar{z}}$ the conservation eqs. (1.9) read

$$\bar{\partial}T = \partial\Theta; \qquad \bar{\partial}\Theta = \partial\bar{T} \tag{1.10}$$

Therefore one can construct the following closed 1-forms

$$\begin{aligned} P &= Tdz-\Theta d\bar{z} \\ \bar{P} &= \bar{T}d\bar{z}-\Theta dz \\ M &= (zT-\bar{z}\Theta)dz-(\bar{z}\bar{T}-z\Theta)d\bar{z} \end{aligned} \tag{1.11}$$

The conserved quantities momentum-energy and angular momentum are defined as integrals of these forms over appropriately chosen closed contours. In particular one defines

$$\langle \oint_{(x)} PA(x)\ldots\rangle = {}^{\partial}/_{\partial z}\langle A(x)\ldots\rangle \tag{1.12}$$

and the spin operator S: $\mathcal{A} \to \mathcal{A}$

$$SA(x) = \oint_{(x)} MA(x) \tag{1.13}$$

<u>Conformal invariance in 2D</u>. Conserved quantities (1.11) are general space-time symmetries of the 2D relativistic field theory. When the theory is scale symmetric the equation

$$\Theta(x) = 0 \tag{1.14}$$

holds. The stress tensor conservation in this case turns to the form

$$\bar{\partial}T = 0 \ ; \quad \partial\bar{T} = 0 \ , \tag{1.15}$$

This means that T and $\bar{T}$ are holomorphic functions of z and $\bar{z}$ respectively. In this case besides the generator of dilations

$$D = \int zTdz + \bar{z}\bar{T}d\bar{z} \tag{1.16}$$

which is also a conserved operator, one in fact has an infinite number of conserved quantities of the form

$$L_{\varepsilon} = \int_C \varepsilon(z)T(z)\frac{dz}{2\pi i} \ , \tag{1.17}$$

for any holomorphic vector field $\varepsilon(z)$ (and the same structures for $\bar{T}$ and $\bar{\varepsilon}(\bar{z})$). The form under integration is closed and L_{ε} is conserved in any region of $\mathbb{R}^2$ where $\varepsilon(z)$ and $T(z)$ are both holomorphic.

In two-dimensional case (as distinct from higher-dimensional spaces) the space of infinitesimal conformal transformations (1.3) is infinite-dimensional and includes all the transformations of the form

$$z \to z + \varepsilon(z) \ , \tag{1.18}$$

with holomorphic $\varepsilon(z)$. In the scale invariant field theory these transformations are generated by operators (1.17). Consequently, the scale invariant 2D field theory has an infinite-dimensional conformal symmetry. In the famous BPZ paper [11] it was explained how to use this symmetry when studying the scale invariant 2D field theory.

Considering an infinitesimal conformal transformation $\varepsilon(z)$ which is regular in the point x where some local field $A \in \mathcal{A}$ is situated one defines the action of this transform on

A

$$\delta_\varepsilon A(x) = \oint_{(x)} \frac{dz}{2\pi i}\, \varepsilon(z)T(z)A(x) \qquad (1.19)$$

This action is supposed to be local in the sense that it results in a finite number of terms containing ε and its derivatives at the point x [11]

$$\delta_\varepsilon A(x) = \sum_{i=0}^{N} B_i \partial^i \varepsilon(x) \qquad (1.20)$$

where B_i are local fields. It is convenient to introduce operators L_n, acting in $\mathcal{A}$ as

$$L_n A(x) = \frac{1}{2\pi i} \oint_{(x)} T(z)A(x) z^{n+1} dz \qquad (1.21)$$

and analogous operators $\bar{L}_n$ with T replaced by $\bar{T}$. In particular, the rotation and dilation operators become

$$S = L_0 - \bar{L}_0 \; ; \quad D = L_0 + \bar{L}_0 \qquad (1.22)$$

Therefore the eigenvalues of $L_0, \bar{L}_0$ are combinations of spin and scale dimension of the corresponding eigenfield $A_{(\Delta,\bar{\Delta})}$. They are called right and left dimensions $(\Delta,\bar{\Delta})$ of this field. Also

$$L_{-1}A = \partial A \; ; \quad \bar{L}_{-1}A = \bar{\partial} A \qquad (1.23)$$

Consider the variation of the field T(z) under infinitesimal conformal transformation ε. The most general form consistent with scaling arguments and unitarity is [11]

$$\delta_\varepsilon T(z) = \varepsilon(z)\partial T(z) + 2\partial\varepsilon(z)T(z) + \frac{c}{12}\partial^3 \varepsilon(z) \qquad (1.24)$$

where the coefficient near $\partial^3 \varepsilon$ should be proportional to the identity operator with the numerical coefficient known as the central charge c. This form of transformation is equivalent to the following general form of OPE of two stress tensor components T [11]

$$T(z)T(z') = \frac{c}{2(z-z')^4} + \frac{2}{(z-z')^2} T(z') + \frac{1}{z-z'} \partial T(z') + \text{reg.} \qquad (1.25)$$

where reg. denotes terms regular as $z \to z'$. It can be verified that in terms of operators L_n introduced above this OPE is equivalent to the following commutation relation for L's

$$[L_m, L_n] = (m-n)L_{m+n} + \frac{c}{12}(m^3-m)\delta_{m+n} \qquad (1.26)$$

To see this one considers the action of product of two operators L on arbitrary field $L_m L_n A$. According to definition (1.21) this can be written down as the double integral

$$\oint_{C_1} d\zeta_1 (\zeta_1 - x)^{m+1} T(\zeta_1) \oint_{C_2} d\zeta_2 (\zeta_2 - x)^{n+1} T(\zeta_2) A(x) \qquad (1.27)$$

over contours C_1 and C_2 surrounding the point x, the second contour being inside C_1. To commute L_m and L_n one is to deform the exterior contour inside the interior one. On doing this one picks up the singularities in the OPE (1.25) of two T's. This just lead to the commutation relation quoted above.

Algebra (1.26) is the central extension of the algebra of vector fields [12], and the constant c is called the central charge. This algebra has long been known in the conformally invariant models that arise in theory of relativistic strings, see i.g., [13-15], and is called the Virasoro algebra. We shall denote it by Vir; Operators $\bar{L}$ form the same algebra (1.26) and commute with L's. The joint algebra of generators L_n, $\bar{L}_n$ is denoted as Vir x $\overline{\text{Vir}}$. We have found that the Virasoro algebra (more strictly Vir x $\overline{\text{Vir}}$) acts in $\mathcal{A}$ and therefore $\mathcal{A}$ can be classified via irreducible representations of Vir.

The above mentioned locality of the action of

conformal transformations on fields means that in fact we deal with the highest weight representations of Vir x $\overline{\text{Vir}}$. In $\mathcal{A}$ there are special fields Φ_ℓ called primary fields. They are annihilated by all L and $\bar{L}$ with positive indices and are eigenstates of L_0 and $\bar{L}_0$ with eigenvalies being right and left dimensions $(\Delta_\ell,\bar{\Delta}_\ell)$ of the primary field Φ_ℓ

$$\begin{aligned} L_n\Phi_\ell &= \bar{L}_n\Phi_\ell = 0 \qquad \text{for} \quad n > 0 \\ L_0\Phi_\ell &= \Delta\Phi_\ell\ , \qquad \bar{L}_0\Phi_\ell = \bar{\Delta}\Phi_\ell \end{aligned} \tag{1.28}$$

This correspond to simple homogeneous transformation of primary fields under conformal transformations

$$\delta_\varepsilon\Phi_\ell(x) = \varepsilon(x)\partial\Phi_\ell(x)+\Delta_\ell\partial\varepsilon(x)\Phi_\ell(x) \tag{1.29}$$

without inhomogeneous higher derivative terms in ε, and the following form of the OPE of T near Φ_ℓ

$$T(z)\Phi_\ell(x,\bar{x}) = \frac{\Delta_\ell}{(z-x)^2}\Phi_\ell(x,\bar{x}) + \frac{1}{z-x}\frac{\partial}{\partial x}\Phi_\ell(x,\bar{x}) + \text{reg.} \tag{1.30}$$

Apart from the primary fields Φ_ℓ, space $\mathcal{A}$ contains all the fields which can be obtained from Φ_ℓ by successive action of operators L_n, $\bar{L}_n$ with $n > 0$, i.e., fields of the form

$$L_{-n_1}L_{-n_2}\ldots L_{-n_N}\bar{L}_{-m_1}\ldots\bar{L}_{-m_M}\Phi_\ell \tag{1.31}$$

where $n_i > 0$, $m_i > 0$. It can be assumed in (1.31) that $n_1 \le n_2 \le \ldots \le n_N$ and $m_1 \le m_2 \le \ldots \le m_N$, since the other orderings are reducible to this by means of the commutation relations (1.26). The fields (1.31) are called "conformal descendants" of the primary field Φ_ℓ. The linear space spanned by all independent vectors of the form (1.31) is called the Verma module of Vir x $\overline{\text{Vir}}$. Because of commutativity of Vir and $\overline{\text{Vir}}$ this space is obviously has a structure of direct product of Verma moduli, corresponding to Vir and $\overline{\text{Vir}}$.

The structure of the Virasoro algebra Verma moduli is well studied in mathematical literature [16-18]. It is found that generally this space is irreducible representation of Vir. When special relations between parameters c and Δ are satisfied null-vectors occur in the Verma module. Null-vector is a vector in module which is different from the highest weight vector (primary field) and satisfies the same conditions (1.28) as a primary field. It gives rise to a submodule in the Verma module. The submodule transforms independently on the rest vectors in the module under the action of Virasoro generators. The conditions for null-vector to emerge are known as the Kac spectrum [16]

$$\Delta_{(m,n)} = \Delta_0 + \frac{1}{4}(m\alpha_+ + n\alpha_-)^2 \tag{1.32}$$

where

$$\alpha_\pm = \frac{\sqrt{1-c} \pm \sqrt{25-c}}{\sqrt{24}} \tag{1.33}$$

This is a set of relations parameterized by two positive integers m and n. When one of these relations holds one finds a null-vector with dimension $\Delta_{(m,n)}+mn$. In this case we call the corresponding representation degenerate and integer number mn the level of degeneracy. The corresponding primary field with dimension $\Delta_{(m,n)}$ is also called degenerate [11].

Once finding a null-vector one can consistently set it to zero (and all the corresponding submodule too) due to homogeneous transformation law (1.29) of primary fields. This is the way to reach the so called degenerate CFT's [11] and then the remarkable series of minimal models of CFT discovered in the pioneering paper [11] that we now turn to discuss.

Instead of considering the degeneration of the Vir representation in general we turn to a special degenerate module which is relevant to all conformal field theories. This is the vacuum module where the identity operator I plays the role of highest weight vector. The degeneration in this module occurs at the level 1. The dimension of the highest weight vector I is zero and one can readily verify that the vector $L_{-1}I$ is a null-vector. Having in mind the interpretation of L_{-1} as a space differentiation (1.23) one should set to zero corresponding submodule due to translation invariance of the vacuum state. The rest space can be built as a linear span of independent w.r.t. commutation relations (1.26) vectors of the form $L_{-n_1}L_{-n_2}\ldots L_{-n_N}I$ with $2 \le n_1 \le n_2 \le \ldots \le n_N$. In other terms the vacuum module can be considered as a space of T-composed fields i.e. fields constructed by merging together several T's and its z-derivatives. For example

$$T = L_{-2}I, \qquad \partial T = L_{-3}I, \qquad \partial^2 T = 2L_{-4}I, \tag{1.34}$$

Next field in this space can be called T^2 and defined as a first regular term in the OPE of two T's

$$T(z)T(z') = \frac{c}{2(z-z')^4} + \frac{2T(z')}{(z-z')^2} + \frac{\partial T(z')}{(z-z')} + T^2(z') + O(z-z') \tag{1.35}$$

i.e.

$$T^2 = L_{-2}^2 I \tag{1.36}$$

In general the vacuum module is an irreducible representation of the Virasoro algebra. In special cases a null-vectors emerge in this module - some composite field turns to be primary. Inspecting the Kac formula one finds that there are extra null vectors (in fact infinite number of them) if the central charge takes one of the discrete series of values

$$c = 1 - \frac{6(p-q)^2}{pq} \tag{1.37}$$

with two coprime integers $p,q > 1$. The first additional null vector has dimension pq. As we shall see, setting this vector to zero one arrives to minimal model one arrives to the minimal model denoted $\mathcal{M}(p/q)$ in [11] where these models were discovered and investigated.

The simplest nontrivial case which we shall use as an example in the following considerations in p/q = 2/5. In this case the central charge is c = -22/5. One finds that the vector

$$(L_{-2}^{2}-3/5L_{-4})I \sim \partial^{2}T-{}^{10}/_{3}T^{2} \tag{1.38}$$

is a null-vector. Equating it to zero leads to equation for the stress tensor

$$\partial^{2}T-{}^{10}/_{3}T^{2} = 0 \tag{1.39}$$

which can be viewed as a defining equation of motion for the minimal model $\mathcal{M}(2/5)$. Similar equations hold for more complicated minimal models. They include more complicated composite fields. E.g., in the model $\mathcal{M}(3/4)$ (c = 1/2) the vacuum module is degenerate at the level 6 and

$$\frac{13}{2}\,\partial^{4}T - 132\partial(T\partial T) + 225(\partial T)^{2}+ 64T^{3} = 0 \tag{1.40}$$

In the model $\mathcal{M}(2/7)$ (c = -68/7) where the vacuum module is also degenerate at the level 6 one has

$$\frac{13}{8}\,\partial^{4}T - \frac{77}{2}\,\partial(T\partial T) + \frac{63}{2}(\partial T)^{2}+ 49T^{3} = 0 \tag{1.41}$$

We call the operator equations of this type the vacuum Ward identities [19]. They can be considered as the equations of motion for the theories $\mathcal{M}(p/q)$. All the analytic content of these theories can be derived from vacuum Ward identities. In particular they contain the spectrum of dimensions of primary fields $\Phi_{(n,m)}$, $0<n<p-1$, $0<m<q-1$ in $\mathcal{M}(p/q)$, derived by BPZ [11]

$$\Delta_{(m,n)} = \frac{(pm-qn)^{2}-(p-q)^{2}}{4pq} \tag{1.42}$$

and the BPZ differential equations for correlation functions [11]. The derivation proceeds as follows. One writes down the the Ward identities for correlation functions of primary fields $\langle\Phi_1(x_1)\ldots\Phi_N(x_N)\rangle$ with several T-insertions (these identities can be derived as a consequences of the OPE (1.30)), then merges together enough T's to produce the vacuum null-vector and equates it to zero.

Consider as an example the theory $\mathcal{M}(2/5)$. In this case it is sufficient to consider the following Ward identities for the correlation functions:

$$\langle T(z)\Phi_1(x_1)\ldots\Phi_N(x_N)\rangle =$$

$$\sum_{i=1}^{N}\left[\frac{\Delta_i}{(z-x_i)^2} + \frac{1}{(z-x_i)}\partial/\partial x_i\right]\langle\Phi_1(x_1)\ldots\Phi_N(x_N)\rangle \qquad (1.43)$$

$$\langle T(z)T(z')\Phi_1(x_1)\ldots\Phi_N(x_N)\rangle = \left[\frac{c}{2(z-z')^4}+\right.$$

$$\sum_{i,j=1}^{N}\left(\frac{\Delta_i}{(z-x_i)^2} + \frac{1}{(z-x_i)}\partial/\partial x_i + \frac{2}{(z-z')^2} + \frac{1}{(z-z')}\partial/\partial z'\right)\cdot$$

$$\left.\left(\frac{\Delta_j}{(z'-x_j)^2} + \frac{1}{(z'-x_j)}\partial/\partial x_j\right)\right]\langle\Phi_1(x_1)\ldots\Phi_N(x_N)\rangle \qquad (1.44)$$

Separating T^2 from the last Ward identity one finds:

$$\langle T^2(z)\Phi_1(x_1)\ldots\Phi_N(x_N)\rangle = \sum_{i,j=1}^{N}\left(\frac{\Delta_i}{(z-x_i)^2} + \frac{1}{(z-x_i)}\partial/\partial x_i\right)\cdot$$

$$\left.\left(\frac{\Delta_j}{(z-x_j)^2} + \frac{1}{(z-x_j)}\partial/\partial x_j\right)\right]\langle\Phi_1(x_1)\ldots\Phi_N(x_N)\rangle \qquad (1.45)$$

The equation of motion (1.39) gives the following relation for the correlation function:

$$\left[\sum_{i,j=1}^{N}\left(\frac{\Delta_i}{(z-x_i)^2} + \frac{1}{(z-x_i)}\partial/\partial x_i\right)\left(\frac{\Delta_j}{(z-x_j)^2} + \frac{1}{(z-x_j)}\partial/\partial x_j\right) - \right.$$

$$\left.\sum_{i=1}^{N}\left(\frac{{}^9/_5\Delta_i}{(z-x_i)^4} + \frac{3}{5(z-x_i)^3}\partial/\partial x_i\right)\right]\langle\Phi_1(x_1)\ldots\Phi_N(x_N)\rangle = 0 \qquad (1.46)$$

The leading singularity as $z \to x_i$ in this eq. $\dfrac{\Delta_i(\Delta_i+1/5)}{(z-x_i)^4}$ exhibits the spectrum of primary field dimensions in $\mathcal{M}(2/5)$. Next to leading singularities terms in (1.46) lead to known differential equations [11] for the correlation function $\langle \Phi_1(x_1)\ldots\Phi_N(x_N)\rangle$.

It is convenient to organize the primary fields $\Phi_{(n,m)}$ in the rectangular table with (p-1)x(q-1) sites. E.g. in the theory $\mathcal{M}(2/5)$ this table has the form

0	-1/5	-1/5	0

(1.47)

It has a central symmetry $n \to p-n$, $m \to q-m$ due to the corresponding symmetry of the spectrum (1.42); one is left with (p-1)(q-1)/2 independent primary fields. In $\mathcal{M}(2/5)$ there are only two primary fields: the identity operator I and scalar field φ with dimensions (-1/5,-1/5). Note the negative sign of the central charge and one of dimensions. This implies that this theory is nonunitary. Nevertheless it is physically meaningful and describes the Lee-Yang singularity in the two-dimensional Ising model [20].

2. Minimal Models on Riemann Surfaces

The constructions of the first part were performed for the field theory on the infinite Euclidean plane, which is equivalent under the conformal invariance to Riemann sphere. In last years it became clear that the conformal field theory is well defined on the surfaces of higher genera g [21-23]. In this case the main object under consideration is the partition function of the theory. Correlation functions can be considered as partition functions on the surfaces with punctures. The partition function depends on the metrics of the surface. The essential dependence is on the moduli of the corresponding complex structure; the dependence on the Weil rescalings has the universal form of the Liouville action.

In this lecture we consider the minimal models on Riemann surfaces. For this problem general results are known in the case of torus (g = 1), where the partition function can be constructed in terms of the characters of irreducible representations of the conformal algebra. In particular, the general classification of the modular invariant toroidal partition functions relevant to unitary minimal models was achieved [24].

Generally these partition functions $Z(m,\bar{m})$, where m represents moduli of the Riemann surface, can be constructed as a modular invariant sesquilinear forms in special functions $X(m)$ (blocks or generalized characters) holomorphically dependent on moduli [21]

$$Z(m,\bar{m}) = e^{S_{Liouv}} \sum R_{ij} X_i(m) X_j(\bar{m}) \tag{2.1}$$

where S_{Liouv} carries the anomalous Weil dependence of the partition function and R_{ij} is Hermitian matrix. Here we shall treat only the holomorphic dependence governed by blocks $X(m)$ and try to use to this purpose the vacuum Ward identities discussed in the previous lecture. This approach to minimal models on the surfaces with nontrivial topology was proposed by Drinfeld (see also [23]). Our strategy will be just the same as in the case of infinite plane. We first construct the correlation functions with several stress tensor insertions. With the use of Ward identities these can be related to derivatives of the partition function in moduli. Then we construct the nontrivial vacuum null-vector and recover linear differential equations in moduli for the partition function (more strictly for block $X(m)$).

Consider first the simplest case of torus. The complex structure of the torus is characterized by the modular parameter τ. Torus may be represented as a parallelogram generated by vectors 1 and τ in the complex ξ plane with pairwise identified sides. Denote

$$q = e^{2i\pi\tau} \tag{2.2}$$

The partition functions of minimal models can be constructed of the characters of the Virasoro algebra irreducible representations, which are blocks in the case of toroidal geometry [25]. Character corresponding to the primary field $\Phi_{(m,n)}$ is defined as

$$\chi_{m,n}(q) = q^{-c/24} \operatorname*{tr}_{[\Delta_{m,n}]} q^{L_0} \tag{2.3}$$

where the trace is over the space of irreducible representation of Vir corresponding to primary field $\Phi_{(m,n)}$. These characters are combined to form a modular invariant partition function [24]. Explicit form of all the characters relevant to minimal models is known [26]:

$$\chi_{m,n}(q) = \frac{q^{\frac{(mp-nq)^2}{4pq}}}{\eta(q)} \sum_{k\in\mathbb{Z}} \left[q^{k^2pq+k(mp-nq)} - q^{k^2pq+(mp+nq)+mn}\right] \tag{2.4}$$

where

$$\eta(q) = q^{1/24} \prod_{k=1}^{\infty} (1-q^k) \tag{2.5}$$

is the Dedekind function.

The set of $(p-1)(q-1)/2$ characters, corresponding to model $\mathcal{M}(p/q)$, forms a finite-dimensional representation of the modular group. This means that they all are linearly independent solutions of a modular-invariant linear differential equation of order $(p-1)(q-1)/2$. We shall see that this differential equation is just the consequence of the vacuum Ward identities on the torus [19,27].

Again take as an example the simplest theory $\mathcal{M}(2/5)$ considered in the first lecture. The translational invariance on the torus requires the correlation function $\langle T(\xi)\rangle = \langle T\rangle$ to be ξ-independent. This expectation value is related to the τ-derivative of the partition function Z of the theory $\mathcal{M}(2/5)$:

$$\langle T\rangle = 2\pi i Z^{-1} \frac{\partial Z}{\partial \tau} \tag{2.6}$$

Similarly, the two-point function $\langle T(\xi)T(\xi')\rangle$ is related to τ-derivative of $\langle T\rangle$:

$$Z\oint_A \langle T(\xi)T(\xi')\rangle d\xi/2\pi i = \frac{\partial(Z\langle T\rangle)}{\partial \tau} \tag{2.7}$$

where the closed contour A can be chosen say to connect

points 0 and 1 in the ξ-plane. The correlation function $\langle T(\xi)T(\xi')\rangle$ is a symmetric elliptic function of the difference of arguments $\xi-\xi'$. The only singularity is at $\xi = \xi'$, where the singular terms are prescribed by the OPE (1.25). We can therefore write in general form

$$\langle T(\xi)T(\xi')\rangle =$$

$$-\frac{c}{12}(\partial/\partial\xi)^4\log\theta_1(\xi-\xi')-2\langle T\rangle(\partial/\partial\xi)^2\log\theta_1(\xi-\xi')+U \qquad (2.8)$$

where U is independent on $\xi-\xi'$ and $\theta_1(\xi)$ is the standard elliptic theta-function

$$\theta_1(\xi) = \sum_{n\in\mathbb{Z}}(-)^n e^{i\pi\tau(n+1/2)^2+2\pi i n\xi} \qquad (2.9)$$

The singular elliptic functions in the first and second terms of (2.8) are chosen to ensure the correct singularity structure at $\xi = \xi'$. In addition the integral of these functions from 0 to 1 (over the A cycle) vanishes. This leads to the following expression of U in terms of derivatives of the partition function Z:

$$U = (2\pi i)^2 Z^{-1}(\partial/\partial\tau)^2 Z \qquad (2.10)$$

On letting the argument ξ tend to ξ' in the correlation function (2.8) and subtracting the singular terms, we can obtain the vacuum expectation value of the operator $T^2(\xi)$ determined by eq. (1.35). This gives

$$(2\pi)^{-4}\langle T^2-\frac{3}{10}\partial^2 T\rangle Z = \qquad (2.11)$$

$$\left[(q\,\partial/\partial q)^2+\frac{1}{6\pi^2}\frac{\theta_1'''}{\theta_1'}\,q\,\partial/\partial q-\frac{c}{8\pi^4}\left(\frac{\theta_1^{(v)}}{\theta_1'}-\frac{1}{72}\left(\frac{\theta_1'''}{\theta_1'}\right)\right)\right]Z$$

where θ_1', θ_1''' and $\theta_1^{(v)}$ are respectively the first, third, and fifth ξ-derivatives of the function (2.9) at $\xi = 0$. In the theory $\mathcal{M}(2/5)$ eq.(1.39) holds. This gives a second order differential equation for the partition function Z. In order to write this equation explicitly in modular invariant form, it is convenient to introduce the function

$$2\pi i P(\tau) = 24\eta^{-1}\partial\eta/\partial\tau \tag{2.12}$$

where $\eta(\tau)$ is the Dedekind function (2.5). Under the basic modular transformations, T: $\tau \to \tau+1$ and R: $\tau \to -1/\tau$, it transforms as follows

$$P(\tau+1) = P(\tau)$$
$$P(-1/\tau) = \tau^2 P(\tau) + \frac{12\tau}{2\pi i} \tag{2.13}$$

A modular form of weight k is the function $G_k(\tau)$ which is regular at $q = 0$ and has the following transformation properties under modular transformations:

$$G_k(\tau+1) = G_k(\tau); \qquad G_k(-1/\tau) = \tau^k G_k(\tau). \tag{2.14}$$

It follows from (2.13) that the differential operator

$$\partial_k = q\partial/\partial q - \frac{k}{12} P \tag{2.15}$$

acts from the space of modular forms of weight k into the space of modular forms of weight k+2. The space of modular forms can be generated as a ring of polynomials in the algebraically independent forms Q and R of weights 4 and 6 respectively [28]:

$$Q = -P^2 - 12\partial_2 P$$
$$R = -3\partial_4 Q \tag{2.16}$$

These functions, like $P(\tau)$, are defined in the way that their q-expansions start with 1. In this notation, after substituting $c = -22/5$, eq. (2.11) takes the modular invariant form

$$(\partial_2\partial_0 - \frac{11}{3600})Z_{(2/5)} = 0 \tag{2.17}$$

This equation has two linearly independent solutions, with the behavior $Z \sim q^{11/60}$ and $Z \sim q^{-1/60}$ as $q \to 0$. This is in accordance with the dimensions $\Delta = 0$ and $\Delta = -1/5$ in the model $\mathcal{M}(2/5)$. It can be verified that the solution of this equation is a combination of the characters $\chi_{1,1}$ and $\chi_{1,2}$

of the theory $\mathcal{M}(2/5)$.

Differential equations of higher order arise in other minimal models. For instance, in the models $\mathcal{M}(3/4)$ and $\mathcal{M}(2/7)$ the vacuum Verma module contains the second null-vector at the sixth level and equations of motion (1.40) and (1.41) hold. To write down the corresponding equations we have to bring into consideration the correlation functions $\langle T(\xi)T(\xi')T(\xi'')\rangle$. Corresponding equations are

$$(\partial_4\partial_2\partial_0-\frac{107}{(48)^2}Q\partial_0+\frac{46}{(48)^3}R)Z_{(3/4)}=0 \tag{2.18}$$

$$(\partial_4\partial_2\partial_0-\frac{5}{252}Q\partial_0+\frac{85}{(42)^3}R)Z_{(2/7)}=0 \tag{2.19}$$

Solutions of the corresponding characteristic equations reproduce the dimensions of primary fields in these minimal theories: (0, 1/2, 1/16) and (0, -2/7, -3/7) respectively.

Let us now turn to the case of higher genera. To obtain the vacuum Ward identities in an explicit form as the differential equations in moduli one needs a convenient parameterization of the moduli space. Every surface with $g = 2$ is hyperelliptic and may be represented as a double-fold covering of the Riemann sphere with six branching points. Hence the $g = 2$ moduli space is parameterized by the positions of these points up to projective transformations and discrete automorphisms. A representation of a surface as a ramified covering of the sphere proved to be very useful when investigating a conformal field theory on the surface [29-31]. In this representation powerful methods of conformal field theory on the sphere can be used. The ramification points behave like conformal fields of special type and the partition function is interpreted as a correlation function of these fields.

We start with the general consideration of the conformal field theory on the double-fold ramified covering of the Riemann sphere and restrict ourselves to right (holomorphic) symmetries only. These are generated by two right stress tensor components T_1 and T_2, each attached to corresponding sheet of the Riemann surface. They are independent and separately satisfy the operator product expansion (1.25) with the same central charge c. The symmetry generated is thus the direct product of two Virasoro algebras and therefore the space of states is the direct product of two independent spaces. This happens everywhere but the ramification points. T_1 and T_2 turn to each other on the way around the point, i.e. we consider them as two branches of a double-valued function. It is convenient to use the following combinations:

$$\begin{aligned} T &= T_1+T_2 \\ \Theta &= T_1-T_2 \end{aligned} \tag{2.20}$$

Here T is single-valued and plays a role of ordinary stress tensor on the sphere. It satisfies (1.25) with $\tilde{c} = 2c$. Θ is primary field w.r.t. T and has the following operator product expansion:

$$\Theta(z)\Theta(z) = \frac{\tilde{c}}{2(z-z')^4} + \frac{2T(z')}{(z-z')^2} + \frac{\partial T(z')}{(z-z')} + \text{reg.} \tag{2.21}$$

It changes in sign around the ramification point, which we therefore call a Ramond point w.r.t. symmetry algebra. To clarify this algebra consider the power expansion of say T_1 around the Ramond point, situated say at z = 0:

$$T_1(z) = \sum_{n\in\mathbb{Z}/2} K_n z^{-n-2} \tag{2.22}$$

Standard calculations described in the first lecture lead to the following commutation relations for operators K_n

$$[K_m,K_n] = \frac{m-n}{2} K_{m+n} + \frac{c}{24}(m^3-m)\delta_{m+n}\ , \qquad m,n\in\mathbb{Z}/2 \tag{2.23}$$

This is of course isomorphic to the Virasoro algebra with central charge c. One can bring it to the standard form (1.26) using combinations

$$\tilde{L}_{2n} = 4K_n - \frac{c}{8}\,\delta_{n,0} \tag{2.24}$$

Clearly, this isomorphism exhibits the usual conformal

symmetry in the local uniformizing coordinate system. Generators (2.24) are just the expansion coefficients of the stress tensor in the coordinate $t = z^{1/2}$ and the second term in (2.24) reflects the inhomogeneous transformation low (1.24) of the stress tensor.

The primary Ramond state σ_ℓ is the state annihilated by all operators K_n, $n \in \mathbb{Z}/2$, with positive n. It is also eigenstate of K_0

$$K_n \sigma_\ell = 0 \quad \text{for } n > 0$$
$$K_0 \sigma_\ell = \delta_\ell \sigma_\ell \tag{2.25}$$

The single valued stress tensor T generates the even subalgebra of the Virasoro algebra (2.23):

$$T = \sum_{n \in \mathbb{Z}} 2K_n z^{-n-2} = \sum_{n \in \mathbb{Z}} L_n z^{-n-2} \tag{2.26}$$

Under the conformal transformations of the sphere σ_ℓ has dimension $\Delta_\ell = 2\delta_\ell$. Now L_{-1} (not $K_{-1/2}$) is interpreted as the differentiation in the Ramond point position

$$L_{-1}\sigma_\ell(x) = \partial/\partial x \, \sigma_\ell(x) \tag{2.27}$$

There is a special primary Ramond state (Ramond vacuum) σ with dimension $\Delta_0 = \frac{c}{16}$

$$L_0 \sigma = 2K_0 \sigma = \frac{c}{16} \sigma \tag{2.28}$$

which is degenerate on the first level and satisfies

$$K_{-1/2}\sigma = 0 \tag{2.29}$$

This state is a "coordinate deformation" of the usual vacuum state and (2.29) reflects its translational invariance in the uniformizing coordinate system. The partition function of the theory on the double-fold covering in our approach is interpreted as a correlation function of Ramond vacuum fields

$$Z(x_1,\ldots x_{2n}) = \langle\sigma(x_1)\ldots\sigma(x_{2n})\rangle \qquad (2.30)$$

Now turn to minimal models. Because of the algebra isomorphism mentioned above one need not to reanalyze the degeneration problem in the Ramond sector. It can be simply written down from the Kac spectrum (1.32) where the parameters $\alpha_\pm$ are in usual relation (1.33) with c (not $\tilde{c}$). If c corresponds to the minimal model spectrum (1.37), in the Ramond vacuum Verma module there are extra null-vectors, different from (2.29). Taking again the theory $\mathcal{M}(2/5)$ one finds

$$(K_{-1}^2 - \frac{3}{20} K_{-2})\sigma = 0 \qquad (2.31)$$

In the uniformizing coordinate this is nothing but the vacuum null-vector (1.38). Now the vacuum Ward identity has the form of the following equation of motion for the field σ:

$$\partial^2\sigma - \frac{3}{5} T\sigma = 0 \qquad (2.32)$$

Note that eq.(2.32) is of the usual form of equation for the second level degenerate field [11]. Using this one immediately writes down the system of second order differential equations for the partition function (2.30) of the model $\mathcal{M}(2/5)$:

$$\left(\frac{\partial^2}{\partial x_i^2} - \frac{3}{10}\sum_{j\neq i}^{2n}\left[-\frac{11}{40(x_i-x_j)^2} + \frac{1}{x_i-x_j}\,{}^{\partial}/\partial x_j\right]\right)Z_{(2/5)} = 0 \qquad (2.33)$$

The situation in other minimal models is not so nice. For example, equations (1.40) and (1.41) are now rewritten as

$$(256\ K_{-1}^3 - 264\ K_{-1}K_{-2} + 105\ K_{-3} + 93\ K_{-3/2}^2)\sigma = 0 \qquad (2.34)$$

for $\mathcal{M}(3/4)$ and

$$(196\ K_{-1}^3 - 77\ K_{-1}K_{-2} + 29\ K_{-3} - 7\ K_{-3/2}^2)\sigma = 0 \qquad (2.35)$$

for $\mathcal{M}(2/7)$. The trouble is with the term $K_{-3/2}^2\sigma$, which has not straightforward translation to differential operators, acting on the partition function. We shall show that the problem can be circumvented in the case of $g = 2$, i.e. the surface with no more then six ramification points.

Let $Z(x_1,\dots x_6)$ be the six-point correlation function (2.30). Our purpose is to rewrite the correlation function $\Gamma(x_1,\dots x_6) = \langle K^2_{-3/2}\sigma(x_1)\sigma(x_2)\dots\sigma(x_6)\rangle$ as a differential operator applied to $Z(x_1,\dots x_6)$. Consider the state $K_{-3/2}\sigma$. One finds the following OPE

$$\Theta(z)K_{-3/2}\sigma(0) =$$

$$\frac{c\sigma(0)}{4z^{7/2}} + \frac{\partial\sigma(0)}{z^{5/2}} + \frac{L_{-2}\sigma}{2z^{3/2}} + \frac{2L^2_{-3/2}\sigma}{z^{1/2}} + O(z^{1/2}) \qquad (2.36)$$

with Θ is defined in (2.20). Consider the correlation function with one field σ replaced by $K_{-3/2}\sigma$ and one $\Theta(z)$ insertion (note that due to the symmetry of the hyperelliptic surface, the correlation $\langle K_{-3/2}\sigma(x_1)\sigma(x_2)\dots\sigma(x_6)\rangle$ vanishes). It has the following general z-dependence

$$\langle\Theta(z)K_{-3/2}\sigma(x_1)\sigma(x_2)\dots\sigma(x_6)\rangle =$$

$$\frac{A_0(x_i)+(z-x_1)A_1(x_i)+(z-x_1)^2A_2(x_i)}{(z-x_1)^3\prod_{i=1}^{6}(z-x_i)^{1/2}} \qquad (2.37)$$

where A_0, A_1 and A_2 are z-independent functions. Comparing the $(z-x_1)$-expansion of the right hand side of (2.37) with the OPE (2.36) order by order one first finds A's in terms of correlation functions containing only "even" Ramond fields $\partial\sigma$ and $L_{-2}\sigma$, and then recovers Γ in terms of A's. Note, that would we take more then six points in the correlation function, there would be more A's and we could not solve these relations. In the six-point case the we get

$$2\Gamma(x_1,x_2,\dots,x_6) = \left[-\frac{1}{4}\sum_{i=2}^{6}\frac{1}{x_1-x_i}L^{(1)}_{-2} + \frac{1}{8}\sum_{i=2}^{6}\frac{1}{(x_1-x_i)^2}\partial/\partial x_1 + \right.$$

$$\left.\frac{c}{32}\left[\sum_{\substack{i,j=2\\ i\neq j}}^{6}\frac{1}{(x_1-x_i)(x_1-x_j)^2} - \frac{1}{2}\sum_{i=2}^{6}\frac{1}{(x_1-x_i)^3}\right]\right]\langle\sigma(x_1)\dots\sigma(x_6)\rangle \qquad (2.38)$$

where the operator $L^{(1)}_{-2}$ in the r.h.s. acts on the $\sigma(x_1)$

inside correlation function. This results in the following equations for g = 2 partition functions in the examples considered above

$$\Big[\partial^3/\partial x_i^3 - \frac{33}{16}\sum_{j\neq i}\frac{1}{x_{ij}}\frac{\partial^2}{\partial x_i\partial x_j} + \frac{3}{128}\sum_{j\neq i}\frac{1}{x_{ij}^2}\big(5\,\partial/\partial x_i + 18\,\partial/\partial x_j\big) -$$

$$\frac{93}{256}\sum_{\substack{k\neq i\\ j\neq i}}\frac{1}{x_{ik}x_{ij}}\partial/\partial x_j - \frac{63}{8192}\sum_{j\neq i}\frac{1}{x_{ij}^3} + \frac{93}{8192}\sum_{\substack{k\neq i\\ j\neq i}}\frac{1}{x_{ik}x_{ij}^2}\Big]Z_{(3/4)}(x_1,\ldots x_6) = 0, \qquad (2.39)$$

$$\Big[\partial^3/\partial x_i^3 - \frac{11}{14}\sum_{j\neq i}\frac{1}{x_{ij}}\frac{\partial^2}{\partial x_i\partial x_j} + \frac{1}{98}\sum_{j\neq i}\frac{1}{x_{ij}^2}\big(45\,\partial/\partial x_i + 19\,\partial/\partial x_j\big) -$$

$$\frac{1}{28}\sum_{\substack{k\neq i\\ j\neq i}}\frac{1}{x_{ik}x_{ij}}\partial/\partial x_j - \frac{1649}{5488}\sum_{j\neq i}\frac{1}{x_{ij}^3} + \frac{17}{748}\sum_{\substack{k\neq i\\ j\neq i}}\frac{1}{x_{ik}x_{ij}^2}\Big]Z_{(2/7)}(x_1,\ldots x_6) = 0, \qquad (2.40)$$

where $x_{ij}=x_i-x_j$. Merging any pair of points x_i in the partition function one recovers the spectrum of primary field dimensions in corresponding minimal theories.

The examples discussed above have demonstrated that vacuum Ward identities can be represented as the closed differential equations on the moduli space. In particular, in the case of g = 2 hyperelliptic surface, represented as a double-fold covering of the Riemann surface, they are of the form of differential equations in the positions of branching points with rational coefficients. On the other hand, when considering the hyperelliptic surface of higher genus, we have not been able to close these identities to obtain differential equations of this type. This seems quite reasonable. In the g > 2 case the hyperelliptic surfaces form a submanifold in the moduli space. It is natural to expect that in general the true differential equations include differentiations in the transverse to hyperelliptic submanifold directions. This is what we can see in the calculations on the double-fold covering. For g > 2 there are odd holomorphic quadratic differentials, contributing in general to the Θ-field correlations, and this is their contribution that can not be specified in terms of differentiations with respect to x_i. Note that precisely this contribution "differentiate" the partition function in the transverse to hyperelliptic submanifold direction.

The theory $\mathcal{M}(2/5)$ is somewhat peculiar in this sense. The differential equations close within the hyperelliptic submanifold. Note that the null-vector (2.31) does not contain operators K_{-n} with half-integer n. This means that in this case the Ramond vacuum σ is degenerate as highest weight vector of the T-generated subalgebra only, without respect of the complete symmetry algebra. This subalgebra of symmetric in two sheets conformal transformations is sufficient to completely determine the holomorphic dependence of the partition function on the hyperelliptic submanifold. A natural question arises: is there any minimal theory suitable for this case? Looking through the family of central charges (1.37) one readily finds the theory $\mathcal{M}(3/10)$ with the central charge -44/5, which is just twice the charge of $\mathcal{M}(2/5)$. The table of primary field dimensions in this theory reads:

2	49/40	3/5	1/8	-1/5	-3/8	-2/5	-11/40	0
0	-11/40	-2/5	-3/8	-1/5	1/8	3/5	49/40	2

(2.41)

This is just the set of T-primary fields one expects for the double-fold $\mathcal{M}(2/5)$. In the dimension 2 operator $\Phi_{(2,1)}$ we recognize the Θ-field. The state $\Phi_{(1,2)}$ of dimension -11/40 is just the Ramond vacuum σ. Other Ramond states are $\Phi_{(1,4)}$ of dimension -3/8 (the "coordinate deformation" of the field φ in the background theory $\mathcal{M}(2/5)$), and fields $\Phi_{(2,2)}= K_{-3/2}\sigma$ and $\Phi_{(2,4)}= K_{-1/2}\Phi_{(1,4)}$, which are primary with respect to symmetric subalgebra. The rest fields in the table (2.41) are single-valued w.r.t. Θ. These are $\Phi_{(1,5)}$, which is just the field φ located on one of the sheets of the covering, and $\Phi_{(1,3)}$ of dimension -2/5. This is interpreted as two φ-fields on different sheets, "glued" together at the same point of the sphere. The field $\Phi_{(2,3)}$ is the only Θ-descendant of $\Phi_{(1,3)}$, which is primary in symmetric subalgebra.

As a consequence of this identification, one can write down all the correlation functions of the theory $\mathcal{M}(2/5)$ on the hyperelliptic surface from these of the theory $\mathcal{M}(3/10)$ on the sphere. In particular, from the results of ref.[32] one finds the following solution to eqs.(2.33) in terms of contour integrals:

$$Z_{(2/5)}(x_1,\dots x_{2g+2}) = \qquad (2.42)$$

$$\int_C dt_1\dots dt_g \frac{\prod_{k\neq l}(t_k-t_l)^{3/5}\prod_{i=1}^{2g+1}[(x_i-x_{2g+2})^{11/20}\prod_{j\neq i}^{2g+1}(x_j-x_i)^{3/20}]}{\prod_{k=1}^{g}[(t_k-x_{2g+2})^{11/10}\prod_{i=1}^{2g+1}(t_k-x_i)^{3/10}]}$$

where the integration is over any nontrivial closed hypersurface C.

3. Liouville Action in Cone Gauge

3.1 Introduction. Recent progress in two-dimensional gravity (Liouville theory) is due to the cone gauge parameterization of a 2D surface, introduced by Polyakov [33]. This gauge is defined by the following form of the space-time distance (x^+ and x^- are coordinates on 2D surface, which are complex in the euclidean case and not necessarily related by the complex conjugation)

$$ds^2 = dx^+(dx^-+h(x^+,x^-)dx^+) \qquad (3.1.1)$$

Contrary to the commonly used conformal gauge, where the space-time distance has the form

$$ds^2 = e^{\varphi}|dz|^2 \qquad (3.1.2)$$

the cone gauge leads to well-defined perturbation theory in the inverse central charge of the theory [33]. Moreover, in this gauge new unexpected symmetry of the theory under sl(2) current algebra shows up [33] which permits one an exact calculation of scaling parameters for $c < 1$ [34]. The connection with the sl(2) current algebra was recently recovered through the path integral quantization of the coadjoint orbits of the Virasoro algebra [35].

Two-dimensional Liouville gravity is essentially an induced gravity and Liouville action arises as an effective action of some massless conformally invariant field theory in curved background space-time. It is characterized by single parameter i.e. central charge c of the inducing field theory, which essentially defines the effective number of massless degrees of freedom. Hence the theory is an effective infrared theory and all fields with finite correlation length for Liouville theory point of view contribute to ultraviolet cutoff only.

In two-dimensional conformal field theory the trace of the stress tensor T_{ab}, which defines the reaction of the effective action on an infinitesimal variation of background metric

$$\delta \log Z = \frac{1}{4\pi} \int T_{ab} \delta g^{ab} \sqrt{g}\, d^2x \tag{3.1.3}$$

is simply related to the space-time geometry

$$T_{ab} g^{ab} = -\frac{c}{12} R \tag{3.1.4}$$

where R is scalar curvature of the background metric. Combined with the stress tensor conservation this relation defines unambiguously (at least in the space with trivial topology) the vacuum expectation value of the stress tensor components and therefore fixes the effective action which is clearly determined by the value of the central charge c only and does not depend on other details of the conformal field theory.

In the conformal gauge the system (3.1.3), (3.1.4) is easily integrated and leads to the Liouville action in the familiar form

$$Z \sim \exp \frac{c}{96\pi}\int(\partial_a\varphi)^2 d^2x \qquad (3.1.5)$$

Here this effective action S_{Liouv} will be considered in the cone gauge, where it appears to be of more complicated form. Now it is functional of h and according to (3.1.3) generates connected correlations of "left" stress components $T = T_{--}$. Related to the Liouville action is another interesting functional that we shall denote as the effective potential W(u). It may be obtained as the Legendre transform of Liouville action and classically may be considered as an effective action in some background stress u. Moreover this correspondence persists on quantum level, i.e. after suitable renormalization of constants W(u) is also the Fourier transform of the cone gauge Liouville action. Interesting observation by Polyakov [33] is that this potential W(u) is tightly related to sl(2) current algebra; in fact it generates connected correlation functions of sl(2) currents. In what follows we consider these two functionals in more detail. Similar approach of the Liouville action was used in [36].

The Liouville action is well known to arise when calculating determinants of conformally covariant Laplace operator acting in the space of differentials of weight λ. It turns out that the potential W(u) also can be recovered as an anomaly in determinants of a family of covariant differential operators of order 2j+1 (for all positive integer or half-integer j). Simplest nontrivial example of this type (second order) is the operator $\partial_-^2 + \frac{u}{2}$. We shall consider this family of operators, analyze their determinants and calculate corresponding anomaly (i.e. the central charge k_j in the sl(2) current algebra).

Consideration of these determinants was motivated by one-loop calculation in cone quantum gravity, performed by Polyakov. This beautiful calculation uses the determinant of the third-order operator of the family. We shall consider this result in sect.3.5.

3.2 Liouville action. In the cone gauge we consider metric tensor of the following form

$$g_{ab} = \begin{pmatrix} h & 1/2 \\ 1/2 & 0 \end{pmatrix}; \quad g^{ab} = \begin{pmatrix} 0 & 2 \\ 2 & -4h \end{pmatrix}; \quad \sqrt{g} = 1/2i \qquad (3.2.1)$$

The corresponding metric connection is as follows

$$\begin{aligned} \Gamma^{+}_{++} &= -\partial_- h \\ \Gamma^{-}_{++} &= \partial_+ h + 2h\partial_- h \\ \Gamma^{-}_{+-} &= \partial_- h \end{aligned} \qquad (3.2.2)$$

all other components are zero. The scalar curvature is

$$R = 4\partial_-^2 h \qquad (3.2.3)$$

Covariant conservation of the stress tensor in the cone coordinates acquires the form

$$(\partial_+ - 2h\partial_- - 3\partial_- h)T_{--} + \partial_- T_{+-} = 0 \qquad (3.2.4a)$$

$$\partial_- T_{++} + (\partial_+ - 2h\partial_- - 2\partial_- h)T_{+-} - \partial_+ hT_{--} = 0 \qquad (3.2.4b)$$

and eq.(3.1.4) is rewritten as

$$T_{+-} - hT_{--} = -\frac{c}{12}\partial_-^2 h \qquad (3.2.5)$$

With (3.2.5) the first of eqs.(3.2.4) turns to the following closed equation for $T = T_{--}$

$$(\partial_+ - h\partial_- - 2\partial_- h)T_{--} = \frac{c}{12}\partial_-^3 h \qquad (3.2.6)$$

In the cone gauge relation (3.1.3) acquires the following form

$$\delta \log Z = \delta S_{Liouv}(h) = -\frac{1}{\pi}\int \langle T_{--}\rangle \delta h \, d^2x \qquad (3.2.7)$$

where by d^2x we denote $\frac{1}{2i}\, dx^+ \wedge dx^-$. This equation combined with eq.(3.2.6) gives closed system of functional relations for $S_{Liouv}(h)$.

Note that the second of eqs.(3.2.4) after substitution of (3.2.5) may be written in the form

$$\partial_-(T_{++} - h^2T_{--} + \frac{c}{6}\, h\partial_-^2h) = \frac{c}{12}(\partial_+\partial_-^2h - h\partial_-^3h) \qquad (3.2.8)$$

This equation leads to the following expression for the component T_{++}

$$T_{++} = h^2T_{--} + \frac{c}{12}\left[\partial_+\partial_-h - 3h\partial_-^2h + \frac{1}{2}(\partial_-h)^2\right] \qquad (3.2.9)$$

which holds up to zero modes of the operator ∂_-.

Considering h as a small perturbation one could solve eq.(3.2.6) iteratively and then integrate (3.2.7) to find $S_{Liouv}(h)$ as a series in h. Alternatively this series may be obtained with the use of formal expansion

$$Z(g+\delta g) = Z(g)\sum_{n=0}^{\infty} \frac{1}{(4\pi)^n n!} \qquad (3.2.10)$$

$$\int <T_{a_1b_1}(x_1)\ldots T_{a_nb_n}(x_n)>\delta g^{a_1b_1}\ldots\delta g^{a_nb_n}\sqrt{g_1}d^2x_1\ldots\sqrt{g_n}d^2x_n$$

In the cone gauge this means that Z(h) generates the stress tensor $T = T_{--}$ correlation functions on the infinite plane

$$Z(h) = \sum_{n=0}^{\infty} \frac{(-)^n}{\pi^n n!} \int <T(x_1)\ldots T(x_n)>h(x_1)d^2x_1\ldots h(x_n)d^2x_n \qquad (3.2.11)$$

All these infinite plane correlations can be recovered from the following Ward identities, which reflect the commutation relations for conformal symmetry generators

$$<T(x^-)T(x_1^-)\ldots T(x_n^-)>=\sum_{i=0}^{n} \frac{c}{2(x^- - x_i^-)^4} <T(x_1^-)\ldots \not{T}(x_i^-)\ldots T(x_n^-)>+$$

$$\sum_{i=0}^{n}\left[\frac{2}{(x^- - x_i^-)^2} + \frac{1}{x^- - x_i^-}\,\partial/\partial x_i^-\right]<T(x_1^-)\ldots T(x_n^-)> \qquad (3.2.12)$$

Hence the Liouville action $S_{Liouv}(h) = \log Z(h)$ generates connected correlation functions $<T(x_1^-)\ldots T(x_n^-)>_c$. Connected correlations are produced by the same Ward identities (3.2.12) with the central term omitted

$$\langle T(x_1^-)T(x_2^-)\rangle_c = \frac{c}{2(x_{12}^-)^4}$$

$$\langle T(x^-)T(x_1^-)\ldots T(x_n^-)\rangle_c = \qquad (3.2.13)$$

$$\sum_{i=1}^{n}\left[\frac{2}{(x^- - x_i^-)^2} + \frac{1}{x^- - x_i^-}\,\partial/\partial x_i^-\right]\langle T(x_1^-)\ldots(x_n^-)\rangle_c$$

Note that besides recurrence relations (3.2.13) these connected correlation functions permit more explicit representation. As was mentioned above the Liouville action in particular results from the anomaly in the determinant of the covariant operator $\bar{\partial}_\lambda$, acting from the space of differentials of weight λ to the space of differential forms of weight $(1,\lambda)$. Corresponding central charge is

$$c_\lambda = -2(1+6\lambda-6\lambda^2) \qquad (3.2.14)$$

In the cone gauge corresponding operators are $\partial_+ - h\partial_- - \lambda\partial_- h$ and one finds

$$\det(\partial_+ - h\partial_- - \lambda\partial_- h) = Z_{c_\lambda}(h) \qquad (3.2.15)$$

Determinant representation leads to the following diagram expansion of the Liouville action

$$S_{Liouv}(h)\Big|_{c=c_\lambda} = -\sum_{n=1}^{\infty}\frac{1}{n\pi^n}\left[\;\cdots\;\right] d^2x_1\ldots d^2x_n \qquad (3.2.16)$$

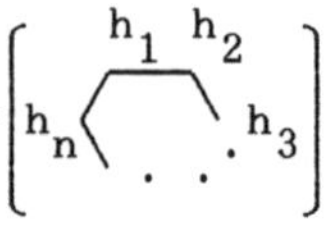

where propagators are $1/x_{ij}^-$ and in each vertex the operator

$$\rangle\cdots = \lambda\overrightarrow{\partial}_- - (1-\lambda)\overleftarrow{\partial}_- \qquad (3.2.17)$$

acts. This is a λ-dependent family of representations for the connected stress tensor correlations. The expansion with $\lambda = 0$ is the most convenient one. Performing vertex differentiation one finds the same diagram expansion as (3.2.16) with propagator $1/(x_{ij}^-)^2$ and unity in the vertex.

Since central charge enters S_{Liouv} simply as a multiplier it is convenient to get rid of it and introduce the standard functional $S_L(h)$, independent on c:

$$S_{Liouv}(h) = \frac{c}{2}\, S_L(h), \qquad (3.2.18)$$

S_L is determined by the following functional relations

$$\delta S_L(h) = -\frac{1}{6\pi}\int u(h)\delta h d^2x$$

$$(\partial_+ - h\partial_- - 2\partial_- h)u = \partial_-^3 h \qquad (3.2.19)$$

and expands in h as follows

$$\delta S_L(h) = \sum_{i=1}^{\infty} \frac{1}{n\pi^n}\Big[\, n$$
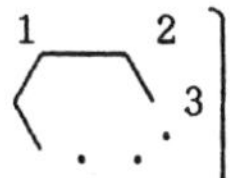
$$\Big]\, h(x_1)\dots h(x_n)d^2x_1\dots d^2x_n \qquad (3.2.20)$$

with propagator $1/(x_{ij}^-)^2$.

In particular the following closed expression for connected stress correlations in conformal field theory holds

$$\langle T_1 \dots T_n\rangle_c = c\sum_{1}^{(n-1)!/2} \frac{1}{(x^-_{C_1C_2}x^-_{C_2C_3}\dots x^-_{C_nC_1})^2} \qquad (3.2.21)$$

where the summation is over all cycles of length n in permutations of n elements (1,2,...,n).

3.3 Potential W(u). On attempting to quantize gravity in the cone gauge one is brought to perform something like functional integral over h weighted with the Liouville action $S_{Liouv}(h)$. Define a generating functional for h-correlations through the following formal expression

$$e^{W(u)} = \int \mathcal{D}h\; e^{S_{Liouv}(h)+\frac{c}{12\pi}\int u(x)h(x)d^2x} \qquad (3.3.1)$$

where u is a classical background stress. In the quasiclassical limit $c \quad -\infty$ one can use the classical equation of motion for h_{cl}

$$\frac{c}{12\pi}\, u + \frac{c}{2}\,\frac{\delta S_L}{\delta h}\,(h_{cl}) = 0 \qquad (3.3.2)$$

Comparing this equation with the first of eqs.(3.2.19) one

finds that it reduces to the second one of eqs.(3.2.19), which should be read now "from right to left", i.e. as the equation for h with u given. Classical action is determined by the relation

$$\delta W_{cl}(u) = \frac{c}{12\pi}\int h(u)\delta u(x)d^2x \qquad (3.3.3)$$

To get rid of the central charge define standard functional $W_L(u)$

$$W_{cl}(u) = \frac{c}{12} W_L(u) \qquad (3.3.4)$$

which is fixed by the c-independent system of relations

$$\begin{aligned} \delta W_L(u) &= \frac{1}{\pi}\int h\delta u \ d^2x \\ (\partial_-^3 + 2u\partial_- + \partial_- u)h &= \partial_+ u \ . \end{aligned} \qquad (3.3.5)$$

Now we are arrived at new functional: potential $W_L(u)$. To get insight to its origin consider the sl(2) current algebra. It contains three currents j^0 and $j^{\pm}$ holomorphically dependent on x^+. Commutation relations are defined by the following operator product expansions

$$\begin{aligned} j^0(x_1^+)j^0(x_2^+) &= \frac{k}{2(x_{12}^+)^2} + \text{reg.} \\ j^+(x_1^+)j^+(x_2^+) &= \text{reg.} \\ j^-(x_1^+)j^-(x_2^+) &= \text{reg.} \\ j^0(x_1^+)j^{\pm}(x_2^+) &= \pm\frac{1}{x_{12}^+} j^{\pm}(x_2^+) + \text{reg.} \\ j^0(x_1^+)j^0(x_2^+) &= \frac{k}{(x_{12}^+)^2} + \frac{2}{x_{12}^+} j^0(x_2^+) + \text{reg.} \end{aligned} \qquad (3.3.6)$$

where k is the central charge of the current algebra. Define the following x^+ and x^- dependent field [33,37]:

$$2H(x) = j^-(x^+) + 2x^-j^0(x^+) - (x^-)^2j^+(x^+) \ . \qquad (3.3.7)$$

This field enjoys the following operator product expansion, which encodes all the commutation relations (3.3.6)

$$H(x)H(0)=-\frac{k}{4}\left(\frac{x^-}{x^+}\right)^2-\frac{x^-}{x^+}\,H(0)-\frac{(x^-)^2}{2x^+}\,\partial_-H(0)\;+\;\text{reg.} \qquad (3.3.8)$$

This expansion can be utilized in the form of Ward identities for H-correlations

$$\langle H(x)H(x_1)\dots H(x_n)\rangle=\sum_{i=1}^{n}-\frac{k}{4}\left(\frac{x^- - x_i^-}{x^+ - x_i^+}\right)^2\langle H(x_1)\dots H(\not{x_i})\dots H(x_n)\rangle-$$

$$\sum_{i=1}^{n}\left(\frac{x^- - x_i^-}{x^+ - x_i^+}+\frac{1}{2}\,\frac{(x^- - x_i^-)^2}{x^+ - x_i^+}\,\frac{\partial}{\partial x_i^-}\right)\langle H(x_1)\dots H(x_n)\rangle \qquad (3.3.9)$$

Using these Ward identities one successively recovers all the multipoint correlation functions of fields H on the infinite plane x^+, x^-. These are nothing but condensed form of the sl(2)-currents correlation functions.

Define the generating functional for these multipoint correlations

$$U(u) =$$

$$\sum_{n=0}^{\infty}\frac{1}{n!\pi^n}\int\langle H(x_1)\dots H(x_n)\rangle u(x_1)\dots u(x_n)d^2x_1\dots d^2x_n. \qquad (3.3.10)$$

It is clear that the functional W(u) = logU(u) generates the "connected" H-correlations $\langle H(x_1)\dots H(x_n)\rangle_c$, which are recovered by Ward identities (3.3.9) without central term

$$\langle H(x_1)H(x_2)\rangle_c = -\frac{k}{4}\left(\frac{x_{12}^-}{x_{12}^+}\right)^2,$$

$$\langle H(x)H(x_1)\dots H(x_n)\rangle_c= \qquad (3.3.11)$$

$$-\sum_{i=1}^{n}\left(\frac{x^- - x_i^-}{x^+ - x_i^+}+\frac{1}{2}\,\frac{(x^- - x_i^-)^2}{x^+ - x_i^+}\,\frac{\partial}{\partial x_i^-}\right)\langle H(x_1)\dots H(x_n)\rangle_c$$

Now it remains to check that the functional obtained this way

$$W(u) =$$

$$\sum_{n=2}^{\infty} \frac{1}{n!\pi^n}\int \langle H(x_1)\ldots H(x_n)\rangle_c u(x_1)\ldots u(x_n) d^2x_1\ldots d^2x_n. \quad (3.3.12)$$

satisfies the same defining relations (3.3.5) as $W_L(u)$. Perturbative derivation proceeds as follows: consider variation

$$\delta W(u) = \frac{k}{2\pi}\int h\delta u\ d^2x \quad (3.\ .13)$$

For h the following diagram expansion holds

$$\frac{k}{2}\ h(x) =$$

$$\sum_{n=1}^{\infty} \frac{1}{n!\pi^n}\int \langle H(x)H(x_1)\ldots H(x_n)\rangle u_1\ldots u_n d^2x_1\ldots d^2x_n. \quad (3.3.14)$$

Differentiating one has

$$\frac{k}{2}\ \partial_-^3 h(x) =$$

$$\sum_{n=1}^{\infty} \frac{1}{n!\pi^n}\int \langle \partial_-^3 H(x)H_1\ldots H_n\rangle u_1\ldots u_n d^2x_1\ldots d^2x_n. \quad (3.3.15)$$

Using Ward identities for connected correlation functions one finds

$$\langle \partial_-^3 H(x)H(x')\rangle_c = \pi\frac{k}{2}\ \partial_+\delta^2(x-x'),$$

$$\langle \partial_-^3 H(x)H(x_1)\ldots H(x_n)\rangle_c =$$

$$-\sum_{i=1}^{n}\left(\pi\partial_-\delta^2(x-x_i) + \pi\delta^2(x-x_i)^{\partial}/_{\partial x_i^-}\right)\langle H(x_1)\ldots H(x_n)\rangle_c \quad (3.3.16)$$

Finally one arrives at

$$\frac{k}{2}\ \partial_-^3 h = \frac{k}{2}\ \partial_+ u - \frac{k}{2}\ (u\partial_- h + \partial_-(uh))$$

This is exactly the equation (3.3.5) and therefore

$$W(u) = \frac{k}{2}\ W_L(u) \quad (3.3.17)$$

To add to the analogy with the Liouville action it is interesting to construct a determinant representation for the potential $W_L(u)$. It turns out that in this case there is a family of covariant differential operators with the anomaly being reducible to the functional $W_L(u)$. The simplest example is the operator

$$\partial_-^2 + \frac{u}{2} \tag{3.3.18}$$

This observation leads in particular to the following diagram expansion for $W_L(u)$

$$W_L(u) = 2\sum_{n=2}^{\infty} \frac{(-)^{n+1}}{n(2\pi)^n} \left[n \right.$$

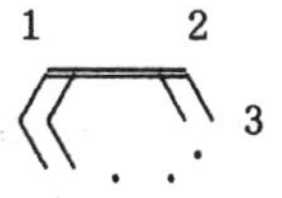

$$\left. \right] u_1 \ldots u_n d^2x_1 \ldots d^2x_n \tag{3.3.19}$$

where the double line corresponds to propagator x_{ij}^-/x_{ij}^+ and the vertex is equal to 1. In the next section several more complicated determinant representations are suggested.

To close the section it is worth to note that recent calculations by KPZ [34] suggest that the abovementioned relation between functionals $S_L(h)$ and $W_L(u)$ holds also in the quantum case. Namely after suitable renormalization of constants the generating functional for quantum correlations of field h reduces to the potential $W_L(u)$. Quantum relation has the form

$$\langle e^{\frac{1}{\pi}\int Th\, d^2x} \rangle_c = e^{\frac{k}{2} W_L(zT)} \tag{3.3.20}$$

where $\langle \ldots \rangle_c$ means the average over the fluctuations of the induced gravity with the central charge c while k and z are in the following relation to c

$$c - 13 = 6\left(k+2 + \frac{1}{k+2}\right) \tag{3.3.21}$$

$$z = \frac{2}{k+2} \ . \tag{3.3.22}$$

3.4. <u>Determinants of covariant operators</u>. Consider the symmetric operator (3.3.18) in the two-dimensional space with coordinates x^+, x^-. To study the local properties of its determinant introduce commutative (bosonic) field $\varphi(x)$. Then

$$\det^{-1/2}(\partial_-^2 + \frac{u}{2}) = \int \mathcal{D}\varphi \; e^{\frac{1}{2\pi}\int \varphi(\partial_-^2 + \frac{u}{2})\varphi \, d^2x} \tag{3.4.1}$$

In the free field theory defined by (3.4.1) the unperturbed propagator (for $u = 0$) is

$$\langle\varphi(x)\varphi(0)\rangle = -\frac{x^-}{x^+} \tag{3.4.2}$$

while the interaction with the external field $u(x)$ is through the current $H(x)$ defined as $\delta\text{Action} = \frac{1}{\pi}\int H(x)\delta u(x) d^2x$, where

$$H(x) = \frac{1}{4}\varphi^2 \tag{3.4.3}$$

In unperturbed theory the two-current operator product expansion is

$$H(x)H(0) = \frac{1}{8}\left(\frac{x^-}{x^+}\right)^2 - \frac{x^-}{x^+}H(0) - \frac{(x^-)^2}{2x^+}\partial_- H(0) + \text{reg.} \tag{3.4.4}$$

Comparing this with (3.3.8) we conclude that the theory under consideration exhibits the current algebra with $k = -1/2$ and the effective action in external field u is generated by the connected current correlations. Therefore

$$\det^{-1/2}(\partial_-^2 + \frac{u}{2}) = e^{-\frac{1}{4}W_L(u)} \tag{3.4.5}$$

In particular the standard determinant expansion lead to diagram representation (3.3.18) for the potential $W_L(u)$.

The nonperturbative derivation of this relation is also possible. In nonzero external $u(x)$ the field φ satisfy the linear equation of motion

$$(\partial_-^2 + \frac{u}{2})\varphi(x) = 0 \tag{3.4.6}$$

Classically this equation lead to the following third order current conservation equation

$$(\partial_-^3 + 2u\partial_- + \partial_- u)H(x) = 0 \qquad (3.4.7)$$

We shall see now that in the quantum case there is an anomalous term in the r.h.s. of (3.4.7) and calculate it. To make expression (3.4.3) sensible in quantum case define current through the operator product expansion

$$H(x) = \frac{1}{4} \overline{\lim_{\varepsilon \to 0}} \left[\varphi(x+\varepsilon)\varphi(x) + {}^{\varepsilon^-}\!/_{\varepsilon^+}\right] \qquad (3.4.8)$$

Consider the two-point expectation value $G(x,x') = \langle\varphi(x)\varphi(x')\rangle$. This symmetric in x and x′ function satisfy the equation

$$(\partial_-^2 + \frac{1}{2} u(x))G(x,x') = \pi\delta^2(x-x') \qquad (3.4.9)$$

Let $\Phi_1(x^-,x^+)$ and $\Phi_2(x^-,x^+)$ by a pair of linear independent zero modes of the operator (3.3.18), non-singularly parameterized by the coordinate x^+ and normalized by the condition $\Phi_1(x)\partial_-\Phi_2(x) - \partial_-\Phi_1(x)\Phi_2(x) = 1$. Then

$$G(x,x') = \frac{\Phi_1(x)\Phi_2(x')-\Phi_2(x)\Phi_1(x')}{(x-x')^+} + \Omega(x,x') \qquad (3.4.10)$$

where $\Omega(x,x')$ is zero mode of (3.3.18) in each variable x^- and $(x')^-$. By definition (3.4.8)

$$4\langle H(x)\rangle = \partial_+\Phi_1(x)\Phi_2(x) - \Phi_1(x)\partial_+\Phi_2(x) + \Omega(x,x) \qquad (3.4.11)$$

With this representation one finds

$$(\partial_-^3 + 2u\partial_- + \partial_- u)\langle H(x)\rangle = -\frac{1}{4}\partial_+ u \qquad (3.4.12)$$

More careful analysis shows that (3.4.12) holds not only for vacuum expectation but in the operator sense, i.e. for every correlation function.

Note that the operator (3.3.18) enjoys the following covariance property. Consider the group of coordinate x^- diffeomorphisms: $x^- \to f(x^-)$ (they may depend on x^+ as a parameter; presently this is not important). If u transforms as the projective connection under these substitutions, i.e.

$$\delta u = \omega\, \partial_- u + 2\partial_-\omega\, u + \partial_-^3\omega \tag{3.4.13}$$

for infinitesimal $f = x^- + \omega$, then this operator acts covariantly from the space of differentials of weight $j = -1/2$, i.e. quantities that transform as

$$\delta\psi_j = \omega\, \partial_-\psi_j + j\partial_-\omega\, \psi_j \tag{3.4.14}$$

to the space of differentials of weight 3/2. In this sense the action in (3.4.1) is invariant under these diffeomorphisms. It is also invariant under substitutions $x^+ \to x^+ + \varepsilon(x^+)$, if one considers φ as a differential of weight 1/2 under transformations of this type.

Classically the current $H(x)$ defined by (3.4.3) transforms as a differential of weight -1 under x^- coordinate substitutions. Like stress tensor in conformal field theory the quantum current (3.4.8) exhibits an inhomogeneous term under these transformations

$$\delta H = \omega\, \partial_- H - \partial_-\omega\, H + \frac{k}{2}\, \partial_-^3\omega \tag{3.4.15}$$

For finite substitutions it has the form

$$H(f) = \frac{\partial f}{\partial x^-}\, H(x) - \frac{k}{2}\, \frac{\partial f}{\partial x^+} \tag{3.4.16}$$

The antisymmetric third order differential operator $\partial_-^3 + 2u\partial_- + \partial_- u$, which we have already met in (3.3.5), enjoys the similar covariance property: it acts from the space of differentials of weight -1 to the space of weight 2 differentials. Anomaly of its determinant also reduces to functional $W_L(u)$. Introduce anticommuting fields Φ and ψ

$$\det(\partial_-^3 + 2u\partial_- + \partial_- u) = \int \mathcal{D}\Phi\mathcal{D}\psi\; e^{\frac{1}{\pi}\int \Phi(\partial_-^3 + 2u\partial_- + \partial_- u)\psi\, d^2x} \tag{3.4.17}$$

The action here is also invariant under x^- diffeomorphisms. Now the unperturbed two point function is

$$\langle\Phi(x)\Psi(0)\rangle = -\frac{1}{2}\frac{(x^-)^2}{x^+} \qquad (3.4.18)$$

and the current has the form

$$H(x) = \Phi\,\partial_-\Psi - \partial_-\Phi\,\Psi \qquad (3.4.19)$$

The operator product expansion of two currents is again of the form (3.3.8) with k = 4. Therefore

$$\det(\partial_-^3 + 2u\partial_- + \partial_- u) = e^{2W_L(u)} \qquad (3.4.20)$$

There is a remarkable family of differential operators in ∂_- with the abovestated covariance property. Consider the space of differential operators of order 2j+1 with j non-negative integer or half-integer number and with coefficients containing homogeneously the only function $u(x^-)$ and its x^--derivatives. They are homogeneous in the following sense: operator $\partial/\partial x_-$ weights 1 and the weight of the function u is 2 (then $\partial_- u$ weights 3, etc.). Then there is a unique operator D_j (up to an overall constant) of this type which act covariantly from the space of differentials of weight -j to the space of weight j+1 differentials.

Requiring this property one finds few simple examples

$$\begin{aligned}
D_{1/2} &= \partial_-^2 + \frac{1}{2}u \\
D_1 &= \partial_-^3 + 2u\partial_- + \partial_- u \\
D_{3/2} &= \partial_-^4 + 5u\partial_-^2 + 5\partial_- u\partial_- + \frac{3}{2}\partial_-^2 u + \frac{9}{4}u^2 \qquad (3.4.21) \\
D_2 &= \partial_-^5 + 10u\partial_-^3 + 15\partial_- u\partial_-^2 + 9\partial_-^2 u\partial_- + 16u^2\partial_- + 2\partial_-^3 u + 16u\partial_- u
\end{aligned}$$

Calculations become rather involved for higher order operators. They can be simplified considerably by the use of the following observation. The differential operators under consideration are related to the null-vectors in the Verma modules of the Virasoro algebra (1.26) in the so called classical limit [11]. Consider the Virasoro Verma module and tend the value of the central charge $c \to -\infty$, keeping the highest weight Δ and the level under consideration to be finite. The series of the Kac curves (1.32) $\Delta_{(1,n)}(c)$ in this limit tend to finite weights $\Delta_{(1,n)}(c \to -\infty) = -(n-1)/2$ and the corresponding null-vectors can be associated with the operators described above.

In this sense the space of differential operators of the type described above, acting on the differential of weight j may be associated with the highest weight representation of the "classical" Virasoro algebra. To reach this classical limit (i.e. the limit $c \to -\infty$) one should substitute in the commutation relations of the Virasoro algebra (1.26) as follows

$$L_m = \frac{c}{12}\,\ell_m \qquad \text{for} \quad m \le 2\,,$$
$$L_m = \ell_m \qquad \text{for} \quad m > 2 \tag{3.4.22}$$

and in the resulting structure constants set $c \to \infty$. The limiting commutation relations are free of parameter c and one results in the following Lie algebra

$$[\ell_m, \ell_n] = (m-n)\ell_{m+n} \quad \text{if } (m > -2 \text{ and } (n > -2 \text{ or } m+n \le -2)) \text{ or } (n > -2 \text{ and } m+n \le -2)$$
$$[\ell_m, \ell_{-m}] = m^3 - m \qquad \text{if } |m| > 1 \tag{3.4.23}$$
$$[\ell_m, \ell_n] = 0 \qquad \text{rest cases}$$

We shall not analyze in detail the structure of this algebra, but simply note, that all the operators ℓ_m with $m \le -2$ are commutative and may be represented as multiplication by functions, for instance $\ell_{-2} \to u(x)$. With the representation of ℓ_{-1} as $\partial/\partial x$ the commutation relations lead to

$$\ell_{-n-2} \to \frac{1}{n!}\partial^n u \qquad n \ge 0$$

and the arbitrary vector in the Verma module becomes the differential operator with the coefficients constructed of u and its derivatives, acting on a differential of the highest weight. The action of ℓ_n's with positive n in the module is defined completely by the following rules

$$\begin{aligned}
\ell_0\Phi_j &= (j+n)\partial^n\Phi_j \\
\ell_1\partial^n\Phi_j &= n(2j+n-1)\partial^{n-1}\Phi_j \\
\ell_2\partial^n\Phi_j &= n(n-1)(3j+n-2)\partial^{n-2}\Phi_j
\end{aligned} \tag{3.4.24}$$

where Φ_j is the highest vector of weight j, and the commutation relations

$$\begin{aligned}
[\ell_0,\partial^n u] &= (n+2)\partial^n u, && \text{for } n \geq 0 \\
[\ell_1,\partial^n u] &= n(n+3)\partial^{n-1}u, && \text{for } n \geq 0 \\
[\ell_2,u] &= 6, \quad [\ell_2,\partial^n u] = n(n-1)(n+4)\partial^{n-2}u, && \text{for } n \geq 1
\end{aligned} \tag{3.4.25}$$

The covariant differential operator of the order 2j+1 is the null-vector in the module of highest weight -j (it is annihilated by all ℓ_n's with positive n.

This approach can be used to calculate the covariant differential operators. In particular one can find in general the linear in u term of the operator D_j

$$D_j = \partial^{2j+1} + \frac{j(j+1)(2j+1)}{3}\sum_{k=0}^{2j-1} \frac{6C_k^{2j-1}}{(k+2)(k+3)}\,\partial^k u\partial^{2j-1-k} + O(u^2) \tag{3.4.26}$$

This leads to the following expression for the zero field (u = 0) current

$$H_j = \sum_{k=0}^{2j-1} \frac{(2j-k)(k+1)}{2}\,(-)^k\partial^k\Phi\,\partial^{2j-1-k}\Psi \tag{3.4.27}$$

with the anticommuting fields Ψ and Φ. In this normalization the two-point function is

$$\langle\Phi(x)\Psi(0)\rangle = -\frac{1}{(2j)!}\,\frac{(x^-)^{2j}}{x^+} \tag{3.4.28}$$

and one finds that the two-current operator product expansion is of the form (3.3.8) with

$$k_j = \frac{2j(2j+1)(2j+2)}{6} \tag{3.4.29}$$

This suggests that

$$\det D_j = \exp \frac{k_j}{2} W_L(u) \qquad (3.4.30)$$

One question naturally arises after the last calculation. It looks like the answer (3.4.30) holds for the operator (3.2.26) with the nonlinear in u terms truncated, i.e.when the expression (3.4.27) is exact. Now, what is the role of the nonlinear terms? The perturbative analyses of few first operators suggests the following answer. For $j > 1$ the two-current operator product expansion (3.3.8) is not correct as the relation between the distributions and holds only up the functions with local support. The δ-functions and their derivatives actually emerge in the r.h.s of (3.3.8). They result in the contact terms in the diagram expansion of the effective potential. The nonlinear in u terms are designed in just the way to cancel all the contact terms contributions. To support this picture one can perform the nonperturbative derivation of the current anomaly, as it was done for the second order operator, and find that the correct form of the anomaly is characteristic exactly for the currents of the covariant operators.

3.5. <u>One-loop quantum gravity</u>. It was shown by A.M.Polyakov (unpublished) that the expression (3.4.20) for the third order determinant can be used to calculate quickly the one-loop quantum renormalization of the constants (3.3.21) and (3.3.22). Up to this order they are

$$\begin{aligned} k &= \frac{c-25}{6} - \ldots \\ z &= \frac{2}{k+2} = 12/c + 156/c^2 + \ldots \end{aligned} \qquad (3.5.1)$$

This is to be compared with the result of the quasiclassical evaluation of the functional integral (3.3.20)

$$\langle e^{\frac{1}{\pi}\int Th\, d^2x} \rangle_c = \int \mathcal{D}h\; e^{S_{Liouv}(h) + \frac{1}{\pi}\int T(x)h(x)d^2x} \tag{3.5.2}$$

In the zero-loop order this is estimated by the classical solution h_{cl}, i.e. the solution of the equation

$$(\partial_-^3 + 2u\partial_- + \partial_- u)h_{cl} = \partial_+ u \tag{3.5.3}$$

with the identification $T = \frac{c}{12}\, u$

$$e^{S_{Liouv}(h_{cl}) + \frac{1}{\pi}\int T(x)h_{cl}(x)d^2x} = e^{\frac{c}{12}\, W(\frac{12}{c}\, T)} \tag{3.5.4}$$

Expanding $h = h_{cl} + f$ one has to the one-loop order the Gaussian integral

$$\int \mathcal{D}f\; e^{\int \frac{\delta^2 S_{Liouv}(h_{cl})}{\delta h(x)\delta h(x')} f(x)f(x')d^2x\, d^2x'} =$$

$$\det^{-1/2}\left[\frac{\delta^2 S_{Liouv}(h_{cl})}{\delta h(x)\delta h(x')}\right] \tag{3.5.5}$$

On the other hand $\delta S_L/\delta h(x) = -\frac{1}{6\pi}\, u(x)$, where u is implicitly determined by h_{cl} through eq.(3.5.3). Therefore, the one-loop determinant is $\det^{-1/2}\left[\delta u(x)/\delta h(x')\right]$. Differentiating eq. (3.5.3) one finds this to be

$$\det^{-1/2}\left[(\partial_+ - h\partial_- - 2\partial_- h)^{-1}(\partial_-^3 + 2u\partial_- + \partial_- u)\right] \tag{3.5.6}$$

Determinants for both operators entering (3.5.6) are ready:

$$\begin{aligned} \det(\partial_+ - h\partial_- - 2\partial_- h) &= e^{-\frac{c_2}{2}\, S_L(h)} = e^{-13 S_L(h)} \\ \det(\partial_-^3 + 2u\partial_- + \partial_- u) &= e^{-\frac{k_1}{2}\, W_L(u)} = e^{2W_L(u)} \end{aligned} \tag{3.5.7}$$

and for one-loop contribution we have

$$\det^{-1/2}(\ldots) = e^{-\frac{13}{2}\, S_L(h_{cl}) - W_L(\frac{12}{c}\, T)} \tag{3.5.8}$$

Combining with the classical contribution one has

$$e^{\frac{c-25}{12} W_L(\frac{12}{c} T)+\frac{13}{\pi c}\int Th_{cl} d^2x} \tag{3.5.9}$$

Recall now that $\delta W/\delta u$ is $\frac{1}{\pi}$ h. So, to this order (3.5.9) reduces to

$$e^{\frac{c-25}{12} W_L\left[\frac{12}{c}(1+\frac{13}{c})T\right]} \tag{3.5.10}$$

in agreement with the expansion (3.5.1).

References

1. Patashinskii A.Z., Pokrovskii V.L., Fluctuation theory of Phase Transitions, Pergamon Press, Oxford, 1979.
2. Kadanoff L.P., Proc. Int. School Phys. "Enrico Fermi", Course L.I., Acad.Press, New York-London, 1971.
3. Gribov V.N., Migdal A.A., ZhETF, 1968, 55, p.1498.
4. Migdal A.A., ZhETF, 1968, 55, p.1964.
5. Polyakov A.M., ZhETF, 1970, 59, p.542.
6. Kadanoff L.P., Phys.Rev.Lett., 1969, 23, p.1430.
8 Polyakov A.M., ZhETF, 1969, 57, p.271.
9. Wilson K.G., Phys.Rev., 1969, 179, p.1499.
10. Polyakov A.M., ZhETF, 1974, 66, p.23.
11. Belavin A.A., Polyakov A.M., Zamolodchikov A.B. Nucl.Phys. 1984, B241, p.333.
12. Gelfand I.M., Fuks D.B., Funkts. analiz i prilozh., 1968, 2, p.92.
13. Mandelstam S., Phys. Rep., 1974, 13C, p.259.
14. Schwarz J.H., Phys.Rep., 1982, 89, N4, p.223.
15. Green M.B., Schwarz J.H., Nucl Phys., 1983, B218, p.43.
16. Kac V., Lect.Notes in Physics, 1979, 94, p.441.
17 Feigin B.D., Fuks D.B., Funkts. analiz i prilozh., 1982, 16, p.47.
18. Feigin B.D., Fuks D.B., Funkts. analiz i prilozh., 1983, 17, p.91.
19. Zamolodchikov Al.B., Preprint ITEP, 146-88, 1988.
20. Cardy J.L., Phys.Rev.Lett., 1985, 54, p.1354.
21. Friedan D., Shenker S.H., EFI preprint, 86-18A, 1986.

22. Friedan D., Martinec E., Shenker S., Nucl.Phys., 1986, B271, p.93.

23. Eguchi T., Ooguri H., Nucl.Phys., 1987, B282, p.308.

24. Itsikson C., Zuber J.B., Nucl.Phys., 1986, B275[FS17], p.580; Capelli A., Itsikson C., Zuber J.B., Sacley Preprint, PhT/86-122, 1986.

25. Cardy J.L., Nucl.Phis., 1986, B270[FS16], p.186.

26. Rocha-Charidi A., in "Vertex Operators in Math. and Phys." ed.by J.Lepousky, S.Mandelstam and I.Singer, Springer Verlag, N.Y., 1985.

27. Eguchi T., Ooguri H., Phys.Lett., 1988, B203, p.44.

28. Leng S. Introduction to Modular Forms. Springer-Verlag, 1976.

29. Zamolodchikov Al.B., Nucl.Phis., 1987, B285[FS19], p.481.

30. Bershadski M.A., Radul A.O., Inter.J.Modern.Phys., 1987, 2A, p.165.

31. Knizhnik V.G., Commun.Math.Phys., 1987, 112, p.567.

32. Dotsenko Vl.S., Fateev V.A., Nucl.Phys., 1984, B240[FS12] p.311; 1985, B251[FS13] p.691.

33. Polyakov A.M., Mod.Phys.Lett., 1987, A2, p.893.

34. Knizhnik V.G., Polyakov A.M., Zamolodchikov A.B., Mod.Phys. Lett., 1988, A3, p.819.

35. Alekseev A., Shatashvili S. LOMI preprint, E-16-88, 1988.

36. Yoshida K. INFN preprint N600, 1988.

37. Zamolodchikov A.B., Fateev V.A., YaF, 1986, 43, p.75.

Alexey Zamolodchikov, Institute of Theoretical and Experimental Physics, Moscow 117259, USSR.

THE COULOMB GAS REPRESENTATION OF MINIMAL MODELS ON RIEMANN SURFACES

Omar Foda

Institute for Theoretical Physics
University of Nijmegen, Nijmegen, NL

Abstract

The Coulomb gas representation of the A_n-series of $c = 1-6/(m(m+1)), m \geq 3$, minimal models is extended to compact Riemann surfaces of genus $g > 1$. An integral representation of the partition functions, for any m and g, is obtained as the difference of two gaussian correlation functions of a background charge = (background charge on sphere) $\times$ ($1 - g$), and screening charges integrated over the surface. The coupling constant $\times$ (compactification radius)^2 of the gaussian expressions are, as on the torus, $m(m+1)$, and $m/(m+1)$.

The partition functions obtained are modular invariant, have the correct conformal anomaly and—restricting the propagation of states to any single handle—one verifies explicitly the decoupling of the null states. Many problems remain open, in particular one would like to see how the partition function on a given Riemann surface relates consistently to the corresponding functions on lower genera under degeneration.

1. Introduction:

Consider the A_n discrete unitary series of minimal conformal field theories [1], characterized by the central charges

$$c = 1 - \frac{6}{m\,(m+1)}\,, \quad m \geq 3 \tag{1}$$

Its Coulomb gas (CG) representation [2] on the Riemann sphere, and the torus (at the level of the partition functions) is well-understood. It can be obtained [3, 4] starting from the underlying restricted-solid-on-solid (RSOS) models [5], and performing a series of mappings, that recast them as special cases of the six-vertex model, the continuum of which is well-known. However, there are open problems already at the level of the torus. In particular, we do not know how to write down expressions for the most general n-point function on the torus, that explicitly satisfies the requirement of double-periodicity.

One way to address this issue is to re-derive the torus correlation functions in a different approach, and compare results. Since operator insertions on a compact Riemann surface Σ of genus g can be considered as punctures on the surface, with appropriate boundary conditions for the quantum fields that live on it, any correlation function can be obtained from a partition function on a higher genus surface through degeneration, and suitable projections. This is *one* reason to consider formulating the minimal models on $g > 1$ surfaces. Another reason is that modular invariance on multi-loop surfaces should give us more information about the structure of the models, in addition to what we learnt on the torus. In this talk, I wish to outline how the CG representation of the simplest series of minimal models is obtained on higher-genus compact Riemann surfaces. Details can be found in [6]. We proceed as follows:

o Start with a RSOS model, based on the Dynkin diagram of a A_n group, formulated on a lattice with the topology of a compact Riemann surface of genus g .

o In order to map on a CG, we need to define the action of the transfer matrix on the topologically-non-trivial lattice. For that we cut the lattice along surface-dividing cycles, into g separate parts. Each part has the topology of a torus with either one or two punctures.

o Map the configurations on each part, called "patch", on a linear combinations of the configurations of cyclic-RSOS models: models that are based on the Dynkin diagrams of the affine groups $\hat{A}_n$.

o Re-attach the punctured tori to recover the original Riemann surface.

o Map the cyclic RSOS configurations, on the multi-loop surface, on a CG.

Our result will be in the form of the difference of two expressions. Each of these is a gaussian functional integral of a scalar field compactified on a circle, and coupled to the curvature of underlying Riemann surface, which results in the appearence of a background charge that should be included in the functional integral. Furthermore, there appear screening charges, integrated over the surface, that ensure charge neutrality.

2. The Coulomb Gas:

Mapping the minimal models, represented as critical RSOS models, on a Coulomb gas has been reviewed in detail in [4]. The basic intermediate steps in mapping on a CG are as follows [7, 3, 4]:

⋆ The Random Cluster Representation: The RSOS configurations, e.g. Fig.1, where the integers assigned to the sites indicate the height variables that satisfy the RSOS conditions [3,4,5], can be mapped on random clusters.

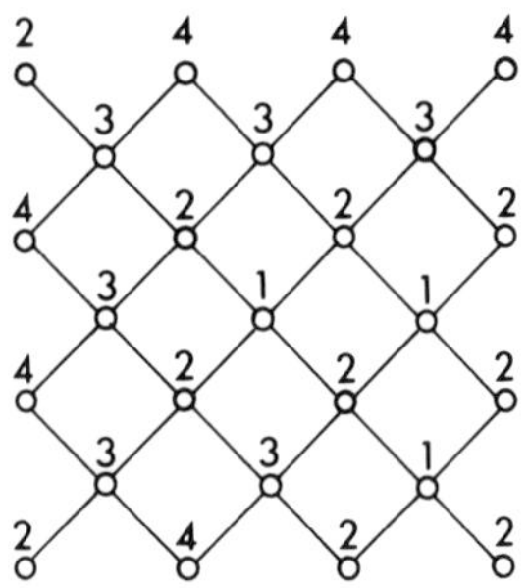

Figure 1: A RSOS configuration.

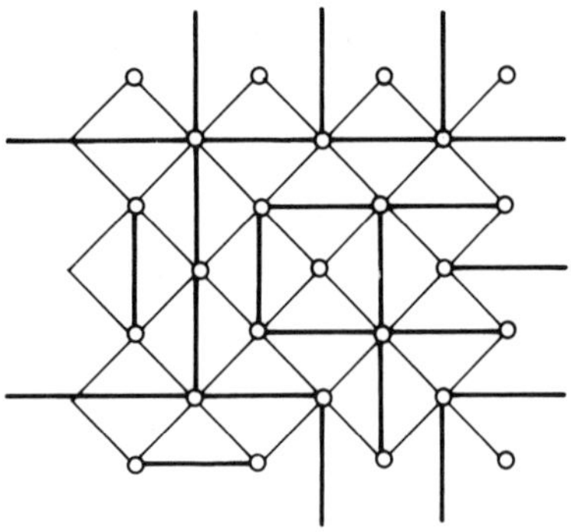

Figure 2: A random cluster configuration.

A typical random cluster configuration is shown in Fig.2. It corresponds to the RSOS configuration in Fig.1. The random clusters inherit the Boltzmann weights of the initial RSOS configurations. They are assigned to them by the action of the transfer matrix, that starts from a given initial one-dimensional state, and generates all possible consistent configurations. Let us call the bonds parallel to the direction of action *horizontal*, and those orthogonal to it *vertical*. One can show [5] that a horizontal bond acquires a Boltzmann weight one. A vertical bond acquires a weight:

$$\left(S_1^{h_i} S_1^{h_{i'}} / S_1^{h_{i-1}} S_1^{h_{i+1}}\right)^{1/2}, \quad h_{i-1} = h_{i+1} \tag{2}$$

where h_i is a height: $1 \leq h_i \leq m$, and $S_1^{h_i} = \sin\left(h_i \pi/(m+1)\right)$ is a component of the eigenvector corresponding to the largest eigenvalue, $2\cos\left(\pi/(m+1)\right)$.

⋆ THE POLYGON REPRESENTATION: The basic step towards a CG representation is to map the clusters on random polygons [7]: Fig.3 :

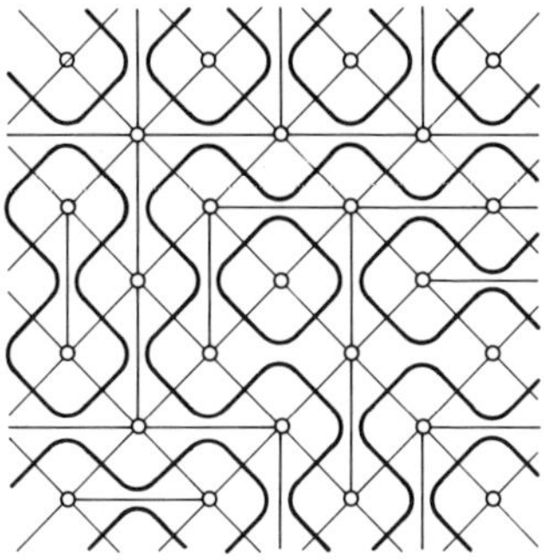

Figure 3: A random polygon configuration.

They eventually become the equipotential lines of the scalar potential of the Coulomb Gas. On arbitrary compact Riemann surfaces, the polygons can be classified, for our purposes, into *three* classes:

- Contractible polygons that enclose no surface curvature.
- Contractible polygons that *do* enclose surface curvature.
- Non-contractible polygons. On multi-loop surfaces, we will distinguish between those that are homologically trivial, i.e. surface-dividing, and homologically non-trivial.

⋆ THE ORIENTED POLYGON REPRESENTATION: One can give the polygons orientations, and sum over all possible combinations of oriented polygons. Regarding a polygon as a domain wall that separates regions of different heights, the orientation of a polygon can be set—by convention—to indicate which region on the side of the polygon is higher: crossing a clockwise-oriented (anti-clockwise oriented) polygons indicates taking a step upwards (downwards).

What we learnt from the lattice approach is that in mapping the minimal models on a CG, on the sphere and the torus, the basic data that characterize the model show up in three different aspects of the final CG representation. These are in one-to-one corespondence with the polygon clases listed above:

- In the specific value of the gaussian coupling constant, or equivalently the compactification radius. This determines the magnitudes of the CG charges.
- In the appearence of extra charges at the conical singularities, on the Riemann sphere. This causes the shift in the central charge.
- In the appearence of additional phase factors that weigh the different soliton

sectors, of the torus partition function, differently. This is the origin of the projection of the null states.

To extend the CG picture to higher genera, we start from an RSOS model defined on a lattice with the topology of a multi-loop Riemann surface, and map that on a CG. We proceed in three steps:

We can readily say something about the first step. Polygons that are both contractible, and do not enclose curvature are "local": they do not know about the curvature of the surface, or its topology. They can be treated exactly as they were on the sphere and the torus, and one ends up with the local vertices of a six-vertex model with a definite coupling constant. In the continuum limit, it renormalizes on a gaussian model with an action

$$A[\varphi] = \frac{\mathrm{g}}{4\pi} \int d^2z \ \sqrt{|\mathfrak{g}|} \ \mathfrak{g}^{z\bar{z}} \ \partial_z \varphi \ \partial_{\bar{z}} \varphi \ , \quad \mathrm{g} = \frac{m}{m+1} \tag{3}$$

$|\mathfrak{g}|$ is the determinant of the metric tensor $\mathfrak{g}^{z\bar{z}}$.

3. The Curvature:

Consider a model defined on a planar region with free boundary conditions. To define it on a cylinder (or a torus) one considers only those configurations with periodic boundary conditions in one direction (or both). On the surface of a cube, one has to do something similar: find a set of boundary conditions, on the plane with boundaries, that simulate the configurations allowed on the surface of the cube.

Take a lattice with RSOS variables on the surface of a cube: Fig.4. :

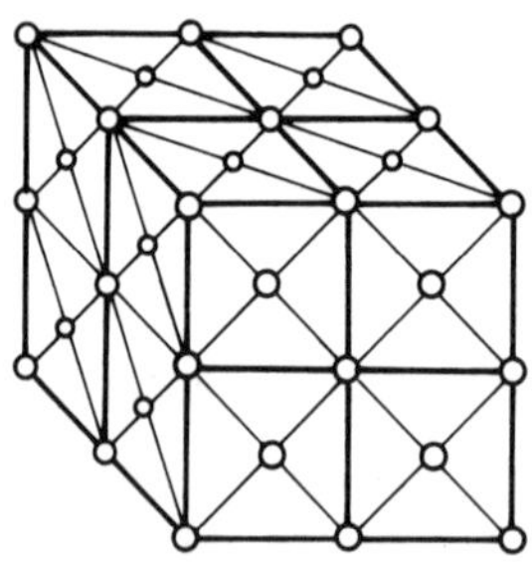

Figure 4: A RSOS lattice on the surface of a cube.

Choose two opposite faces, and open the surface along their diagonals, to obtain a surface with the topology of a cylinder. This can then be opened flat on the plane. To do that, certain sites are *split* as shown in Fig.5:

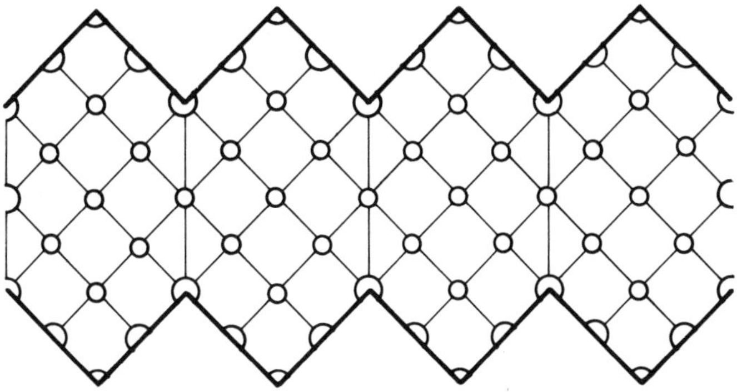

Figure 5: The RSOS lattice on the surface of
the cube opened on the plane, with no *inert* sites.

All information about the shape of the original surface is now in the fact that all parts of a "split-site"share the same height variable. Fill the gaps with *inert* sites: sites connected by *horizontal bonds only*—they do not independently fluctuate: they are bound to carry the same height as the split-sites connected to them by the horizontal bonds, and therefore do not contribute to the dynamics. These are shown in Fig.6:

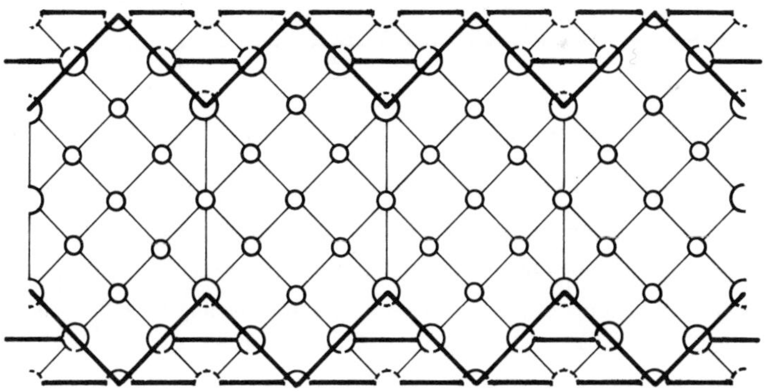

Figure 6: The same surface, as above, with inert sites, and bonds.

Since the horizontal bonds have Boltzmann weight one, the weight of the entire configuration remains unchanged. In a sense, we could have started with free boundary conditions, generated all possible configurations, and retained only those which satisfy the boundary conditions we require. We will refer to the compact surfaces we start with, in general, as Σ, and to the corresponding open surfaces, with inert sites, as Σ_c.

Consider the polygons that live on the lattice of Fig.6. (They are not drawn in the figure.) Computing their weights, one finds that each one has exactly the same weight: $2\cos\Big(2\pi/(m+1)\Big)$ [3,4]. This includes polygons that, on Σ, would enclose a corner, where curvature is concentrated. But these are the weights of *entire* polygons. To obtain local expressions, in order to get closer to a local field theory, we distribute them on the segments that constitute the polygons. To distribute the weights, one goes to the oriented-polygon representation. A clockwise-oriented polygon gets a weight exp $\Big(i\pi/(m+1)\Big)$, counter-clockwise-oriented polygons get exp $\Big(-i\pi/(m+1)\Big)$.

These phases can be split into a product of smaller phases and distributed on the polygon segments in proportion to their line curvature [6]. This must be done on those segments that live on Σ, and not on the fill-up regions $\Sigma_c - \Sigma$. Going back to Σ, and distributing the weights on the *available* segments, one is left with extra phase factors, that translate, in the CG picture, to charge insertions at the conical singularities. The fact that all polygons—including those that enclose corners—have the same weight, is the origin of the background charge introduced in [2]. Consider a square lattice on a genus-two surface Σ as in Fig.7:

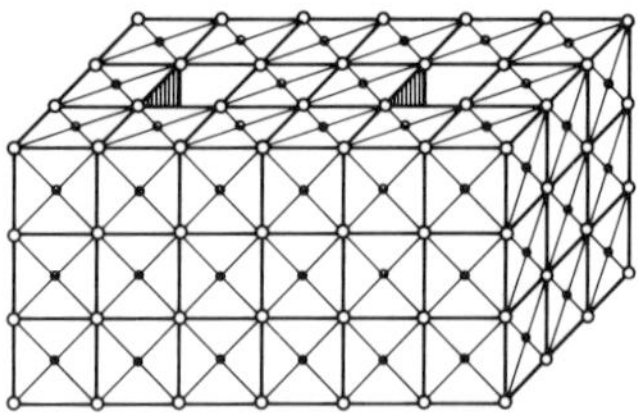

Figure 7: A diagonal square lattice with the topology of a genus-two surface.

Our discussion extends readily to $g > 1$ surfaces. Cut the lattice open in two parts : "patches ", along a surface-dividing cycle. Consider either patch: it is basically a lattice with the topology of a torus, but with one face missing. (Going through this excercise on $g > 2$, some of the patches will have *two* missing faces.) Each patch can now be opened the way one would open a torus:

- Open the patch into a cylinder with missing regions.
- Open the cylinder onto a planar surface. Notice that the latter has *two* types of missing regions:
  - i : regions with boundaries that are due to the way we cut the original surface along its surface-dividing cycles.
  - ii : regions due to opening the cylinder-like surfaces on the plane.

Since regions of the first type are rectangular, one can still define a direction for the action of the transfer matrix in their presence. Initially, we will assign them free boundary conditions. To recover the configurations allowed on Σ, we have to restrict the configurations to match along the boundaries where two patches should be "glued" together in the sense that domain walls should run continuously across the boundaries: "glue-lines". Regions of the second type have triangular parts, and we will fill them with inert sites connected by bonds that are horizontal with respect to the action of the transfer matrix, on that patch, as showm in Fig.8:

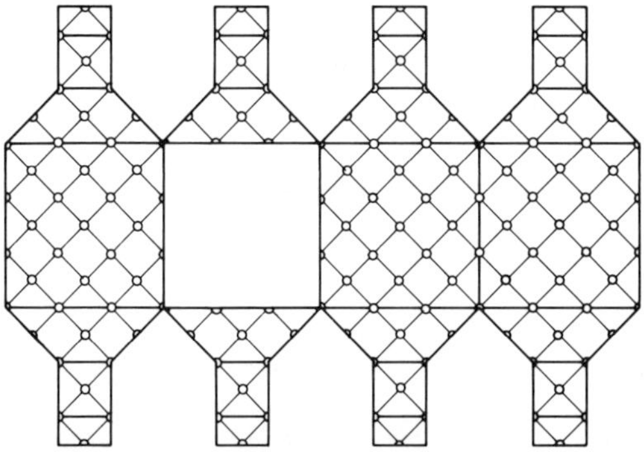

Figure 8: A RSOS model on an open patch of a $g > 1$ Riemann surface.

- Allow the transfer matrix to generate all possible configurations, and retain only those consistent with the Σ boundary conditions.

It is straightforward now to show—following the procedure detailed in [6]—that all contractible loops do indeed get the same weight, and that, on Σ, one is left with extra phase factors, that translate to charges proportional to the background curvature.

The background charges can be included *via* a modification of the action [6]. The modified action is

$$A[\varphi] = \frac{\mathrm{g}}{4\pi} \int d^2 z \, \sqrt{|\mathfrak{g}|} \, \left(\mathfrak{g}^{z\bar{z}} \, \partial_z \varphi \, \partial_{\bar{z}} \varphi \; - \; \frac{i}{4\pi(m+1)} R \varphi \right) \tag{4}$$

Where $\mathrm{g} = m/(m+1)$. The functional integral based on (7) will vanish unless we include screening charges that neutralize the background charge. Their positions $\{z_i\}$ will be integrated over the surface.

4.The Soliton Sectors:

Finally, we come to the non-contractible polygons. As shown in [3], each configuration of non-contractible polygons with a fixed assignment of heights has weight one. To obtain the full weight of a configuration, one has to sum over all height assignments, that are consistent with the RSOS rules. We can think of a height as a colour, and refer to the number of ways one can assign consistent heights as the number of colourings. Let us outline how the mapping of the non-contractible polygons on the torus goes:

- Since the non-contractible polygons do not intersect, one can always perform a set of modular transformations to rearrange them in a convenient way. We will start from configurations that wind only around the a-cycle of the torus: Fig. 8. This set is complete—in the sense that we include all solitons that wind around the a-cycle only—but not modular invariant.
- Re-express the number of allowed RSOS height-assignments in terms of linear combinations of gaussian height-assignments.
- Perform modular transformations to obtain a modular invariant result.

A typical configuration on the torus is given in Fig.9:

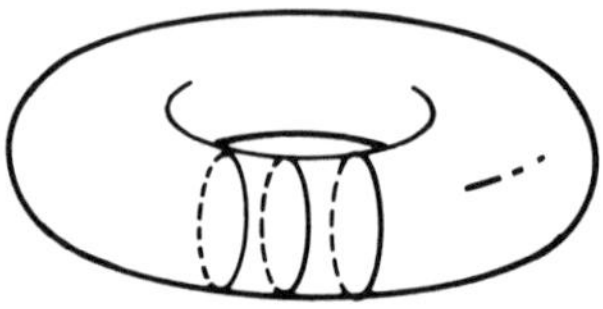

Figure 9: Non-contractible polygons winding on the a-cycle of the torus.

The total number of height assignments is given by the number of closed walks of length N on the corresponding Dynkin diagram. This can be computed using the adjacency matrix of the Dynkin diagram. The answer is the trace of the matrix raised to power$N$:

$$\sum_{m_j=1}^{m} \Big(\cos m_j\pi/(m+1)\Big)^N \tag{5}$$

Next, we follow [3] and re-express eq.(5) in terms of the number of walks, of length N, on the Dynkin diagrams of gaussian models. These are the diagrams of the affine $\hat{\mathrm{A}}_{2n-1}$ groups, Fig. 10:

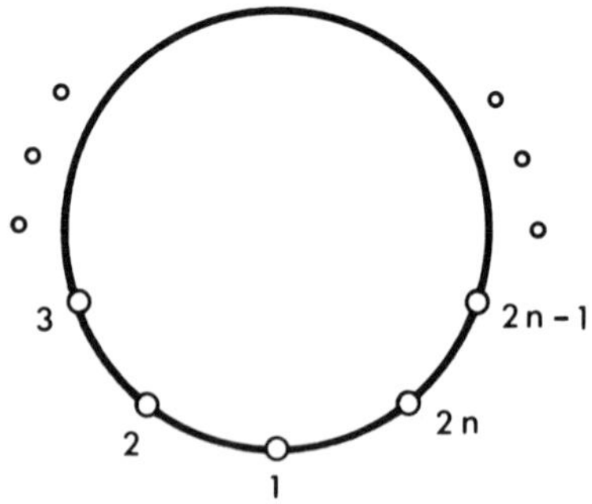

Figure 10: The Dynkin diagram of $\hat{A}_{2n-1}$.

they have the topology of a circle with $2n$ nodes. Now we can consider the non-contractible polygons on multi-loop surfaces. We begin with genus two.

o We choose to start with a set of configurations that resembles the one-loop case most. Let us draw our Riemann surface in a way that emphasizes the fact that it is basically two *punctured* tori, with a connecting cylinder. We start from configurations that are restricted in two ways:

o No polygons wind around the b-cycles of the Riemann surface: This is not a drastic restriction.

o All polygons that wind around the connecting-cylinder to be homologically-trivial. This is the restriction that we will be most interested in.

We are left with configurations of the type shown in Fig. 11:

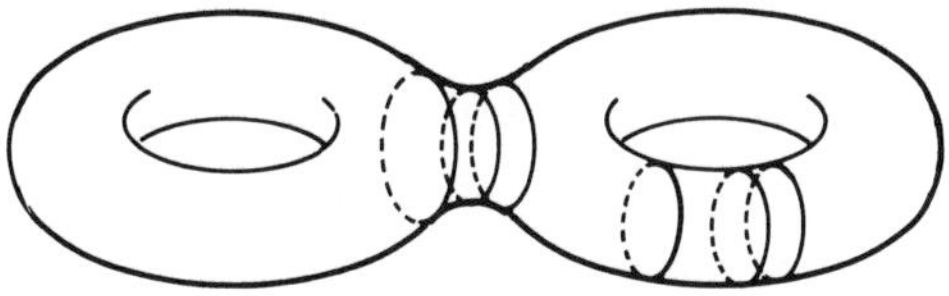

Figure 11: Non-contractible polygons on a $g = 2$ Riemann surface.

These describe closed walks on the Dynkin diagrams that have the topologies shown in Fig.12:

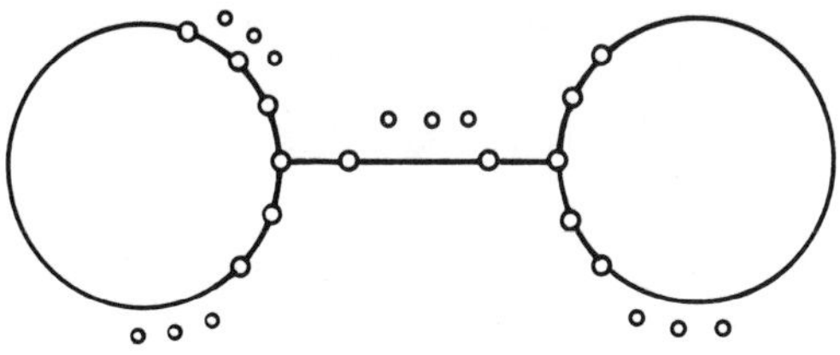

Figure 12: The topology of the walks on Fig. 11, as walks on the Dynkin diagram.

In particular, we have not yet included the configurations in Fig. 13:

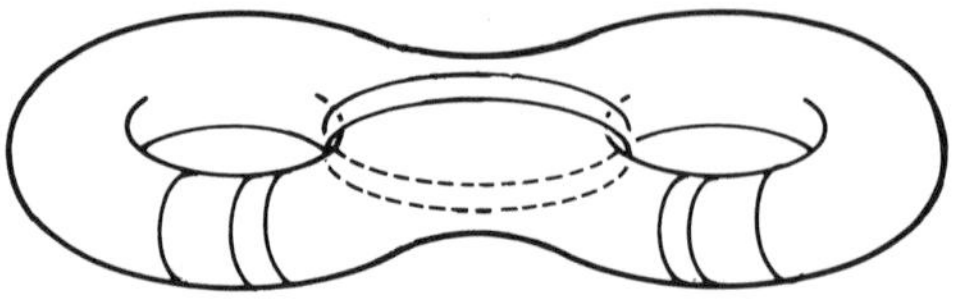

Figure 13: Another class of non-contractible polygons on a $g = 2$ Riemann surface.

They correspond to walks on the Dynkin diagrams, that have the topology shown in Fig.14:

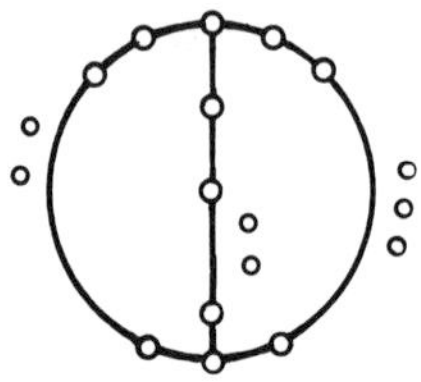

Figure 14: The topology of the walks on Fig. 13, as walks on the Dynkin diagram.

We will show that each of these restrictions can be removed by modular transformations.

Consider the configurations in Fig.11. We wish to express these in terms of walks on affine diagrams. The polygons that originally wind around the connecting cylinder cannot contribute to the discontinuities that form the solitons. The reason is that they are homologically trivial. The fluctuations associated with them will contribute only to the quantum part of the partition function, but not to the classical part that we are now interested in. Thus one can factor them out, and concentrate on the polygons that wind around the a-cycles.

But these are exactly the configurations that appear on the torus. Therefore, we can go ahead and re-express their number in terms of the soliton configurations of gaussian models: the set of all soliton configurations on a given a-cycle can be expressed in terms of the difference between the soliton configurations of two gaussian theories, just as in eq.(8). But here we do not have a single a-cycle, but two. Once we glue the patches, we end up with *four* different types of soliton configurations:

$$\frac{1}{4}\Big\{[\text{ both handles carry solitons of } r = m(m+1)\,]$$
$$-[\text{the first handle carries solitons of } r = m(m+1),$$

the second has solitons of $r = m/(m+1)$]
− [the reverse of the previous situation]
+ [both handles carry solitons $r = m/(m+1)$] } (6)

Is this division of configurations modular invariant? On a multi-loop Riemann surface, we can classify all modular transformations into two types:

a. those which mix cycles that belong to the *same* handle.

b. those which mix the cycles that belong to *different* handles. Since the modular group is generated by the Dehn twists, it has five generators on genus two.

Four of these are of type-a mentioned above. Acting with these on the solitons generated by the polygon configurations in Fig.11, we obtain the set of all possible solitons that live only on one handle or the other, while the polygons on the connecting cylinders remain unchanged. This removes our first restriction on the polygons.

However, the result obtained this way is not modulr invariant. The reason is that we have not yet included those configurations generated by the cycle-mixing fifth generator: we have not yet removed the second restriction on the configurations we started with.

To obtain the fully modular invariant result we should include the orbits generated by the fifth modular transformation. The result is

$$\left(\frac{1}{2}\right)^{g}(\det \mathrm{Im}\tau)^{1/2}\times$$

$$\left\{\sum_{\alpha\beta\gamma}\Theta\begin{pmatrix}\frac{1}{2}\alpha+\frac{1}{2}\beta+\gamma\\0\end{pmatrix}(0|2m(m+1)\tau)\overline{\Theta}\begin{pmatrix}\frac{1}{2}\alpha+\frac{1}{2}\beta+\gamma\\0\end{pmatrix}(0|2m(m+1)\tau)\right.$$

$$\left.-\sum_{\alpha(-\beta)\gamma}\Theta\begin{pmatrix}\frac{1}{2}\alpha+\frac{1}{2}\beta+\gamma\\0\end{pmatrix}(0|2m(m+1)\tau)\overline{\Theta}\begin{pmatrix}\frac{1}{2}\alpha-\frac{1}{2}\beta+\gamma\\0\end{pmatrix}(0|2m(m+1)\tau)\right\} \tag{7}$$

where we have performed a Poisson resummation, and the theta function is defined as

$$\Theta\begin{pmatrix}\alpha\\\beta\end{pmatrix}(z|\tau)=\sum_{n\in\mathbb{Z}^g}\exp[\,i\pi(n+\alpha)\cdot\tau\cdot(n+\alpha)+2\pi i(n+\alpha)\cdot(z+\beta)\,]$$

The characteristics α and β are g-dimensional vectors, and τ is the $g\times g$ period matrix of the genus g Riemann surface. α,β, and γ take the values: $\alpha\in\mathbb{N}_p/p$, $\beta\in\mathbb{N}_q/q$, $\gamma\in\mathbb{N}_2/2$,

where $\mathbb{N}_n$ stands for the set $\{\mathbb{N} \bmod n\}$. The first term in (7) is the multi-loop classical part of a gaussian theory that has g $r^2 = m(m+1)$, the second has g$r^2 = m/(m+1)$

5. The Partition Functions:

We are finally in a position to put write down the partition functions. A minimal RSOS model is mapped, for any m and g, on a linear combination of cyclic RSOS models. Each of these describes a scalar field compactified on a circle, of a prescribed radius, and coupled to the background curvature, in the presence of screening charges V_{α_+} and V_{α_-} that neutralize the background charge. The partition function of the minimal model is given by a linear combination of functional integrals that describe the scalar theories:

$$Z_{min}(m) = \int [d\varphi] \int \cdots \int \prod V_{\alpha_+} \prod V_{\alpha_-}$$
$$\left\{ \exp -(\frac{1}{4\pi}) \Big(m(m+1)\Big) \left(\varphi'\varphi' + i\frac{R}{4\pi(m+1)}\varphi\right) \right\}$$
$$- \int [d\varphi] \int \cdots \int \prod V_{\alpha_+} \prod V_{\alpha_-}$$
$$\left\{ \exp -(\frac{1}{4\pi}) \Big(m/(m+1)\Big) \left(\varphi'\varphi' + i\frac{R}{4\pi(m+1)}\varphi\right) \right\} \tag{8}$$

These functional integrals are precisely of the type encountered in studies of the bosonized ghost system. They can be evaluated following [8–11] as

$$Z^g(m) = \left(\frac{1}{2}\right)^g \delta\left(\sum \alpha_i + 2\alpha_0(g-1)\right) |\det\bar{\partial}_0|^{-1}$$
$$\int \cdots \int d^2z_1 \ldots d^2z_N \prod_{i<j} |E(z_i,z_j)|^{2\alpha_i\alpha_j} \prod_{n=1}^{N} |\sigma(z_i)|^{-4\alpha_0\alpha_i} |F_*|^{-8\alpha_0^2}$$
$$\left\{ \sum_{\alpha\beta\gamma} A^g_{\alpha\beta\gamma}\overline{A}^g_{\alpha\beta\gamma} - \sum_{\alpha\beta\gamma} A^g_{\alpha\beta\gamma}\overline{A}^g_{\alpha(-\beta)\gamma} \right\} \tag{9}$$

where

$$A^g_{\alpha\beta\gamma} = \Theta \begin{pmatrix} \frac{1}{2}\alpha + \frac{1}{2}\beta + \gamma \\ 0 \end{pmatrix} \left\{ \sum_i \alpha_i z_i + 2\alpha_0 \Delta | 2m(m+1)\tau \right\} \tag{10}$$

and Δ is a divisor of degree $(g-1)$: the location of he zeros and the poles of the metric, and the position of the background charges $2\alpha_0$.

This is our result for the partition function fot any m and g. It consists of three components:

∘ A quantum part due to the contractible polygons, and the topologically-trivial deformations of the non-contractible ones. This is proportional to $|\det\bar{\partial}_0|^{-1}$,where $\bar{\partial}_0 = \partial_{\bar{z}}$. They can be consistently absorbed in the quantum part since it is r-independent.

∘ The classical part due to the non-contractible loops. It has a structure that decouples of the null states.

o Interactions terms due to the background charge left over by the contractible polygons that enclose curvature, and the screening charges required for charge neutrality.

The interactions of the screening charges are given by the prime form $E(z_i, z_j)$, and $h(z)$ is a half-differential, with zeros at a set of points $\{R_1, \ldots, R_{p-1}\} \equiv \{R_\alpha\}$. The mapping on a CG is defined up to insertions of these charges. The correct number of insertions is determined by charge neutrality. Since the insertions can be anywhere on the surface, they should be integrated over. The interactions of the screening charges and the background charge are given by σ: a section of a trivial bundle, with no zeros, or poles:

$$\sigma(z) = \frac{h(z)}{\prod_{\alpha=1}^{g-1} E(z, R_\alpha)} \tag{11}$$

For details see [8—11]. Since our lattice naturally defines a metric that is flat everywhere, up to singularities at isolated points, we included a term $|F_*|^{-4\alpha_0^2}$ that describes the interaction of the isolated background charges. This has been given in [11], where a metric similar to our lattice metric has been discussed:

$$F_* = \frac{\prod_{\alpha<\beta} E(R_\alpha, R_\beta)}{\prod_\alpha h'(R_\alpha)} \tag{12}$$

We wish to check the consistency of eq.(9). We immediately see the following:

o It is manifestly modular invariant: It is the difference between two gaussian-like correlation functions. Each of these is seperately manifestly modular invariant: it contains a complete set of solitons, all with the same weight.

o It does *not* have $c = 1$. This can be seen by looking at the ground state. The lowest energy state: the state with no solitons cancels out in the subtraction of the two terms in eq.(9). Thus the true ground state is one with a non-trivial soliton, it has a higher energy, which can be absorbed in a shift of the central charge, just as on the torus . But what is the value of the central charge? Since the central charge can be understood as an anomaly, it is independent of the surface one formulates the theory on. Since the background charge is

(the background charge on the sphere) × (1 – genus)

we see that the energy momentum tensor is the same as on the sphere, and we have the correct conformal anomaly. In [6], a check of the decoupling of the null-states was performed by restricting the propagation of states to an arbitrary single handle, and "shrinking "the rest. The expression obtained this way is identical to the coresponding one-loop partition function, up to an overall normalization. Since the one-loop partition function has decoupled null states, and since one can repeat the same procedure for each handle independently, one verifies explicitly the decoupling of the null states.

ACKNOWLEDGEMENTS:

I would like to thank the organizers Professors Z. Haba, and J. Sobczyk for hospitality at Karpacz, A. Coste, J. Gates, Al. Morozov, H. Römer, B. Schroer, Al. Zamolodchikov, R. Zuchini, and all my Polish friends for their pleasent company and discussions, and R. van Doorn for assistence with the preparation of this TEXscript. This work was supported by'De Stichting voor Fundamenteel Onderzoek der Materie.'

REFERENCES:

[1] C. Itzykson, H. Saleur, and J.-B. Zuber eds., *Conformal Field Theory and Applications in Statistical Mechanics*, World Scientific, 1988.

[2] Vl. S. Dotsenko and V. A. Fateev, **B240** [FS 12] (1984) 312; *ibid* **B251** [FS 13] (1985) 691; Phys. Lett. **154B** (1985) 291.
B.L. Feigin and D.B. Fuchs, Functs. Anal. Prilozhen **16** (1982) 47; *ibid* **17** (1983) 91.

[3] V. Pasquier, J. Phys. **A20** (1987) L1229.
Di Francesco, H. Saleur, and J.-B. Zuber, J. Stat. Phys. **49** (1987) 57.

[4] O. Foda and B. Nienhuis, Utrecht preprint THU-88-34, to appear in Nucl. Phys. B.

[5] G. A. Andrews, R. J. Baxter, and P. J. Forrester, J. Stat. Phys. **35** (1984) 193.

[6] O. Foda, Nijmegen preprint 1989.

[7] B. Nienhuis, in *Phase Transitions and Critical Phenomena* , Vol. 11, C.C. Domb, and J.L. Lebowitz, eds. Academic Press, 1987.
M.P.M. den Nijs, Phys. Rev. **B27** (1983) 1674.

[8] L. Alvarez-Gaumé et. al., Comm. Math. Phys. **106** (1986) 40; *ibid* **112** (1987) 503.

[9] E. Verlinde and H. Verlinde, Nucl. Phys. **B288** (1987) 357.

[10] T. Eguchi and H. Ooguri, Phys. Lett. **187B** (1987) 127.

[11] A. Gerasimov, A. Marshakov, A. Morozov, M. Olshanetsky, and S. Shatashvili, ITEP preprint 1989.

Omar Foda
Institute for Theoretical Physics
University of Nijmegen, Toernooiveld,
6525 ED Nijmegen, The Netherlands.

NON-COMPACT WZW MODELS AND $(QCD)_2$

Zbigniew Haba
Institute of Theoretical Physics, University of Wroclaw

The $(QCD)_2$ has been studied for a long time [1]-[17] as a simplified model of quark interaction. The massless $(QCD)_2$ can be bosonized through the evaluation of the determinant of the Dirac (or Weyl) operator in an external gauge field [7]-[17]. It is equal to the exponential of the Wess-Zumino-Witten (WZW) action [18]. The theories in Minkowski and Euclidean space differ substantially concerning the compactness (resp. non-compactness) of the invariance group. This is a result of a peculiar non-compactness of the chiral transformations of massless fermions noted earlier e.g. by Zumino [19]. We shall restrict ourselves to the Euclidean formulation[20]. At infinite coupling the gluon self-interaction $g^{-2}F^2$, which depends on a dimenional parameter g, is absent. So, what remains is a conformal invariant WZW action (this has been discussed earlier in refs.[10],[21]-[23]).

It appears that the $(QCD)_2$ at infinite coupling corresponding to the interaction of non-abelian gauge fields with left and right handed massless fermions represents a general case of unitary minimal series of conformal field theories. This follows from the description of coset WZW models in terms of fermions discussed in refs.[21]-[23]. Then, the realization of unitary minimal series by coset models goes through the Goddard-Kent-Olive construction [24]-[25] as discussed in refs.[23],[26]-[28].

We restrict ourselves here to Euclidean Dirac fermions. In such a case the functional integral over fermions can be expressed as the WZW model on G^C/G , where G^C is the complexification of the compact gauge group G. We introduce the coordinates ND on G^C/G resulting from the Iwasawa decomposition g=NDK. Then, the functional integral in the WZW G^C/G model can be computed explicitly (this has been discussed for SL(2,C)/SU(2) in our earlier papers [29]-[31], whereas the solubility of the general G^C/G model has been pointed out by Gawedzki and Kupiainen [28]). We calculate the fermionic Green's functions showing that they are scale invariant with a non-canonical scaling dimension .

The $(QCD)_2$ fulfills the Osterwalder-Schrader positivity condition. It can be constructed in Hilbert space. The Hamiltonian has positive spectrum. The reflection positivity is preserved in the limit $g \to \infty$. In this limit the fermionic QCD correlation functions are products of the correlation functions of G^C/G-valued fields, which do not fulfill the reflection positivity, and free Fermi fields. The relation of non-compact coset models to QCD could explain the operator realization of non-compact current algebras and their relation to compact algebras (see refs.[33],[35],[37],[38]).

Let us assume that the gauge field A_r (r=1,2) defined on the sphere (or compactified plane) takes values in an antihermitian representation g of the algebra of a gauge group G (we shall furtheron denote the unitary representation of G also by G). Let G^C be the complexification fo G. Then, we can express A_r by elements of G^C

$$A=A_1-iA_2 = g^{-1}\partial g \qquad g\in G^C \tag{1}$$

where $\partial=\partial_1-i\partial_2$. Then, $\bar{A}=A_1+iA_2= g^+\bar{\partial}(g^+)^{-1}$.

The gauge transformation corresponds to the right translation

$$g \to gU \qquad U\in G \tag{2}$$

The Dirac operator is

$$\not{D}=\gamma^r(\partial_r+A_r)=\begin{bmatrix} 0 & \bar{\partial}+\bar{A} \\ \partial+A & 0 \end{bmatrix} \tag{3}$$

Its Green's function

$$\not{D}^{-1}(z,z')=\begin{bmatrix} 0 & S(z,z') \\ \bar{S}(z,z') & 0 \end{bmatrix} \tag{4}$$

where

$$S(z,z') = g(z)^{-1}g(z')S_0(z,z') \quad \text{and} \quad \bar{S} = g(z)^+(g(z')^+)^{-1}\overline{S_0}(z,z') \tag{5}$$

with

$$S_0(z,z') =\partial^{-1}(z,z') = -(2\pi)^{-1}(\bar{z}-\bar{z}')^{-1} \quad \text{and} \quad \overline{S_0}(z,z') =\bar{\partial}^{-1}(z,z') \tag{6}$$

The determinant of the operator (3) was computed in refs. [13]-[14],[32]

$$\det\not{D} = \exp -W_{WZW}(R^2) \tag{7}$$

where

$$R^2 = gg^+ \tag{8}$$

and

$$W_{WZW}(R^2) = (2\pi)^{-1}\mathrm{Tr}\int R^{-1}\partial R\ R^{-1}\bar{\partial}R\ -i(3\pi)^{-1}\mathrm{Tr}\int(R^{-1}dR)^3 \tag{9}$$

where the second term is integrated over a 3-manifold with the sphere as its boundary.

We are going to compute the functional integral over R using a proper choice of coordinates. It is known (see e.g. ref.[34]) that any positive matrix $R \in G^C$ can be expressed in the form

$$R = NDN^{+} \tag{10}$$

where N is upper triangular and D is positive and diagonal. We shall discuss only SL(2,C) and Sp(n,C). In the first case

$N = \{Z_{ij}\}$ is $n \times n$ matrix where $Z_{ii}=1$ and $Z_{ij}=0$ if $i>j$, and $D=\{\exp\phi_j\}$ (11)

is the diagonal $n \times n$ matrix with the diagonal elements fulfilling the relation $\phi_n=-\phi_1-\ldots-\phi_{n-1}$ (for SL(n,C)).

In the second case

$$N = \begin{bmatrix} 1 & Z \\ 0 & 1 \end{bmatrix} \qquad D = \begin{bmatrix} Y^{-1} & 0 \\ 0 & Y \end{bmatrix} \tag{12}$$

where N is $2n \times 2n$ matrix with $Z \in GL(n,C)$ and Y is a positive $n \times n$ matrix.

Using the transformation property

$$W_{wzw}(gh^{-1}) = W(g) + W(h^{-1}) + 2\mathrm{Tr}\int g^{-1}\partial g\; h^{-1}\bar{\partial}h$$

we get

$$W_{wzw}(R^2) = (2\pi)^{-1}\mathrm{Tr}\int D^{-1}\partial D\; D^{-1}\bar{\partial}D + (2\pi)^{-1}\mathrm{Tr}\int (D^{-1}N^{-1}\bar{\partial}N)^{+}\; N^{-1}\bar{\partial}N\; D \tag{13}$$

Now, in eq.(1)

$$g = RU \tag{14}$$

We increase the number of fermions coupled to the gauge field A_r k-times (flavors). Then, the $(QCD)_2$ action reads

$$W'_{QCD} = -(2g^2)^{-1}\mathrm{Tr}\int F_{12}F_{12} + k\, W_{wzw} \tag{15}$$

where

$$F_{12} = \partial_1 A_2 - \partial_2 A_1 - [A_1, A_2] \tag{16}$$

The conventional functional integral is

$$dA \exp\text{-}W(A)$$

We wish to change the integration variables from A to g (1). Then, the Jacobian J of the transformation $A \rightarrow g$ is (see refs.[11][23][28])

$$J = |\det(\,\partial + [A,\ \ ]\,)|^2 \tag{17}$$

J is equal to $\exp\text{-}W^{ad}_{wzw}$, where the gauge field A is taken in the adjoint representation. Denoting the relative weight of the adjoint representation to the G-representation by c(G) we get for the functional measure the expression

$$d\mu = dUdR \exp\text{-}W_{QCD} \tag{18}$$

where W_{QCD} is equal to W'_{QCD} (15) with k replaced by k+c(G)

SU(2) gauge group

For G=SU(2) ([29]-[31],[38])

$$N=\begin{bmatrix} 1 & Z \\ 0 & 1 \end{bmatrix} \qquad D=\begin{bmatrix} \exp\text{-}\phi & 0 \\ 0 & \exp\phi \end{bmatrix}$$

$$W_{WZW}=\frac{1}{\pi}\int\left[\partial\phi\bar{\partial}\phi + \exp(-2\phi)\,\partial\bar{Z}\,\bar{\partial}Z\right] \tag{19}$$

Next

$$F_{12}= U^{-1}\partial(R^{-1}\bar{\partial}R)U \tag{20}$$

So, that

$$TrF_{12}F_{12}= Tr\left[\partial(R^{-1}\bar{\partial}R)\right]^2 \tag{21}$$

Let us introduce new coordinates for R

$$v = \exp\text{-}\phi\ \bar{\partial}Z \tag{22}$$

The WZW action (19) in the variables (22) becomes quadratic

$$W_{WZW}=\frac{1}{\pi}\int\left[\partial\phi\bar{\partial}\phi + v\bar{v}\ \right) \tag{23}$$

Now, let us compute the matrix entering the Green's function (5)

$$R(z)^{-1}R(z') = \begin{bmatrix} \alpha & \beta \\ \nu & \mu \end{bmatrix} \tag{24}$$

where

$$\alpha = \exp(\phi(z)-\phi(z')\ ) \quad - Z(z)\left[\bar{Z}(z')-\bar{Z}(z)\right] \exp(-\phi(z)-\phi(z')\ )$$
$$\mu = \exp(\phi(z')-\phi(z)\ ) \ + Z(z')\left[\bar{Z}(z')-\bar{Z}(z)\right] \exp(-\phi(z)-\phi(z')\ )$$
$$\nu = \exp(-\phi(z)-\phi(z'))\ \left[\bar{Z}(z')-\bar{Z}(z)\right]$$
$$\beta = \exp(\phi(z)-\phi(z'))\ Z(z') - \exp(\phi(z')-\phi(z))\ Z(z) - \exp(-\phi(z)-\phi(z'))\ Z(z)Z(z')\left[\bar{Z}(z')-\bar{Z}(z)\right] \tag{25}$$

We are going to compute the correlation functions of Fermi fields. These correlation functions depend on the gauge degrees of freedom U. The dU integral of gauge dependent variables is equal to zero. So, we should either fix the gauge or work with gauge invariant variables. We shall do the latter computing expectation values not of $\bar{\psi}$'s themselves but of $U\bar{\psi}$ e.g. consider (we suppress the color and flavor indices)

$$\left\langle U(z)\bar{\psi}_1(z)\ \psi_2(z')U(z')^{-1}\right\rangle = \left\langle U(z)S(z,z')U(z')^{-1}\right\rangle = S_0(z,z')\left\langle R(z)^{-1}R(z')\right\rangle \tag{26}$$

and

$$\left\langle U(z_1)\bar{\psi}_1(z_1)\psi_2(z'_1)U(z'_1)^{-1}U(z_2)\bar{\psi}_1(z_2)\psi_2(z'_2)U(z'_2)^{-1}\right\rangle =$$
$$=S_0(z_1,z'_1)S_0(z_2,z'_2)\left\langle R(z_1)^{-1}R(z'_1)R(z_2)^{-1}R(z'_2)\right\rangle$$
$$- S_0(z_1,z'_2)S_0(z_2,z'_1)\left\langle R(z_1)^{-1}R(z'_2)R(z_2)^{-1}R(z'_1)\right\rangle \tag{27}$$

Furtheron for computation of the expectation values it is convenient to change the variables from Z to v (eq.(22)). This change of variables brings the Jacobian (see refs.[28]-[31])

$$J=|\det[\exp(-\phi)\partial]|^2= \exp\tfrac{1}{2\pi}\int\partial\phi\bar{\partial}\phi \tag{28}$$

The efffect of J is to change the coefficient in front of the $\partial\phi\bar{\partial}\phi$ term in the action (19)

$$k \rightarrow k^{*}= k-\tfrac{1}{2}$$

We consider now the fermionic correlation functions (26)-(27) first in QCD and next in the soluble conformal invariant infinite coupling model (we have computed some correlation functions in this model in refs.[29]-[31]). First, consider the expectation value (26). We need to compute the expectation values of the matrix elements in eq.(25)

$$\left\langle \exp(-\phi(z)-\phi(z'))\ Z(z)\left[\bar{Z}(z')-\bar{Z}(z)\right]\right\rangle = \int \left[\bar{S}_0(z',w')-\bar{S}_0(z,w')\right]S_0(z,w)\ \left\langle v(w)\overline{v(w')}\right\rangle$$
$$\left\langle \exp\left[\phi(w)+\phi(w')-\phi(z)-\phi(z')\right]\right\rangle \tag{29}$$

where we assumed that the expectation values of ϕ and v are independent. This is true in the leading order of $\frac{1}{\gamma^2}$ expansion . Now, because ϕ is Gaussian

$$\left\langle \exp\left[\phi(w)+\phi(w')-\phi(z)-\phi(z')\right]\right\rangle = \exp\tfrac{1}{2}\left\langle\left[\phi(w)+\phi(w')-\phi(z)-\phi(z')\right]^2\right\rangle \tag{30}$$

Then, in eq.(25) we have also

$$\left\langle \exp\left[\phi(z)-\phi(z')\right]\right\rangle = \exp\left[\langle\phi(0)^2\rangle - \langle\phi(z)\phi(z')\rangle\right] \tag{31}$$

In QCD $\langle\phi(z)\phi(z')\rangle$ tends to a constant, when $z \rightarrow z'$ (because F^2 contains $(\Delta\phi)^2$) and

$$\left\langle\phi(z)\phi(z')\right\rangle \sim -(4k+2)^{-1}\ln|z-z'| \tag{32}$$

for large $|z-z'|$ (independently of g).

So, the term (31) behaves as $|z-z'|^{(4k+2)^{-1}}$ for large $|z-z'|$. A simple power counting shows that the term (29) behaves as $|z-z'|^{3(4k+2)^{-1}}$ for large $|z-z'|$. Hence,

$$\left\langle\bar{\psi}_1(z)\psi_2(z')\right\rangle \sim (\bar{z}-\bar{z}')^{-1}\ |z-z'|^{3(4k+2)^{-1}} \tag{33}$$

for large $|z-z'|$.

Now, at $g=\infty$ we shall have $\exp\langle\phi(0)^2\rangle$ diverging to infinity in eq.(31), whereas in eq.(29) the divergence is $\exp2\langle\phi(0)^2\rangle$. If we wish to obtain a finite result by means of a multiplicative renormalization of the correlation functions (26)-(27) we must divide

$\langle\bar{\psi}(z)\psi(z')\rangle$ by $\exp 2\langle\phi(0)^2\rangle$. In such a case the term (31) becomes zero (this has been pointed out in ref.[35], where some similar calculations have been performed). The r.h.s. of eq.(33) is the exact renormalized two-point function for $(QCD)_2$ with the gauge group SU(2) and $g=\infty$.

The calculation of the four-point function (27) is more involved. We discuss only the terms, which do not vanish (after renormalization) in the limit $g \to \infty$. Then,

$$R(z_1)^{-1}R(z'_2)R(z_2)^{-1}R(z'_1) = \begin{bmatrix} \alpha & \beta \\ \nu & \mu \end{bmatrix} \tag{34}$$

where

$$\alpha = -\exp\left[-\phi(z_1)-\phi(z_2)-\phi(z'_1)-\phi(z'_2)\right] Z(z_1)\left[Z(z'_2)-Z(z_2)\right]\left[\bar{Z}(z'_2)-\bar{Z}(z_1)\right]\left[\bar{Z}(z'_1)-\bar{Z}(z_2)\right] \tag{35}$$

$$\mu = \exp\left[-\phi(z_1)-\phi(z_2)-\phi(z'_1)-\phi(z'_2)\right] Z(z'_1)\left[Z(z'_2)-Z(z_2)\right]\left[\bar{Z}(z'_2)-\bar{Z}(z_1)\right]\left[\bar{Z}(z'_1)-\bar{Z}(z_2)\right] \tag{36}$$

$\nu=\beta=0$

The expectation value of α (35) is

$$\langle\alpha\rangle = (2\pi)^{-4}(z'_2-z_1)(z'_1-z_2)(\bar{z}'_2-\bar{z}_2)\int(z_1-w_1)^{-1}(z'_2-w_2)^{-1}(z_2-w_2)^{-1}(\bar{z}'_2-\bar{w}'_2)^{-1}(\bar{z}_1-\bar{w}'_2)^{-1}$$
$$(\bar{z}'_1-\bar{w}'_1)^{-1}(\bar{z}_2-\bar{w}'_1)^{-1}\langle v(w_1)v(w_2)\bar{v}(w'_1)\bar{v}(w'_2)\rangle$$
$$\exp\tfrac{1}{2}\left\langle\left[\phi(w_1)+\phi(w_2)+\phi(w'_1)+\phi(w'_2)-\phi(z_1)-\phi(z_2)-\phi(z'_1)-\phi(z'_2)\right]^2\right\rangle \tag{37}$$

In the limit $g \to \infty$ the v-expectation value is

$$\left(\frac{2\pi}{2k+1}\right)^2\left[\delta(w_1-w'_1)\delta(w_2-w'_2)+\delta(w_2-w'_1)\delta(w_1-w'_2)\right] \tag{38}$$

whreas the ϕ-expectation value is determined by the two-point function (32)

$$\exp\tfrac{1}{2}\langle....\rangle = |w_1-w_2|^{-4r}|z_1-z_2|^{-r}|z'_1-z'_2|^{-r}|z'_1-z_2|^{-r}|z'_2-z_1|^{-r}|z'_1-z_1|^{-r}|z'_2-z_2|^{-r}|z_1-w_1|^{2r}|w_2-z_1|^{2r}$$
$$|w_1-z_2|^{2r}|w_2-z_2|^{2r}|z'_1-w_1|^{2r}|z'_1-w_2|^{2r}|z'_2-w_1|^{2r}|z'_2-w_2|^{2r} + w_1\leftrightarrow w_2 \tag{39}$$

where

$$r= (4k+2)^{-1}$$

So, we get for the four-point function of the Fermi fields ψ an integral representation involving the integrals $d^2w_1d^2w_2$ (see also ref.[29]). A simple power counting in eq.(39) shows that ψ in eq.(27) has the scale dimension

$$d = \tfrac{1}{2} - \tfrac{3}{2}r \tag{40}$$

in accordance with eq.(33). The power counting argument can be extended to arbitrary n-point functions showing the scale invariance of the n-point functions.

U(n) gauge group

We apply now the formula (13) for GL(n,C). Let us change variables from N to B

$$N^{-1}\bar{\partial}N = \bar{\partial}B \tag{41}$$

where B is an upper triangular matrix (with zeros on the diagonal).
Then, eq.(13) reads

$$W_{WZW} = \frac{1}{2\pi}\mathrm{Tr}\int\left[D^{-1}\partial D\, D^{-1}\bar{\partial}D + (D^{-1}\bar{\partial}B)^{+}\,\bar{\partial}B\, D\right] \tag{42}$$

It is easy to see that if we express the matrices in the coordinates (11) , then eq.(42) takes the form

$$W_{WZW} = \frac{1}{2\pi}\int\left(\sum_{i=1}^{n}\partial\phi_i\bar{\partial}\phi_i + \sum_{k<m}^{n}\exp(\phi_m-\phi_k)\,\partial\bar{b}_{km}\bar{\partial}b_{km}\right) \tag{43}$$

if G^C=GL(n,C)(the gauge group U(n)). For SL(n,C) (complexification of SU(n)) we should replace ϕ_n by $-\sum_{k=1}^{n-1}\phi_k$.

Let us make the next change of variables

$$\bar{\partial}b_{km} = \exp\tfrac{1}{2}(\phi_k-\phi_m)\,\bar{\partial}\Phi_{km} \tag{44}$$

Then,

$$W_{WZW} = \frac{1}{2\pi}\int\left(\sum_{i=1}^{n}\partial\phi_i\bar{\partial}\phi_i + \sum_{k<m}^{n}\bar{\partial}\Phi_{km}\partial\bar{\Phi}_{km}\right) \tag{45}$$

Hence, ϕ are real and Φ complex massless free fields.

The Jacobian of the transformation (41) is equal to 1.The Jacobian of the transformation (44) is equal to

$$J = \prod_{k<m}^{n}|\det[\exp\tfrac{1}{2}(\phi_m-\phi_k)\partial]|^2 = \exp\sum_{k<m}^{n}\frac{1}{8\pi}\int\bar{\partial}(\phi_k-\phi_m)\partial(\bar{\phi}_k-\bar{\phi}_m) \tag{46}$$

Hence, together with the determinant (17) the effective action reads

$$W = \frac{k+c}{2\pi}\int\left(\sum_{i=1}^{n}\partial\phi_i\bar{\partial}\phi_i + \sum_{k<m}^{n}\bar{\partial}\Phi_{km}\partial\bar{\Phi}_{km} - \frac{1}{8\pi}\int\sum_{k<m}^{n}\bar{\partial}(\phi_k-\phi_m)\partial(\bar{\phi}_k-\bar{\phi}_m)\right) \equiv$$

$$\equiv \frac{k+c}{2\pi}\int\sum_{i,k=1}^{n}\partial\phi_i\, M_{ik}\bar{\partial}\phi_k + \sum_{k<m}^{n}\bar{\partial}\Phi_{km}\partial\bar{\Phi}_{km} \tag{47}$$

with a certain invertible matrix M.
Now, in eq.(26)

$$\langle R(z)^{-1}R(z')\rangle = \langle (N^{+})^{-1}(z)\, D^{-1}(z)\, N^{-1}(z)\, N(z')D(z')N^{+}(z')\rangle \tag{48}$$

In order to compute the expectation value (48) we must solve first eqs.(41)

$$(N^{-1}\bar{\partial}N)_{km} = \exp\tfrac{1}{2}(\phi_k-\phi_m)\,\bar{\partial}\Phi_{km} \tag{49}$$

This is easy, because N is nilpotent (see ref.[36]). Then, the expectation value in eq.(48) is reduced to a Gaussian integral. The calculations although straightforward are quite tedious if $n \geq 3$. The gauge groups SU(n) and USp(n) will be discussed in more detail in forthcoming publications.

I would like to thank Sergio Albeverio and Krzysztof Gawedzki for interesting discussions.

References

[1] J. Schwinger, Phys.Rev.128,2425(1962)

[2] G. 't Hooft, Nucl.Phys.B75,461(1974)

[3] C.G. Callan,Jr., N. Coote and D.J. Gross, Phys.Rev.D13,1649(1976)

[4] A. Casher, J.B. Kogut and L. Susskind, Phys.Rev.D10,732(1974)

[5] I.V.Belvedere, J.A. Swieca, K.D. Rothe and B. Schroer, Nucl.Phys.B153,112(1979), N.K. Nielsen,K.D. Rothe and B. Schroer, Nucl.Phys.B160,330(1979)

[6] K. Furuya, R.E. Gamboa Saravi and F.A. Schaposnik, Nucl.Phys.B208,159(1982), R.E. Gamboa Saravi, F.A. Schaposnik and J.E. Salomin, Nucl.Phys.B185,239(1981)

[7] R.E. Gamboa Saravi, F.A. Schaposnik and J.E. Salomin, Phys.Rev.D30,1353(1984)

[8] E. Fradkin, C.M. Naon and F.A. Schaposnik, Phys.Rev.D36,3809(1987)

[9] K.D. Rothe, Nucl.Phys.B269,269(1986)

[10] H. Falomir and E.M. Santangelo, Phys.Rev.Lett.56,1659(1986)

[11] A. Polyakov and P.B. Wiegmann, Phys.Lett.B131,121(1983), B141,223(1984)

[12] E. Witten, Comm.Math.Phys.92,455(1984)

[13] O. Alvarez, Nucl.Phys.B238,61(1984)

[14] P. di Vecchia, B. Durhuus and J.L. Petersen, Phys.Lett.144B,245(1984)

[15] R.Roskies, in Symmetries in Particle Physics, edited by I.Bars et al,Plenum,New York,1984

[16] O. Gonzalez and A.N. Redlich, Phys. Lett.147B,150(1984)

[17] A.Balachandran,G. Marmo, V.B. Nair and C.G. Trahern, Phys.Rev.D25,2713(1982)

[18] J. Wess and B. Zumino, Phys.Lett.37B,95(1971)

[19] B. Zumino, Phys.Lett.69B,369(1977)

[20] K. Osterwalder and R. Schrader, Helv.Phys.Acta 46,277(1973)

[21] G.D. Date, Y. Frishman and J. Sonnenschein, Nucl.Phys.B283,365(1987)

[22] Y. Frishman and J. Sonnenschein, Nucl.Phys.B294,804(1987)

[23] K. Bardakci, E. Rabinovici and B. Saring, Nucl.Phys.B299,151(1988)

[24] P. Goddard, A. Kent and D. Olive, Phys.Lett.B152,88(1985)

[25] P. Goddard, A. Kent and D. Olive, Comm.Math.Phys.103,105(1986)

[26] D. Karabali, Q.-H.Park, H.J. Schnitzer and Z. Yang, Phys.Lett.216B,307(1989)

[27] P. Bowcock, Nucl.Phys.B316,86(1989)

[28] K.Gawedzki and A. Kupiainen, Phys.Lett.215B,119(1988), Nucl.Phys.B320,625(1989)

[29] Z. Haba, Int.J.Mod.Phys.A4,267(1989)

[30] Z. Haba, BiBoS preprint No.364,1989(to appear in Mod.Phys.Lett.A)

[31] Z. Haba, BiBoS preprint No.378,1989(subm. to Int.J.Mod.Phys.)

[32] A. Kupiainen and J. Mickelsson, Phys.Lett.185B,107(1987)

[33] Z. Haba, Mod.Phys.Lett. A(in print) and BiBoS preprint No.381,1989

[34] A. Terras, Harmonic Analysis on Symmetric Spaces II.,Springer,1987

[35] K. Gawedzki, these Proceedings and IHES preprint 89/32

[36] S. Albeverio, T. Arede and Z. Haba, BiBoS preprint No.330,1988(to appear in J.Math.Phys.) and paper in preparation

[37] A. Morozov, Lectures in Karpacz

[38] A. Morozov,JETPLett.49,305(1989)(inRussian),A.V.Marshakov, Phys.Lett.224B,141(1989),D. Nemeschansky, Phys.Lett.B224,121(1989), A. Bilal, preprint CERN-TH.5370/1989, J. Distler and Z. Qiu, Cornel preprint, CLNS 89/911, L.J. Dixon, J. Lykken and M.E. Peskin, preprint SLAC-PUB-4884,1989

IRREDUCIBLE AUTOMORPHIC FUNCTIONS AND CONFORMAL FIELD THEORY

W. Rühl

Universität Kaiserslautern, Fachbereich Physik, Kaiserslautern, FRG

1. Introduction

The topics treated in this talk concerns an application of real analysis on a Riemann surface. We consider the Riemann surface as represented by the fundamental domain $\mathcal{F}_\Gamma$ of the discrete group Γ acting on the unit disk $\mathcal{D}$. Then a correlation (or N-point) function is obtained by summing the corresponding planar function over Γ. This is known in potential theory as the method of mirror charges. The standard approach to this problem is by means of complex analysis and is based on the use of Θ-functions. This approach is so powerful due to the Jacobian map which maps the Riemann surface on a Jacobian variety. After that map details on the concrete Riemann surface are superfluous. They have been incorporated in the Abelian differentials of the first kind.

Though less powerful our approach tries to avoid also using explicit knowledge of the surface as far as possible. We use
(1) automorphic, in particular irreducible automorphic functions;
(2) the Casimir operator (= Laplace operator with spin) and its spectrum

to obtain expansions on correlation functions. The mathematical methods used are related to Selberg's trace formula[1,2].

2. Automorphic and irreducible automorphic functions

Consider the unit disk

$$\mathcal{D} = \{z \mid |z| < 1\}$$

with Poincaré measure

$$d\nu(z) = \frac{1}{\pi}(1-|z|^2)^{-2}\, d^2z$$

SU(1,1) acts on $\mathcal{D}$ by

$$z \longrightarrow z_g = \frac{az + b}{cz + d}$$

$$a = \bar{d},\ b = \bar{c}, \qquad |a|^2 - |b|^2 = 1$$

Define a unitary representation in $\mathcal{L}^2_\nu(\mathcal{D})$ by

$$T_g^{(m)} f(z) = \exp[im \arg(cz+d)]\, f(z_g) \qquad m \in \mathbb{Z}$$

We use the definition: $\psi(z)$ is automorphic with respect to $T^{(m)}$ and the discrete group Γ iff it is an invariant vector (in $\mathcal{L}^2_\nu(\mathcal{D})$ or an appropriately extended space) so that

$$T_\gamma^{(m)} \varphi(z) = \varphi(z) \text{ for all } \gamma \in \Gamma.$$

For simplicity we consider compact Riemann surfaces and hyperbolic groups Γ.

In order to obtain an irreducible function we need an irreducible representation. They can be realized in a space $\mathcal{L}^2(S)$ where S is a double covering of $\partial\mathcal{D}$

$$S = \{\xi \mid \xi = e^{i\frac{\psi}{2}},\ 0 \leq \psi < 4\pi,\ \lim_{|z|\to 1} z = e^{i\psi}\}$$

SU(1,1) acts on S by

$$\xi \xrightarrow[g]{} \xi_g$$

$$e^{i\psi_g} = \frac{ae^{i\psi} + b}{ce^{i\psi} + d}$$

Let us denote

$$\chi = \{\sigma, \epsilon\}, \qquad \epsilon \in \{0,1\}$$

$$\alpha_\sigma(g,\xi) = |c\xi^2 + d|^{\sigma-1}$$

Then

$$T_g^\chi F(\xi) = \alpha_\sigma(g,\xi)\ F(\xi_g)$$

where

$$F(\xi) = (-1)^\epsilon F(-\xi)$$

defines a representation which is unitary and irreducible (the principal series) if

σ purely imaginary

$F \in \mathcal{L}^2_{\mu,\epsilon}(S)$

$d\mu(\psi) = \frac{1}{4\pi}\, d\psi$

After enlarging $\mathcal{L}^2_{\mu,\epsilon}(S)$ to the space of hyperfunctions $\mathcal{A}'_\epsilon(S)$, automorphic functions can be defined with respect to T^χ. They can be extended from S into the interior of $\mathcal{D}$ by the Poisson kernel

$$\kappa_{(\sigma,m)}(z,\psi) = e^{i\frac{m}{2}\phi}\left[\frac{1 - ze^{-i\phi}}{1 - \bar{z}e^{+i\psi}}\right]^{\frac{m}{2}}\left|\frac{1 - ze^{-i\phi}}{(1 - |z|^2)^{\frac{1}{2}}}\right|^{-\sigma-1}$$

where $m-\epsilon \in 2\mathbb{Z}$. It maps $\mathcal{A}'_\epsilon(S)$ invertibly onto a space $\mathcal{E}_{(\sigma,m)}$ of C^∞

functions over $\mathcal{D}$. Provided $\sigma \notin \mathbb{Z}$ this map

(1) intertwines T^{χ} on $\mathcal{L}^2_{\mu,\epsilon}(S)$ with $T^{(m)}$ on $\mathcal{E}_{(\sigma,m)}$;

(2) has as image $\mathcal{E}_{(\sigma,m)}$ a space of eigenfunctions of the Casimir operator $-\Delta^{(m)}$ with eigenvalue $\lambda = 1-\sigma^2$

$$\Delta^{(m)} = + m^2(1-|z|^2) - 2m\,(1-|z|^2)(z\partial_z - \bar{z}\partial_{\bar{z}}) + 4(1-|z|^2)^2\,\partial_z\partial_{\bar{z}}$$

It follows that

(a) $\mathcal{E}_{(\sigma_1,m_1)}$ $\mathcal{E}_{(\sigma_2,m_2)}$ carry equivalent representations iff $m_1-m_2 \in 2\mathbb{Z}$;

(b) $T^{(m)}$ is irreducible on $\mathcal{E}_{(\sigma,m)}$ and unitary if σ is purely imaginary and is equivalent to T^{χ};

(c) if $F \in \mathcal{A}'_{\epsilon}(S)$ is irreducible and automorphic then its image φ_λ in $\mathcal{E}_{(\sigma,m)}$ is als irreducible automorphic.

3. The spectrum of the Casimir operator.

The Casimir operator $\Delta^{(m)}$ reduces to the well known Laplacian for $m = 0$. But $-\Delta^{(m)}$ is not positive definite, its spectrum has a finite discrete negative part. Let the Riemann surface be compact and have genus g. Then the spectrum of $\Delta^{(m)}$ is discrete throughout and belongs to different series of representations of SU(1,1)

(i) $\sigma_i = i\rho_i$, the principal series;

$$\lambda_i = 1 + \rho_i^2 > 1$$

(ii) $0 < \sigma_i < 1$, the supplementary series $\lambda_i = 1-\sigma_i^2$;

(iii) the discrete holomorphic and antiholimorphic series

$$\lambda_i = -4n(n-1) \leq 0$$

$$1 \leq n \leq |\tfrac{m}{2}|$$

$$m \text{ even: } n \in \{1, 2, \ldots \tfrac{|m|}{2}\}$$

$$m \text{ odd: } n \in \{\tfrac{3}{2}, \tfrac{5}{2}, \ldots \tfrac{|m|}{2}\}$$

(iv) the trivial representation with $\lambda = 0$.

These eigenvalues are degenerate. For the discrete series the degeneracy is

$$k_n = (2n-1)\ (g-1)$$

in each branch. This follows from the Riemann-Roch theorem or Selberg's trace formula. Note that by the Gauss-Bonnet theorem

$$g-1 = \nu(\mathcal{F})$$

Any automorphic function w.r.t. $T^{(M)}$ can be expanded in the eigenfunctions $\varphi_\lambda(z)$

$$f(z) = \sum_{\lambda_i} c_{\lambda_i} \varphi_{\lambda_i}(z)$$

$$c_{\lambda_i} = \int_{F_\Gamma} f(z)\ \overline{\varphi_{\lambda_i}}(z)\ d\nu_i(z)$$

4. Two-point functions.

A 2-point function of a (quasi-primary) field in conformal field theory can be defined so that it is automorphic in both arguments with respect to $T^{(m)}$, $\frac{1}{2}|m|$ being the spin of the field. Let $k(z_1, z_2)$ be the two-point invariant

$$k(z_1, z_2) = \frac{|z_1 - z_2|}{[(1-|z_1|^2)(1-|z_2|^2)]^{\frac{1}{2}}}$$

$$k((z_1)_g, (z_2)_g) = k(z_1, z_2)$$

and define

$$\tilde{K}_\chi(z_1,z_2) = \exp[im \arg(z_1-z_2)]\; k(z_1,z_2)^{-2\delta}$$

where δ is the dimension of the field. Then we assume

$$\tilde{W}_\chi(z_1,z_2) = \sum_{\gamma_i \in \Gamma} \exp[im \arg (c_i z_1 + d_i)] \cdot \tilde{K}_\chi((z_1)\gamma_i, z_2)$$

We want to expand

$$\tilde{W}_\chi(z_1,z_2) = \sum_{\text{spec}} \varphi_{\lambda_i}(z_1)\, c_{\lambda_i}(z_2) = \sum_{\text{spec}} c'_{\lambda_i}(z_1)\, \varphi_{\lambda_i}(z_2)$$

This expansion can be performed explicitly without knowledge of the $\varphi_{\lambda_i}(z)$ due to the sum over Γ representation. The arguments go as follows.

Step 1:

$$c_\lambda(z_2) = \int_{\mathcal{F}_\Gamma} \tilde{W}_\chi(z_1,z_2)\, \overline{\varphi_\lambda(z_1)} d\nu(z_1)$$

$$= \int_{\mathcal{D}} \tilde{K}_\chi(z_1,z_2)\, \overline{\varphi_\lambda(z_1)}\, d\nu(z_1)$$

Step 2: Let $s(z) \in SU(1,1)$ be such that

$$(0)_{s(z)} = z$$

Then

$$c_\lambda(z_2) = \int_{\mathcal{D}} \tilde{K}_\chi(z_1,0)\, \overline{F(z_2;z_1)}\, d\nu(z_1)$$

$$F(z_2;z_1) = T^{(m)}_{s(z_2)} \varphi_\lambda(z_1)$$

Here we can integrate over $\varphi = \arg z_1$.

Step 3: Using the fact that $F(z_2;z_1)$ is an eigenfunction of $-\Delta^{(m)}$ with the eigenvalue λ, we obtain

$$c_\lambda(z_2) = H_\chi(\lambda) \begin{cases} \overline{a_{m,0}(z_2)}, & m \geq 0 \\ \overline{a_{0,|m|}(z_2)} & m \leq 0 \end{cases}$$

$$a_{m,0}(z_2) = \frac{1}{m!} \left[\frac{\partial}{\partial z}\right]^m F(z_2;z) \Big|_{z=0}, \quad m \geq 0$$

$$a_{0,|m|}(z_2) = \frac{1}{|m|!} \left[\frac{\partial}{\partial \bar{z}}\right]^{|m|} F(z_2;z) \Big|_{z=0}, \quad m \leq 0$$

Step 4: For the determination of $H_\chi(\lambda)$ it is essential that $F(z_2;z_1)$ is an eigenfunction of $-\Delta^{(m)}$ with respect to z_1 and eigenvalue λ. $H_\chi(\lambda)$ is otherwise independent of F. Replace F by the Poisson kernel. If follows easily that

$$H_\chi(\lambda) = \begin{bmatrix} -\frac{1}{2}(\bar{\sigma}+1) + \frac{1}{2}|m| \\ |m| \end{bmatrix}^{-1} \cdot \int_0^1 dt\, F(t)\, \phi\left[\frac{t}{1-t}\right] (1-t)^{\frac{1}{2}(\bar{\sigma}+1)-2}$$

where F(t) is a hypergeometric series

$$F(t) = \sum_{r,s=0}^{\infty} \delta_{m,r-s} \begin{bmatrix} -\frac{1}{2}(\bar{\sigma}+1) + \frac{1}{2}m \\ r \end{bmatrix} \begin{bmatrix} -\frac{1}{2}(\bar{\sigma}+1) - \frac{1}{2}m \\ s \end{bmatrix} \cdot t^{\frac{1}{2}(r+s)}$$

If σ is purely imaginary then H_χ and ϕ are connected by a Fourier integral transform. If

$$\phi(\xi) = \xi^{-\delta}$$

then $H_\chi(\lambda)$ can be calculated explicitly. The final result can be expressed in the form

$$\widetilde{W}_{\chi}(z_1,z_2) = \sum_{\text{spec}} H_{\chi}(\lambda)\ \varphi_{\lambda}(z_1)\ \overline{D^{(m)}(z_2)\ \varphi_{\lambda}(z_2)}$$

where $D^{(m)}$ is a differential operator of degree $|m|$, e.g.

$$m > 0:\ D^{(m)}(z) = \sum_{k=0}^{m-1} \frac{1}{(m-k)!} \begin{bmatrix} \frac{m}{2} - 1 \\ k \end{bmatrix} \cdot (-\bar{z})^{k}\ (1-|z|^2)^{m-k} \left[\frac{\partial}{\partial z}\right]^{m-k}$$

5. Three-point functions.

We can proceed similarly in the case of three-point functions for three (quasi-primary) fields of covariance

$$\chi_i = (m_i, \delta_i)\ ,\ i \in \{1,2,3\}$$

We define

$$k_{\delta_1\delta_2\delta_3}(z_1,z_2,z_3) = \prod_{i<j} k(z_i,z_j)^{-2\delta_{ij}}$$

$$\begin{bmatrix} \delta_{23} \\ \delta_{31} \\ \delta_{12} \end{bmatrix} = \frac{1}{2} \begin{bmatrix} -1 & 1 & 1 \\ 1 & -1 & 1 \\ 1 & 1 & -1 \end{bmatrix} \begin{bmatrix} \delta_1 \\ \delta_2 \\ \delta_3 \end{bmatrix}$$

and $\{m_{ij}\}$ analogously. Then the three-point function is (up to normalization)

$$\widetilde{W}_{\chi_1\chi_2\chi_3}(z_1,z_2,z_3) = \sum_{\gamma_i\gamma_j\in\Gamma} \exp i\,[m_1 \arg(c_i z_1 + d_i) + m_2 \arg(c_j z_2 + d_j)] \cdot \widetilde{K}_{\chi_1\chi_2\chi_3}((z_1)_{\gamma_i},\ (z_2)_{\gamma_j},\ z_3)$$

where $\widetilde{K}_{\chi_1\chi_2\chi_3}$ is the corresponding planar function

$$\tilde{K}_{\chi_1\chi_2\chi_3}(z_1,z_2,z_3) = \prod_{i<j} \exp[im_{ij} \arg(z_i-z_j)]$$

$$\cdot\ k_{\delta_1\delta_2\delta_3}(z_1,z_2,z_3)$$

The function $\tilde{W}$ is automorphic with respect to $T^{(m_1)}$, $T^{(m_2)}$, $T^{(m_3)}$ in the respective arguments. Of course we can replace $k(z_i,z_j)^{-2\delta_{ij}}$ by $\phi_{ij}(k(z_i;z_j)^2)$. Then one calculates

$$c_{\lambda_1\lambda_2}(z_3) = \int_{\mathcal{F}_\Gamma \times \mathcal{F}_\Gamma} \tilde{W}_{\chi_1\chi_2\chi_3}(z_1,z_2,z_3)\ \overline{\varphi^{(m_1)}_{\lambda_1}(z_1)\ \varphi^{(m_2)}_{\lambda_2}(z_2)}\ d\nu(z_1)\ d\nu(z_2)$$

Going through similar steps as before we arrive at

$$\tilde{W}_{\chi_1\chi_2\chi_3} = \sum_{\lambda_1\lambda_2} \varphi_{\lambda_1}(z_1)\varphi_{\lambda_2}(z_2) \sum_{k\in\mathbb{Z}} H_{\chi_1\chi_2\chi_3,k}(\lambda_1,\lambda_2)$$
$$\cdot\ \overline{D^{(k)}\varphi^{(m_1)}_{\lambda_1}(z_3)}\ \overline{D^{(M-k)}\varphi^{(m_2)}_{\lambda_2}}$$

where

$$M = \frac{1}{2}(m_1 + m_2 + m_3) \in \mathbb{Z}$$

and H is a Fourier transform over the product of the ϕ_{ij}. For more details see [3].

REFERENCES

1. I.M. Gelfand, M.I. Graev, I.I. Pyatetskii-Shapiro, Representation Theory and Automorphic Functions, Saunders, Philadelphia - London - Toronto 1969.

2. D.A. Hejhal, Duke Math. J. 43 (1976), 441, contains also a complete list of references to Selberg's original works.

3. W. Rühl, Irreducible automorphic functions and the N-point functions of conformal field theory, Z. Phys. C, to appear.

(IN)CONSISTENCY OF NON-COMPACT WZNW-MODELS

Andreas Wipf

Max Planck Institut für Physik und Astrophysik, München, FRG

It is generally believed that two-dimensional conformally invariant sigma models for fields $X^\mu \in M$ coupled to a background metric, antisymmetric tensor field and dilaton field with vanishing β-functions, are consistent [CALLAN et.al 1985]. Such models can be interpreted as string propagation on the target manifold M. A well studied example is the propagation on $M_4 \times G_c$, with bosonic string action on the Minkowski space M_4 and Wess-Zumine-Novikov-Witten action on a compact Lie group G_c. The conformally invariant ones ($c = 26$) are indeed unitary for certain classes of vacuum representations [GEPNER and WITTEN 1986].

Clearly one would like to go beyond the $M_4 \times G_c$ compactification scheme, in particular on shouldn't assume apriori that exactly four dimensions do not curl up. As a first step in this direction we study the exactly soluble models where the target manifold is a non-compact group G with WZNW action

$$S = \frac{-k}{16\pi} \int \mathrm{tr}\Big(\partial_a g\, \partial^a g^{-1}\Big) + \frac{k}{24\pi} \int d^3\xi \epsilon^{abc}\, \mathrm{tr}\Big((g^{-1}\partial_a g)(g^{-1}\partial_b g)(g^{-1}\partial_c g)\Big). \quad (1)$$

These G-models are halfway between the bosonic string and the general σ- models. We shall see that the only viable candidate is $G = SU(1,1) \times G_c$, where G_c should be taken such that the central charges add up to 26. For this models the metric can be chosen to have signature $(-,+,...,+)$, so that the target space has one time direction. Also one can show that the β-functions vanish to all orders, similarly as for the compact WZNW models [ALLEN et.al 1988] (there exists a parallelizing torsion). However, although these G-models are all conformally invariant, none of them is unitary [BALOG et.al 1988]. This shows that only a subclass of the conformally invariant σ-models can be consistently quantized.

1. Classical Propagation

We consider the propagation on the only viable group (see below) $G = SU(1,1)\times G_c$, and may leave out G_c in what follows. Topologically $SU(1,1)$ is $R^2 \times S^1$, where R^2 and S^1 are generated by the two non-compact and the compact generators, respectively. Clearly we must interpret the 'compact coordinate' on the circle as time to get the correct signature. To avoid causality problems associated with a periodic time we shall take the universal covering $\widetilde{SU(1,1)}$ as space-time for the string. Now one can show that this model is classically completely consistent, that is the general solutions of the equation of motion following from (1), with causal initial conditions at time $\tau = 0$ and fulfilling the reparametrization constraints $T_{\pm\pm} = 0$, moves causal at any later time. A string moves causal if every point on the string moves forward in time with velocity $v < c$. After having shown consistency on the classical level, we now turn to the related quantum mechanical problem.

2. Quantized G-Models

The algebraic part of the quantization of G-models is a rather straightforward generalization of the similar problem for compact groups [GEPNER and WITTEN 1986]. The conserved currents

$$J(z) = -\frac{k}{2}\,\partial_z g\,g^{-1} \qquad \bar{J}(\bar{z}) = -\frac{k}{2}\,\partial_{\bar{z}} g\,g^{-1} \qquad (z = e^{i(\tau+\sigma)})$$

can be expanded into Laurent series and the current algebra is equivalent to the Kac-Moody (KM) algebra for the expansion coefficients

$$[J^a_n, J^b_m] = i f^{ab}{}_c J^c_{n+m} - K g^{ab} n \delta_{n+m,0} \qquad (n, m \in Z).$$

The expansion coefficients of the Sommerfield-Sugawara energy-momentum tensor define a Virasoro algebra with central charge

$$c = \frac{2K}{2K - Q} D \qquad \text{where} \qquad Q = f^{ax}{}_y f^{yb}{}_x \ , \ D = \text{dim} G$$

and form a semi-direct product with the KM generators

$$[L_n, J^a_m] = -m\,J^a_{n+m}.$$

The Hilbert space is the Fock space generated by the creation operators $J^a_{-n}, n \geq 0$ from vacuum states $|\alpha\rangle$. We shall only consider highest weight representations such that L_0 has a lowest eigenvalue on the vacuum sector and thus all $J^a_n, n > 0$, annihilate the vacuum states. The vacuum sector transforms according to some representaion of Lie(G), spanned by the J_0^a. For non-compact KM algebras the Fock space is never unitary, since

$$| J_{-1}^a \, | \alpha\rangle |^2 = -K g^{aa} \langle \alpha \, | \alpha \rangle,$$

and the Killing metric is indefinite. This is the analogue of the situation in the case of the free bosonic string on Minkowski space. The real question is whether the 'physical' subspace on which the reparametrization constraints hold

$$\begin{aligned} L_n \, | phys\rangle = \bar{L}_n \, | phys\rangle = 0 \qquad \text{for} \qquad n > 0 \\ (L_0 - \bar{L}_0) \, | phys\rangle = (L_0 + \bar{L}_0 - 2\alpha_0) \, | phys\rangle = 0 \end{aligned} \tag{2}$$

is free of negative norm states.

3. The Unitarity Problem

A crucial difference to the compact case is that a unitary vacuum representation is already infinite dimensional in the non-compact case. Thus the Fock space is always built over infinitely degenerate vacuum sectors. It is not too difficult to show that there are always ghosts in the 'physical' subspace of the grade 1 states $\{J_{-1}^a \, | \alpha\rangle\}$, unless $G = SU(1,1)$. To see this one notes that any non-compact Lie algebra with rank> 1 contains both a $su(2)$ and a $su(1,1)$ subalgebra. By labelling the vacuum states as $| jh\rangle$, where j and h denote the quantum numbers corresponding to one of these subalgebras, one can show [BALOG et.al 1988] that the 'physical' states

$$| \Phi\rangle = x \, J_{-1}^+ \, | jh-1\rangle + y \, J_{-1}^- \, | jh+1\rangle + z(x,y) \, J_{-1}^3 \, | jh\rangle,$$

where $z(x,y)$ is determined by the physical state condition $L_1 \, | \Phi\rangle = 0$, have only positive norms if $K < 0$ for $su(2)$ and $K > 0$ for $su(1,1)$. Since K cannot be both positive and negative we conclude that for any non-compact $G \neq SU(1,1)$ there are always negative norm states fulfilling the constraints. So we are left with the only candidate $SU(1,1)$.

To proceed one observes that the L_0 constraint implies that the grade N states are built over their own vacuum sector which carries a $\widetilde{SU(1,1)}$ representation with Casimir

$$C_N = 2(K-1)(N - \alpha_0).$$

In deriving this condition we have used the Sugawara form of L_0. It should be compared with the mass shall condition for the free bosonic string

$$p^2 = m^2 = \frac{1}{\alpha}(N - \alpha_0)$$

with $K-1$ playing the role of the sting tension.

For negative string tension $K-1$ the Casimirs C_N become negative for large N and the corresponding vacua carry principal representation of $\widetilde{SU(1,1)}$ with

$$C_N = j_N(j_N+1) \qquad j_N = -\frac{1}{2} + i\omega_N \quad , \quad \omega_N \quad \text{real.}$$

These grade N states are generated out of their vacuum representation by acting N times with the vector operators J^a_{-1} and thus transform according to the reducible representation

$$\theta_{j_N} \otimes \left(\theta_1\right)^{\otimes N} = \theta_{j_N+N} \oplus \ldots \oplus \theta_{j_N-N}. \tag{3}$$

The states in the extremal representations $\theta_{j_N \pm N}$ can be shown to be 'physical', that is they are annihilated by the constraints $L_{n>0}$. However, since $j_E = j_N + N$, their Casimirs are complex, which shows that the extremal representations , although fulfilling the reparametrization constraints, are non-unitary.

In the most promising case of positive string tension $K-1$ the Casimirs C_N become positive for large N and the corresponding vacua carry highest weight representations of $\widetilde{SU(1,1)}$. Then all irreducible representations θ_{j_N+n} in the Clebsch-Gordan decomposition (3) are highest weight ones. In particular the highest weight state of the extremal (and 'physical') representation is explicitly given by $|E_N^+\rangle = (J_{-1}^+)^N \, |j_N \, j_N\rangle$. Since the vacuum Casimir is proportional to N for large N, one has $j_N \sim \sqrt{N}$ and the magnetic quantum number of $|E_N^+\rangle$ becomes positive for large N, $h^{(N)} = j_N + N \sim N$. On the other hand

$$\langle E_N^+ | J_0^+ J_0^- \, | E_N^+\rangle = -2h_N \, \langle E_N^+ \, | E_N^+\rangle,$$

showing that either $|E_N^+\rangle$ or $J_0^- \, |E_N^+\rangle$ has negative norm. Hence the case of positive string tension is excluded as well and we have shown that the 'physical' subspace always posesses negative norm ghost states, even for the group $\widetilde{SU(1,1)}$.

4. Conclusions and Remarks

We have shown that in the G-models not all ghost states are eliminated by the reparametrization constraints (2). In the string context this means that truly consistent string propagation rules out certain target manifolds, hence restricting the possible vacua of the string.

A natural question to ask is whether one can impose additional constraints, such that one recovers unitarity after imposing these constraints. First one might be interested in group invariant constraints and a natural set of such constraints would be the analogue of the Virasoro operators for higher order Casimirs. However, such operators would probably not produce unitarity for non-compact higher rank groups, since the number of Casimirs is always less than the number of time like directions.

Recently it has been realized that the Liouville and Toda field theories can be regarded as WZNW models, with certain natural, conformally invariant, constraints. These constraints break the $SL(2,R) \sim SU(1,1)$ invariance in the Liouville case or the G invariance in the Toda case, but they respect the conformal invariance [FORGACS et.al 1989]. The constraints are very restrictive and it is likely (at least in the Liouville case) that the physical subspace of the Hilberspace is positive.

References

Allen R.W., Jack I. and Jones D. (1988) Chiral σ-models and the dilaton β-function, preprint SHEP 14

Balog J., O'Raifeartaigh L., Forgacs P. and Wipf A. (1988) Consistency of string propagation on curved spacetimes: an SU(1,1) based counterexample, preprint DIAS-STP-45 and MPI-PAE/PTh 66

Callan C.G., Friedan D., Martinec E.J. and Perry M.J. (1985) Strings in background fields, Nucl. Phys. B262, 593-609

Forgacs P., Wipf A., Balog J., Feher L. and O'Raifeartaigh L. (1989) Liouville and Toda theories as conformally reduced WZNW theories, preprint MPI-PAE/PTh 29 and DIAS-STP-41

Gepner D. and Witten E. (1986) Strings on group manifolds, Nucl. Phys. B278, 493-549

Andreas Wipf, *Max-Plack-Institut für Physik und Astrophysik, - Werner-Heisenberg-Institut für Physik, P.O.Box 40 12 12, Munich (Fed. Rep. Germany)*

IV. QUANTIZATION OF CONSTRAINED SYSTEMS

Aspects of BRST-Quantization

J.W. van Holten
NIKHEF-H
Amsterdam

Abstract

Quite frequently the time-evolution of physical systems is subject to constraints. This happens in particular if one tries to keep certain symmetries in the description of a system manifest. The relation between symmetries and constraints is explained. A convenient method of dealing with constrained systems in quantum theory is provided by the BRST-procedure. This involves new degrees of freedom (ghosts) and a new symmetry (BRST-invariance). It is shown how the physical Hilbert space of states can be identified with the cohomology structure of the BRST-operator. The relation of this algebraic description to the usual Faddeev-Popov path-integral formulation of gauge theories is described. The construction of theories with local BRST-invariance is addressed next. The interpretation of this local symmetry is given and its application in the construction of a covariant quantum theory of the 4-dimensional superparticle is discussed as an example.

Chapter 1

Constrained Systems

1.1 Introduction

The time-evolution of physical systems is usually described in mathematical language by a differential equation such as Newton's equation in classical mechanics, Maxwell's equations in the theory of electro-magnetism or the Schrödinger equation in quantum mechanics. It happens quite frequently though, that these dynamical equations admit too many solutions, in the sense that not all solutions can be realized as physical motions of the system. The correct physical solutions are then obtained by imposing some constraints on the dynamical variables in addition to the equations of motion.

As an example, consider the relativistic point-particle, which is described by the particle's worldline giving its position x^i as a function of time t. Mathematically, the worldline is a map from the real line R^1 into d-dimensional Minkowski space-time M^d. The time-like co-ordinate τ parametrizing the line is called parameter-time, whilst the time co-ordinate of space-time is $x^0 = ct$. In the absence of external forces the equation of motion of the particle is

$$\frac{d^2x^\mu}{d\tau^2} = 0, \tag{1.1}$$

with the solution

$$x^\mu(\tau) = x^\mu(0) + u^\mu\tau. \tag{1.2}$$

The space-time velocity u^μ is constant and its square is negative because of the time-like direction of the worldline:

$$u_\mu u^\mu = -c^2. \tag{1.3}$$

An important physical observation is, that c is a *universal* constant, equal to the velocity of light, for all particles irrespective of their mass, spin, charge or other physical properties. This imposes an extra condition on the physical solutions of eq.(1.1), which restricts the allowed boundary values of the solution (1.2) to those obeying the relation (1.3). The observation above is true not only for free particles, but for interacting particles as well. Indeed, if a particle is subject to an external force f^μ, the equation of motion becomes

$$\frac{dp^\mu}{d\tau} = f^\mu, \tag{1.4}$$

where $p^\mu = m\dot{x}^\mu$ and the external force has to satisfy

$$p \cdot f = 0. \tag{1.5}$$

This guarantees, that condition (1.3) can still be implemented, which defines the physical mass-shell:

$$p^2 + m^2c^2 = 0. \tag{1.6}$$

This is an invariant hypersurface in momentum space for any relativistic particle. Thus, an electron in an accelerator subject to electric and magnetic fields obeys eq.(1.6) at any point of its orbit, even though its momentum p^μ is changing continuously. This illustrates quite clearly, how the motions of physical systems may be subject to restrictions, thus selecting among all possible solutions of the dynamical equations those satisfying appropriate additional requirements.

Another example of redundant solutions to the equations of motion of a physical system is provided by Maxwell's theory of the electro-magnetic field. The dynamical equations describing the time-development of electric and magnetic fields in vacuo are

$$\frac{1}{c}\frac{\partial \vec{E}}{\partial t} = \vec{\nabla} \times \vec{B}, \qquad \frac{1}{c}\frac{\partial \vec{B}}{\partial t} = -\vec{\nabla} \times \vec{E}. \tag{1.7}$$

Each of the fields $(\vec{E}, \vec{B})$ has three components, but physical electromagnetic waves are transversely polarized. The longitudinal part can be eliminated by imposing constraints on the spatial variation of the fields:

$$\vec{\nabla} \cdot \vec{E} = 0, \qquad \vec{\nabla} \cdot \vec{B} = 0. \tag{1.8}$$

As in the case of the relativistic particle these constraints donot follow from the dynamical equations (1.7), but are compatible with them and serve to select the physically relevant solutions of the evolution equations. Applying eqs.(1.7) twice we can separate them, and taking into account conditions (1.8) we find the wave equations

$$\Box \vec{E} = 0, \qquad \Box \vec{B} = 0, \tag{1.9}$$

where $\Box = \partial^\mu \partial_\mu$ denotes the relativistic Laplacian. Physically, these equations imply that electro-magnetic waves are transversely polarized and travel at constant speed (by definition the speed of light), independent of their frequency. The quantum-mechanical interpretation is, that photons are massless particles with only two helicity states, $\lambda = (+1, -1)$.

Other examples of constrained motion of physical systems are easy to find; they include such systems as the gravitational field described by Einsteins equations, the strong color-forces between quarks described by the Yang-Mills equations and the motion of relativistic strings, which may play a role in the description of strong interactions and/or the quantum theory of gravity. In all these cases the constraints arise mathematically because of some special symmetry of the theory. Therefore the next section is devoted to a discussion of the relation between constraints and symmetries.

1.2 The Noether Theorems

A well-known theorem by Noether [1] states, that a continuous symmetry of the action for a physical system implies the existence of a conserved quantity, the Noether charge. A second less-known but equally important theorem states, that for a continuous gauge symmetry —by which we mean a symmetry the parameter of which may depend on time in an arbitrary way— the Noether charge is not only conserved, but actually vanishes.

[1] See for example ref.[1].

This imposes an additional constraint on the physical solutions of the equations of motion; therefore systems with gauge symmetries are always subject to constraints.

In order to show this, we consider a system described by generalized co-ordinates $q^i, i = 1, ..., N$, where N may be infinite (as in field theory), with the action

$$S = \int_1^2 dt\, L(q^i, \dot{q}^i). \tag{1.10}$$

We restrict ourselves to systems for which the equations of motion are at most of second order in time-derivatives, hence the action is a functional only of the co-ordinates q and the velocities $\dot{q}$. An arbitrary variation of the co-ordinates leads to a variation of the action of the form

$$\delta S = \int_1^2 dt\, \delta q^i \left(\frac{\partial L}{\partial q^i} - \frac{d}{dt}\frac{\partial L}{\partial \dot{q}^i} \right) + \left[\delta q^i \frac{\partial L}{\partial \dot{q}^i} \right]_1^2. \tag{1.11}$$

The Euler-Lagrange equations:

$$\frac{dp_i}{dt} = \frac{\partial L}{\partial q^i}, \tag{1.12}$$

where the generalized momenta are defined as

$$p_i = \frac{\partial L}{\partial \dot{q}^i}, \tag{1.13}$$

represent the equations of motion. As may be seen from eq.(1.11), the Euler-Lagrange equations imply that the action is stationary under variations δq^i vanishing at the end-points of the system's path in configuration space. Hence for *arbitrary* variations δq^i around a physical solution $q^i(t)$ we find

$$\delta S = \left[\delta q^i p_i \right]_1^2. \tag{1.14}$$

We now define a *symmetry* of the action as a variation δq^i such that the Lagrangian $L(q, \dot{q})$ transforms into a total time-derivative:

$$\delta L = \frac{dB}{dt}, \tag{1.15}$$

with $B(q, \dot{q})$ some function of the co-ordinates and velocities. Then the variation of the action is a pure boundary term, and comparing eqs.(1.14), (1.15) we establish the result

$$[\delta q^i p_i - B]_1^2 = 0, \tag{1.16}$$

provided $q^i(t)$ is a solution of the equations of motion. Since the times (t_1, t_2) are arbitrary, the quantity inside the brackets must be time-independent.

If there are k linearly independent symmetries, the variations δq^i depend on k independent parameters $\xi^\alpha, \alpha = 1, ..., k$:

$$\delta q^i = R^i[\xi] = \xi^\alpha R_\alpha^{(0)i} + \dot{\xi}^\alpha R_\alpha^{(1)i} + ..., \tag{1.17}$$

and the Lagrangian transforms into the derivative of a quantity $B[\xi]$:

$$B[\xi] = \xi^\alpha B_\alpha^{(0)} + \dot{\xi}^\alpha B_\alpha^{(1)} + ... \tag{1.18}$$

Now define

$$G[\xi] = R^i[\xi]p_i - B[\xi]$$
$$= \xi^\alpha G_\alpha^{(0)} + \dot{\xi}^\alpha G_\alpha^{(1)} + ... \tag{1.19}$$

By the arguments presented above, $G[\xi]$ is time-independent:

$$\frac{dG}{dt}[\xi] = \xi^\alpha \dot{G}_\alpha^{(0)} + \dot{\xi}^\alpha (G_\alpha^{(0)} + \dot{G}_\alpha^{(1)}) + ... = 0. \tag{1.20}$$

From this equation the Noether theorems follow directly.

First consider the case, that the variations $R^i[\xi]$ define a symmetry only for constant ξ^α. Then $\dot{\xi}^\alpha = 0$, and eq.(1.20) immediately reduces to

$$\frac{dG_\alpha}{dt} = 0, \qquad \alpha = 1, ..., k. \tag{1.21}$$

Thus there is one constant of motion for each (continuous) symmetry of the action. This is Noether's first theorem and the $G_\alpha = G_\alpha^{(0)}$ are called *Noether charges.*

On the other hand, if the ξ^α are allowed to depend on time in an arbitrary way, then $\xi^\alpha, \dot{\xi}^\alpha, ...$ are independent and eq.(1.20) leads to the following set of equations

$$\begin{aligned} \dot{G}_\alpha^{(0)} &= 0, \\ G_\alpha^{(0)} &= -\dot{G}_\alpha^{(1)}, \\ &... \\ G_\alpha^{(N)} &= 0, \end{aligned} \tag{1.22}$$

where we suppose $G[\xi]$ depends at most on the Nth derivative of ξ^α; in practical examples only the first derivative of ξ^α occurs. It follows, that all the components of $G[\xi]$ have to vanish separately:

$$G_\alpha^{(r)} = 0, \qquad r = 0, ..., N. \tag{1.23}$$

This proves the second theorem.

In these lectures I will mostly discuss systems for which the commutator algebra of the infinitesimal symmetry transformations (1.17) closes:

$$[\delta(\xi_1), \delta(\xi_2)]\, q^i = \delta(\xi_3) q^i, \tag{1.24}$$

where

$$\xi_3^\alpha = \xi_1^\beta \xi_2^\gamma f_{\gamma\beta}^{\ \ \alpha}. \tag{1.25}$$

Then the Poisson-brackets of the Noether charges close as well:

$$\{G_\alpha, G_\beta\} = \frac{\partial G_\alpha}{\partial q^i}\frac{\partial G_\beta}{\partial p_i} - \frac{\partial G_\beta}{\partial q^i}\frac{\partial G_\alpha}{\partial p_i} = f_{\alpha\beta}^{\ \ \gamma} G_\gamma. \tag{1.26}$$

In Dirac's terminology such algebraically closed systems of constraints are classified as ***first-class*** [2,3]. For the purpose of these lectures the restriction to systems of first-class constraints is one of practical convenience. All examples mentioned in the introduction fall in this category. However, extension to more complicated systems of constraints is sometimes necessary and I will provide an example of how one can deal with some of them in chapter 3.

The main advantage of systems with first-class constraints is, that the corresponding quantum theories are relatively easy to construct. Strictly speaking, the quantization prescription using the correspondence between Poisson brackets and quantum commutators can be used only for physical variables. For constrained systems this would imply that one must first solve the constraints, identify the physical degrees of freedom and their conjugate momenta, compute their Poisson brackets and replace those by operator commutation relations. In practice this is often difficult. However, if the constraints are first-class, one can work in an extended classical configuration space (or phase space), by using auxiliary non-physical degrees of freedom, and after solving the dynamical equations for the extended system impose constraints to select the physical solutions. The same can be done in the quantum theory: one uses an extended Hilbert space, describing solutions of a generalized Schrödinger equation, which in principle allows for states to depend on non-physical auxiliary degrees of freedom, and then selects the physical states by applying extra constraints. For example, the states of a relativistic particle are required to satisfy the Klein-Gordon equation:

$$\left(p^2 + m^2c^2\right)|phys> = 0, \tag{1.27}$$

the quantum version of the mass-shell condition (1.6). Similarly, systems with local gauge symmetries may generally be quantized in a Hilbert-space containing gauge degrees of freedom. The physical states are then selected by the constraints

$$\mathcal{G}_\alpha|phys> = 0, \tag{1.28}$$

where the $\mathcal{G}_\alpha$ now denote the quantum operator version of the constraints as derived from Noether's theorem. However, the consistency of this procedure requires the constraints to be strictly first-class, since from (1.28) it follows immediately that

$$[\mathcal{G}_\alpha, \mathcal{G}_\beta]\,|phys> = 0. \tag{1.29}$$

This is only true for arbitrary physical states, if the commutator can again be expanded into a linear combination of constraints, as in (1.26):

$$[\mathcal{G}_\alpha, \mathcal{G}_\beta] = i f_{\alpha\beta}^{\;\;\gamma}\, \mathcal{G}_\gamma. \tag{1.30}$$

Note, that the physical interpretation of the constraints is quite obvious: physical states are required to be gauge-invariant. The first-class property merely certifies that such a requirement can be imposed consistently.

In concluding this section, it should be remarked that the boundary between constraints and equations of motion is not always as sharp as suggested here. In particular in reparametrization-invariant theories there often is more than one variable which can play the role of time, and an equation which looks like a constraint in one formulation can become an equation of motion in another. As an example, consider again the relativistic particle. The equation of motion (1.1) is a differential equation in terms of the parameter time τ. The mass-shell condition (1.6) then is a *constraint* on the physical four-momentum p_μ. However, in the quantum theorie the four-momentum becomes an operator, which is realized as $p_\mu = -i\partial_\mu$ in the co-ordinate representation. Then the constraint (1.27) becomes the Klein-Gordon equation, which is a second-order partial differential equation with respect to the co-ordinate time $x^0 = ct$. The mass-shell condition may therefore also be regarded as an *equation of motion*. The Klein-Gordon equation (1.27) can itself be derived from an action

$$S_{KG} = \frac{1}{2} < \psi|\left(p^2 + m^2c^2\right)|\psi> . \tag{1.31}$$

This is the procedure called second quantization, and forms the starting point for the quantum field theory of relativistic particles.

1.3 The Yang-Mills Field

With the general theorems and procedures established, it is useful to study an example to illustrate the details. As the prime example in these lectures I have chosen non-Abelian gauge theory [4]. However, all results can be generalized without much difficulty to reparametrization-invariant theories like gravity, point-particles or strings.

We introduce a set of (real) vector fields $A_\mu^a(x)$ transforming under infinitesimal local gauge transformations with parameter $\Lambda^a(x)$ as

$$\delta A_\mu^a = D_\mu \Lambda^a = \partial_\mu \Lambda^a - f_{bc}{}^a A_\mu^b \Lambda^c, \tag{1.32}$$

where $f_{bc}{}^a$ denote the structure constants of the gauge group. The field strength $F_{\mu\nu}^a$ is defined as

$$F_{\mu\nu}^a = \partial_\mu A_\nu^a - \partial_\nu A_\mu^a - f_{bc}{}^a A_\mu^b A_\nu^c, \tag{1.33}$$

and transforms under the local gauge transformations according to the adjoint representation:

$$\delta F_{\mu\nu}^a = f_{bc}{}^a \Lambda^b F_{\mu\nu}^c. \tag{1.34}$$

The Lagrangian for this field theory is (in three space dimensions):

$$L = -\frac{1}{4g^2} \int d^3x \, (F_{\mu\nu}^a)^2. \tag{1.35}$$

It is fully invariant under the local gauge transformations (1.32),(1.34), therefore the corresponding quantity B, eq.(1.15), vanishes. It is now easy to construct the Noether charges. First we define the canonical momenta:

$$E_a^\mu = g \frac{\partial L}{\partial \dot{A}_\mu^a} = \begin{cases} 0, & \mu = 0; \\ \frac{1}{g} F_a^{0i}, & \mu = i = (1,2,3). \end{cases} \tag{1.36}$$

which equal the 'electric' components of the field strength. Observe, that in the canonical formulation the time component of the gauge field A_0^a plays the role of a Lagrange multiplier and is not a dynamical degree of freedom.

The expression for the conserved charges now becomes after inserting (1.32) and (1.36) into eq.(1.19):

$$G[\Lambda] = \frac{1}{g} \int d^3x \, D_\mu \Lambda^a E_a^\mu, \tag{1.37}$$

and we can write the constraints after partial integration as

$$G_a = \frac{\delta G[\Lambda]}{\delta \Lambda^a} = -\frac{1}{g} \vec{D} \cdot \vec{E}_a \approx 0, \tag{1.38}$$

where the last equality holds only upon using the field equations. In the canonical formulation of the theory one introduces the Poisson brackets

$$\{A_i^a(\vec{x}, t), E_{jb}(\vec{y}, t)\} = g \, \delta_b^a \delta_{ij} \delta^3(\vec{x} - \vec{y}), \tag{1.39}$$

and the Hamiltonian

$$\begin{aligned} H &= \int d^3x \left\{ \frac{1}{2}\vec{E}_a^2 - \frac{1}{g}A_0^a\, \vec{D}\cdot\vec{E}_a + \frac{1}{4g^2}(F_{ij}^a)^2 \right\} \\ &= \int d^3x\, \tfrac{1}{2}\left\{ \vec{E}_a^2 + \vec{B}_a^2 \right\}, \end{aligned} \tag{1.40}$$

where $B_i^a = (1/2g)\, \varepsilon_{ijk} F_{jk}^a$ represents the 'magnetic' components of the field strength and the terms containing the time-component of the gauge field A_0^a vanish after applying the constraints.

From the fundamental Poisson brackets (1.39), it is straightforward to check, that the constraints generate the gauge-transformations on the space-like vector fields:

$$\{A_i^a(\vec{x},t), G[\Lambda]\} = D_i\Lambda^a(\vec{x},t), \tag{1.41}$$

and that the algebra of constraints is closed:

$$\{G_a(\vec{x},t), G_b(\vec{y},t)\} = f_{ab}{}^c G_c(\vec{x},t)\delta^3(\vec{x}-\vec{y}). \tag{1.42}$$

Thus the constraints are first-class. Finally, the dynamical equations are obtained by computing the brackets of the fields with the Hamiltonian:

$$\dot{A}_i^a = \{A_i^a, H\}, \qquad \dot{E}_{ia} = \{E_{ia}, H\}. \tag{1.43}$$

From this classical formulation of Yang-Mills theory the quantum theory is obtained using the correpondence principle, replacing the fields by operators $(\mathcal{A}, \mathcal{E})$, whose equal-time commutators are i times the corresponding Poisson brackets:

$$[\mathcal{A}_i^a(\vec{x},t), \mathcal{E}_{jb}(\vec{y},t)] = ig\, \delta_b^a \delta_{ij} \delta^3(\vec{x}-\vec{y}). \tag{1.44}$$

In the Schrödinger picture the physical states are now represented by wave-functionals $\Psi[A]$ of classical fields $A_i^a(\vec{x})$ which are eigenstates of the Hamiltonian and satisfy the constraints

$$\vec{D}\cdot\vec{\mathcal{E}}_a|\psi> = -ig\, D_i \frac{\delta}{\delta A_i^a}\Psi[A] = 0. \tag{1.45}$$

Obviously such states are gauge-invariant. Note that our analysis can be applied directly to the Abelian theory, which is Maxwell's theory of the electro-magnetic field discussed in the introduction. The constraint on the divergence of the magnetic $\vec{B}$-field appearing there reduces to an identity upon writing $\vec{B}$ as the curl of the vector-potential $\vec{A}$ (the Bianchi-identity) and therefore doesn't lead to more constraints on the physical states.

Chapter 2

BRST-Quantization

2.1 Ghosts

The BRST-formalism for the construction of quantum gauge theories makes use of ghost-variables [5,6,7]. At the classical level ghosts are unphysical degrees of freedom described by co-ordinates c^α with Grassmann parity opposite to that of the corresponding gauge variable: ghosts are anti-commuting variables when the gauge degrees of freedom are commuting and *vice versa*. In most examples in this chapter the gauge variables are commuting (bosonic), hence we take ghost-variables c^α to be generically anti-commuting. However, there are some important exceptions to this rule, such as local supersymmetry, which is a fermionic gauge-invariance not discussed in these lectures, Siegel's κ-invariance of superparticles and -strings, which is closely related to supersymmetry on the world-line or world-surface of these objects, and local BRST-invariance, about which I will have more to say later. To avoid confusion, commuting ghosts will be distinguished by a different notation when they appear.

Historically, ghost variables were introduced mainly in order to develop a covariant quantization procedure for gauge theories [5]-[9]. The point of view taken here is that ghosts are required to define BRST-transformations [1] [10,11]. This point of view is somewhat more general, in the first place because in principle BRST-invariance is a purely group-theoretical concept, independent of the specific dynamics of a quantum field theory, and secondly because it is of practical importance in establishing the unitarity of quantum gauge theories not only in covariant, but also in non-covariant gauges. In the notation of sect.(1.2) a BRST-transformation of the variables q^i is defined as

$$\delta_\Lambda q^i = R^i[c] = c^\alpha R_\alpha^{(0)i} + \dot{c}^\alpha R_\alpha^{(1)i} + ..., \tag{2.1}$$

cf. eq.(1.17), where the parameters ξ^α have been replaced by ghost variables c^α and the variational derivative δ_Λ is a Grassmann-odd operator. The transformations of the ghost variables themselves are defined by requiring the BRST-transformations of the q^i to be nilpotent:

$$\delta_\Lambda^2 = 0. \tag{2.2}$$

This result is obtained if

$$\delta_\Lambda c^\alpha = -\frac{1}{2} c^\beta c^\gamma f_{\gamma\beta}^{\ \ \alpha}. \tag{2.3}$$

because of the closure of the algebra, eq.(1.24). Surprisingly, the ghost transformations themselves, eq.(2.3), are now nilpotent as well, as may be checked using the Jacobi identity

[1]For reviews of BRST-quantization see for example refs. [12]-[14]

for the structure functions $f_{\gamma\beta}^{\ \alpha}$. Note, that this identity is another way of expressing the closure of the algebra, and thus the closure is seen to guarantee the existence of a nilpotent Grassmann-odd type of transformations, first introduced in [10,11].

A canonical generator for the BRST-transformations is constructed by introducing ghost momenta π_α, having (symmetric) Poisson-brackets

$$\{c^\alpha, \pi_\beta\} = \{\pi_\beta, c^\alpha\} = \delta_\beta^\alpha, \tag{2.4}$$

and defining the BRST-charge

$$Q = c^\alpha G_\alpha - \frac{1}{2} c^\alpha c^\beta f_{\alpha\beta}^{\ \gamma} \pi_\gamma. \tag{2.5}$$

Here the G_α are the constraints satisfying eq.(1.26). As a result, the Poisson brackets of the BRST-charge take the form

$$\left\{q^i, Q\right\} = \delta_\Lambda q^i, \qquad \{c^\alpha, Q\} = \delta_\Lambda c^\alpha, \tag{2.6}$$

and

$$\{Q, Q\} = 0. \tag{2.7}$$

The last equation states that the BRST-algebra is abelian.

2.2 Quantum BRST-Invariance

In the canonical formulation of quantum field theory the ghosts and their momenta become operators obeying the equal-time anti-commutation relation [2]

$$\{c^\alpha, \pi_\beta\} = -i\, \delta_b^a. \tag{2.8}$$

The BRST-charge $\mathcal{Q}$ is given by an expression similar to (2.5),

$$\mathcal{Q} = c^\alpha \mathcal{G}_\alpha - \frac{1}{2} c^\alpha c^\beta f_{\alpha\beta}^{\ \gamma} \pi_\gamma, \tag{2.9}$$

provided no operator-ordering problems arise. For the ghost-term this is guaranteed in all cases in which the structure functions are traceless: $f_{\alpha\beta}^{\ \beta} = 0$. Note, that in principle operator ordering ambiguities can always be absorbed in a redefinition of $\mathcal{G}_\alpha$. However, when this happens, the conditions on quantum states imposed by the BRST-cohomology generally differ from what is expected naively from the classical theory. Also, the algebra of the constraints $\mathcal{G}_\alpha$ may be changed and one must check that it still closes.

The BRST-charge (2.9) for a closed algebra of constraints with structure functions $f_{\alpha\beta}^{\ \gamma}$ is nilpotent as an operator:

$$\mathcal{Q}^2 = \frac{1}{2}\{\mathcal{Q}, \mathcal{Q}\} = 0, \tag{2.10}$$

as is clear from the defining property (2.2) and the Poisson-bracket relation (2.7). This property is the most important characteristic of the BRST-algebra, since it allows us to extend the notion of gauge invariance to the state space of the quantum theory. This is an immediate consequence of the following property of the space of states, which is explained below:

---

[2] We will continue to use the same notation for the ghosts operators as for the classical variables, as confusion is not likely to occur here.

• In the extended state space constructed from physical, gauge and ghost degrees of freedom, the physical states are represented by the cohomology classes of the BRST-operator $\mathcal{Q}$ [15,16,17].

By cohomology classes I mean the equivalence classes of states which are BRST-invariant and differ at most by a BRST-transformation. In mathematical notation: $|\psi>$ represents a physical state if

$$\mathcal{Q}|\psi>=0, \tag{2.11}$$

and we identify $|\psi>$ with any state $|\psi'>$ such that

$$|\psi'>=|\psi>+\mathcal{Q}|\chi>. \tag{2.12}$$

In analogy with the terminology in the theory of differential forms we call a state which is BRST-invariant *BRST-closed*, whilst it is called *BRST-exact* if it is the BRST-transform of another state: $|\psi>=\mathcal{Q}|\chi>$. Because $\mathcal{Q}$ is nilpotent, BRST-exact states are always BRST-closed. The converse is not true however, and this is precisely what allows non-trivial cohomology classes to exist. Note also, that BRST-exact states have vanishing norm:

$$\|\mathcal{Q}|\chi>\|^2=<\chi|\mathcal{Q}^2|\chi>=0. \tag{2.13}$$

Therefore the gauge-transformations in the space of quantum-states, as induced by the BRST-transformations (2.12), correspond to adding a zero-norm state to a physical state. Such transformations do not alter physical quantities; for example, the zero-norm states are orthogonal to the physical states:

$$<\chi|\mathcal{Q}|\psi>=0, \tag{2.14}$$

and the norm of physical states is invariant under BRST-transformations:

$$\||\psi>+\mathcal{Q}|\chi>\|^2=\||\psi>\|^2. \tag{2.15}$$

The properties just mentioned check the consistency of the definition of physical states as given above. We must now show that the physical states defined in this way are the same as those defined by the first-class constraints (1.28). To this end we expand an arbitrary state in terms of ghosts:

$$|\Psi>=|\psi^{(0)}>+c^\alpha|\psi_\alpha^{(1)}>+\frac{1}{2}c^\alpha c^\beta|\psi_{\alpha\beta}^{(2)}>+\ldots \tag{2.16}$$

where the components $|\psi^{(n)}>$ satisfy

$$\pi_\alpha|\psi^{(n)}_{\alpha_1\ldots\alpha_n}>=0. \tag{2.17}$$

Then

$$\begin{aligned}\mathcal{Q}|\Psi>= \; & c^\alpha\mathcal{G}_\alpha|\psi^{(0)}> \\ & +\tfrac{1}{2}c^\alpha c^\beta\left[\mathcal{G}_\alpha|\psi_\beta^{(1)}>-\mathcal{G}_\beta|\psi_\alpha^{(1)}>+if_{\beta\alpha}^{\ \ \gamma}|\psi_\gamma^{(1)}>\right]+\ldots\end{aligned} \tag{2.18}$$

The expression on the right-hand side vanishes if

$$
\begin{aligned}
&\mathcal{G}_\alpha|\psi^{(0)}>=0,\\
&|\psi_\alpha^{(1)}>=\mathcal{G}_\alpha|\chi^{(0)}>,\\
&\dots,
\end{aligned}
\tag{2.19}
$$

where the states $|\chi^{(n)}>$ have zero ghost-momentum:

$$
\pi_\alpha|\chi^{(n)}>=0. \tag{2.20}
$$

Inserting this solution back into the original expansion of $|\Psi>$ gives

$$
\begin{aligned}
|\Psi> &=|\psi^{(0)}>+\left[c^\alpha\mathcal{G}_\alpha-\tfrac{1}{2}c^\alpha c^\beta f_{\beta\alpha}^{\ \ \gamma}\pi_\gamma\right]|\chi^{(0)}>+\dots\\
&=|\psi^{(0)}>+\mathcal{Q}|\chi>,
\end{aligned}
\tag{2.21}
$$

with

$$
|\chi>=|\chi^{(0)}>+c^\alpha|\chi_\alpha^{(1)}>+\dots \tag{2.22}
$$

Hence $|\Psi>$ has precisely the form (2.12), with $|\psi^{(0)}>$ a BRST-closed state which manifestly satisfies the first-class constraints, eq.(2.19). This establishes the one-to-one correspondence of the BRST-cohomology classes with the physical states as defined by the first-class constraints.

Before concluding this discussion of the BRST-condition on physical states let us recall, that in reparametrization-invariant theories sometimes constraints can be turned into equations of motion, as in the example of second quantization of relativistic particles at the end of sect.(1.2). Since the BRST-condition represents the first-class constraints, it can similarly be turned into a field equation in theories of this kind. Thus we can construct a second-quantized version of a theory in which first-class constraints become equations of motion, by postulating the action

$$
S_{BRST}=\frac{1}{2}<\psi|\mathcal{Q}|\psi>, \tag{2.23}
$$

where the inner-product is supposed to be well-defined on the space of $\mathcal{Q}$-cohomology classes. Note, that this action indeed has the gauge-invariance

$$
|\psi>\rightarrow|\psi'>=|\psi>+\mathcal{Q}|\chi>, \tag{2.24}
$$

so the action (2.23) generates not only the correct equations of motion, but also the cohomology of BRST-equivalence classes. This procedure is the starting point for certain formulations of string field theory [18,19].

2.3 Operator Cohomology

In the extended Hilbert-space in which the ghost degrees of freedom live, physical states are only defined up to a BRST gauge-transformation, and therefore they correspond to BRST-*equivalence classes* of state-vectors, rather than to a particular vector. This peculiarity is reflected in the definition of physical observables. In principle an observable must transform physical states into physical states: it cannot mediate transitions between physical and unphysical states. This implies, that observables must have ghost number $N_{gh}=0$, and *commute* with the BRST-operator in the physical sector of the state-space: if $\mathcal{O}$ is a map from the physical Hilbert-space into itself ($\mathcal{O}: H_{phys}\longrightarrow H_{phys}$), then

$$\mathcal{Q}\mathcal{O}|\Psi> \equiv \mathcal{Q}|\Psi'> = 0, \tag{2.25}$$

for any physical state $|\Psi>$ with $\mathcal{Q}|\Psi> = 0$. Therefore

$$[\mathcal{Q}, \mathcal{O}]\,|\Psi> = 0, \tag{2.26}$$

for any state $|\Psi>$ in H_{phys}. However, this condition has trivial solutions of the type $\mathcal{O} = \{\mathcal{Q}, \Phi\}$, where Φ is an operator with ghost number $N_{gh} = -1$. We conclude, that observables are only defined modulo a BRST-transformation:

$$\mathcal{O}' = \mathcal{O} + \{\mathcal{Q}, \Phi\}, \tag{2.27}$$

and therefore they really correspond to BRST-cohomology *classes* of operators in the extended space of states, not to particular operators.

A particular example of an observable is the Hamiltonian, which generates the time-evolution of the states. Clearly, a physical state must remain a physical state in time, in order for the theory to be unitary. From the above discussion it follows, that any two Hamiltonians differing by a BRST-exact operator:

$$\mathcal{H}' = \mathcal{H} + \{\mathcal{Q}, \Phi\}, \tag{2.28}$$

now must be considered physically equivalent, since they have the same matrix elements between physical states. Choosing a particular Hamiltonian then amounts to choosing a *gauge*. Indeed, in the Lagrangian formulation of the theory, the arbitrariness (2.28) implies the physical equivalence of two Lagrangians differing by a BRST-exact term:

$$L = \dot{q}^i p_i - \mathcal{H} \quad \Leftarrow \quad L' = L - \{\mathcal{Q}, \Phi\}. \tag{2.29}$$

Now adding to the classical (BRST-invariant) action a term which is the BRST-transform of an expression Φ is precisely the Faddeev-Popov prescription for defining a gauge-fixed Lagrangian [7,20], the gauge being defined by the choice of Φ. Thus we see how the Faddeev-Popov procedure is justified from the canonical point of view.

2.4 Quantum Yang-Mills Theory

The above procedures can be applied directly to the example of the Yang-Mills field of sect.(1.3). In order to define BRST-transformations it is convenient to introduce a constant Grassmann-odd parameter Λ and write

$$\begin{aligned} \delta(\Lambda)A_\mu^a &= iD_\mu\left(\Lambda c^a\right) \\ &= i\partial_\mu\left(\Lambda c^a\right) - i f_{bc}{}^a A_\mu^b \Lambda c^c, \\ \delta(\Lambda)c^a &= -\tfrac{i}{2}\,\Lambda c^b c^c f_{cb}{}^a, \end{aligned} \tag{2.30}$$

where the $c^a(x)$ are the Faddeev-Popov ghost fields. Clearly, $\delta(\Lambda)A_\mu^a$ is just a gauge transformation on A_μ^a with $\xi^a = \Lambda c^a$. It is easy to verify that the transformations are nilpotent: $\delta(\Lambda_2)\delta(\Lambda_1) = 0$. This simple algebraic property is a crucial element of the BRST-procedures. Other representations of this algebra exist, the simplest one being a two-dimensional representation consisting of two fields $(\bar{c}, b)$ of opposite Grassmann-parity transforming as

$$\delta(\Lambda)\bar{c} = \Lambda b, \qquad \delta(\Lambda)b = 0. \tag{2.31}$$

Such representations also play a role in the quantization of gauge theories, because they provide the variables of ghost number $N_{gh} = -1$ necessary to construct the gauge-fixing term in the Hamiltonian/Lagrangian, cf. eqs.(2.28),(2.29). These variables are usually called the *anti-ghosts*, not because they represent the ghost anti-particles (there is no relation with the ghost variables through complex conjugation: both the ghosts and the anti-ghosts are real), but because they have opposite ghost number. Clearly, if $\bar{c}$ has $N_{gh}(\bar{c}) = -1$, then b has $N_{gh}(b) = 0$, as follows from the transformation rules (2.31). These variables b play the role of Lagrange multipliers used to impose the gauge. Because the number of gauge conditions must equal the number of gauge degrees of freedom, the number of anti-ghosts and Lagrange multipliers equals the number of ghosts.

To show how the procedure works, we introduce anti-ghost BRST-multiplets $(\bar{c}_a, b_a)$ for the Yang-Mills theory and choose a particular gauge, for example the physical temporal gauge $A_0^a = 0$. This gauge is obtained by choosing $\Phi = \bar{c}_a A_0^a$, which has ghost number equal to -1 as required. In the Lagrangian BRST-formalism the corresponding action is

$$\begin{aligned} S' &= \int d^4x \left\{ -\frac{1}{4g^2}\left(F_{\mu\nu}^a\right)^2 - \delta_\Lambda\left(\bar{c}_a A_0^a\right)\right\} \\ &= \int d^4x \left\{ -\frac{1}{4g^2}\left(F_{\mu\nu}^a\right)^2 - b_a A_0^a + i\bar{c}_a D_0 c^a\right\} \end{aligned} \tag{2.32}$$

We have seen, that in the classical theory A_0^a plays the role of a Lagrange multiplier for the constraints (1.38). In the presence of the ghost fields, eq.(2.32), these equations are modified and we find

$$b_a - \frac{1}{g}\vec{D}\cdot\vec{E}_a + i f_{ab}{}^c \bar{c}_c c^b = 0, \tag{2.33}$$

which is no longer a constraint on the vector fields, but an equation to eliminate b_a. Because of this modification the constraints (1.38),(1.45) are no longer guaranteed to be satisfied in this version of the theory. However, they are now reobtained by applying the BRST-conditions (2.11) and (2.12). Clearly only the combination of the action (2.32) *plus* the BRST-conditions is equivalent to the original version of the theory.

Note also, that in (2.32) the anti-ghost $\bar{c}_a$ plays the role of the ghost-momentum, because of the special choice of gauge. Therefore the canonical anti-commutation relations for the (anti-)ghosts read in the temporal gauge:

$$\{c^a(\vec{x},t), \bar{c}_b(\vec{y},t)\} = \delta_b^a \delta^3(\vec{x}-\vec{y}). \tag{2.34}$$

As usual in field theory the conserved BRST-charge can be obtained from a divergenceless BRST-current J_μ:

$$\mathcal{Q} = \int d^3x\, J_0, \qquad \partial\cdot J = 0, \tag{2.35}$$

where J_μ is found by considering a BRST-variation of the action with a local (space-time dependent) parameter $\Lambda(x)$. The rigid BRST-invariance of S' then guarantees this variation to be of the form [3]:

$$\delta[\Lambda(x)]S' = -i\int d^4x\, \partial^\mu\Lambda\, J_\mu + \textit{boundary terms}. \tag{2.36}$$

This is a special instance of the algorithm known as the Noether procedure. Eqs. (2.35), (2.36) are valid in any BRST-invariant formulation of the theory. However, only with our special choice of temporal gauge the current assumes the simple form

[3] See for example ref.[1]

$$\mathcal{J}_0 = -\tfrac{1}{g}c^a \vec{D}\cdot\vec{\mathcal{E}}_a + \tfrac{i}{2}c^a c^b f_{ba}{}^c \bar{c}_c,$$
$$\vec{\mathcal{J}} = 0. \tag{2.37}$$

Clearly all the relevant information is in the time-component and hence in the BRST-charge $\mathcal{Q}$. The expression for $\mathcal{Q}$ thus obtained is the same one as follows directly from the prescription (2.9) using (1.38). However, the present derivation in terms of the BRST-current is useful in later considerations.

Chapter 3

Local BRST-Invariance

3.1 The Noether Construction

In the discussion of quantum Yang-Mills theory I made use of an algorithm for the construction of the BRST-current known as the Noether procedure. This construction of the conserved current associated with a rigid symmetry of an action is of quite general validity: it is correct whenever under local transformations the variation of the action modulo boundary terms can be written in a form which contains only first derivatives of the parameters [1]. To see this, suppose that the degrees of freedom of the theory are continuous space-time fields $q^i(x)$ transforming as in (1.17):

$$\delta q^i(x) = R^i[\xi] \tag{3.1}$$

with space-time dependent parameters $\xi^\alpha(x)$. Let the action transform as

$$\delta S = \int d^n x \, \partial_\mu \xi^\alpha \, J^\mu_\alpha + \textit{boundary terms}. \tag{3.2}$$

for some J^μ_α, depending on the form of the R^i_α. We first observe, that under rigid transformations ($\partial_\mu \xi^\alpha = 0$), the action is invariant modulo boundary terms. Thus these rigid transformations do constitute a symmetry of the theory. On the other hand, after using the equations of motion the action is invariant under ***any*** transformation of the $q^i(x)$ which vanishes at the boundary. This is the content of the variational principle. Therefore, after a partial integration of equation (3.2) and throwing away boundary terms we find that

$$\partial \cdot J_\alpha = 0, \tag{3.3}$$

upon substitution of the equations of motion. It follows, that the quantity

$$G = \int d^{n-1}\vec{x} \, J_0(\vec{x}, t) \tag{3.4}$$

is conserved. This expression agrees with the previous one (1.19) for the Noether charge, modulo terms which vanish after using the equations of motion.

A corollary of this result is, that one can compensate the local variation of the action (3.2) to first order by a term

$$\Delta S = \int d^n x \left(-A^\alpha_\mu J^\mu_\alpha \right), \tag{3.5}$$

provided A^α_μ transforms as

$$\delta A^\alpha_\mu = \partial_\mu \xi^\alpha + \delta' A^\alpha_\mu. \tag{3.6}$$

Thus A_μ^α is seen to be the gauge field of the local symmetry to be realized. The problem of constructing an action which is fully invariant under the local symmetries (3.1),(3.6) now amounts to finding the form of the higher-order terms $\delta' A_\mu^\alpha$ in (3.6) such that the variations

$$\delta' A_\mu^\alpha J_\alpha^\mu + A_\mu^\alpha \delta J_\alpha^\mu \tag{3.7}$$

can be compensated by a term of the form

$$\Delta' S = \int d^n x \, \frac{1}{2} \, A_\mu^\alpha H_{\alpha\beta}^{\mu\nu} A_\nu^\beta \tag{3.8}$$

etc. This Noether construction of actions with local symmetries is used in the following section to construct actions with local BRST-invariance.

3.2 Gauging BRST-Symmetry

BRST-invariance is the translation of the local gauge-invariance of a classical theory into a symmetry of the quantum mechanical Hilbert space. It corresponds to a rigid symmetry of the effective gauge-fixed action, as for example in eq.(2.30). It is possible to turn this symmetry itself into a local gauge-symmetry [20]-[23], for example using the Noether construction sketched above. Actually, this turns out to provide some of the simplest non-trivial examples of the Noether procedure possible: in all known cases the construction is complete already after the first step, i.e. adding the Noether term eq.(3.5) [23]. The reason is, that the Noether current of BRST-invariance has itself extremely simple transformation properties; for example, the BRST Noether-current (2.35) of Yang-Mills theory in the temporal gauge is itself locally BRST-*invariant*:

$$\delta[\Lambda(x)] \, J^\mu = 0. \tag{3.9}$$

The same statement is true also in the covariant gauge $\partial \cdot A^a = 0$ [23]. In the case of quantum-gravity the BRST-current is not invariant under local BRST-transformations (except in $d = 1$), but transforms precisely in such a way that it can be compensated for by the variation $\delta' \psi_\mu$ of the anti-commuting BRST-gauge field ψ_μ, leading to the vanishing of the corresponding combination of terms in eq.(3.7) [23]:

$$\delta' \psi_\mu \, J^\mu + \psi_\mu \, \delta J^\mu = 0. \tag{3.10}$$

Thus in both cases we find that with the correct transformation rule for the BRST gauge-field the action

$$S = S' + i \int d^n x \, \psi_\mu J^\mu \tag{3.11}$$

is invariant under *local* BRST-transformations.

For the example of the Yang-Mills theory in the temporal or covariant gauge this result is extremely simple to prove: since the current is invariant under local BRST-transformations, and the variation of the gauge-fixed action S' takes the form (2.36), the action S of eq.(3.11) is locally BRST-invariant (modulo boundary terms) if we define

$$\delta[\Lambda(x)] \, \psi_\mu = \partial_\mu \Lambda. \tag{3.12}$$

Analoguous results hold for the BRST-extended version of reparametrization-invariant classical theories, like point-particles and strings.

3.3 Physical Content of Local BRST-Invariance

Having constructed theories with local BRST-invariance, we must now address the question what physical content such theories have. The answer turns out to be surprisingly simple:

• Theories with gauged BRST-symmetry have the *same* physical content as the theories with rigid BRST-invariance from which they are derived [23].

At the classical level this is not difficult to prove: the locally BRST-invariant theory reduces to the original theory with rigid BRST-invariance by choosing an appropriate gauge-condition. Again, the Yang-Mills theory in the temporal gauge provides a simple example: because the spatial components of the BRST-current vanish: $\vec{\mathcal{J}} = 0$, the Noether term in the action contains only the time-components of the current and of the BRST gauge-field. Eq.(3.11) then reads

$$\mathcal{S} = \mathcal{S}' + i \int d^n x \, \psi_0 \, \mathcal{J}^0. \tag{3.13}$$

Choosing the Weyl gauge (temporal gauge)

$$\psi_0 = 0, \tag{3.14}$$

the action $\mathcal{S}$ is seen to reduce to $\mathcal{S}'$. In more complicated gauges, in which also the space components of the BRST current and the BRST gauge-field come into play, one can prove the same result, provided both the scalar and the vector part of the gauge field vanish. This is possible if one combines gauge freedom and field equations (equations of motion):

$$\begin{aligned} \partial \cdot \psi &= 0, \qquad && (\textit{gauge condition}), \\ \partial_\mu \psi_\nu - \partial_\nu \psi_\mu &= 0, \qquad && (\textit{equation of motion}). \end{aligned} \tag{3.15}$$

The equation of motion proposed here, which asserts that there is no long-range BRST-field contributing to the BRST-charge at some closed surface very far away, is fully consistent with the coupling of the BRST gauge-field defined by the Noether term [23]. It is also necessary for the BRST-condition $Q = 0$ to have the same meaning here as in the original theory.

One can show, that all these arguments can be generalized to the case of gravity and other reparametrization-invariant theories. This confirms the validity of our general statement about equivalence of local and rigid BRST-invariance at the classical level. It remains to check, that the result holds in the case of quantum theories as well.

To interpret local BRST-symmetry at the quantum level, we must consider the first-class constraint following from BRST gauge-invariance. Since the Noether charge of the local BRST-symmetry is the same as that associated with the rigid transformations (because of the simple structure of the BRST gauge-couplings and our choice of field equation), the first class constraint on the physical states becomes

$$\mathcal{Q}|phys> = 0. \tag{3.16}$$

This is precisely the BRST-condition (2.11) and shows, that the *normalizable* physical states in this theory are the same as those of the original theory [1].

[1] Note, that we have not removed here the degeneracy associated with the spurious zero-norm states; this can be done by gauging the ghost-number current, see ref. [26]

A convenient way of obtaining the condition (3.16) is to perform the BRST-quantization of the theory with local BRST-invariance. I will call this procedure the 'second-level' or '$BRST_2$-quantization'. First note, that the quantum algebra of first-class constraints of our theory is very simple:

$$\{\mathcal{Q}, \mathcal{Q}\} = 2\mathcal{Q}^2 = 0. \tag{3.17}$$

All structure constants vanish and we need to introduce only one *commuting* ghost variable $s(x)$. Applying eq.(2.9) then gives the second-level $BRST_2$-operator Ω:

$$\Omega = \int d^{n-1}\vec{x}\, s(x)\, \mathcal{J}_0(x), \tag{3.18}$$

which is nilpotent as expected:

$$\Omega^2 = \frac{1}{2}\, \{\Omega, \Omega\} = 0. \tag{3.19}$$

It is now straightforward to verify, that the cohomology classes of Ω reproduce exactly the states (3.16):

$$\Omega|\psi> = 0, \qquad and \qquad |\psi> \sim |\psi> + \Omega|\chi>, \tag{3.20}$$

imply $\mathcal{Q}|\psi>= 0$. The proof follows the lines of sect.(2.2), but with power series expansions in the variable $s(x)$ instead of polynomials in anti-commuting variables c^α. I leave the details of the proof as an exercise. However, I wish to draw attention to one notable property of Ω. Expanding the $BRST_2$-ghost $s(x)$ into Fourier components:

$$s(x) = s_0 + \sum_{n \neq 0} \left(s_n e^{2\pi i n x} + s_n^* e^{-2\pi i n x} \right), \tag{3.21}$$

we observe, that Ω is of the form:

$$\Omega = s_0 \mathcal{Q} + ... \tag{3.22}$$

Therefore rescaling the original ghosts by a constant factor s_0 we can write

$$\Omega' = \mathcal{Q} + ... \tag{3.23}$$

and the relation between Ω and $\mathcal{Q}$ becomes transparent: Ω is just a rescaled version of $\mathcal{Q}$ with extra terms, containing the higher-order Fourier modes of the BRST-current $\mathcal{J}^\mu$ and needed to eliminate the dependence of states on the corresponding modes of the $BRST_2$-ghosts $s(x)$.

3.4 Gupta-Bleuler Constraints

The reader who has followed the discussion so far may well wonder what is the point in all this. We had an elegant way of reformulating first-class constraints in terms of a cohomology problem, by introducing the notion of BRST-transformations. Local BRST-invariance then leads to another, slightly more complicated, cohomology problem with the same solutions, and clearly this process can be continued indefinitely without any clear advantage, except that each time the BRST-condition at one level becomes a first-class constraint on the next one.

In the folowing sections I want to show, that there are situations in which local BRST-symmetry can accomplish something which is not possible in the original rigid BRST-formulation of a theory. However, the theories I have in mind are somewhat more complicated than the ones we have been discussing so far. Namely, we will now consider theories in which not all constraints are of first-class type. Then the (anti-)commutator algebra of the constraints in the quantum theory takes the general form

$$[\mathcal{G}_\alpha, \mathcal{G}_\beta\} = i f_{\alpha\beta}^{\ \ \gamma} \mathcal{G}_\gamma + 2c_{\alpha\beta}, \tag{3.24}$$

where the last term is *not* a constraint and does not vanish identically. Note that in our notation we have explicitly allowed constraints of either Grassmann-parity (even or odd) with a view to later applications. With the algebra of constraints (3.24) it is no longer possible to define physical states in an extended Hilbert space by the condition (1.28), because eq.(1.29) cannot be fulfilled. However, it is often possible to separate the constraints in such a way, that even with the algebra (3.24) they become 'almost' first class. More precisely, we can often construct a set of complex (holomorphic) constraints Φ_A and their (anti-holomorphic) conjugates $\Phi_A^\dagger$ as complex linear combinations of the $\mathcal{G}_\alpha$ with the (anti-)commutator algebra [24,25]

$$\begin{aligned}
[\Phi_A, \Phi_B\} &= i f_{AB}^{\ \ C} \Phi_C, \\
\left[\Phi_A^\dagger, \Phi_B^\dagger\right\} &= i\, \Phi_C^\dagger \bar{f}^C_{\ AB}, \\
\left[\Phi_A, \Phi_B^\dagger\right\} &= 2\, Z_{AB},
\end{aligned} \tag{3.25}$$

where all the non-closure of the algebra is contained in the last (anti-)commutator. Thus the algebra of holomorphic constraints is first-class by itself, and so is the set of anti-holomorphic ones. Only the mixed (anti-)commutators do not close in the set of constraints.

The physical states in theories of this kind can now be defined by the holomorphic first-class constraints in ket-space

$$\Phi_A|phys> = 0. \tag{3.26}$$

the anti-holomorphic ones being realized on the bra-states:

$$< phys|\Phi_A^\dagger = 0. \tag{3.27}$$

This procedure is known as the Gupta-Bleuler method of imposing constraints in quantum theory. In the following it is shown how to construct a BRST-extension of the Gupta-Bleuler method for handling holomorphic first-class constraints [25].

3.5 Non-Abelian BRST-Invariance

Because the holomorphic first-class constraint algebra (3.25) can be realized in Hilbert space by the Gupta-Bleuler procedure (3.26).(3.27) one might expect that it is possible to generalize the BRST-procedure to this case, provided one uses a complex holomorphic BRST-operator $\mathcal{Q}$ with conjugate $\mathcal{Q}^\dagger$, and defines the physical states by

$$\mathcal{Q}|phys> = 0, \qquad < phys|\mathcal{Q}^\dagger = 0. \tag{3.28}$$

This complex BRST-charge should be constructed out of the holomorphic first-class constraints only, and its cohomology classes should correspond to the physical states. To achieve this, one simply introduces a set of complex holomorphic ghosts c^A with conjugates $\bar{c}^A$, and their momenta $(\pi_A, \bar{\pi}_A)$:

$$\left[\pi_A, c^B\right\} = -i\delta_A^B. \tag{3.29}$$

Then one defines:

$$\mathcal{Q} = c^A \Phi_A - \tfrac{1}{2}(-)^A \, c^A c^B f_{BA}^{\ \ C} \pi_C,$$
$$\mathcal{Q}^\dagger = \bar{c}^A \Phi^\dagger_A - \tfrac{1}{2}(-)^A \, \bar{\pi}_C \bar{f}^C_{\ BA} \bar{c}^B \bar{c}^A \tag{3.30}$$

These operators are nilpotent: $\mathcal{Q}^2 = 0$, and their cohomology classes do correspond to the solutions of the (anti-)holomorphic Gupta-Bleuler constraints (3.26) indeed. However, this procedure does not lead to a satisfactory definition of physical states. The reason is, that the BRST-exact states have non-zero norm; thus they are not spurious and donot decouple from physical matrix elements, which may lead to a violation of unitarity. To see this, consider the algebra of $\mathcal{Q}, \mathcal{Q}^\dagger$ and Z, defined by

$$Z = c^A \, Z_{AB} \, \bar{c}^B. \tag{3.31}$$

The (anti-)commutator algebra reads:

$$\begin{aligned} &\mathcal{Q}^2 = 0, \qquad\qquad \mathcal{Q}^{\dagger 2} = 0, \\ &\left\{ \mathcal{Q}, \mathcal{Q}^\dagger \right\} = 2\, Z, \\ &[\mathcal{Q}, Z] = 0, \qquad \left[\mathcal{Q}^\dagger, Z \right] = 0. \end{aligned} \tag{3.32}$$

In other words, we have a non-Abelian super-algebra of holomorphic and anti-holomorphic BRST-operators with a central charge Z. But since the central charge is not a constraint (it doesn't annihilate all physical states), and $\mathcal{Q}^\dagger$ doesn't annihilate physical ket-states, it now follows that

$$\begin{aligned} \| \, \mathcal{Q}|\chi> \|^2 \; &=< \chi|\mathcal{Q}^\dagger \mathcal{Q}|\chi> \\ &= 2 \; < \chi|Z|\chi> - < \chi|\mathcal{Q}\mathcal{Q}^\dagger|\chi> . \end{aligned} \tag{3.33}$$

Both terms on the right-hand side are non-zero, and the two terms donot cancel identically. Therefore the norm of $\mathcal{Q}|\chi>$ is non-zero. In addition we observe, that the starting point for second quantization in the BRST-formalism given by the action formula (2.23) is no longer available, because this action is not real. We conclude, that naive BRST-quantization of systems with holomorphic first-class constraints in the operator formulation is not consistent.

3.6 Gauging the non-Abelian BRST-Algebra

A solution to the problem of performing the BRST-quantization on a system with holomorphic first-class constraints is provided by the gauging of the non-abelian BRST-algebra constructed above. The key to this solution is the observation that even though the BRST-charges are complex and their algebra is non-abelian, the BRST$_2$-charge Ω is real and nilpotent. We construct Ω by introducing two commuting ghosts $(s, s^\dagger)$ for local complex BRST-transformations, and two canonically conjugate anti-commuting ghosts (γ, β) for the central charge Z:

$$\{\gamma, \beta\} = 1. \tag{3.34}$$

Now we use the general expression (2.29) for the BRST-operator which holds for an arbitrary abelian or non-abelian algebra. This gives:

$$\Omega = s^\dagger \mathcal{Q} + \mathcal{Q}^\dagger s + \gamma\, Z - 2 s^\dagger s\, \beta \tag{3.35}$$

It is easy to check the reality and nilpotency of this operator. Therefore it can be used to construct $BRST_2$-cohomology classes, with states in the same equivalence class differing only by a state *of norm zero*. It follows, that $BRST_2$-exact states are spurious in the usual sense and decouple from physical matrix elements.

It only remains to investigate the physical content of this cohomology structure. The physical states now correspond to classes of states in the doubly-extended Hilbert space containing also $(s, s^\dagger, \beta, \gamma)$, the classes being defined by

$$\Omega|\psi>= 0, \tag{3.36}$$

with the equivalence relation

$$|\psi>\sim|\psi'>=|\psi>+\Omega|\chi>. \tag{3.37}$$

From the symmetry between $\mathcal{Q}$ and $\mathcal{Q}^\dagger$ and their respective ghosts in the expression for Ω, it is easy to understand that one gets two sets of physical states: those corresponding to the cohomology classes of $\mathcal{Q}$ and those of $\mathcal{Q}^\dagger$. Thus the spectrum is doubled as compared to what was expected naively from the holomorphic first-class constraints:

- the solution of the $BRST_2$-cohomology consists of the states $|\psi>$ which are solutions of

$$\mathcal{Q}|\psi>= 0 \qquad or \qquad \mathcal{Q}^\dagger|\psi>= 0, \tag{3.38}$$

and these may be identified with the solutions of

$$\Phi_A|\psi>= 0 \qquad or \qquad \Phi_A^\dagger|\psi>= 0, \tag{3.39}$$

modulo a holomorphic/anti-holomorphic BRST-transformation [2]. This result is straightforward to derive by explicit solution of eqs.(3.36), (3.37) along the lines of sect.(2.2).

3.7 The Superparticle

To conclude this overview of BRST-quantization and local BRST-invariance I will give an example of a theory with holomorphic first-class constraints, illustrating the previous general results [27]. This example is the superparticle [28,29], a point-particle theory having both bosonic and fermionic states of equal mass-energy in its spectrum. In the co-ordinate representation the wave-functions are superfields depending on the position co-ordinates x^μ and anti-commuting spinor co-ordinates θ. The equations of motion and constraints describing this system are [30]

$$\begin{aligned} &D_\mp\Phi(x,\theta) = 0, \qquad \gamma\cdot p\, D_\pm\Phi(x,\theta) = 0, \\ &p^2\Phi(x,\theta) = 0, \end{aligned} \tag{3.40}$$

where

$$D_\pm = i\left(\frac{\partial}{\partial\bar{\theta}_\pm} + \gamma\cdot p\,\theta_\mp\right), \tag{3.41}$$

is the super-covariant derivative and $\theta_\pm$ denotes the chiral components of the spinor co-ordinate: $\gamma_5\theta_\pm = \pm\theta_\pm$. Note that the equations (3.40) with opposite chirality are complex conjugates. Therefore we can separate them according to the Gupta-Bleuler prescription,

[2] Again, to remove the degeneracy due to the (anti-)holomorphic BRST-transformations one should actually gauge also the ghost-number operators of the BRST-algebra [26]

and work either with only the upper or the lower signs in eq.(3.40). The difference is, that in one case the theory describes a massless scalar and a helicity $+1/2$ fermion, in the other case a massless scalar and a helicity $-1/2$ fermion.

In order to achieve the BRST-quantization we introduce two sets of *commuting* chiral spinor ghosts $(\omega_{\mp}, \phi_{\pm})$ for the first two spinorial constraints in (3.40), and an anti-commuting scalar ghost c for the real mass-shell constraint [27]. The (anti-)holomorphic BRST-operators become

$$\begin{aligned} \mathcal{Q} &= \bar{\omega}_- D_- + \bar{D}_+ \gamma \cdot p\, \phi_- + c p^2 - 2\bar{\omega}_- \phi_- b, \\ \mathcal{Q}^\dagger &= \bar{D}_+ \omega_+ + \bar{\phi}_+ \gamma \cdot p D_- + c p^2 - 2\bar{\phi}_+ \omega_+ b, \end{aligned} \tag{3.42}$$

where b is the ghost operator conjugate to c. The system described here is truely *holomorphic* first-class only, since the anti-commutator of the two conjugate BRST-operators has a non-vanishing central charge Z:

$$\frac{1}{2}\left\{\mathcal{Q}, \mathcal{Q}^\dagger\right\} = Z = \left(\bar{\omega}_- - \bar{\phi}_+ \gamma \cdot p\right) \gamma \cdot p \left(\omega_+ - \gamma \cdot p \phi_-\right). \tag{3.43}$$

We can now insert the expressions into Ω, eq.(3.35), and check that this operator is real and nilpotent. The solutions of the BRST_2-cohomology problem defined by Ω now describe the two supermultiplets with opposite helicity referred to above, i.e. we find *two* massless scalars and the two helicity states of a massles fermion. Together these states describe the supermultiplet known in field theory as the chiral or scalar multiplet. Our theorem stating that we can only obtain this *combined* set of states in a properly BRST-quantized theory is seen to be equivalent here with the CPT-theorem. This is not surprising, since a real BRST-operator like Ω can be used to construct a consistent second quantized field theory model with *real* action (2.23), in which $\mathcal{Q}$ is to be replaced by Ω. For such theories the CPT-theorem is known to hold. This provides at the same time a consistency check and an interpretation of the general results as applied to the superparticle.

Acknowledgement.

I am indebted to S. Aoyama, J. Kowalski-Glikman, Z. Hasiewicz and J. Lukierski for long and stimulating discussions, especially on some of the topics discussed in chapter 3.

Bibliography

[1] H. Georgi: Weak Interactions and Modern Particle Theory (Benjamin/Cummings, 1984)

[2] P.A.M. Dirac, Proc. Roy. Soc. A246 (1958), 326

[3] A. Hanson, T. Regge and C. Teitelboim, Proc. Acc. Naz. d. Lincei (1976)

[4] C.N. Yang and R.C. Mills, Phys. Rev. 96 (1954), 191

[5] R.P. Feynman, Acta Phys. Pol. 26 (1963), 697

[6] B. DeWitt, Phys. Rev. 162 (1967), 1195

[7] L.D. Faddeev and V.N. Popov, Phys. Lett. B25 (1967), 29

[8] G. 't Hooft, Nucl. Phys. B35 (1971), 167

[9] G. 't Hooft and M. Veltman, Nucl. Phys. B50 (1972), 318

[10] C. Becchi, A. Rouet and R. Stora, Ann. Phys. 98 (1976), 28

[11] I.V. Tyutin, Lebedev preprint FIAN 39 (1975), unpublished

[12] T. Kugo and I. Ojima, Supp. Prog. Theor. Phys. 66 (1979), 1

[13] M. Henneaux Phys. Rep. 126 (1985), 1

[14] L. Baulieu, Phys. Rep. 129 (1985), 1

[15] T. Kugo and I. Ojima, Phys. Lett. B73 (1978), 459; Prog. Theor. Phys. 60 (1978), 1869

[16] G. Curci and R, Ferrari, Nuovo Cim. 35A (1976), 273

[17] K. Fujikawa, Prog. Theor. Phys. 59 (1978), 2045; Prog. Theor. Phys. 63 (1980), 1364

[18] W. Siegel, Phys. Lett. B149 (1984), 157; *ibid.* 162; *corrected version*: Phys. Lett. B151 (1985), 391;*ibid.* 396

[19] E. Witten, Nucl. Phys. B268 (1986), 79; Nucl. Phys. B276 (1986), 291

[20] F.R. Ore and P. van Nieuwenhuizen, Nucl. Phys. B204 (1982), 317

[21] L. Baulieu, B. Grossman and R. Stora, Phys. Lett. B180 (1986), 95

[22] J.P. Ader and J.C. Wallet, Phys. Lett. B192 (1987), 103

[23] J.W. van Holten, Nucl. Phys. B315 (1989), 740

[24] J. Lukierski, in: Supersymmetry and Supergravity 1983, ed. B. Milewski (World Scientific, Singapore, 1983)

[25] Z. Hasiewicz, J. Kowalski-Glikman, J. Lukierski and J.W. van Holten, Phys. Lett. B217 (1989), 95

[26] J. Kowalski-Glikman, J.W. van Holten, Z. Hasiewicz and J. Lukierski, *in preparation*

[27] S. Aoyama, J. Kowalski-Glikman, J.Lukierski and J.W. van Holten, Phys. Lett. B216 (1989), 133

[28] R. Casalbuoni, Phys. Lett. B62 (1976), 49

[29] L. Brink and J. Schwarz, Phys. Lett. B100 (1981), 310

[30] S.J. Gates, M.T. Grisaru, M. Rocek and W. Siegel: Superspace (Benjamin/Cummings, 1983)

CAN PERTURBATIVE FIELD THEORY BE MADE FINITE?

Iwo Bialynicki-Birula

Institute for Theoretical Physics
Polish Academy of Sciences
Warsaw, Poland

I will argue in this lecture that the answer to the question posed in the title is a qualified 'yes'.

Almost forty years ago Dyson[1] pointed out that the perturbation expansion of quantum electrodynamics can not be analytic at the point $e^2 = 0$. If the convergence radius of the perturbation series were finite, argued Dyson, one could replace e^2 by $-e^2$ and still obtain a convergent expansion. However, the negative value of e^2 corresponds to attraction between like charges so that the states with a very large number of electrons would have the energy lower than the vacuum. The vacuum will, therefore, become unstable and the theory must become inconsistent.

This lack of analyticity at $e^2 = 0$ has also been found when various specific calculations in quantum electrodynamics were carried out. For example, the hyperfine splitting in positronium is given by the expression[2]

$$\Delta\nu_{pos} = \frac{1}{2}\alpha^4 m\left[\frac{7}{6} - \left(\frac{16}{9} + \ln 2\right)\frac{\alpha}{\pi} + \frac{1}{2}\alpha^2 \ln\frac{1}{\alpha} - 1.937\alpha^2\right], \tag{1}$$

which contains the logarithm of the fine structure constant.

Similar logarithmic terms are present in the formulas for the Lamb shift and for the hyperfine splitting in hydrogen.[2]

A logarithm of the electric charge also appears in the formula for the correlation energy of the degenerate electron gas[3], which for for small values of e takes on the form

$$E_{corr}/N = (me^4/\pi^2\hbar^2)\ (1 - \ln 2)\ \ln(me^2/r_o\hbar^2), \qquad (2)$$

where r_o is the packing radius of the electrons.

In all these cases logarithms of the fine structure constant are multiplied by powers of the electric charge.

The purpose of this lecture is to show that logarithmic terms of a similar type appear in the lowest order expressions for the renormalization constants in quantum field theory if a modified perturbation theory is used.

Standard perturbation expansion in relativistic quantum field theories leads to divergent integrals over momenta of virtual particles. Luckily, in a large class of theories - called renormalizable field theories - all divergencies can be hidden in a finite number of constants. We must fix the values of these constants from the outside and the standard perturbation theory, as a matter of principle, can not answer questions involving the renormalization constants. Apart from that, renormalizable field theories have (at least in perturbation theory) full predictive power. The appearance of infinite expression at the intermediate steps, however, is disconcerting not only from the philosophical point of view (we believe that Nature abhors infinities), but from the practical point of view as well. In order to avoid the dangers so succinctly expressed by John von Neumann - "Just because it's infinite, it doesn't meant it's zero, it might be finite" (quoted by E.T.Jaynes) one develops an elaborate calculational procedure - the regularization scheme - which enables one to artificially make everything finite and keep it finite until all the infinities are completely hidden from sight. Despite all the successes of such an approach, it would be obviously preferable to develop a calculational scheme that would not require ad hoc regularization procedures and still render all expressions finite. I would like to argue in this lecture that such a scheme can be extracted from the original theory, provided one abandons the requirement that the propagation of free particles determines the behavior of interacting particles at high momenta in perturbation theory.

The most popular regularization scheme, before the advent of dimensional regularization, was the scheme introduced by Pauli and Villars.[4,5] In this scheme, each free particle propagator in momentum space,

$$(m - \gamma\cdot p)^{-1} \quad \text{or} \quad (\mu^2 - k^2)^{-1}, \tag{3}$$

is replaced by a sum of such expressions with the coefficients and masses chosen in such a way that the propagator falls of at high momenta as fast as is needed to avoid divergences. For example, for Dirac particles this is achieved by replacing the propagator (3) by the sum

$$(m - \gamma\cdot p)^{-1} + c_1(M_1 - \gamma\cdot p)^{-1} + c_2(M_2 - \gamma\cdot p)^{-1}, \tag{4}$$

where the c_i's and the M_i's obey the conditions

$$1 + c_1 + c_2 = 0, \qquad m + c_1M_1 + c_2M_2 = 0, \tag{5}$$

which guarantee the falling off of the propagator as p^{-3} that is sufficient for convergence of all the momentum space integrals for a typical theory with the Yukawa coupling.

The modification (4) of the fermion propagator in the Pauli-Villars scheme is introduced ad hoc and the extra terms are removed at the end, after renormalization, by taking the limit $M_i \to \infty$. As has been observed, however, in my earlier papers,[6,7] one may obtain the same modification of the propagator as in the Pauli-Villars scheme just by a different rearrangement of the terms in the original Lagrangian without changing its value. However, this observation left much to be desired since the rearrangement was arbitrary and I did not give any clues as to what are the principles that govern the reshuffling of the terms in the Lagrangian.

In this lecture I would like to go beyond my old observation and add an additional requirement - that of the self-consistency - to make the splitting of the Lagrangian unique. The requirement of self-consistency simply means that the modified propagator for low values of momenta must agree with the true propagator to a higher degree of accuracy than its free counterpart. More precisely, I shall require that the inverse of the propagator (4), which is a third order polynomial in $\gamma\cdot p$, coincides with the first three terms of the expansion in powers of $\gamma\cdot p$ of the true propagator for interacting particles. This requirement fixes the values of the (previously arbitrary) constants M_i.

The power expansion of the true propagator can be derived from its spectral representation,

$$G(p) = Z_2\left[\frac{1}{m - \gamma\cdot p} + \int \frac{dM\ \rho(M)}{M - \gamma\cdot p}\right]. \tag{6}$$

The integration in this formula extends from minus to plus infinity (with the mass gap from $-m$ to m). The expansion of the inverse of the propagator around the point m is

$$G^{-1}(p) = Z_2^{-1}(m - \gamma\cdot p)\left[1 + (m - \gamma\cdot p)\int \frac{dM\ \rho(M)}{M - \gamma\cdot p}\right]^{-1}$$

$$= Z_2^{-1}(m - \gamma\cdot p)[1 + a(m - \gamma\cdot p) + b(m - \gamma\cdot p)^2 + \ldots], \tag{7}$$

where the coefficients a and b are determined by the spectral function ρ. In perturbation theory, the renormalization constant and the physical mass are expressed in terms of divergent integrals, but the renormalized spectral function and hence the coefficients a and b are given as power series in the physical coupling constant with finite coefficients. Each series begins with the second order term. The first three terms of the expansion of $G^{-1}(p)$ give the Pauli-Villars form (4) of the propagator, which is sufficient to render all momentum space integrals finite.

In contradistinction to the standard Pauli-Villars procedure, in my approach the masses M_i are determined from the self-consistency conditions and they are not set equal to infinity at the end of the calculation. For that reason, all renormalization constants will also remain finite. They will be given as well defined functions of the coupling constant. For small values of the coupling constant one can easily find their form from the knowledge of their dependence on the cutoff parameter in the standard perturbation theory. The logarithmic dependence on the cutoff means the logarithmic dependence on the Pauli-Villars masses. These masses, in turn, can be shown from (4) and (7) to behave as 1/g for small values of the coupling constant. This gives the behavior $g^2\ln(g)$ for the renormalization constants that I have discussed at the beginning.

The procedure outlined above can be cast into a more systematic form by taking the Lagrangian of the theory as the starting point and by insisting that only the reshufflings of the terms in the Lagrangian are allowed, but not the modifications that would change its value.

I shall illustrate this in a simple case of the field theory of spinor and scalar fields interacting via the Yukawa coupling. The Lagrangian in this case is of the form

$$L = \bar{\psi}(i\gamma\cdot\partial - m_0)\psi + \tfrac{1}{2}(\partial\phi\cdot\partial\phi - \mu_0^2) + g_0\bar{\psi}\,\psi\,\phi, \qquad (8)$$

where all the field operators would be called the bare field operators in the standard terminology, but in my approach this distinction is not so crucial. The Lagrangian (8) is split into the "free" Lagrangian L_0, which is treated without approximations, and the part L_2 describing the interaction treated by perturbation theory. This splitting is accomplished by adding to the first two terms of (8) and subtracting from its last term a third order polynomial in $(i\gamma\cdot\partial)$ sandwiched between the operators $\bar{\psi}$ and ψ,

$$L = L_0 + L_1,$$

$$L_0 = \bar{\psi}(A + Bi\gamma\cdot\partial + C(i\gamma\cdot\partial)^2 + D(i\gamma\cdot\partial)^3)\psi + \tfrac{1}{2}(\partial\phi\cdot\partial\phi - \mu_0^2) \qquad (9)$$

The propagators obtained from L_0 have the Pauli-Villars form (4) and, therefore, all momentum space integrals will converge. The self-consistency conditions in this approach are introduced through the requirement that the correction terms to the inverse propagator of the spinor field,

$$S^{-1}(p) = A + B\gamma\cdot p + C(\gamma\cdot p)^2 + D(\gamma\cdot p)^3, \qquad (10)$$

that are introduced by the interaction L_1, must vanish up to the terms of the third order in $\gamma\cdot p$. This requirement is completely analogous to the standard one that the value of the mass and the normalization of the propagator are correctly given by the free propagator. However, I go two steps further requiring that the quadratic and the cubic term of the inverse propagator are also correctly reproduced by the new free theory. All that one must do to satisfy this requirement is to expand the self energy part $\Sigma(p)$, calculated with the use of the propagator (10), around the value $\gamma \cdot p = m$ and to cancel the first four terms by the appropriate choice of the four constants A,B,C and D.

There are three remarks that should be made here.

First, the masses M_1 and M_2 obtained from the self-consistency condition turn out to be complex. As has been shown, however, by Lee and Wick[8] such complex masses, when they very large, should not cause any problems.

The second remark is concerned with the propagator of the scalar field. This propagator is finite, but I have not added and subtracted any terms to make the mass of the scalar particle equal to its physical mass. This can be done by the usual resummation of the geometric series that gives the propagator in terms of the self-energy part. It is awkward to do this resummation for the scalar particle just by adding and subtracting the appropriate counter terms because in the standard theory the mass correction for the scalar particle is quadratically divergent which in my approach leads to its independence of the coupling constant ($g^2 * (1/g)^2 = 1$). Such a mass correction when subtracted from the interaction Lagrangian would put into question the whole perturbative treatment since L_1 would then contain terms which are independent of the coupling constant.

The third remark is about the absence of the $\lambda\phi^4$ term in the original Lagrangian. This term is not needed because there are no infinite contributions to the four-boson vertex but there will be, of course, an induced term of this type, whose dependence on the coupling constant will again be of the logarithmic type, $\lambda \simeq g^4 \ln(g)$.

Finally, I would like to comment on the feasibility of extending my approach to gauge theories. There are two, obviously related problems with such an extension: the preservation of gauge invariance and infrared divergencies. The first problem can be overcome by a generalization of the splitting of the Lagrangian into L_0 and L_1, which employs the covariant derivatives of the spinor field involving also the gauge field. This technique has already been successfully used in the past[9] in conjunction with Pauli-Villars regularization. The presence of infrared divergencies, however, makes the power expansion of the propagator around the point $p^2 = m^2$ invalid since the propagator develops, in general, a branch point there. Perhaps the use of the Fried-Yennie gauge[10], in which infrared divergencies disappear, can solve this problem.

To summarize, my proposal to develop a new scheme of calculations in quantum field theory is based on the observation that the Pauli-Villars regularization method can be combined with the requirement of self-consistency which enables one to determine the the Pauli-Villars propagators. As a consequence of this requirement, all results are not just finite, but also unique. The renormalization constants become nonanalytic functions of the coupling constant.

This research was supported by the Grant CPBP 01.03. Part of this work was carried out in collaboration with Tomasz Radożycki.

References

1. Dyson, F.J. (1952) Phys.Rev.85, 631.
2. Jauch, J.M. and F.Rohrlich (1976) The theory of photons and electrons (Springer, New York).
3. Fetter, A.L. and J.D.Walecka (1971) Quantum theory of many-particle systems (McGraw-Hill, New York).
4. Pauli, W. and F.Villars (1949) Rev.Mod.Phys.21, 434.
5. Schweber, S.S. (1961) An introduction to relativistic quantum field theory (Row, Peterson and Co., Evanston).
6. Bialynicki-Birula, I. (1960) Phys.Rev.Lett.5, 533.
7. Bialynicki-Birula, I. (1961) Phys.Rev.122, 1942.
8. Lee, T.D. and G.C.Wick (1970) Phys.Rev.D2, 1033.
9. Faddeev, L.D. and A.A.Slavnov (1980) Gauge fields, introduction to quantum theory (Benjamin/Cummings, Reading).
10. Fried, H.M. and D.R.Yennie (1958) Phys.Rev.112, 1391.

Iwo Białynicki-Birula, Institute for Theoretical Physics, Lotników 32, 02-668 Warszawa, Poland.

N=6 FROM CENTRAL EXTENSION OF DOUBLY INFINITE SUPERALGEBRAS*

by

M.Chaichian
Department of High Energy Physics, University of Helsinki
Siltavuorenpenger 20c, SF-00170 Helsinki, Finland,

D.Leites**
Institute for Advanced Study, Princeton, NJ08540, USA
and

J.Lukierski
Institute for Theoretical Physics, University of Wroclaw,
ul.Cybulskiego 36, 50205, Wroclaw, Poland

* Talk presented by J.Lukierski

** One leave of absence from Department of Mathematics, University of Stockholm, Box 6701, S-113 85 Stockholm Sweden

1. Introduction.

For the two-dimensional world, in particular string theories, the N-extended superconformal algebra (for all N) is infinite-dimensional. Usually one constructs the locally superconformal two-dimensional theories by gauging only their finite $OSp(2|N)\oplus OSp(2|N)$ superalgebra [1]. Such a result can be understood as imposing the constraints, which trivialize the local invariance with respect to the remaining infinite number of degrees of freedom [2]. In this letter we shall consider generalized Kac-Moody algebras with full N-extended Virasoro superalgebras as describing the local gauge degrees of freedom. It is known (see, e.g. [3]) that the central extension of Kac-Moody algebras is due to the existence of the nondegenerate even symmetric invariant bilinear form, which is the Killing form. Interestingly enough, among simple Lie superalgebras of vectorfields on a supercircle, only one, i.e. the N=6 Neveu-Schwarz conformal superalgebra, has nonvanishing Killing form. We shall show below, after some mathematical preliminaries, that the "quantum" version, with nonvanishing central extension obtained through the Killing form, exists only for N=6 super Virasoro gauge algebra (we consider here only the Neveu-Schwarz boundary conditions).

2. Infinite-dimensional superalgebras of vector fields on D=1 superspace.

Let us recall briefly the infinite-dimensional superalgebras of vector fields on the superline $R^{1/N}$ and the supercircle $S^{1|N}$ [4-7]:

i) Superline $\mathcal{R}^{1|N}$

We denote the polynomial algebra by $R(N) = R[x,\theta_1,...,\theta_N]$. We define

$$W_N = vect(1|N) = \left\{ der\ R(N) : D = F\frac{\partial}{\partial x} + F_i\frac{\partial}{\partial\theta_i} \right\} \qquad (1a)$$

with $F, F_1, \ldots, F_N \in (N)$, $i=1,\ldots,N$ and summation over the repeated indices is understood;

$$S_N = svect(1|N) = \{D \in vect(1|N), div\ D = 0\} \qquad (1b)$$

(special or supervolume - preserving superalgebra),

where $div(F \frac{\partial}{\partial x} + F_i \frac{\partial}{\partial \theta_i}) = \frac{\partial F}{\partial x} + (-1)^{P(F_i)} \frac{\partial F_i}{\partial \theta_i}$ and $P(F_i)$ is the parity of the superfunction;

$$K_N = K(1|N) = \left\{ D_F = \Delta(F) \frac{\partial}{\partial x} + \frac{\partial F}{\partial x} \theta_i \frac{\partial}{\partial \theta_i} + (-1)^{P(F)} \frac{\partial F}{\partial \theta_i} \frac{\partial}{\partial \theta_i} \right\} \qquad (1c)$$

(contact superalgebra$\equiv O(N)$-conformal superalgebra),

where $\Delta(F) = 2F - \theta_i \frac{\partial F}{\partial \theta_i}$.

We observe that the vector fields (1c) can be written in the form

$$D_F = 2F \frac{\partial}{\partial x} + (-1)^{P(F)} D_i F \cdot D_i, \qquad (2a)$$

where $\mathcal{D}_i = \frac{\partial}{\partial \theta_i} + \theta_i \frac{\partial}{\partial x}$. It is easy to calculate that the superalgebra $K^{NS}(1|N)$ takes the form [6]

$$[D_F, D_G] = D_{\{F,G\}}, \qquad (2b)$$

where

$$\{F,G\} = 2\left(F \frac{\partial G}{\partial x} - G \frac{\partial F}{\partial x}\right) + (-1)^{P(F)} D_i F \cdot D_i G. \qquad (2c)$$

The superalgebras (1a)-(1c) and their formal or smooth versions (i.e. $R(N)$ replaced by formal power series or smooth functions) are the Lie superalgebras of vector fields on the superline.

ii) Supercircles $S^{1|N}$ and its Möbius double covering $S^{1|N-1,M}$

We describe the superspace of functions on $S^{1|N}$ by Laurent power series $R^{1}(N)=C[x^{-1},x,\theta_1,\dots,\theta_N]$, i.e. by Fourier series ($x=e^{i\varphi}$, where φ is the angle parameter).

If we consider $S^{1|N}$ as a trivial boundle of rank N over S^1, and replace $R(N)$ by $R^{1}(N)$ in the above definitions, we get the superalgebras $vect^{1}(1|N)$, $svect^{1}(1|N)$ and $K^{1}(1|N)$. Unlike $svect(1|N)$, the superalgebras $svect^{1}(1|N)$ admit a deformation, $svect^{1}_{\lambda}(1|N)=\{D\in vect^{1}(1|N),\ div\ x^{\lambda}D = 0\}$, found recently by Schwimmer and Seiberg [8].

The contact superalgebras $K^{1}(1|N)$ are known in the physics literature as Neveu-Schwarz (NS) [9] conformal superalgebras ($K^{1}(1|N)\equiv K^{NS}(1|N)$).

One can also consider $S^{1|N-1,M}$ with the Grassmann sector describing a nontrivial bundle over S^1. It appears that by change of variables, such nontrivial bundles can be described by the Whitney sum of the trivial bundle of rank $N-1$ with the Möbius bundle over S^1 with the coordinate, say, θ_N in the fibre. The vector fields (1c) in such a case are modified as follows:

$$K^{R}(1|N) = \left\{D^{R}_{F} = \Delta(F)D_x + D_xF + (-1)^{P(F)}\frac{\partial F}{\partial\theta_i}\frac{\partial}{\partial\theta_i}\right\}, \tag{3}$$

where $D_x = \frac{\partial}{\partial x} - \frac{\theta_i}{2x}\frac{\partial}{\partial\theta_i}$.

The superalgebra of vector fields (1d) can be called Ramond [10] conformal superalgebra $K^{R}(1|N)$.

Further in this paper we shall discuss only trivial bundles over S^1 (the NS sector of conformal superalgebras).

Recently it has been shown [6,7,11] that the following infinite-dimensional superalgebras admit central extension:

$vect^{1}(1|N)$: $N=1,2$

$svect^{1}_{\lambda}(1|N)$: $N=1,2$

$K^{NS}(1|N)$: $N=1,2,3,4$.

3. Kac-Moody superalgebras with finite and infinite dimensional target superalgebras.

Our aim here is to propose a new "doubly infinite" superalgebra which admits a central extension due to the existence of Killing form for the infinite-dimensional target superalgebra.

In order to construct such an example, let us consider the superextensions of loop algebras, describing the smooth mappings $S^1 \rightarrow \tilde{G}$, where $\tilde{G}$ is a supergroup and $\tilde{g}$ is the corresponding superalgebra. If we introduce the corresponding $D=2$ current algebra with the superalgebra $\tilde{g}$-valued currents $J(\tau,\sigma) \equiv J^a(\tau,\sigma)T_a$, where $[T_a,T_b\}=f_{ab}^{\ c}\ T_c$ describes the adjoint matrix representation of g, the "quantum" extension of the superloop algebra, which is a supersymmetric extension of the Kac-Moody algebra[1], looks as follows:

$$[J_a(\sigma),J_b(\sigma')]=i\delta(\sigma-\sigma')f_{ab}^{\ c}\ J_c(\sigma') +$$

$$\frac{i}{2\pi}\ \varkappa\ \frac{\partial}{\partial\sigma}\ \delta(\sigma-\sigma')g_{ab} \quad (4)$$

where $J_a(\sigma) \equiv J_j(0,\sigma)$ and $g_{ab}=str(T_aT_b)$ describes the Killing metric on the superalgebra $\tilde{g}$.

Below we shall consider the generalized superloop algebras with $\hat{g}$ describing the *infinite-dimensional* superalgebras of vector fields, discussed in i). Our aim here is to consider the central extensions for such generalized superloop algebras, in particular, for the choice $\tilde{g} = K^{NS}(1|N)$.

There are two ways of writing down our superextension:

i) To consider the supergroup indices a,b and c in (2)

as describing the infinite discrete basis of $vect^1(1|N)$, $svect^1_\lambda(1|N)$, or $K^{NS}(1|N)$, obtained e.g. by taking the Laurent expansions on the supercircle $S^{1|N}$. For example, for $K^{NS}(1|N)$ the supergroup indices are described by 2^N-triples of integer numbers $(n_1,\ldots,n_r)$ with $r = 2^N$.

ii) One can equivalently replace the discrete indices (a,b,c) in eq.(2) by a continuous index $X=(x,\theta_1,\theta_2,\ldots,\theta_N)$ describing the parametrization of the supercircle $S^{1|N}$, e.g. in the case of $\tilde{g} = K^{NS}(1|N)$ by a substitution

$$J_a(\sigma) \to J(\sigma;X) \equiv J(\sigma;x,\theta_1,\ldots,\theta_N) \qquad (5)$$

in eq.(2). with the structure constants $f_{ab}{}^c$ replaced by the structure distributions of the $K^{NS}(1|N)$ superalgebra written in a local supercurrent form (see, e.g. [7,12]). We propose the following superalgebra for the supercurrents $J(\sigma;X)$,

$$[J(\sigma;X),J(\sigma';X')]$$

$$=i\delta(\sigma-\sigma')\{D_i\delta(X-X')D_i J(\sigma';X')-2\delta(X-X')\cdot J'(\sigma',x')$$

$$-(4-N)\delta'(X-X')J(\sigma',X')\}+\frac{i}{2\pi}\varkappa\frac{\partial}{\partial\sigma}\delta(\sigma-\sigma')g(X,X'), \qquad (6)$$

where $J'=\frac{\partial}{\partial x}J$, $\delta'=\frac{\partial}{\partial x}\delta$.

4. Unique choice: N=6.

It is easy to see that the supercurrents $J(\sigma,X)$ in eq.(6) have the conformal dimension[2)] $d=3-\frac{N}{2}$, while the dimension of the symmetric metric $g(X,X')$ is equal to 4-N. There are the following two local choice of $g(X,X')$ consistent with the choice of dimensionless parameter $\varkappa$,

$$N=2: \qquad g(X,X')=\delta''(X-X'), \qquad (7a)$$

N=6: $g(X,X')=\delta(X-X')$. (7b)

One can show, however that only the choice (7b) is consistent, i.e. defines an invariant bilinear symmetric nondegenerate (Killing) form on $K^{NS}(1|N)$. Indeed, if we introduce for N=6 the scalar product

$$\langle F,G\rangle=\int dv^{N}_{x,\theta}F(X)G(X) \ , \qquad (8)$$

where $dv^{N}_{x,\theta}$ is the volume element in the coordinates $x,\theta_1,\ldots,\theta_N$ [15], one can show that

$$\langle\delta F,G\rangle + \langle F,\delta G\rangle=0, \qquad (9a)$$

or equivalently,

$$\langle\{F,G\},H\rangle=\langle F,\{G,H\}\rangle, \qquad (9b)$$

where { , } is defined by (4b).

It follows from our construction that for N=6 the superalgebra (6) with $g(X,X')$ given by (7b) satisfies the Jacobi identity, or equivalently, the central extension term satisfies the 2-cocycle integrability condition.

The superalgebra (6) for N=6 can be described by an infinite sequence of N=6 superconformal currents $J_n(X)$ ($n=0,\pm1,\pm2,\ldots$; $X=(x,\theta_1,\ldots,\theta_6)$), where

$$J(\sigma:X)=\frac{1}{2\pi}\sum_{n=-\infty}^{+\infty} e^{-in\sigma}\ J_n(X) \qquad (10)$$

and

$$[J_n(X),J_m(X')]=$$

$$i\left\{D_i\delta(X-X')D_i J_{n+m}(X')-2\delta(X-X')J'_{n+m}(X')+2\delta'(X-X')J_{n+m}(X')\right\}$$

$$+\ \varkappa n\delta_{n+m,0}\,\delta(X-X'), \qquad (11)$$

(with $J'\equiv\frac{\partial}{\partial x}J$ and $\delta'\equiv\frac{\partial}{\partial x}\delta$).

If we expand the supercurrent $J_n(X)$ in the Laurent modes on $S^{1|N}$, we obtain the description of our superalgebra in a discrete basis (see *i)* in this section).

5. N=6 superalgebra as Poisson-Lie bracket.

It is known that the Poisson-Lie brackets described by the N-extended super Virasoro algebra provides the supersymmetric extensions of the KdV equation [16,12]. In order to describe an analogous Hamiltonian flow on coadjoint orbits of our new superalgebra (6) for N=6, one should classify the Hamiltonians with conformal dimension three, corresponding to the bilinear Hamiltonian of the second KdV structure. As the simplest choice, we obtain

$$H=\int dv^{6}_{\sigma,x,\theta}\ \partial_\sigma J(\sigma;X)\partial_\sigma J(\sigma;X). \qquad (12)$$

It is interesting to observe that for the N=6 superalgebra (6) one can write the physical Hamiltonian, of dimension one, generalizing in the simplest way the Sugawara Hamiltonian $H=\int d\sigma\, j^a j_a$, as follows:

$$H=\frac{\pi}{\varkappa}\int dv^{6}_{\sigma,x,\theta}\ J(\sigma;X)J(\sigma,X). \qquad (13)$$

The supersymmetric equation of motion, with one-dimensional N=6 supersymmetry (in the variable X), describing the time evolution in the space with three bosonic variables (τ,σ,x) and six Grassmann variables, which corresponds to the choice of Poisson-Lie bracket given by (6) (with N=6) and the Hamiltonian (13), looks surprisingly simple as follows:

$$\partial_\tau\mathcal{J}(\tau,\sigma;X)+\partial_\sigma\mathcal{J}(\tau,\sigma;X)=\frac{2\pi}{\varkappa}\,(6-N)\mathcal{J}\partial_\sigma\mathcal{J}=0, \qquad (14)$$

where $\mathcal{J}(\tau,\sigma;X)|_{\tau=0} \equiv J(\sigma;X)$ and the relation $(\mathcal{D}_i\mathcal{J})^2=0$ has been used. The general solution of (14) is given by $\mathcal{J}=f(\tau-\sigma;X)$, where f is an arbitrary $D=2$, $N=6$ superfield.

6. Conclusions.

Our basic superalgebra (6) describes the supersymmetric two-dimensional local gauge theory with infinite-dimensional N-extended superconformal algebra as the graded gauge algebra. We see the following several possible applications:

a) Following [2], if one restricts the representations of the superalgebra (6) by suitable constraints, one can obtain $D=2$, N-extended superconformal gravity, as well as other locally N-extended superconformal field theories in two dimensions. In such an approach, only a finite number of $D=2$ gauge fields are not pure gauge degrees of freedom.

b) One can also consider the models with nontrivial gauging of infinite number of supercurrents $J_n(X)$ (see 10)). In such a case, our basic superalgebra (6) describes "membrane-like" supersymmetry, with asymmetric treatment of two continuous variables: one, denoted by x is supersymmetrized and carries the representations of $K^{NS}(1|N)$ (N-extended superconformal symmetry), while the second one, denoted by σ is invariant under supersymmetry and describes the local density of the internal symmetry charges. We recall that in the light-cone gauge the (super)membrane with spherical topology has been described as a one-dimensional (super) Yang-Mills theory with infinite-dimensional gauge group of symplectic (area-preserving) diffeomorphisms [17,18]. Our choice, described by eq.(11), corresponds to a different type of supermembrane solutions, describing the $N=6$ supermembrane with the topology $S^1\times S^{1|6}$ as an infinite family of N=6 superstrings.

It should also be mentioned that:

c) One can show that for other infinite-dimensional superalgebras of vector fields on $S^{1|N}$, i.e. for $vect^{1}(1|N)$ and $svect^{1}_{\lambda}(1|N)$, the Killing form does not exist for any N(N=0,1,2,...). We see, therefore, that the generalized superloop algebras with $\tilde{g}$ described by $vect^{1}(1|N)$ and $svect^{1}_{\lambda}(1|N)$ do not permit a central extension via the Killing form. In that sense the choice of $K^{NS}(1|6)$ corresponding to our algebra (6) , is unique.

d) The central extension of generalized superloop algebra can be also obtained if the target algebra has an extension. The choice of superalgebras $\tilde{g}$ for which such a central extension exists are listed at the end of sect.1.

e) The existence of the nondegenerate symmetric form for our N=6 superalgebra permits to consider the solutions of Yang-Baxter equation [19,20] with values in the N=6 super Virasoro algebra (see [21]).

f) One can show that contrary to the case of conventional Kac-Moody superalgebras the superextension $J(\sigma,X)\rightarrow J(\sigma,\eta;X)=\Psi(\sigma,X)+\eta J(\sigma,X)$ $(\eta^2=0)$ for the N=6 superalgebra (11) is not consistent with $\varkappa\neq 0$.

The results presented in our talk shall be soon published [22].

Footnotes

[1] It should be mentioned that the Kac-Moody algebra, describing the smooth mapping $S^1 \rightarrow \tilde{G}$, where $\tilde{G}$ is a compact Lie group, can be extended in two ways:
i) by replacing the circle S^1 by a supercircle $S^{1|N}$, parametrized by S^1 and N real Grassmann coordinates. It can be argued [12] that such supersymmetric extension of Kac-Moody algebras *with central extension exists only for* N=1;
ii) by replacing the Group G by a supergroup $\tilde{G}$, i.e. by introducing in the loop algebra the structure constants of the Lie superalgebra $\tilde{g}$ (see, e.g. [13,14]. The supersymmetric extension described by the formula (2) is of this type.

[21] We recall that in the mass dimensions, *dim* $\delta(\sigma)$, $\delta(x)$, $\frac{\partial}{\partial\sigma}$, $\frac{\partial}{\partial x}$ = 1, *dim* $\theta = -\frac{1}{2}$, *dim* $d\,\theta = \frac{1}{2}$ and *dim* $D_i = \frac{1}{2}$.

References

[1] P.van Nieuwenhuizen, J.Mod.Phys. *A1* (1986) 155; E.A.Bergshoeff, E.Sezgin and H.Nishino, Phys.Lett. *B166* (1986) 141; J.W.van Holten, Nucl.Phys. *B277* (1986) 429.

[2] K.Schoutens, Nucl.Phys.*B292*(1987)150; A.Sevrin and J.W.van Holten, *ibid* p.474

[3] P.Goddard and D.Olive, Intern.J.Mod.Phys.*A1*(1986)303

[4] V.G.Kac, Adv.Math.26(1977)8

[5] D.A.Leites, Theory of Supermanifolds, Izdat. Karel'sk. Filiala AN SSSR, Petrozavodsk, 1983 (in Russian)

[6] B.L.Feigin and D.A.Leites, in the Proc.of the Int. Seminar on "Group-Theor. Methods in Physics", Zvenigorod, 1983, vol.I, Nauka, Moscow, 1984 (in Russian); (English transl., Harwood, 1986)

[7] K.Schoutens, Nucl.Phys.*B295*(1988)634

[8] A.Schwimmer and N.Seiberg, Phys.Lett.*B184*(1987)191

[9] A.Neveu and J.H.Schwarz, Nucl.Phys.*B31*(1971)86

[10] P.Ramond, Phys.Rev.*D3*(1971)2415

[11] V.G.Kac and J.W.van de Leur, On classification of superconformal Algebras, University of Utrecht preprint, No 549 (1988)

[12] M.Chaichian and J.Lukierski, Phys.Lett.*B212*(1988)461

[13] B.L.Feigin, D.A.Leites and V.V.Serganova, as in [6]

[14] P.Goddard, D.Olive and G.Waterson, Commun. Math. Phys. *112* (1987) 591

[15] I.N.Bernshtein and D.A.Leites, Funct.Anal.and Appl. *11* (1977) 45; I.M.Gel'fand, V.V.Minkhin and V.N.Shander, *ibid, 20* (1986) 310

[16] B.A.Kupershmidt, Phys.Lett.*A109*(1985)417

[17] J.Hoppe, B.deWit and H.Nicolai, Nucl.Phys.*B305* (FS 23)(1988)545

[18] E.Bergshoeff, E.Sezgin, Y.Tanii and P.K.Townsend, Super p-branes as gauge theories of volume preserving diffeomeorphisms, ICTP preprint IC/88/186 (October 1988)

[19] V.G.Drinfel'd, in the Proc.Intern.Congress of Mathematicians (MSRI, Berkeley, 1986)

[20] J.M.Maillet, Phys.Lett.*B167*(1986)401

[21] D.Leites and H.Ooguri, IAS, Princeton preprint, in preparation

[22] M.Chaichian, D.Leites and J.Lukierski, "New Infinite--dimensional Superalgebra with Central Extension", Phys.Lett.B, in press

ON THE CLASSICAL BRST COMPLEX

Zbigniew Hasiewicz
Institute of Theoretical Physics.
University of Wroclaw
Cybulskiego 36
PL-50-205 Wroclaw.POLAND

In this note we collect some facts on the classical BRST complex for the constraints forming the centrally extended Lie algebra. Firt section is devoted to the presentation of the scheme of symplectic reduction in terms of the momentum mapping. All the statements are valid also for reducible system. In the next section the reduction of the ring of functions is described cohomologically. Finally we construct the Hamiltonian representations of the cohomological operation . One of them gives the global version of the procedure introduced in [B-F]. The second generalizes and gives the cohomological description of the Gupta-Bleuler approach. (see also the lectures of J.W. van Holten - this volume). The detailed presentation of the geometry of symplectic reduction can be found in [G], while the systematic superspace description of BRST method is presented in [H].

I. CONSTRAINTS, MOMENTUM MAPPING AND REDUCED PHASE SPACE.

1. It happens very often that the constraints in Hamiltonian systems have their origin in the symmetry described by the action on phase space of some Lie group. Assume that we are given the set $\{\phi_i\}_1^k$ of k functions defining the constraints $\phi_i=0; 1\le i\le k$ on some symplectic manifold (phase space) (M,ω). Assume moreover that the

Hamiltonian vector fields $\{h_i\}_1^k$ corresponding to the constraints; $i(h_i)\omega + d\phi_i = 0$, form the representation of k-dmensional Lie algebra $\mathfrak{G}$:

$$[h_i, h_j] = c^k_{ij} h_k, \tag{1}$$

and are complete on M.

Then we can say that they are the fundamental vector fields corresponding to the action on M of some Lie group G ,whose Lie algebra is $\mathfrak{G}$ (for definiteness we can think about universal covering group).

The functions $\{\phi_i\}$ can be regarded as the components of the mapping

$$\Phi : M \longrightarrow \mathfrak{G}^* \text{ (dual of } \mathfrak{G}\text{)} \tag{2}$$

called the momentum for the Hamiltonian action of G on M.

It is easy to see that they satisfy the following Poisson algebra:

$$\{\phi_i, \phi_j\} = c^k_{ij} \phi_k + \Omega_{ij} \tag{3a}$$

where Ω_{ij} is constant on M.

In the coordinate free notation, (3a) can be rewritten as follows:

$$\{\phi_x, \phi_y\} - \phi_{[x,y]} = \Omega(x,y) \tag{3b}$$

where $x,y \in \mathfrak{G}$ and $\phi_z = \langle \phi, z\rangle$(the value of covector on vector).

Notice that Ω on the r.h.s of (3b) is the cocycle i.e defines the central extension of $\mathfrak{G}$.

2. We describe now briefly the procedure called the symplectic reduction, which associates to the constraints $\Phi = 0$ the reduced phase space.

It can be illustrated by the following diagram:

$$\begin{array}{ccc} M_\Phi & \xrightarrow{I} & (M,\omega) \\ \pi \downarrow & & \\ (B_\Phi, \omega_0) & & \end{array} \qquad \pi^*\omega_0 = I^*\omega \tag{4}$$

whose elements are described below.

The set M_Φ is just the solution to the constraints

$$M_\Phi := \Phi^{-1}(0) = \{ p \; ; \; \Phi(p) = 0 \} \tag{5}$$

and $I: M_\Phi \longrightarrow M$ is the canonical inclusion.

In order to have the symplectic reduction well defined we

assume that $0\in\mathfrak{G}^*$ is weakly regular value of Φ [B] ,i.e, M_Φ is the submanifold of M and $\mathcal{T}M_\Phi=\ker d\Phi\,|M_\Phi$. This means in particular that the constraints are not assumed to be independent, which would correspond to the special case when $0\in\mathfrak{G}^*$ is regular value of Φ. The 2-form $I^*\omega$ induced on M_Φ has null directions (along the coordinates having no their canonically conjugated variables on M_Φ) forming integrable ($I^*\omega$ is closed) kernel distribution $K(I^*\omega)$ on M_Φ. The space B_Φ is the set of leaves of $K(I^*\omega)$(equivalence classes of the points joined by the integral curves of $K(I^*\omega)$;"gauge transformations"). We assume that B_Φ does admit the manifold structure, such that the canonical projection $\pi : M_\Phi \longrightarrow B_\Phi$ is smooth. Φ being the momentum mapping for the action of a Lie group guarantees automatically, that $K(I^*\omega)$ has constant dimension, which is neccessary, but not always sufficient condition for π to be smooth. Finally, there is unique symplectic form ω_0 on B_Φ satisfying $\pi^*\omega_0=I^*\omega$. (B_Φ,ω_0) is called the reduced phase space corresponding to the constraint(s)Φ.

3. We have to comment on the equivariance properties of Φ [SO] and describe in more details the structure of the leaves of $K(I^*\omega)$ ("gauge orbits"). [MW]

We assumed that Φ is the momentum mapping for the Hamiltonian action ψ_g: M ⟶ M ; $g\in G$ of the Lie group G (the Lie algebra of G is (1)). The domain M of Φ is thus the representation space of G (acting by symplectomorphisms). On the other hand the target $\mathfrak{G}^*$, carries also the natural , coadjoint representation of G. We may ask if Φ intertwines them. The answer is positive when the cocycle Ω is absent in (3). The general situation is described in the theorem which states that there exsists unique affine action $J_g : \mathfrak{G}^* \longrightarrow \mathfrak{G}^*$ of G ,such that Φ is equivariant

$$\Phi\circ\psi_g = J_g\circ\Phi \quad ; \; g\in G. \tag{6a}$$

Moreover, the action J can be expressed as follows:

$$J_g(\xi) = Ad_g^*(\xi)+\theta(g) \quad ; \; g\in G,\; \xi\in\mathfrak{G}^*, \tag{6b}$$

where Ad_g^* (the linear part of J_g) is coadjoint representation and the $\mathfrak{G}^*$-valued cocycle on G is given by

$$\Theta(g) = \Phi\circ\psi_g - Ad_g\Phi \ . \qquad (6c)$$

The above expression contains Φ explicitely and we can expect that it depends on the points of M. It can be shown however that (6c) is constant on M and defines the object living on G only. The cocycle Ω is the "infinitesimal version" of Θ. More precisely :

$$\Omega(x,y) = -\langle (d_G\Theta)(x), y\rangle \quad ;x,y\in\mathfrak{G} \ , \qquad (6d)$$

where d_G denotes the exterior derivative on G.
Let us now define the Lie subalgebra $\mathfrak{G}_0$ of $\mathfrak{G}$:

$$\mathfrak{G}_0 := \{x\in\mathfrak{G} \ ; \ \Omega(x,\cdot)=0 \ \} \qquad (7)$$

($\mathfrak{G}_0$ is subalgebra due to the cocycle property of Ω).
The Lie group $G_0\subset G$ corresponding to $\mathfrak{G}_0$ is the stabilizer subgroup of $0\in\mathfrak{G}^*$ under the affine action (6b). Using the equivariance (6a) of Φ one can show that the leaves of the kernel distribution $K(I^*\omega)$ are the homogenous spaces of. G_0. In fact G_0 is the maximal connected subgroup which leaves M_Φ invariant.
In general the action of G_0 is not neccesarilly (even locally) free (this would correspond to $0\in\mathfrak{G}^*$ being regular value of Φ -independent constraints). At the moment the locally free action is not needed, as we will use only the fact that B_Φ can be expressed as the quotient of M_Φ by the action of G_0:

$$B_\Phi = M_\Phi \ / \ G_0 \qquad (8)$$

II. THE RINGS OF FUNCTIONS & pre-BRST COMPLEX.

1. Associated with the diagram (4) describing the symplectic reduction, there is the corresponding diagram for the rings of functions:

$$\begin{array}{ccc} \mathcal{F}(M_\Phi) & \overset{I^*}{\longleftarrow} & \mathcal{F}(M) \\ {\scriptstyle \pi^*}\uparrow & & \\ \mathcal{F}(B_\Phi) & & \end{array} \tag{9}$$

where I^* and π^* are the pull-backs induced by inclusion I and projection π.

Under certain regularity assumptions (M_Φ closed submanifold of M) we can identify $\mathcal{F}(M_\Phi)$ with the quotient algebra:

$$\mathcal{F}(M_\Phi) \simeq \mathcal{F}(M) \;/\; I(\Phi) \tag{10}$$

where the (multiplicative) ideal $I(\Phi) = \mathcal{F}(M)\langle\Phi,\mathfrak{G}\rangle$ is identified with that containing all functions on M, vanishing on M_Φ. The generator $\langle\Phi,\mathfrak{G}\rangle$ of $I(\Phi)$ is overcomplete in general.

The function on M_Φ defines the function on B_Φ iff it is constant along the orbits of G_o. Since , by assumption G_o is connected this is the same as to say that it is annihilated by all the vector fields corresponding to the Lie algebra $\mathfrak{G}_o$ of G_o. In other words:

$$\mathcal{F}(B_\Phi) = \mathcal{F}(M_\Phi)^{G_o} = \{f\in\mathcal{F}(M_\Phi); x_{M\Phi}f=0, x\in\mathfrak{G}_o\} \tag{11}$$

2. The reduction of the ring of functions can be described cohomologically. Here the assumptions about the irreducibility of constraints play the crucial role. In what follows we assume that $0\in\mathfrak{G}^*$ is the regular value of Φ. On the general case we comment briefly at the end of this section.

Define the complex:

$$\mathcal{E}(M,\mathfrak{G},\mathfrak{G}_o) := \mathcal{F}(M) \otimes \wedge\mathfrak{G} \,\hat{\otimes}\, \wedge\mathfrak{G}_o^* \tag{12a}$$

(denoted by $\mathcal{E}$ below), which is naturally bigraded

$$\mathcal{E}^{m,n} := \mathcal{F}(M) \otimes \wedge^m \mathfrak{G} \,\hat{\otimes}\, \wedge^n \mathfrak{G}^* \tag{12b}$$

($\hat{\otimes}$ denotes the graded tensor product) and the differential $\mathcal{D}$ is the sum of two anticommuting, nilpotent operators:

$$\mathcal{E}^{m-1,n} \overset{\delta_\Phi}{\longleftarrow} \mathcal{E}^{m,n} \overset{d_{\mathfrak{G}o}}{\longrightarrow} \mathcal{E}^{m,n+1} \tag{12c}$$

They are defined canonically:

$$\delta_\Phi(f\otimes x_1\wedge\ldots\wedge x_m\hat{\otimes}\,\omega) := \sum_{i=1}^{m} f\langle\Phi,x_i\rangle\otimes x_1\wedge\ldots\wedge\hat{x}_i\wedge\ldots\wedge x_m\hat{\otimes}\,\omega \;.$$

$$d_{\mathfrak{G}o}(f\otimes x_1\wedge\ldots\wedge x_m\hat{\otimes}\,\omega) := d_{\mathfrak{G}o}(f\otimes x_1\wedge\ldots\wedge x_m) + $$
$$+(-)^m(f\otimes x_1\wedge\ldots\wedge x_m)\hat{\otimes}\, d_{\mathfrak{G}o}\omega \qquad (12d)$$

where $\mathcal{F}(M)\otimes\bigwedge^m\mathfrak{G}$ -valued 1-form $d_{\mathfrak{G}o}(f\otimes x_1\wedge\ldots\wedge x_m)$ is given by

$$\langle d_{\mathfrak{G}o}(\ldots),y\rangle := y_M f\otimes x_1\wedge\ldots\wedge x_m + f\otimes ady(x_1\wedge\ldots\wedge x_m) \qquad (12e)$$

for any $y\in\mathfrak{G}o$. In the above formula y_M is the fundamenthal vector field corresponding to $y\in\mathfrak{G}o$ and ady is the adjoint representation extended canonically (by derivation property) to $\bigwedge\mathfrak{G}$.

The differential $d_{\mathfrak{G}o}\omega$ is ordinary exterior differential of left invariant form on Go.

One introduces the grading (the ghost number)

$$\mathcal{E}^k := \oplus_{n-m=k}\mathcal{E}^{m,n} \qquad (12e)$$

such that $\mathcal{D}: \mathcal{E}^k \longrightarrow \mathcal{E}^{k+1}$ becomes the differerential of degree 1.

Using the technique of spectral sequences [B-T] one shows that:

$$\mathcal{H}^m_{\mathcal{D}}(\mathcal{E}) = \begin{cases} 0 \;;\; m<0 \\ \mathcal{H}^m_d(\mathcal{H}_{o\delta}(M,\mathfrak{G}),\mathfrak{G}^*_o) \;;\; m>0 \end{cases} \qquad (13a)$$

The above result means in particular:

$$\mathcal{H}^0_{\mathcal{D}}(\mathcal{E}) = \mathcal{F}(B_\Phi) \qquad (13b)$$

i.e. the 0-th cohomology gives the ring of functions on reduced symplectic manifold. The higher cohomology classes represent the invariant cohomologies of the leaves which are locally diffeomorphic with G.

In the general (weakly regular) case of reducible constraints the result (13a) is not true. Its validity depends crucially on the vanishing of all(higher than 0-th) homologies of the complex formed by first two factors of (12a) with differential δ_Φ. In order to achieve this acyclicity one is forced to apply the Tate procedure [TA] and to introduce the ghosts for ghosts [F]. Though the modification follows some canonical rule, the simple structure of (12a) is spoiled. In all further considerations we assume that the constraints are irreducible. It is not obvious that all the constructions are directly applicable in general case. This question is under investigation.

2. The space $\mathcal{C}$ is the Grassmann algebra and $\mathcal{D}$ is the antiderivation of degree 1. It is then obvious that $\mathcal{D}$ can be thaught as an odd vector field on the supermanifold

$$\mathcal{S} := M \times (\mathfrak{G} \oplus \overset{*}{\mathfrak{G}o}) \qquad (14)$$

In this interpretation the space of complex is identified with the ring of functions on $\mathcal{S}$: $\mathcal{C} = \mathcal{F}(\mathcal{S})$.

In order to keep the correspondence with the complex, we introduce the even vector field on $\mathcal{S}$, which splits $\mathcal{F}(\mathcal{S})$ into the direct sum of eigenspaces in accordance with (12e).

Denote the vector fields corresponding to $\mathcal{D}$ and the ghost number derivation by $V_{\mathcal{D}}$ and $V_{\mathcal{N}}$ respectively. They form the algebra:

$$\begin{aligned} V_{\mathcal{D}}V_{\mathcal{D}} &= 0 \\ [V_{\mathcal{N}}, V_{\mathcal{D}}] &= V_{\mathcal{D}} \end{aligned} \qquad (15)$$

Choose the coordinate system $(\mathcal{P}_i)_1^k$, (η^α) in the odd part of $\mathcal{S}$, corresponding to the choice of bases $\{1_i\}_1^k$ and $\{\theta^\alpha\}_1^l$ of $\mathfrak{G}$ and $\overset{*}{\mathfrak{G}o}$. Assume that the basis of $\mathfrak{G}$ contains the dual to that of $\overset{*}{\mathfrak{G}o}$. We can write:

$$\begin{aligned} V_{\mathcal{N}} &= \sum_1^l \eta^\alpha \, \partial/\partial\eta^\alpha - \sum_1^k \mathcal{P}_i \, \partial/\partial\mathcal{P}^i \\ V_{\mathcal{D}} &= \sum_1^k \langle\Phi, 1_i\rangle \, \partial/\partial\mathcal{P}^i + \\ &+\sum_1^l \eta^\alpha (h_\alpha + \sum_{i,j} c^j_{\alpha i} \mathcal{P}_j \, \partial/\partial\mathcal{P}_i - \tfrac{1}{2} \sum_{b,c} c^c_{ab} \eta^b \partial/\partial\eta^c) \end{aligned} \qquad (16)$$

In the above formula h_α is the Hamiltonian vector field of $\phi_\alpha = \langle\Phi, 1_\alpha\rangle$, and c^k_{ij} are the structure constants of $\mathfrak{G}$ in the basis $\{1_i\}_1^k$.

III. HAMILTONIAN REPRESENTATION OF COHOMOLOGY BRST COMPLEX.

1. When the constraints Φ are of first class (l=k), there exsists the Hamiltonian realisation of cohomological operations. We introdoce the canonical orthosymplectic form on the superspace $\mathcal{S}$:

$$\hat{\omega} := \omega + \Sigma_1^k \, d\mathscr{P}_i \wedge d\eta^i \quad (\omega\text{-symplectic form on } M) \tag{1}$$

such that the vector fields $V_{\mathscr{N}}$ and $V_{\mathscr{D}}$ are Hamiltonian:

$$\begin{aligned} i(V_{\mathscr{N}})\hat{\omega} + d\mathfrak{N} = 0 \quad &; \quad V_{\mathscr{N}}\psi = \{\mathfrak{N}, \Psi\} \\ i(V_{\mathscr{D}})\hat{\omega} + d\mathfrak{C} = 0 \quad &; \quad V_{\mathscr{D}}\Psi = \{\mathfrak{C}, \Psi\} \quad ; \quad \Psi \in \mathscr{F}(\mathscr{P}) \end{aligned} \tag{2}$$

The Hamiltonians $\mathfrak{N}$ ond $\mathfrak{C}$ have the standard form:

$$\begin{aligned} \mathfrak{N} &= \Sigma_i \eta^i \mathscr{P}_i \\ \mathfrak{C} &= \Sigma_i \langle \Phi, 1_i \rangle \eta^i - \tfrac{1}{2} \Sigma_{ijk} c^k_{ij} \eta^i \eta^j \mathscr{P}_k \end{aligned} \tag{3}$$

In the case when the constraints are mixed or of second class (l<k), the Hamiltonian realisation (2-3) of cohomological operations in $\mathfrak{G}$ does not exist; the corresponding superspace $\mathscr{P}$ is asymmetric - there are more antighosts than ghosts*). This is the manifestation of the geometrical property of the constraint submanifold which is not coisotropic in M i.e. not all the Hamiltonian vector fields, but only those corresponding to Go, are tangent to M_Φ.

The original idea, proposed by Dirac [D], was to introduce smaller symplectic manifold M, where all the second class constraints are solved. Then B_Φ can be obtained as the symplectic reduction associated to the first class constraints algebra $\mathfrak{G}$o on M_D. However solving the constraints is outside the spirit of the BRST methods. Below we present the two approaches to the general system (3a-I) without solving the second class part. The closer look at the bundle $\mathscr{T}M|M_\Phi$ suggests the ways of producing the first class system out of (3-I). We can enlarge (in some canonical way) the manifold M and impose all the constraints or reduce the system of constraints on M

2. Let us describe the first method briefly. It corresponds to that presented in [B-F]

The affine action (6b) of G on $\mathfrak{G}^*$ defines unique orbit $\mathscr{O}$ of $0 \in \mathfrak{G}^*$, which can be identified with the coset G/Go. The orbit $\mathscr{O}$ carries the canonical symplectic structure. The

*) Even if $[\Omega]=0$ in H $(\mathfrak{G})$ we are not allowed to kill it by the shift $\Phi \longrightarrow \Phi - \Sigma$ $(d_{\mathfrak{G}}\Sigma = \Omega)$ since the reduced phase spaces of Φ and $\Phi - \Sigma$ are different in general.

easiest way to describe it, is via the corresponding Poisson bracket on $\mathfrak{G}^*$:

$$\{f,g\}(\xi) := \xi([df,dg]) + \Omega(df,dg) \qquad (4)$$

where $f,g \in \mathcal{F}(\mathfrak{G}^*)$ and the differentials df, dg are canonically identified with the elements of $\mathfrak{G}^{**} \sim \mathfrak{G}$. The Poisson bracket (4), defines the symplectic form $\omega_{\mathcal{O}}$ when restricted to $\mathcal{O}$. The momentum mapping $\Phi_{\mathcal{O}}$ corresponding to the canonical (Hamiltonian) affine action of G on $\mathcal{O}$ is just the inclusion of $\mathcal{O}$ into $\mathfrak{G}^*$.

The momentum on $\mathcal{M}$ is given by the formula: $\Phi_{\mathcal{M}} = \Phi_M - \Phi_{\mathcal{O}}$.

We define the submanifold of $\mathcal{M}$:

$$\mathcal{M}_{\Phi} := \Phi_{\mathcal{M}}^{-1}(0) = \{ (p,\xi) \in M \times \mathcal{O} \ ; \ \Phi_M(p) = \xi \}. \qquad (5)$$

From $\{\Phi_{\mathcal{M}}x, \Phi_{\mathcal{M}}y\} = \Phi_{\mathcal{M}}[x,y]$ it follows that $\mathcal{M}_{\Phi}$ is coisotropic. Consequently, the null foliation on $\mathcal{M}_{\Phi}$ is generated by all the fundamenthal vector fields of $\mathfrak{G}$.

The reduced phase space has the form: $\mathcal{B}_{\Phi} = \mathcal{M}_{\Phi}/G$. It is not difficult to see the (set theoretical) identification

$$\mathcal{B}_{\Phi} \sim B_{\Phi} \qquad (6)$$

The Hamiltonian description of corresponding complex is now straightforward as the superspace $\mathcal{S} = \mathcal{M} \times (\mathfrak{G} \oplus \mathfrak{G}^*)$ admits the canonical orthosymplectic form .

3. The idea of the approach presented below came from the observation that under some favourable circumstances (the strict conditins are far from being clear) one can get the reduced phase space B_{Φ} and the ring of functions $\mathcal{F}(B_{\Phi})$ by extracting from (3a) the maximal first class subalgebra.

Let us fix the direct sum decomposition

$$\mathfrak{G} = \mathfrak{G}o \oplus \mathfrak{m} \qquad (7)$$

Then Ω defines the non degenerate form on $\mathfrak{m}$. The maximal first class subalgebra corresponds to the choice of the Lagrangian subspace $\mathfrak{l}$ of $\mathfrak{m}$ forming together with $\mathfrak{G}o$ the subalgebra $\mathfrak{h} = \mathfrak{G}o \oplus \mathfrak{l}$. (if in addition $[\mathfrak{G}o, \mathfrak{l}] \subseteq \mathfrak{l}$ then $\mathfrak{l}$ defines the real polarization of the orbit $\mathcal{O}$). However, in general the real subalgebra $\mathfrak{h}$ does not exist. We are forced to look for $\mathfrak{h}$ in the complexification $\mathfrak{G}^c$ of $\mathfrak{G}$.

The general definition of the subalgebra with all desired properties was given in [K](in slightly different context), where it was called the polarization of $\mathcal{O}$:

i) $\mathfrak{h} \supset \mathfrak{G}o$

ii) $\dim_c \mathfrak{G} / \mathfrak{h} = \dim \mathfrak{G}/\mathfrak{G}o$

iii) $\Omega | \mathfrak{h}\times\mathfrak{h} = 0$

iv) $\mathfrak{h} + \bar{\mathfrak{h}}$ is the subalgebra of $\mathfrak{G}$

We will describe less general case of the Kaehler polarization. It corresponds to the decomposition of $\mathfrak{m}^c$ into the direct sum: $\mathfrak{m}^c = \mathfrak{m}_+ \oplus \mathfrak{m}_-$ of eigenspaces (belonging to the eigenvalues $+i$ and $-i$) of $\mathfrak{G}o$ invariant complex structure operator $J(J^2=-1)$. The algebra $\mathfrak{h}=\mathfrak{G}o \oplus \mathfrak{m}_+$ is then of first class We can introduce the complex superspace $\mathcal{P}=Mx(\mathfrak{h}\oplus\mathfrak{h}^*)$ and the Hamiltonian representation of the cohomological operator associated to the complex subalgebra $\mathfrak{h}$. Introducing the bases $\{m_\alpha\}$ in $\mathfrak{m}_+$ and $\{l_i\}$ in $\mathfrak{G}^c_0$ we define the structure constants and complex BRST operator $\mathfrak{S}_{\mathfrak{h}} = \overline{\mathfrak{S}_{\mathfrak{h}}}$.

The above construction is purely algebraic. We should ask about its geometrical meaning. The main question is if the 0-th cohomology space gives the functions on B_Φ. It is not hopeless to expect that $\mathcal{H}^0(\mathfrak{S}_{\mathfrak{h}}) = \mathcal{F}(B_\Phi)^c$. It is confirmed in several concrete cases [L^3],[FR],[A].. The general geometric mechanism wchich enables us to prove the desired result is under investigation.

References

[G] V.Guillemin,S.Sternberg SYMPLECTIC TECHNIQUES IN PHYSICS Cambridge University Press 1985

[H] M.Henneaux CLASSICAL FOUNDATIONS OF BRST SYMMETRY Bibliopolis Napoli 1988.

[SO]J.-M.Souriau STRUCTURE DES SYSTEMES DYNAMIQUES Dunod Paris 1969

[B]R.Bott Ann.Math.60(1954)248-261

[MW]J.E.Marsden,A.Weinstein Rep.Math.Phys.5(1974)121-130

[B-T]R.Bott,L.W.Tu DIFFERENTIAL FOFMS IN ALGEBRAIC GEOMETRY Springer Verlag 1984

[T] J.Tate Illinois J.Math.1(1957)14-21

[F] J,Fish,M.Henneaux,J.Stasheff,C.Teitelboim Commun.Math.Phys.120(1989)379-407

[D] P.A.M.Dirac Can.J.Math.2(1950)129

[B-F] I.A.Batalin,E.S.Fradkin Nucl.Phys.B279(1987)514-528

[K] B.Kostant Actes,Congres intern.Math.2(1970)395-400

[L^3] Lectures of J.A.de Azcarraga, J.Lukierski and L.Lusanna in Proc.of XXth Winter School in Karpacz SUPERSYMMETRY AND SUPERGRAVITY 1983 ed.B.Milewski World Scientific Publ.Co.Singapore 1983

[FR] A.Frydryszak Phys.Rev.D30(1984)2172

[A] S.Aoyama,J.Kowalski-Glikman,J.Lukierski,J.W.van Holten Phys.Lett.201B(1988)487, ibid.216B(1989)133

STRINGS AND INSTANTONS IN CHIRAL MODELS

Michael I. Monastyrsky

Institute of Theoretical and Experimental Physics, Theoretical Department, Moscow, USSR.

1. Introduction

In the fruitful union between mathematics and physics topology dominates in the mathematical aspects. Having started about 15 years ago with the discovery of gauge monopoles by applying homotopic methods, topological methods now cover practically all branches of physics from gravitation to condensed matter. This union has also influenced topology itself.

One of the most extraordinary recent achievments in topology is the discovery by S. Donaldson of different differentiable structures on simply connected four-dimensional manifolds (including R^4) by applying ideas from Yang-Mills theory (instanton connections). Another striking discovery is the construction of new topological invariants of knots (the so-called Jones polynomials), closely connected with such beautiful constructions as von Neumann factors and ice models in exactly soluble two-dimensional statistical systems. The investigation of these unpredictable and deep connections between new mathematical concepts and physical models, such as string and superstring theories, suggests a long and fruitful path for this part of mathematical physics.

The purpose of our paper is more precise. I should like to pay attention to two items appearing in the investigation of concrete physical phenomena, which may also be interesting from a mathematical point of view.

2. Strings as links in condensed matter.

By strings, we actually mean cords (line singularities or defects in condensed matter). The following models appear in the context of a real problem in condensed matter : let us consider a set of closed, non-self intersecting curves $l = l_1,...l_n)$ in M (e.g.M : S^3 or R^3). l is called a link and M is the physical space of a system. Let us consider a continuous map $\Phi:M/l \to V$ where V is the order parameter space. In this situation, l is called the defect Σ of Φ. It is interesting to consider the following problem: What is the system of topological invariants which allows us to decide whether it is possible to unlink this system of linked closed strings ? The well-known invariant of this type is the classical Gauss coefficient k_G. But knowledge of this coefficient is not sufficient to solve the linkability problem. A well-known example of the so-called Whitehead link (see e.g. ROLFSEN) shows that $k_G(l_1,l_2) = 0$ is only a necessary condition for unlinkability. The Whitehead link is an example of an isotopically unlinked link with $k_g = 0$. Another example is the Borromean rings (ROLFSEN). This link consists of three circles (l_1, l_2, l_3) that can be split by cutting one of them. Every k_G $(l_i, l_j) = 0$, but it is impossible to unlink them just homotopically (with intersections). In our paper with V.S. Retakh (MONASTYRSKY, RETAKH), we have solved the problem of linkability in terms of high-order linking coefficients.

Another interesting problem involving knot theory is the following : we consider a map Φ : $S^3 \backslash \Sigma$ V with the defect Σ : 1. We study the homotopy classes of the images $\Phi(\bar{l}_i)$, where $\bar{l}_i$ is the parallel of l_i, in connection with properties of $\Phi(m_i)$, where m_i is a meridian of l_i, using the so-called Milnor coefficients (see details in MONASTYRSKY & RETAKH).

Definition. The defect Σ is called m-essential if every $\Phi(m_i)$ is not contractible to a point in V for all m_i $(l < i < n)$.

Theorem. Let $\pi_1(V) = Z_2$, and the defect Σ be m-essential. The loop $\Phi(l_i)$ can be contracted iff

$$\sum_{j=1}^{i-1} \mu(i,j) + \sum_{i=j+1}^{n} \mu(i,j) = 0 \pmod 2$$

Here $\mu(i,j)$ is the first Milnor coefficient of two loops l_i, l_j. For the definition of Milnor coefficients see e.g. MONASTYRSKY & RETAKH).

This result is only connected with homotopical properties of space V and $(M\backslash 1)$. So it is impossible to distinguish between the defects in two different physical situations having equal homotopical properties of V (e.g. $V = RP^2$ or SO(3)).

Let me finish this discussion by posing the following question : Is there any connection between linking invariants such as Milnor coefficients and Jones-type polynomials ?

3. Instantons in supersymmetrical chiral models.

The following problems arose in studies of instantons in QCD.

There are two approaches to the definition of superinstantons in four-dimensional Yang-Mills theory.

One consists in the generalization of the duality equation to superduality (SEMIKHATOV). The other uses the following construction : let us consider the ordinary instanton, i.e. the instanton solution of an even part of super Yang-Mills equations. Let us act on the instanton by the superconformal group G. The algebra of this group is defined in the paper (FAYET & FERRARA). We thus obtain a superinstanton. This definition is used in practical calculations in QCD in the paper (NOVIKOV at all.).

The natural question is whether these two definitions of instantons are equivalent. The answer is not so evident. If it is true, we can apply all the results obtained in an ordinary case to this situation. It is worthwhile noting that in the two-dimensional case the answer is affirmative. We give a short outline using the results of the paper (MIKHAYLOV & PERELOMOV). The action of two-dimensional supersymmetrical sigma model with a compact gauge group G is the following :

$$S = (1/2 \int d^2\theta d^2 x \{D\hat{\Phi}, D\hat{\Phi}) - ([\hat{\Phi}, \overline{D}\hat{\Phi}],[\hat{\Phi},D\hat{\Phi}])\}$$

Here $x = (x^1, x^2)$ is an ordinary space coordinate, $\theta = \theta^1, \theta^2)$ is a Grassmann coordinate, $\theta^2 = \theta^{1*}$, where * denotes the involution in Grassmann algebra, $\hat{\Phi}(x,\theta)$ is a chiral superfield with even part Φ taking its value in an orbit of the adjoint representation of G. We denote by (,) the usual scalar product in the Lie algebra of G.

Differential operators D and $\overline{D}$ are defined as usual

$$D = \partial/\partial\theta + i\theta\partial,\ \overline{D} = \partial/\partial\theta^* - i\theta^*\partial,\ \partial = \partial/\partial z,\ \overline{\partial} = \partial/\partial\overline{z},\ z = x_1 + ix_2$$

In this notation, the topological charge is given by

$$Q = \int d^2\theta d^2 x(\Phi,[\overline{D}\hat{\Phi}, D\hat{\Phi}]_+)$$

where $[\ ,\ ]_+$ means even part of the commutator. The duality equations can be represented in the form

$$D\hat{\Phi} = [\hat{\Phi}, D\hat{\Phi}];\ \text{or}\ \overline{D}\hat{\Phi} = -[\hat{\Phi}, \overline{D}\hat{\Phi}]$$

The solution of these equations in our case is

$$f(z) + \theta\tilde{f}(z)$$

where $f(z)$, $\tilde{f}(z)$ are analytical functions taking their values, respectively, on the even and odd part of Grassmann algebra. As the conformal group in this case is the infinite-dimensional group of conformal transformations, our conjecture is valid.

Acknowledgements

Completing this report in Laboratoire de Physique des Solides, Orsay, France, I would like to thank John BECHHOEFER for his kind assistance in helping to correct the final version and Françoise Degrave for typing the manuscript.

4. References

Fayet P., Ferrara S. (1977) Supersymmetry. Phys. Reports 32, 249-334

Mikhaylov A.V., Perelomov A.M. (1979) Instantons type solutions in super-symmetrical chiral models, JETP Lett. 29, 403-407

Monastyrsky M.I., Retakh V.S. (1986) Topology of linked defects in condensed matter. Commun. Math. Phys. 103, 445-459

Novikov V.A., Shifman M.A., Vainstein A.I., Voloshin M.B., Zakharov V.I. (1983) Supersymmetry transformations of instantons. Nucl. Phys. B229, 394-406

Rolfsen D. (1976) Knots and links (Publish or Perish, Wilmington)

Semikhatov A.M. (1982) Supersymmetrical instanton, JETP Lett. 35, 560-563

Prof. Michael Monastyrsky
Theoretical Department
Institute of Theoretical and Experimental Physics,
Moscow, 117259, USSR

Orbifold Singularities in Classical and Quantum Mechanics

H. Römer
C. Emmrich

Fakultät für Physik, Universität Freiburg
Hermann-Herder-Str. 3, D-7800 Freiburg, FRG

Consider a Lie group G acting on a manifold M. It is well known that the quotient M/G, the set of orbits of G in M is, in general, not a manifold, but only an orbifold which contains singular points. In fact, singularities in M/G occur, whenever there is a jump in the conjugacy class (H) of the isotropy groups $H \subset G$ of points in M.

Our aim is to investigate the effect of such singular points in configuration and phase spaces of classical and quantum systems. We shall restrict ourselves to a description of the main ideas and refer to the original publications [1,2,3] for technical details.

We work within the framework of Lagrangean mechanics with a (possibly infinite dimensional) manifold Q, the configuration space, and its cotangent bundle $P=T^*Q$ as phase space. T^*Q carries an exact symplectic structure $\omega=-d\Theta$ in a natural way.

Let a Lie group G with Lie algebra $\mathfrak{g}$ act on Q. There is an induced action on $P=T^*Q$, and for every $\xi \in \mathfrak{g}$ a fundamental vector field

$$\xi_p(p) = \frac{d}{dt} e^{t\xi} p \Big|_{t=0}$$

on P. Moreover, there is an equivariant momentum map [4] $J: P \to \mathfrak{g}^*$, defined by

$$\langle J(p), \xi \rangle = J_\xi(p) = \Theta(\xi_p)(p).$$

and fulfilling

$$J(gp) = Ad^*_{g^{-1}} J(p) .$$

For a common symmetry group like rotations in space, which transforms a physical configuration into a different one, it is not absolutely necessary to go over to the quotient P/G.

The situation is quite different for symmetry transformations like gauge transformations in Yang-Mills theory or coordinate transformations for gravity theory. This time the transformations are not between different physical states but between different descriptions of the same physical states. So, the transition to the quotient M/G becomes a conceptual necessity. We shall call symmetries of this kind redundancy symmetries.

For these symmetries, there is a freedom to fix redundant variables at every time t separately, and typically the Lagrangean will be invariant under time dependent transformations g(t). Hence, Noether's theorem gives

$$0 = \frac{d}{dt} \sum p_i \, \xi_Q \, q^i \varepsilon(t) = \varepsilon(t) \frac{dJ_\xi}{dt} + \dot{\varepsilon} J_\xi ,$$

and therefore

$$J_\xi \equiv 0 \text{ for all } \xi \varepsilon \mathfrak{g}.$$

The system is constrained to the zero set of the momentum map. The singular points of the zero set $J^{-1}(0)$ are points in P with non generically large isotropy group $H \subset G$. Under quite general conditions [5] the nature of these singularities is quite simple and explicitely known: In the neighbourhood of a singularity, $J^{-1}(0)$ is locally equivalent to a quadratic cone.

For Yang-Mills and gravity theory the contraint $J \equiv 0$ is identical with Gauß's law and the ADM constraints respectively. The singular points are gauge potentials which are left fixed by some groups $H \neq \{e\}$ of gauge transformations or metrics which admit some isometry. The quadratic conical nature of these singularities has been established also in this infinite dimensional situation [5].

The symplectic form ω is degenerate on the level set $J^{-1}(0)$, which is, hence, not a good candidate for a symmetry reduced phase space. Rather one has to consider the quotient

$$\tilde{P}_0 := J^{-1}(0)/G \; .$$

More generally, for $\mu \, \varepsilon \, \mathfrak{g}^*$ one defines

$$\tilde{P}_\mu := J^{-1}(\mu)/G_\mu \; ,$$

where G_μ is the isotropy group of $\mu \varepsilon \mathfrak{g}^*$ with respect to the coadjoint representation.

This procedure is known as Marsden-Weinstein reduction [4]. If $\tilde{P}_\mu$ is a manifold (which is guaranteed if $\mu \varepsilon \mathfrak{g}^*$ is a weakly regular value of J and if G_μ acts properly with conjugate isotropy groups on $J^{-1}(\mu)$), then $\tilde{P}_\mu$ has a natural symplectic structure, $\tilde{\omega}_\mu$ uniquely defined by

$$\pi_\mu{}^*\tilde{\omega}_\mu = i_\mu{}^*\omega \quad ,$$

where i_μ is the injection $J^{-1}(\mu)\to P$ and π_μ the projection $J^{-1}(\mu)\to\tilde{P}_\mu$. Unfortunately, in general there is more than one orbit type in $J^{-1}(\mu)$ and $\tilde{P}_\mu$ fails to be a manifold.

As far as classical mechanics is concerned, this difficulty can be circumvented in the following way:
We first define $P_{(H)}\subset P$ as the set of points with orbit type (H),i.e., isotropy groups conjugate to $H\subset G$. $P_{(H)}$ is a (not necessarily symplectic) submanifold of P. Then take for arbitrary $\mu\in\mathfrak{g}^*$

$$P_{(H)\mu} := P_{(H)} \cap J^{-1}(0).$$

$P_{(H)\mu}$ is invariant under G_μ.

$P=T^*Q$ carries an almost Kähler structure and the action of G on Q can be lifted to an action on T^*Q by almost Kaehler automorphisms with equivariant momentum map. Then the following theorem of M. Otto [2] becomes applicable:

Let G act properly on (P,ω) by almost Kaehler automorphisms.
Then

$$\tilde{P}_{(H)\mu} := P_{(H)} \cap J^{-1}(0)/G_\mu$$

is a symplectic manifold whose symplectic form $\tilde{\omega}_\mu$ is uniquely determined by

$$i^*\omega = \pi^*\tilde{\omega}_\mu.$$

So, reduction of phase space is performed in two steps:

1. decomposition into orbit types: $P = \bigcup_{(H)} P_{(H)}$.
2. reduction on $P_{(H)}$.

For a G-invariant Hamiltonean, the classical trajectories will never leave $J^{-1}(0)\cap P_{(H)}$. Hence, for classical mechanics, it will always suffice to take $\tilde{P}_{(H)\mu}$ as a reduced phase space.

The situation is more involved for quantum mechanics for at least two reasons:
First, quantum mechanics uses configuration space rather

than phase space or at least a polarization of phase space. Secondly, and more importantly, there is no justification for restricting oneself to a given orbit type (H) because quantum transitions between classical configurations of different orbit types are always possible, as there is no reason to assume that the wave function(al) has support in only one orbit type.

Restriction to the zero set $J^{-1}(0)$ and going over from $J^{-1}(0)$ to the reduced phase space $\tilde{P}_0 = J^{-1}(0)/G$ by a procedure of "gauge fixing" both amount to imposing constraints. Hence, we are facing a problem of quantization of a constrained system with the additional complication that the solution set of the constraints does not form a manifold.
Two different approaches are conceivable

(1) Extrinsic quantization: One first quantizes an unconstrained system, which still contains redundant degrees of freedom. Then, one translates the constraint functions into operators and constructs a Hilbert space of physical states by imposing vanishing conditions, when (annihilation parts of) the constraint operators are applied. The Gupta-Bleuler and the BRST formalisms are two examples of extrinsic quantization.

(2) Intrinsic quantization: One first solves for the classical constraints by introducing appropriate coordinates and then only quantizes in terms of these coordinates.

The equivalence of these two alternative quantization schemes is not guaranteed. In particular, extrinsic quantization, at least on intermediate stages, allows for fluctuations into configurations, which violate the constraint equations. The equivalence problem is even more serious [1,3], if the constraint set fails to be a manifold. An example of non equivalence in quantum gravity, which is of particular relevance in the context of this work, is given in [6].

In any case, for a system with a redundancy symmetry the intrinsic approach seems to be conceptionally preferable. Unfortunately, it is in general hard if not impossible to solve for the constraints even locally.

One very promising and attractive method of intrinsic quantization has been announced by Śniatycki and Weinstein [7], who show, that, even in the singular case, one can define a classical Poisson algebra of functions on $\tilde{P}_0 = J^{-1}(0)/G$ by identifying functions on P_0 with G-invariant elements in $C^\infty(P)/\langle J\rangle$, where $\langle J\rangle$ is the ideal generated by the momentum map J. However, the difficult and crucial step of quantizing this Poisson algebra has, so far, only been performed in simple special cases.

The strategy of intrinsic quantization we are going to follow in this work is the formulation and solution of a Schrödinger equation on an orbifold.

For this approach, it is a crucial observation, that, at least for the cases of interest to us, it is possible to identify a substitute for a reduced configuration space. Indeed, if $P=T^*Q$ is a cotangent bundle and if G acts on Q with only one orbit type, then, under suitable technical conditions (always met in physically interesting cases), there is a natural identification of the reduced phase space with a cotangent bundle:

$$\tilde{P}_0 := J^{-1}(0)/G \simeq T^*(Q/G).$$

More precisely, the following theorem holds:

If Q has only one orbit type, then
$J^{-1}(0) \subset T^*Q$ contains only one one orbit type and
$J^{-1}(0) \subset T^*Q$ is a G-invariant subbundle with
$\pi(J^{-1}(0))=Q$ and there is a natural symplectic isomorphism
$J^{-1}(0)/G \simeq T^*(Q/G)$.

This theorem is to be found in [4] (Thm 4.3.3) for free action of G on Q and was generalized in [3].
In general, of course, there are several orbit types in Q, but there is a generic orbit type of lowest symmetry, which is dense and open in Q.

Orbifold singularities in Q/G correspond to points of higher symmetry. In this situation, the above identification is true for a dense open subset, which leaves out singular points of Q/G. Then it is possible to define a Hamilton operator H and a Schrödinger equation first on the regular points in Q/G. (For the quantization of classical operators there may still be ordering problems.) The Hamilton operator and the Schrödinger equation on the orbifold Q/G are then constructed by self-adjoint extensions of H. The behaviour of the wave function at the orbifold points of Q/G is of particular physical interest.

Singular points in Q/G are linearization unstable. In

[8] it was shown, that for these points the linearized constraints have to be supplemented by additional quadratic constraint conditions, which suppress spurious solutions of the linearized constraint equations. The effect of these additional constraints is a suppression of transitions to configurations of lower symmetry.

This leads to the conjecture, formulated in [1], that in the Schrödinger picture, the wave function(al) shows a particular concentration at the orbifold points. A very peculiar quantum mechanism for the enhancement of symmetric configurations would emerge this way. We test this conjecture by a detailed treatment of some finite dimensional models.

In the vicinity of an orbifold singularity one expects Q/G to have an iterated cone structure ("cones over cones"), the prototype being the set $\mathfrak{g}/G$ of adjoint orbits of a semisimple group G, which is just given by a Weyl chamber.

This provides a motivation to consider Schrödinger equations on a cone C over a (compact Riemannian) manifold M of dimension n. [1,3]

The metric over the (double) cone C over M is given by

$$ds^2 = dr^2 + r^2 g_M(x) dx \otimes dx,$$

where g_M is the Riemannian metric on M, x are coordinates on M and $0 \leq r < \infty$ for the cone and $-\infty < r < \infty$ for the double cone. Rescaling the metric g_M on M corresponds to a variation of the "opening angle" of C.

The Schrödinger equation on C has the form

$$\left\{ \frac{\partial^2}{\partial r^2} + \frac{n}{r}\frac{\partial}{\partial r} + \frac{1}{2}\Delta_M + 2(E-V) \right\} \psi = 0,$$

where $n = \dim M$ and Δ_M is the Laplace operator on M with respect to the metric g_M.

We shall consider two different potentials V

(i) $V \equiv 0$: free motion

(ii) $V = \frac{1}{2}\omega^2 r^2$: oscillator potential.

After separation of the radial coordinate r and the azimuthal coordinates x we obtain the radial equation

$$-2h_\ell \varphi := \left\{ \frac{d^2}{dr^2} + \frac{n}{r}\frac{d}{dr} - \frac{\ell^2}{r^2} - 2V \right\} \varphi = -2E\varphi .$$

Here $l^2 \geq 0$ is the eigenvalue of $-\Delta_M$.

The behaviour of the formal regular and singular solution of this equation near the origin r=0 is

$$\varphi_r(r) \sim r^{\frac{1-n}{2}+\nu(\ell)}$$

and

$$\varphi_s(r) \sim r^{\frac{1-n}{2}-\nu(\ell)}$$

with

$$\nu(\ell) = \sqrt{\frac{(n-1)^2}{4} + \ell^2} \,.$$

Already at this stage we notice

- a tendency towards a concentration around r=0 of the regular wave function for decreasing opening angle (decreasing l^2)
- The regular solution is always locally square integrable around r=0 , and the singular solution is locally square integrable for $\nu(l)<1$. For $n\geq 3$ we always have $\nu(l)\geq 1$.

These features are also born out if we now consider the problem of self-adjoint extensions for the radial Hamiltoneans h_l.

We first define h_l on the domain $D=C_0^\infty(\mathbb{R}\setminus\{0\})$ of smooth functions with compact support in $\mathbb{R}\setminus\{0\}$. Clearly, h_l is symmetric on D.

For $n\geq 3$, h_l is essentially self-adjoint. The unique self-adjoint extension is obtained by closure. The spectrum of h_l turns out to be

for $V=\frac{1}{2}\omega^2r^2$: $E_{j\ell}=(2j+1+\nu(\ell))\,\omega$ (purely discrete)
and for $V\equiv 0$: $E\geqslant 0$ (purely continuous).

More complicated and also more interesting is the case $n\lesssim 2$. It is relevant for our considerations, since we expect low dimensions to enter because of the orbifold structure (cones over cones) of Q/G.

We shall restrict ourselves to extensions such that the eigenstates factor into radial and a azimuthal part, and for the double cone we shall assume interchange of its two halves to be a symmetry operation. (These two conditions imply time reversal invariance.)

For $\nu(l)<1$, h_l is not essentially self-adjoint. The defect indices $n_\pm(l)$ are

$n_\pm(l) = d_l$ for the single cone and

$n_\pm(l) = 2d_l$ for the double cone.

d_l is the degeneracy of the eigenvalue l^2 of $-\Delta_M$.

The admissible self-adjoint extensions are non-unique and depend on d_l parameters for the single cone and $2d_l$ parameters for the double cone. Leaving out some marginal cases which are discussed in [3], the results [3] can be summarized as follows:

- The spectrum is
  $E_{jl} = (2j + 1 \pm \nu(l))\omega + \varepsilon_j$ (discrete) for the oscillator potential and
  $E \geq 0$ plus in general exactly one bound state with negative energy for each $\nu(l)<1$ for the vanishing potential.
- For both potentials, by a suitable choice of the self-adjoint extensions, one can achieve that the ground state of H has arbitrarily low eneryy $E<0$ and arbitrary l^2 with $\nu(l)<1$.
- There is a clear tendency towards a concentration of the wave function around the symmetric configuration r=o, which confirms the conjecture described above.

There is no evident physical principle which should fix the choice of the self-adjoint extension. One could think of a regularization by rounding off the tip of the (double) cone. This would correspond to a self-adjoint extension which only keeps the regular solution. But for a redundancy symmetry the singularity in configuration space is physical and such a regularization procedure does not seem to be well motivated.

One could also consider more general self-adjoint extensions, which are non diagonal in the radial and azimuthal part. They would provide a very peculiar symmetry breaking mechanism.

References

[1] B. de Barros Cobra, Doctoral Thesis Freiburg 1986 (unpublished)
B. de Barros Cobra, H. Römer
Lett.Math.Phys. 13, (1987), 189

[2] M. Otto
J. Geom.Phys. Vol. 4, (1987), 101.

[3] C. Emmrich, MSc Thesis, Freiburg 1989 (unpublished).
C. Emmrich, H. Römer, Freiburg THEP 89/11, to be

published.

[4] For a standard reference
R. Abraham, J.E. Marsden Foundations of Mechamics, Second Edition, Updated 1985 Printing. The Benjamin/ Cunnings Publishing Company (1978).

[5] J. M. Arms, J.E. Marsden, V. Moncrief
Comm.Math.Phys. 78 (1981), 455.

[6] A. Ashtekar, G.T. Horowitz
Phys.Rev. D26 (1982), 3342.

M.J. Gotay
Class.Quantum Grav. 3, (1986), 487.

[7] J. Śniatycki, A. Weinstein
Lett.Math.Phys. 7 (1983), 155.

[8] J. Arms
J.Math.Phys. 20 (1979), 443.

List of participants

S. Albeverio
J. Antikainen
H. Arodz
J. Balog
B. Barbashov
C. Becker
I. Bialynicki-Birula
A. Borowiec
B. Broda
R. Budzynski
S. Carlip
M. Chaichian
D. Chruscinski
F. Colomo
A. Coste
J. Derezinski
P. Dorey
F. Ferrari
O. Foda
A. Frydryszak
Z. Galasiewicz
N. Ganoulis
P. Garbaczewski
W. Garczynski
S.J. Gates, Jr.
K. Gawedzki
B. Geyer
R. Gielerak
K. Golec
J. Gora
M. Gorzelanczyk
D. Grech
P. Gusin
Z. Haba
P. Hajac
Z. Hasiewicz
E. Hellmud
E. Hellwig
A. Hietamaki
M. Hubner
J. Krawczynski
L. Jakobczyk
Z. Jaskolski
A. Jevicki
J. Jezierski
M. Kaku
A. Kargol
W. Karwowski
M. Klimek
A. Kryszen
R. Leandre
F. Leitenberger
Z. Lipinski
A. Lisiecki
A. Lugo
J. Lukierski
R. Madsen
A. Majewski
I. Mazur
K. Meissner
R. Mendy
M. Monastyrski
A. Morozov
M. Mozrzymas
W. Mulak
N. Nesterenko
A. Nowicki
Th. Ohrndorf
Z. Oziewicz
J. Pawelczyk

S. Paycha
A. Perelomov
W. Piechocki
Z. Popowicz
J. Potthoff
K. Redlich
M. Röckner
H. Römer
W. Rühl
F. Ruiz
J. Russo
S. Sannan
B. Schroer
H. Sierra
J. Sobczyk
B. Spokoiny
A. Thielmann
B. Tirozzi
B. Torresani
W. Troost
W. Urbanik
J. van Holten
M. Vasiliev
G. Weigt
A. Wipf
Al. Zamolodchikov